D0895558

Cognitive Technologies

Managing Editors: D. M. Gabbay J. Siekmann

Editorial Board: A. Bundy J. G. Carbonell
M. Pinkal H. Uszkoreit M. Veloso W. Wahlster
M. J. Wooldridge

Advisory Board:

Luigia Carlucci Aiello
Franz Baader
Wolfgang Bibel
Leonard Bolc
Craig Boutilier
Ron Brachman
Bruce G. Buchanan
Anthony Cohn
Artur d'Avila Garcez
Luis Fariñas del Cerro
Koichi Furukawa
Georg Gottlob
Patrick J. Hayes
James A. Hendler
Anthony Jameson
Nick Jennings
Aravind K. Joshi
Hans Kamp
Martin Kay
Hiroaki Kitano
Robert Kowalski
Sarit Kraus
Maurizio Lenzerini
Hector Levesque
John Lloyd
Alan Mackworth
Mark Maybury
Tom Mitchell
Johanna D. Moore
Stephen H. Muggleton
Bernhard Nebel
Sharon Oviatt
Luis Pereira
Lu Ruqian
Stuart Russell
Erik Sandewall
Luc Steels
Oliviero Stock
Peter Stone
Gerhard Strube
Katia Sycara
Milind Tambe
Hidehiko Tanaka
Sebastian Thrun
Junichi Tsujii
Kurt VanLehn
Andrei Voronkov
Toby Walsh
Bonnie Webber

Wolfgang Wahlster (Ed.)

SmartKom: Foundations of Multimodal Dialogue Systems

With 193 Figures and 63 Tables

Editor:

Prof. Dr. Wolfgang Wahlster
German Research Center for AI
DFKI GmbH
Stuhlsatzenhausweg 3
66123 Saarbrücken
Germany

Managing Editors:

Prof. Dov M. Gabbay
Augustus De Morgan Professor of Logic
Department of Computer Science, King's College London
Strand, London WC2R 2LS, UK

Prof. Dr. Jörg Siekmann
Forschungsbereich Deduktions- und Multiagentensysteme, DFKI
Stuhlsatzenweg 3, Geb. 43, 66123 Saarbrücken, Germany

Library of Congress Control Number: 2006928318

ACM Computing Classification (1998): H.5, H.1.2, I.2, I.3, I.6.

ISSN 1611-2482
ISBN-10 3-540-23732-1 Springer Berlin Heidelberg New York
ISBN-13 978-3-540-23732-7 Springer Berlin Heidelberg New York

This work is subject to copyright. All rights are reserved, whether the whole or part of the material is concerned, specifically the rights of translation, reprinting, reuse of illustrations, recitation, broadcasting, reproduction on microfilm or in any other way, and storage in data banks. Duplication of this publication or parts thereof is permitted only under the provisions of the German Copyright Law of September 9, 1965, in its current version, and permission for use must always be obtained from Springer. Violations are liable for prosecution under the German Copyright Law.

Springer is a part of Springer Science+Business Media
springer.com

© Springer-Verlag Berlin Heidelberg 2006
Printed in Germany

The use of general descriptive names, registered names, trademarks, etc. in this publication does not imply, even in the absence of a specific statement, that such names are exempt from the relevant protective laws and regulations and therefore free for general use.

Cover Design: KünkelLopka, Heidelberg
Typesetting: by the Editor
Production: LE-TEX Jelonek, Schmidt & Vöckler GbR, Leipzig

Printed on acid-free paper 45/3100/YL 5 4 3 2 1 0

Foreword

For about 25 years research and development projects in the area of human-computer interaction (HCI) have been pursued with the objective to adapt the communication and interaction with the machine to the needs of the human user, and not vice versa. But it was only within the past ten years that significant and substantial progress in the practical realization of the results in this area was achieved with the development of individual forms of interaction like speech processing or visualization. The resulting question, then, was whether it is possible to develop easy-to-use multimodal user interfaces with an attractive market potential.

This was the starting point for the inter-disciplinary research activities in human-computer interaction conducted by six large strategic cooperative projects with 102 partners from science and industry. In 1999, these six so-called lead projects came out ahead of 89 proposals overall in an ideas competition launched by the German federal government. These recently finished research projects were supposed to allow human users in both their private and professional environments to multimodally control and diversely use technical systems via natural modalities of interaction like speech, gestures, facial expressions, tactile and graphical input. Ergonomics and user acceptance of these forms of interaction were the key criteria for the development of prototypes that were supposed to have both a strong scientific attractiveness and a high market potential.

One of these lead projects is SMARTKOM. Coordinated by the German Research Center for Artificial Intelligence in Saarbrücken, a consortium of four well-known industrial companies, two small companies and two middle-sized ones, one research institute and three universities was formed. The objective of SMARTKOM was to conduct fundamental research in the area of robust multimodal interaction under realistic conditions, i.e., interaction has to be possible even if the input is underspecified, ambiguous or partially incorrect. The basic idea was to consider and integrate several modes of interaction — in addition to speech especially gestures and facial expressions — instead of only a single modality, and thereby to achieve a substantially better interpretation of the user's intention. This assessment has been confirmed at international conferences worldwide. Thanks to the dedication of all project partners SMARTKOM's ambitious objectives have been more than accomplished, as for instance:

- the situation-dependent recognition of underspecified, ambiguous or partially incorrect input on both a syntactic and a pragmatic level was demonstrated successfully,
- a multimodal semantic representation language was developed (M3L) that substantially contributes to a worldwide standardization, and last but not least
- speech-based dialogic Web services for car drivers and pedestrians were developed.

Moreover, the know-how gained in the project was protected for the German economy through 52 patent applications, 29 spin-off products and six spin-off companies so far. In the scientific area, the SMARTKOM project resulted in 255 publications, 66 diploma theses, Ph.D. and habilitation theses, State doctorates as well as six appointments to professorships. This makes SMARTKOM the most successful of all 29 lead projects of the Federal Ministry of Education and Research started since 1998. SMARTKOM was funded with 16.8 million € between September 1999 and September 2003. The overall financial means including the matching funds from industry amounted to 25.7 million €.

This book provides a comprehensive overview of the broad spectrum of results of the research conducted in SMARTKOM. I thank and give credit to everyone involved in the project but especially to Professor Wolfgang Wahlster's professional project management and his competent scientific leadership of the distinguished team of researchers.

Bonn, May 2006

Dr. Bernd Reuse
Head of the Software Systems Division
German Federal Ministry of Education and Research

Acknowledgment

A book such as this one could obviously not be put together without the help and cooperation of many people.

I am particularly indebted to the authors who graciously made their contributions available in a timely fashion.

I would like to thank Dr. Anselm Blocher for his excellent editorial assistance and the production of the final camera-ready copy. Special praise goes to Leivy Michelly Kaul and Alexander Kowalski for their assistance in formatting and copy-editing the book. Special thanks go to Ronan Nugent from Springer for his continuous copy-editing and production support.

The SMARTKOM project was made possible by funding from the German Federal Ministry of Education and Research (BMBF) under contract number number 01 IL 905. I would like to thank Dr. Bernd Reuse, Head of the Software Systems Division at BMBF, for his constant and tireless support of the SMARTKOM project.

Saarbrücken, May 2006

Wolfgang Wahlster
Scientific Director of the SMARTKOM Project
CEO of the
German Research Center for Artificial Intelligence
(DFKI)

List of Contributors

J. Adelhardt
Friedrich-Alexander-University
Erlangen-Nuremberg
Martenstraße 3
D-91058 Erlangen
adelhard@
informatik.uni-erlangen.de

J. Alexandersson
DFKI GmbH
Stuhlsatzenhausweg 3
D-66123 Saarbrücken
janal@dfki.de

H. Aras
European Media Lab
Schloß-Wolfsbrunnenweg 33
D-69118 Heidelberg
Hidir.Aras@eml.villa-bosch.de

M. Baudis
Mundwerk AG
Immanuelkirchstraße 3-4
D-10405 Berlin
berlin@excelsisnet.com

A. Batliner
Friedrich-Alexander-University
Erlangen-Nuremberg
Martenstraße 3
D-91058 Erlangen
batliner@
informatik.uni-erlangen.de

T. Becker
DFKI GmbH
Stuhlsatzenhausweg 3
D-66123 Saarbrücken
becker@dfki.de

A. Berton
DaimlerChrysler AG
Wilhelm-Runge-Straße 11
D-89081 Ulm
andre.berton@
daimlerchrysler.com

A. Blocher
DFKI GmbH
Stuhlsatzenhausweg 3
D-66123 Saarbrücken
blocher@dfki.de

N. Braunschweiler
IMS, University of Stuttgart
Azenbergstraße 12
D-70174 Stuttgart
braunnt@ims.uni-stuttgart.de

J. Bryant
International Computer Science Institute
1947 Center Street
Berkeley, CA, USA 94704-1105
jbryant@icsi.berkeley.edu

D. Bühler
University of Ulm
Albert-Einstein-Allee 43
D-89081 Ulm
dirk.buehler@.uni-ulm.de

V. Chandrasekhara
quadox AG
Ohmstraße 2
D-69190 Walldorf
vasu.chandrasekhara@quadox.de

G. Dogil
IMS, University of Stuttgart
Azenbergstraße 12
D-70174 Stuttgart
dogil@ims.uni-stuttgart.de

R. Engel
DFKI GmbH
Stuhlsatzenhausweg 3
D-66123 Saarbrücken
ralf.engel@dfki.de

M. Emele
Sony International (Europe) GmbH
Heinrich-Hertz-Straße 1
D-70327 Stuttgart
emele@gmx.net

C. Frank
Friedrich-Alexander-University
Erlangen-Nuremberg
Martenstraße 3
D-91058 Erlangen
frank@
informatik.uni-erlangen.de

D. Gelbart
International Computer Science Institute
1947 Center Street
Berkeley, CA, USA 94704-1105
gelbart@icsi.berkeley.edu

S. Goronzy
3SOFT GmbH
Frauenweiherstraße 14
D-91058 Erlangen
Silke.Goronzy@3SOFT.de

S. Grashey
Siemens AG
Otto-Hahn-Ring 6
D-81730 Munich
stephan.grashey@siemens.com

I. Gurevych
Technical University Darmstadt
Hochschulstraße 10
D-64289 Darmstadt
Gurevych@
tk.informatik.tu-darmstadt.de

U. Haiber
DaimlerChrysler AG
Wilhelm-Runge-Straße 11
D-89081 Ulm
udo.haiber@daimlerchrysler.com

J. Häußler
European Media Lab
Schloß-Wolfsbrunnenweg 33
D-69118 Heidelberg
jochen.haeussler@
eml.villa-bosch.de

G. Herzog
DFKI GmbH
Stuhlsatzenhausweg 3
D-66123 Saarbrücken
Gerd.Herzog@dfki.de

A. Horndasch
Sympalog Speech Technologies AG
Karl-Zucker-Straße 10
D-91052 Erlangen
horndasch@sympalog.de

C. Hying
Sony International (Europe) GmbH
Heinrich-Hertz-Straße 1
D-70327 Stuttgart
christian.hying@
ims.uni-stuttgart.de

M. Jöst
European Media Lab
Schloß-Wolfsbrunnenweg 33
D-69118 Heidelberg
joest@eml.org

A. Kaltenmeier
DaimlerChrysler AG
Wilhelm-Runge-Straße 11
D-89081 Ulm
alfred.kaltenmeier@
daimlerchrysler.com

A. Kellner
Philips GmbH
Weißhausstraße 2
D-52066 Aachen
Andreas.Kellner@philips.com

T. Klankert
IMS, University of Stuttgart
Azenbergstraße 12
D-70174 Stuttgart
klankert@ims.uni-stuttgart.de

S. Krüger
SAP AG
Dietmar-Hopp-Allee 16
D-69190 Walldorf
sven.krueger@sap.com

Y.H. Lam
Sony International (Europe) GmbH
Heinrich-Hertz-Straße 1
D-70327 Stuttgart
lam@sony.de

M. Löckelt
DFKI GmbH
Stuhlsatzenhausweg 3
D-66123 Saarbrücken
loeckelt@dfki.de

M. Lützeler
Siemens AG
Otto-Hahn-Ring 6
D-81730 Munich
Michael.Luetzeler@siemens.com

R. Malaka
University of Bremen
Bibliothekstraße 1, MZH
D-28359 Bremen
malaka@informatik.uni-bremen.de

M. Merdes
EML Research gGmbH
Schloß-Wolfsbrunnenweg 33
D-69118 Heidelberg
matthias.merdes@
eml-r.villa-bosch.de

W. Minker
University of Ulm
Albert-Einstein-Allee 43
D-89081 Ulm
wolfgang.minker@uni-ulm.de

B. Möbius
IMS, University of Stuttgart
Azenbergstraße 12
D-70174 Stuttgart
moebius@ims.uni-stuttgart.de

G. Möhler
IMS, University of Stuttgart
Azenbergstraße 12
D-70174 Stuttgart
moehler@gmx.de

E. Morais
Faculdade de Eng. Elétrica e de Computação
State University of Campinas, Brazil
emorais@decom.fee.unicamp.br

N. Morgan
International Computer Science Institute
1947 Center Street
Berkeley, CA, USA 94704-1105
morgan@icsi.berkeley.edu

A. Ndiaye
DFKI GmbH
Stuhlsatzenhausweg 3
D-66123 Saarbrücken
ndiaye@dfki.de

H. Niemann
Friedrich-Alexander-University
Erlangen-Nuremberg
Martenstraße 3
D-91058 Erlangen
niemann@
informatik.uni-erlangen.de

E. Nöth
Friedrich-Alexander-University
Erlangen-Nuremberg
Martenstraße 3
D-91058 Erlangen
noeth@
informatik.uni-erlangen.de

D. Pfisterer
University Lübeck
Ratzeburger Allee 160
D-23538 Lübeck
pfisterer@itm.uni-luebeck.de

N. Pfleger
DFKI GmbH
Stuhlsatzenhausweg 3
D-66123 Saarbrücken
Norbert.Pfleger@dfki.de

P. Poller
DFKI GmbH
Stuhlsatzenhausweg 3
D-66123 Saarbrücken
poller@dfki.de

T. Portele
Philips GmbH
Weißhausstraße 2
D-52066 Aachen
Thomas.Portele@philips.com

R. Porzel
European Media Lab
Schloß-Wolfsbrunnenweg 33
D-69118 Heidelberg
porzel@eml.org

S. Rabold
University of Munich
Schellingstraße 3
D-80799 Munich
rabold@phonetik.uni-muenchen.de

J. Racky
Siemens AG
Otto-Hahn-Ring 6
D-81730 Munich
Jens.Racky@siemens.com

H. Rapp
MediaInterface Dresden GmbH
Washingtonstraße 16/16a
D-01139 Dresden
rapp@mediainterface.de

S. Rapp
Sony International (Europe) GmbH
Heinrich-Hertz-Straße 1
D-70327 Stuttgart
rapp@conante.com

N. Reithinger
DFKI GmbH
Stuhlsatzenhausweg 3
D-66123 Saarbrücken
Norbert.Reithinger@dfki.de

B. Reuse
BMBF
Heinemannstraße 2
D-53175 Bonn
Bernd.Reuse@BMBF.bund400.de

H. Röttger
Siemens AG
Otto-Hahn-Ring 6
D-81730 Munich
Hans.Roettger@siemens.com

B. Säuberlich
IMS, University of Stuttgart
Azenbergstraße 12
D-70174 Stuttgart
bettina.saeuberlich@
ims.uni-stuttgart.de

F. Schiel
University of Munich
Schellingstraße 3
D-80799 Munich
schiel@phonetik.uni-muenchen.de

O. Schreiner
Technical University Berlin
Straße des 17. Juni 135
10623 Berlin

M. Schuster
Siemens AG
Otto-Hahn-Ring 6
D-81730 Munich
Matthias.Schuster@siemens.com

A. Schweitzer
University of Stuttgart
Azenbergstraße 12
D-70174 Stuttgart
Antje.Schweitzer@
ims.uni-stuttgart.de

R.P. Shi
Friedrich-Alexander-University
Erlangen-Nuremberg
Martenstraße 3
D-91058 Erlangen
shi@informatik.uni-erlangen.de

S. Steininger
University of Munich
Schellingstraße 3
D-80799 Munich
steins@phonetik.uni-muenchen.de

A. Stolcke
International Computer Science Institute
1947 Center Street
Berkeley, CA, USA 94704-1198
stolcke@icsi.berkeley.edu

M. Streit
Rue de la Gare 108
F-57150 Creutzwald

M. Thomae
Technical University of Munich
Arcisstr. 21
D-80333 Munich
thomae@ei.tum.de

S. Torge
Sony International (Europe) GmbH
Heinrich-Hertz-Straße 1
D-70327 Stuttgart
sunna.torge@web.de

V. Tschernomas
TNM Software GmbH
Grubenstraße 107
D-66540 Neunkirchen
tschernomas@tnmsoft.com

U. Türk
University of Munich
Schellingstraße 3
D-80799 Munich
tuerk@phonetik.uni-muenchen.de

Z. Valsan
Sony International (Europe) GmbH
Heinrich-Hertz-Straße 1
D-70327 Stuttgart
valsan@sony.de

J. te Vrugt
Philips GmbH
Weißhausstraße 2
D-52066 Aachen
Juergen.te.Vrugt@philips.com

W. Wahlster
DFKI GmbH
Stuhlsatzenhausweg 3
D-66123 Saarbrücken
wahlster@dfki.de

V. Zeissler
Friedrich-Alexander-University
Erlangen-Nuremberg
Martenstraße 3
D-91058 Erlangen
zeissler@
informatik.uni-erlangen.de

Contents

Part I Introduction

Part II Multimodal Input Analysis

Part IV Multimodal Output Generation

Part V Scenarios and Applications

Part VI Data Collection and Evaluation

Part I

Introduction

Dialogue Systems Go Multimodal: The SmartKom Experience

Wolfgang Wahlster

DFKI GmbH, Saarbrücken, Germany
wahlster@dfki.de

Summary. Multimodal dialogue systems exploit one of the major characteristics of human-human interaction: the coordinated use of different modalities. Allowing all of the modalities to refer to and depend upon each other is a key to the richness of multimodal communication. We introduce the notion of symmetric multimodality for dialogue systems in which all input modes (e.g., speech, gesture, facial expression) are also available for output, and vice versa. A dialogue system with symmetric multimodality must not only understand and represent the user's multimodal input, but also its own multimodal output. We present an overview of the SMARTKOM system that provides full symmetric multimodality in a mixed-initiative dialogue system with an embodied conversational agent. SMARTKOM represents a new generation of multimodal dialogue systems that deal not only with simple modality integration and synchronization but cover the full spectrum of dialogue phenomena that are associated with symmetric multimodality (including crossmodal references, one-anaphora, and backchannelling). We show that SMARTKOM's plug-and-play architecture supports multiple recognizers for a single modality, e.g., the user's speech signal can be processed by three unimodal recognizers in parallel (speech recognition, emotional prosody, boundary prosody). We detail SMARTKOM's three-tiered representation of multimodal discourse, consisting of a domain layer, a discourse layer, and a modality layer. We discuss the limitations of SMARTKOM and how they are overcome in the follow-up project SmartWeb. In addition, we present the research roadmap for multimodality addressing the key open research questions in this young field. To conclude, we discuss the economic and scientific impact of the SMARTKOM project, which has led to more than 50 patents and 29 spin-off products.

1 The Need for Multimodality

In face-to-face situations, human dialogue is not only based on speech but also on nonverbal communication including gesture, gaze, facial expression, and body posture. Multimodal dialogue systems exploit one of the major characteristics of human-human interaction: the coordinated use of different modalities. The term *modality* refers to the human senses: vision, audition, olfaction, touch, and taste. In addition, human communication is based on socially shared code systems like natural languages, body languages, and pictorial languages with their own syntax, semantics,

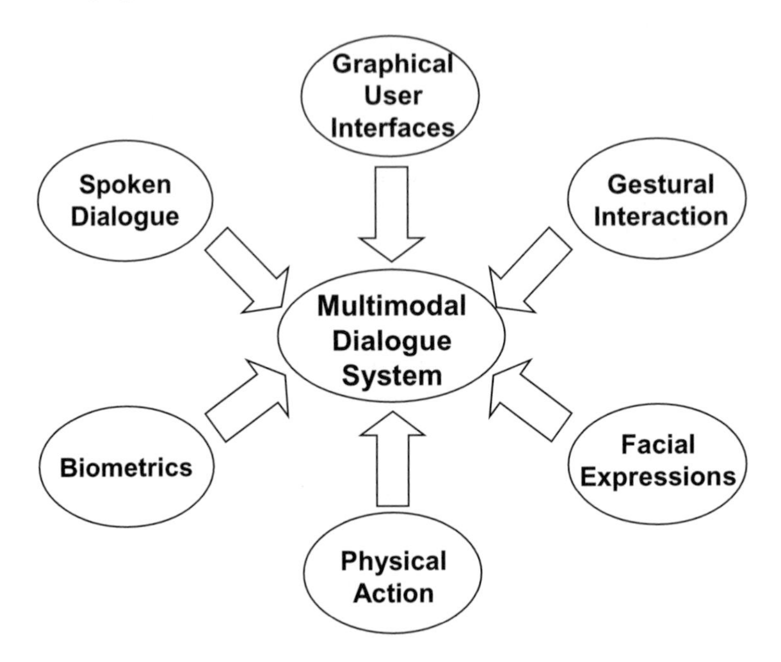

Fig. 1. Merging various dialogue and interaction paradigms into multimodal dialogue systems

and pragmatics. A single semiotic code may be supported by many modalities (Maybury and Wahlster, 1998). For instance, language can be supported visually (i.e., written language), aurally (i.e., spoken language) or tactilely (i.e., Braille script) – in fact, spoken language can have a visual component (e.g., lipreading). Allowing all of the modalities to refer to and depend upon each other is a key to the richness of multimodal communication (see Fig. 1).

Unlike traditional keyboard and mouse interfaces or unimodal spoken dialogue systems, multimodal dialogue systems permit flexible use of input and output modes (Oviatt, 2003). Since there are large individual differences in ability and preference to use different modalities, a multimodal dialogue system permits diverse user groups to exercise control over how they interact with application systems. Especially for mobile tasks, multimodal dialogue systems permit the modality choice and switching that is needed during the changing situational conditions.

With the proliferation of embedded computers in everyday life and the emergence of ambient intelligence, multimodal dialogue systems are no longer only limited to traditional human-computer communication, but become more generally a key component for advanced human-technology interaction and even multimodal human-environment communication (Wahlster and Wasinger, 2006).

Self-service systems, online help systems, Web services, mobile communication devices, remote control systems, smart appliances, and dashboard computers are providing ever more functionality. However, along with greater functionality, the user must also come to terms with the greater complexity and a steeper learning curve.

This complexity is compounded by the sheer proliferation of different systems lacking a standard user interface. The growing emphasis on multimodal dialogue systems is fundamentally inspired by the aim to support natural, flexible, efficient, and powerfully expressive means of human-computer communication that are easy to learn and use. Advances in human language technology and in perceptive user interfaces offer the promise of pervasive access to online information and Web services. The long-term goal of the research in multimodal dialogue systems is to allow the average person to interact with computerized technologies anytime and anywhere without special skills or training, using such common devices as a smartphone or a PDA.

We begin by describing the scientific goals of the SMARTKOM project in Sect. 2, before we introduce the notion of symmetric multimodality in Sect. 3. In Sect. 4, we introduce SMARTKOM as a flexible and adaptive multimodal dialogue shell and show in Sect. 5 that SMARTKOM bridges the full loop from multimodal perception to physical action. SMARTKOM's distributed component architecture, realizing a multiblackboard system, is described in Sect. 6. Then in Sects. 7 and 8, we describe SMARTKOM's methods for multimodal fusion and fission. Section 9 discusses the role of the three-tiered multimodal discourse model in SMARTKOM. Section 10 gives a brief introduction to SmartWeb, the follow-up project to SMARTKOM, which supports open-domain question answering. Open research questions in the young field of multimodal dialogue systems are presented in Sect. 11 in a research roadmap for multimodality. Finally, we discuss the economic and scientific impact of the SMARTKOM project in Sect. 12.

2 SmartKom: A Massive Approach to Multimodality

In this book, we present the theoretical and practical foundations of multimodal dialogue systems using the results of our large-scale project SMARTKOM as the background of our discussion. Our SMARTKOM system (http://www.smartkom.org) is designed to support a wide range of collaborative and multimodal help dialogues that allow users to intuitively and efficiently access the functionalities needed for their task. The application of the SMARTKOM technology is especially motivated in non-desktop scenarios, such as smart rooms, kiosks, or mobile environments. SMARTKOM features the situated understanding of possibly imprecise, ambiguous or incomplete multimodal input and the generation of coordinated, cohesive, and coherent multimodal presentations. SMARTKOM's interaction management is based on representing, reasoning, and exploiting models of the user, domain, task, context, and modalities. The system is capable of real-time dialogue processing, including flexible multimodal turn-taking, backchannelling, and metacommunicative interaction. The four major scientific goals of SMARTKOM were to:

- explore and design new symbolic and statistical methods for the seamless fusion and mutual disambiguation of multimodal input on semantic and pragmatic levels
- generalize advanced discourse models for spoken dialogue systems so that they can capture a broad spectrum of multimodal discourse phenomena

- explore and design new constraint-based and plan-based methods for multimodal fission and adaptive presentation layout
- integrate all these multimodal capabilities in a reusable, efficient and robust dialogue shell that guarantees flexible configuration, domain independence and plug-and-play functionality

3 Towards Symmetric Multimodality

SMARTKOM provides *full symmetric multimodality* in a mixed-initiative dialogue system. Symmetric multimodality means that all input modes (speech, gesture, facial expression) are also available for output, and vice versa. A dialogue system with symmetric multimodality must not only understand and represent the user's multimodal input, but also its own multimodal output.

In this sense, SMARTKOM's modality fission component provides the inverse functionality of its modality fusion component, since it maps a communicative intention of the system onto a coordinated multimodal presentation (Wahlster, 2002). SMARTKOM provides an anthropomorphic and affective user interface through an embodied conversational agent called Smartakus. This life-like character uses coordinated speech, gesture and facial expression for its dialogue contributions.

Thus, SMARTKOM supports face-to-face dialogic interaction between two agents that share a common visual environment: the human user and Smartakus, an autonomous embodied conversational agent. The "i"-shape of Smartakus is analogous to that used for *information* kiosks (see Fig. 2). Smartakus is modeled in 3D Studio Max. It is a self-animated interface agent with a large repertoire of gestures, postures and facial expressions. Smartakus uses body language to notify users that it is waiting for their input, that it is listening to them, that is has problems in understanding their input, or that it is trying hard to find an answer to their questions.

Most of the previous multimodal interfaces do not support symmetric multimodality, since they focus either on multimodal fusion (e.g., QuickSet, see Cohen et al. (1997), or MATCH, see Johnston et al. (2002)) or multimodal fission (e.g., WIP, see Wahlster et al. (1993)). But only true multimodal dialogue systems like SMARTKOM create a natural experience for the user in the form of daily human-to-human communication, by allowing both the user and the system to combine the same spectrum of modalities.

SMARTKOM is based on the situated delegation-oriented dialogue paradigm (SDDP, see Fig. 2): The user delegates a task to a virtual communication assistant (Wahlster et al., 2001). This cannot however be done in a simple command-and-control style for more complex tasks. Instead, a collaborative dialogue between the user and the agent elaborates the specification of the delegated task and possible plans of the agent to achieve the user's intentional goal. The user delegates a task to Smartakus and helps the agent, where necessary, in the execution of the task. Smartakus accesses various digital services and appliances on behalf of the user, collates the results, and presents them to the user.

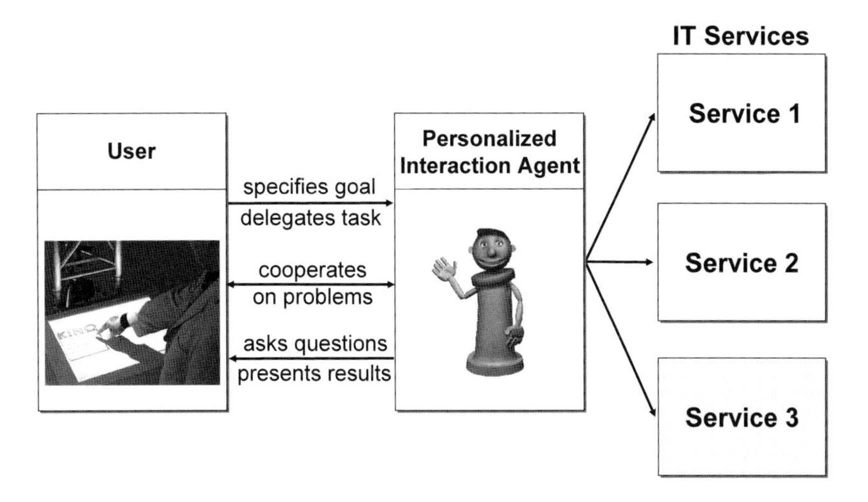

Fig. 2. SMARTKOM's SDDP interaction metaphor

SMARTKOM represents a new generation of multimodal dialogue systems that deal not only with simple modality integration and synchronization but cover the full spectrum of dialogue phenomena that are associated with symmetric multimodality. One of the technical goals of our research in the SMARTKOM project was to address the following important discourse phenomena that arise in multimodal dialogues:

- mutual disambiguation of modalities
- multimodal deixis resolution and generation
- crossmodal reference resolution and generation
- multimodal anaphora resolution and generation
- multimodal ellipsis resolution and generation
- multimodal turn-taking and backchannelling

Symmetric multimodality is a prerequisite for a principled study of these discourse phenomena.

4 Towards a Flexible and Adaptive Shell for Multimodal Dialogues

SMARTKOM was designed with a clear focus on flexibility, as a transmutable system that can engage in many different types of tasks in different usage contexts. The same software architecture and components are used in various roles that Smartakus can play in the following three fully operational experimental application scenarios (see Fig. 3):

- a *communication companion* that helps with phone, fax, email, and authentication tasks

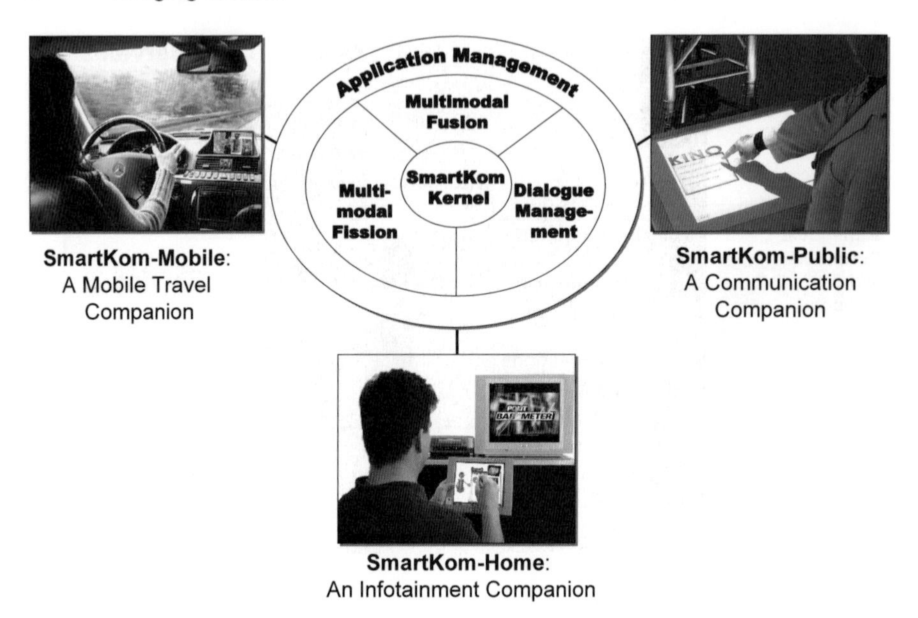

Fig. 3. The three application scenarios of SMARTKOM

- an *infotainment companion* that helps to select media content and to operate various TV appliances (using a tablet computer as a mobile client)
- a *mobile travel companion* that helps with navigation and point-of-interest information retrieval in location-based services (using a PDA as a mobile client)

Currently, the user can delegate 43 types of complex tasks to Smartakus in multimodal dialogues. The SMARTKOM architecture supports not only simple multimodal command-and-control interfaces, but also coherent and cooperative dialogues with mixed initiative and a synergistic use of multiple modalities. SMARTKOM's plug-and-play architecture supports easy addition of new application services.

Figure 4 shows a three-camera configuration of SMARTKOM that can be used as a multimodal communication kiosk for airports, train stations, or other public places where people may seek information on facilities such as hotels, restaurants, and movie theatres. Users can also access their personalized Web services. The user's speech input is captured with a directional microphone. The user facial expressions of emotion are captured with a CCD camera and his gestures are tracked with an infrared camera. A video projector is used for the projection of SMARTKOM's graphical output onto a horizontal surface. Two speakers under the projection surface provide the speech output of the life-like character. An additional camera that can automatically tilt and pan is used to capture images of documents or 3D objects that the user would like to include in multimedia messages composed with the help of SMARTKOM.

As a resource-adaptive multimodal system, the SMARTKOM architecture supports a flexible embodiment of the life-like character that is used as a conversational

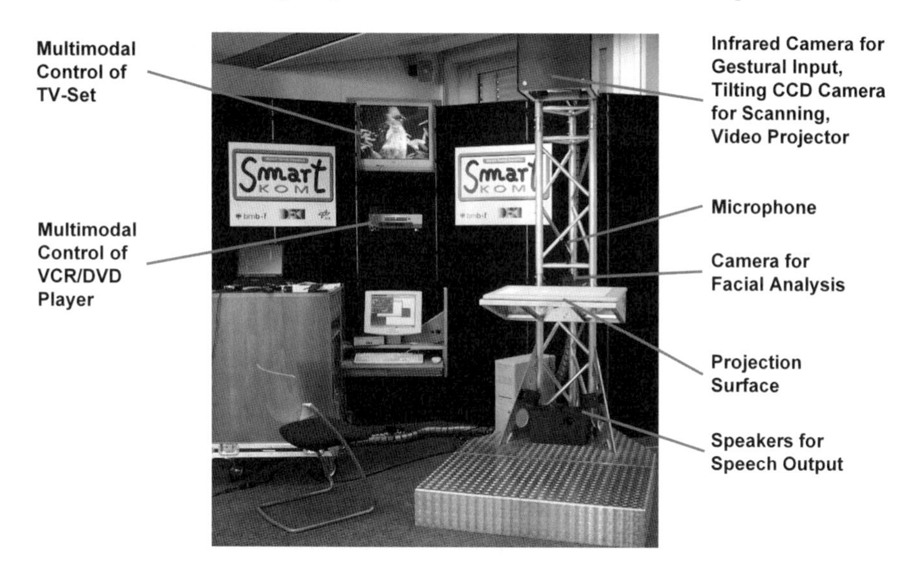

Fig. 4. Multimodal input and output devices of SMARTKOM-Public

partner in multimodal dialogue. The Smartakus agent is visualized either simply as a talking head together with an animated hand, when screen space is scarce, or as a full-body character, when enough screen space is available (see Fig. 2). Thus, Smartakus is embodied on a PDA differently than on a tablet computer or on the large top-projected screen used in the public information kiosk.

5 Perception and Action Under Multimodal Conditions

SMARTKOM bridges the full loop from multimodal perception to physical action. Since the multimodal interaction with Smartakus covers both communicative and physical acts, the mutual understanding of the user and the system can often be validated by checking whether the user and the system "do the right thing" for completing the task at hand.

In a multimodal dialogue about the TV program, the user may browse a TV show database, create a personalized TV listing, and finally ask Smartakus to switch on the TV and tune to a specific program. Smartakus can also carry out more complex actions like programming a VCR to record the user's favourite TV show. Moreover, it can scan a document or a 3D object with its camera and then send the captured image to another person as an email attachment. Fig. 5 shows Dr. Johannes Rau, the former German Federal President, using SMARTKOM's multimodal dialogue capabilities to scan the "German Future Award" trophy and send the scanned image via email to a colleague. This example shows that, on the one hand, multimodal dialogue contributions can trigger certain actions of Smartakus. On the other hand, Smartakus

Fig. 5. The former German Federal President e-mailing a scanned image with the help of Smartakus

Fig. 6. Interactive biometric authentication by hand contour recognition

may also ask the user to carry out certain physical actions during the multimodal dialogue.

For example, Smartakus will ask the user to place his hand with spread fingers on a virtual scanning device, or to use a write-in field projected on the screen for his signature, when biometric authentication by hand contour recognition or signature verification is requested by a security-critical application. Fig. 6 shows a situation in which Smartakus has found an address book entry for the user, after he has introduced himself by name. Since the address book entry, which is partially visualized

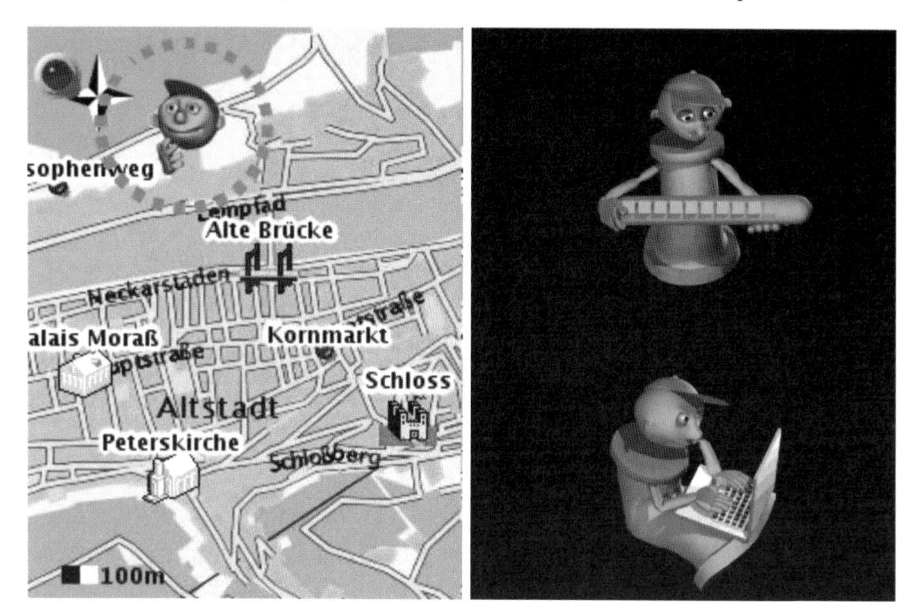

Fig. 7. Adaptive perceptual feedback on the system state

by SMARTKOM on the left part of the display, requests hand contour authentication for this particular user, Smartakus asks the user to place his hand on the marked area of the projected display, so that the hand contour can be scanned by its camera (see Fig. 6).

Since quite complex tasks can be delegated to Smartakus, there may be considerable delays in replying to a request. Our WOZ (Wizard-of-Oz) experiments and user tests with earlier prototypes of SMARTKOM showed clearly that users want simple and fast feedback on the state of the system in such situations. Therefore, a variety of adaptive perceptual feedback mechanisms have been realized in SMARTKOM.

In the upper-left corner of a presentation, SMARTKOM can display a "magic eye" icon, that lights up while the processing of the user's multimodal input is proceeding (see the left part of Fig. 7). "Magic eye" is the common name applied to the green-glow tubes used in 1930s radio equipment to visually assist the listener in tuning a radio station to the point of greatest signal strength. Although SMARTKOM works in real-time, there may be some processing delays caused by corrupted input or complex disambiguation processes.

An animated dashed line (see the left part of Fig. 7) circles the Smartakus character, while the system is engaged in an information retrieval task (e.g., access to maps, EPG (Electronic Program Guide), Web sites). This type of feedback is used when screen space is scarce. When more screen space is available, an animation sequence that shows Smartakus working on a laptop is used for the same kind of feedback. When Smartakus is downloading a large file, it can show a progress bar to indicate to the user how the data transfer is going (see the right part of Fig. 7).

6 A Multiblackboard Platform with Ontology-Based Messaging

SmartKom is based on a distributed component architecture, realizing a multi-blackboard system. The integration platform is called MULTIPLATFORM (**Mu**ltiple **L**anguage **T**arget **I**ntegration **Plat**form **for M**odules, see Herzog et al. (2003)) and is built on top of open source software. The natural choice to realize an open, flexible and scalable software architecture is that of a distributed system, which is able to integrate heterogeneous software modules implemented in diverse programming languages and running on different operating systems. SmartKom includes more than 40 asynchronously running modules coded in four different programming languages: C, C++, Java, and Prolog.

The MULTIPLATFORM testbed includes a message-oriented middleware. The implementation is based on PVM, which stands for *parallel virtual machine*. On top of PVM, a message-based communication framework is implemented based on the so-called publish/subscribe approach. In contrast to unicast routing known from multiagent frameworks, which realize a direct connection between a message sender and a known receiver, MULTIPLATFORM is based on the more efficient multicast addressing scheme. Instead of addressing one or several receivers directly, the sender publishes a notification on a named message queue, so that the message can be forwarded to a list of subscribers. This kind of distributed event notification makes the communication framework very flexible as it focuses on the data to be exchanged and it decouples data producers and data consumers. Compared with point-to-point messaging used in multiagent frameworks like OAA (Martin et al., 1999), the publish/subscribe scheme helps to reduce the number and complexity of interfaces significantly.

GCSI, the **G**alaxy **C**ommunicator **S**oftware **I**nfrastructure (Seneff et al., 1999) architecture is also fundamentally different from our approach. The key component of GCSI is a central hub, which mediates the interaction among various servers that realize different dialogue system components. Within MULTIPLATFORM there exists no such centralized controller component, since this could become a bottleneck for more complex multimodal dialogue architectures.

In order to provide publish/subscribe messaging on top of PVM, we have added another software layer called PCA (**P**ool **C**ommunication **A**rchitecture). In MULTIPLATFORM, the term *data pool* is used to refer to named message queues. Every single pool can be linked with a pool data format specification in order to define admissible message contents. The messaging system is able to transfer arbitrary data contents, and provides excellent performance characteristics (Herzog et al., 2003).

In SmartKom, we have developed M3L (**M**ulti**m**odal **M**arkup **L**anguage) as a complete XML language that covers all data interfaces within this complex multimodal dialogue system. Instead of using several quite different XML languages for the various data pools, we aimed at an integrated and coherent language specification, which includes all substructures that may occur on the different pools. In order to make the specification process manageable and to provide a thematic organization, the M3L language definition has been decomposed into about 40 schema specifications. The basic data flow from user input to system output continuously adds further processing results so that the representational structure will be refined, step-by-step.

The ontology that is used as a foundation for representing domain and application knowledge is coded in the ontology language OIL. Our tool OIL2XSD (Gurevych et al., 2003) transforms an ontology written in OIL (Fensel et al., 2001) into an M3L compatible XML Schema definition. The information structures exchanged via the various blackboards are encoded in M3L. M3L is defined by a set of XML schemas. For example, the word hypothesis graph and the gesture hypothesis graph, the hypotheses about facial expressions, the media fusion results, and the presentation goal are all represented in M3L. M3L is designed for the representation and exchange of complex multimodal content. It provides information about segmentation, synchronization, and the confidence in processing results. For each communication blackboard, XML schemas allow for automatic data and type checking during information exchange. The XML schemas can be viewed as typed feature structures. SMARTKOM uses unification and a new operation called OVERLAY (Alexandersson and Becker, 2003) of typed feature structures encoded in M3L for discourse processing.

Application developers can generate their own multimodal dialogue system by creating knowledge bases with application-specific interfaces, and plugging them into the reusable SMARTKOM shell. It is particularly easy to add or remove modality analyzers or renderers, even dynamically while the system is running. This plug and play of modalities can be used to adjust the system's capability to handle different demands of the users, and the situative context they are currently in. Since SMARTKOM's modality analyzers are independent from the respective device-specific recognizers, the system can switch in real-time, for example, between video-based, pen-based or touch-based gesture recognition. SMARTKOM's architecture, its dialogue backbone, and its fusion and fission modules are reusable across applications, domains, and modalities.

MULTIPLATFORM is running on the SMARTKOM server that consists of 3 dual Xeon 2.8 GHz processors. Each processor uses 1.5 GB of main memory. One processor is running under Windows 2000, and the other two under Linux. The mobile clients (an iPAQ Pocket PC for the mobile travel companion and a Fujitsu Stylistic 3500X webpad for the infotainment companion) are linked to the SMARTKOM server via WaveLAN.

7 Reducing Uncertainty and Ambiguity by Modality Fusion

The analysis of the various input modalities by SMARTKOM is typically plagued by uncertainty and ambiguity. The speech recognition system produces a word hypothesis graph with acoustic scores, stating which word might have been spoken in a certain time frame. The prosody component generates a graph of hypotheses about clause and sentence boundaries with prosodic scores. The gesture analysis component produces a set of scored hypotheses about possible reference objects in the visual context. Finally, the interpretation of facial expressions leads to various scored hypotheses about the emotional state of the user. All the recognizers produce time-stamped hypotheses, so that the fusion process can consider various temporal constraints. The key function of modality fusion is the reduction of the overall uncertainty and the mutual disambiguation of the various analysis results. By fusing symbolic and statistical information derived from the recognition and analysis components for speech, prosody, facial expression and gesture, SMARTKOM can correct various recognition errors of its unimodal input components and thus provide a more robust dialogue than a unimodal system.

In principle, modality fusion can be realized during various processing stages like multimodal signal processing, multimodal parsing, or multimodal semantic processing. In SMARTKOM, we prefer the latter approach, since for the robust interpretation of possibly incomplete and inconsistent multimodal input, more knowledge sources become available on later processing stages. An early integration on the signal level allows no backtracking and reinterpretation, whereas the multimodal parsing approach has to prespecify all varieties of crossmodal references, and is thus unable to cope robustly with unusual or novel uses of multimodality. However, some early fusion is also used in SMARTKOM, since the scored results from a recognizer for emotional prosody (Batliner et al., 2000) are merged with the results of a recognizer for affective facial expression. The classification results are combined in a synergistic fashion, so that a hypothesis about the affective state of the user can be computed.

In SMARTKOM, the user state is used, for example, in the dialogue-processing backbone to check whether the user is satisfied or not with the information provided by Smartakus. It is interesting to note that SMARTKOM's architecture supports multiple recognizers for a single modality. In the current system, prosody is evaluated by one recognizer for clause boundaries and another recognizer for emotional speech. This means that the user's speech signal is processed by three unimodal recognizers in parallel (speech recognition, emotional prosody, boundary prosody).

The time stamps for all recognition results are extremely important since the confidence values for the classification results may depend on the temporal relations between input modalities. For example, experiments in SMARTKOM have shown that the results from recognizing various facial regions (like eye, nose, and mouth area) can be merged to improve recognition results for affective states like anger or joy. However, while the user is speaking, the mouth area does not predict emotions reliably, so that the confidence value of the mouth area recognizer must be decreased. Thus, SMARTKOM's modality fusion is based on adaptive confidence measures that can be dynamically updated depending on the synchronization of input modalities.

One of the fundamental mechanisms implemented in SMARTKOM's modality fusion component is the extended unification of all scored hypothesis graphs and the application of mutual constraints in order to reduce the ambiguity and uncertainty of the combined analysis results. This approach was pioneered in our XTRA system, an early multimodal dialogue system that assisted the user in filling out a tax form with a combination of typed natural language input and pointing gestures (Wahlster, 1991). QuickSet uses a similar approach (Cohen et al., 1997).

In SMARTKOM, the intention recognizer has the task to finally rank the remaining interpretation hypotheses and to select the most likely one, which is then passed

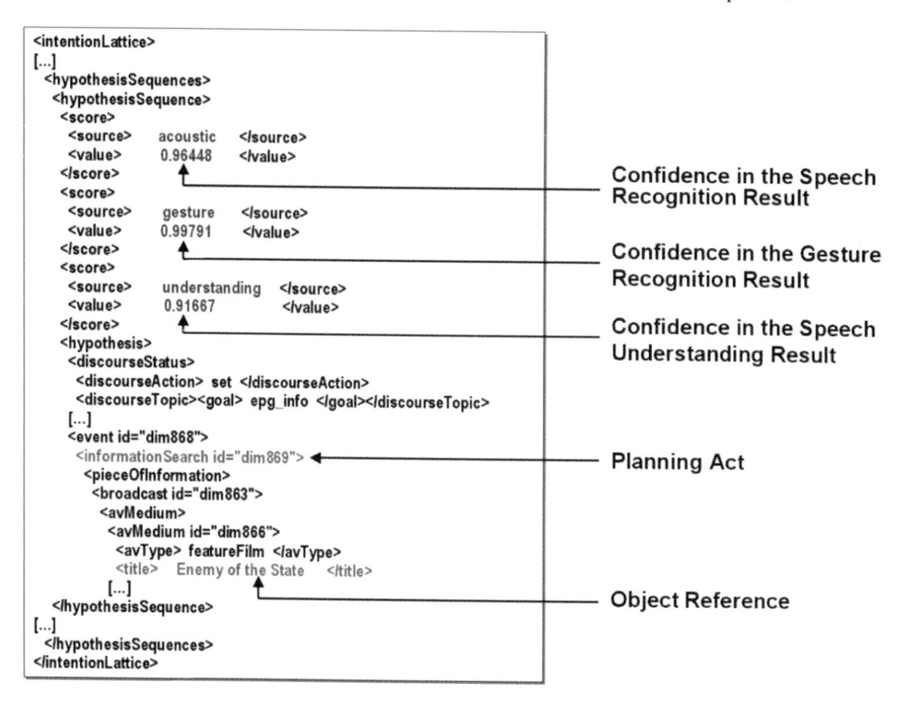

Fig. 8. M3L representation of an intention lattice fragment

on to the action planner. The modality fusion process is augmented by SMARTKOM's multimodal discourse model, so that the final ranking of the intention recognizer becomes highly context sensitive. The discourse component produces an additional score that states how good an interpretation hypothesis fits to the previous discourse (Pfleger et al., 2002). As soon as the modality fusion component finds a referential expression that is not combined with an unambiguous deictic gesture, it sends a request to the discourse component asking for reference resolution. If the resolution succeeds, the discourse component returns a completely instantiated domain object.

Figure 8 shows an excerpt from the intention lattice for the user's input "I would like to know more about this [deictic pointing gesture]". It shows one hypothesis sequence with high scores from speech and gesture recognition. A potential reference object for the deictic gesture (the movie title "Enemy of the State") has been found in the visual context. SMARTKOM assumes that the discourse topic relates to an electronic program guide and the intended action of Smartakus refers to the retrieval of information about a particular broadcast.

8 Plan-Based Modality Fission in SmartKom

In SMARTKOM, modality fission is controlled by a presentation planner. The input to the presentation planner is a presentation goal encoded in M3L as a modality-free representation of the system's intended communicative act. This M3L structure is generated by either an action planner or the dynamic help component, which can initiate clarification subdialogues. The presentation planning process can be adapted to various application scenarios via presentation parameters that encode user preferences (e.g., spoken output is preferred by a car driver), output devices (e.g., size of the display), or the user's native language (e.g., German vs. English). A set of XSLT (Extensible Stylesheet Language Transformations) stylesheets is used to transform the M3L representation of the presentation goal, according to the actual presentation parameter setting. The presentation planner recursively decomposes the presentation goal into primitive presentation tasks using 121 presentation strategies that vary with the discourse context, the user model, and ambient conditions. The presentation planner allocates different output modalities to primitive presentation tasks, and decides whether specific media objects and presentation styles should be used by the media-specific generators for the visual and verbal elements of the multimodal output.

The presentation planner specifies presentation goals for the text generator, the graphics generator, and the animation generator. The animation generator selects appropriate elements from a large catalogue of basic behavioral patterns to synthesize fluid and believable actions of the Smartakus agent. All planned deictic gestures of Smartakus must be synchronized with the graphical display of the corresponding media objects, so that Smartakus points to the intended graphical elements at the right moment. In addition, SMARTKOM's facial animation must be synchronized with the planned speech output. SMARTKOM's lip synchronization is based on a simple mapping between phonemes and visemes. A viseme is a picture of a particular mouth position of Smartakus, characterized by a specific jaw opening and lip rounding. Only plosives and diphthongs are mapped to more than one viseme.

One of the distinguishing features of SMARTKOM's modality fission is the explicit representation of generated multimodal presentations in M3L. This means that SMARTKOM ensures dialogue coherence in multimodal communication by following the design principle "no presentation without representation". The text generator provides a list of referential items that were mentioned in the last turn of the system. The display component generates an M3L representation of the current screen content, so that the discourse modeler can add the corresponding linguistic and visual objects to the discourse representation. Without such a representation of the generated multimodal presentation, anaphoric, crossmodal, and gestural references of the user could not be resolved. Thus, it is an important insight of the SMARTKOM project that a multimodal dialogue system must not only understand and represent the user's multimodal input, but also its own multimodal output.

Figure 9 shows the modality-free presentation goal that is transformed into the multimodal presentation shown in Fig. 10 by SMARTKOM's media fission component and unimodal generators and renderers. Please note that all the graphics and layout shown in Fig. 10 are generated on the fly and uniquely tailored to the dialogue

```
<presentationTask>
 <presentationGoal>
  <inform><informFocus>
   <RealizationType>list</RealizationType>
  </informFocus></inform>
  <abstractPresentationContent>
      <discourseTopic><goal>epg_browse</goal></discourseTopic>
      <informationSearch id="dim24"><tvProgram id="dim23">
          <broadcast><timeDeictic id="dim16">now</timeDeictic>
                 <between>2003-03-20T19:42:32 2003-03-20T22:00:00</between>
            <channel><channel id="dim13"/></channel>
           < /broadcast></tvProgram>
       < /informationSearch>
   <result><event>
       <pieceOfInformation>
        <tvProgram id="ap_3">
          <broadcast><beginTime>2003-03-20T19:50:00</beginTime>
                              <endTime>2003-03-20T19:55:00</endTime>
                             <avMedium><title>Today's Stock News</title></avMedium>
                              <channel>ARD</channel>
          < /broadcast>........</event>
   < /result>
< /presentationGoal>
< /presentationTask>
```

Fig. 9. A fragment of a presentation goal, as specified in M3L

situation, i.e., nothing is canned or preprogrammed. The presentation goal shown in Fig. 9 is coded in M3L and indicates that a list of broadcasts should be presented to the user. Since there is enough screen space available and there are no active constraints on using graphical output, the strategy operators applied by the presentation planner lead to a graphical layout of the list of broadcasts. In an eyes-busy situation (e.g., when the user is driving a car), SMARTKOM would decide that Smartakus should read the list of retrieved broadcasts to the user. This shows that SMARTKOM's modality fission process is highly context aware and produces tailored multimodal presentations.

The presentation planner decides that the channel should be rendered as an icon, and that only the starting time and the title of the individual TV item should be mentioned in the final presentation.

In the next section, we show how the visual, gestural and linguistic context stored in a multimodal discourse model can be used to resolve crossmodal anaphora. We will use the following dialogue excerpt as an example:

1. User: I would like to go to the movies tonight.
2. Smartakus: [displays a list of movie titles] This is a list of films showing in Heidelberg.

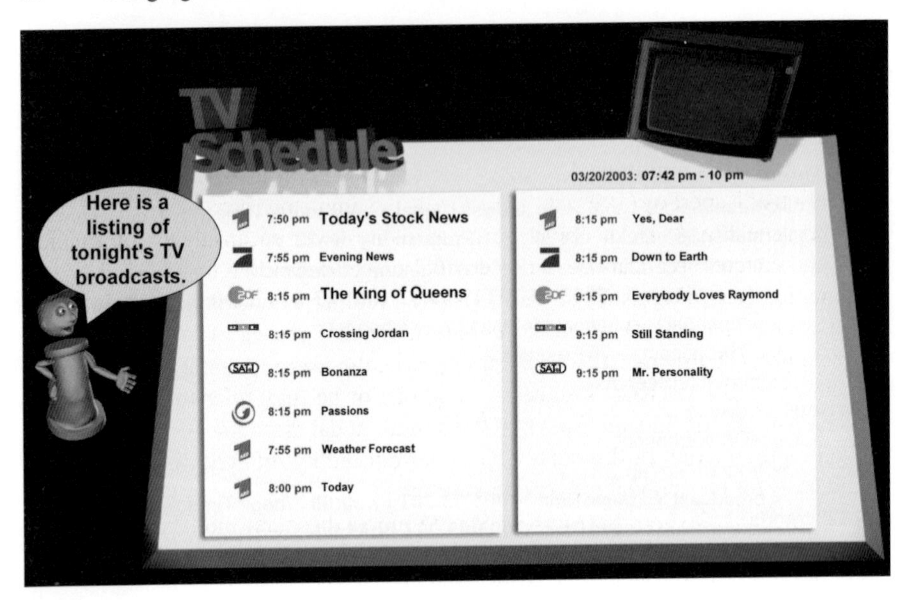

Fig. 10. A dynamically generated multimodal presentation based on a presentation goal

3. User: Hmm, none of these films seem to be interesting ... Please show me the TV program.
4. Smartakus: [displays a TV listing] Here [points to the listing] is a listing of tonight's TV broadcasts (see Fig. 10).
5. User: Please tape the third one!

9 A Three-Tiered Multimodal Discourse Model

Discourse models for spoken dialogue systems store information about previously mentioned discourse referents for reference resolution. However, in a multimodal dialogue system like SMARTKOM, reference resolution relies not only on verbalized, but also on visualized information. A multimodal discourse model must account for entities not explicitly mentioned (but understood) in a discourse, by exploiting the verbal, the visual and the conceptual context. Thus, SMARTKOM's multimodal discourse representation keeps a record of all objects visible on the screen and the spatial relationships between them.

An important task for a multimodal discourse model is the support of crossmodal reference resolution. SMARTKOM uses a three-tiered representation of multimodal discourse, consisting of a domain layer, a discourse layer, and a modality layer. The modality layer consists of linguistic, visual, and gestural objects that are linked to the corresponding discourse objects. Each discourse object can have various surface realizations on the modality layer. Finally, the domain layer links discourse objects with instances of SMARTKOM's ontology-based domain model (Löckelt et al.,

2002). SMARTKOM's three-tiered discourse representation makes it possible to resolve anaphora with nonlinguistic antecedents. SMARTKOM is able to deal with multimodal one-anaphora (e.g., "the third one") and multimodal ordinals ("the third broadcast in the list").

SMARTKOM's multimodal discourse model extends the three-tiered context representation of LuperFoy (1991) by generalizing the linguistic layer to that of a modality layer (see Fig. 11). An object at the modality layer encapsulates information about the concrete realization of a referential object depending on the modality of presentation (e.g., linguistic, gestural, visual). Another extension is that objects at the discourse layer may be complex compositions that consist of several other discourse objects (Salmon-Alt, 2001). For example, the user may refer to an itemized list shown on SMARTKOM's screen as a whole, or he may refer to specific items displayed in the list. In sum, SMARTKOM's multimodal discourse model provides a unified representation of discourse objects introduced by different modalities, as a sound basis for crossmodal reference resolution.

The modality layer of SMARTKOM's multimodal discourse model contains three types of modality objects:

- Linguistic Objects (LOs): For each occurrence of a referring expression in SMARTKOM's input or output, one LO is added.
- Visual Objects (VOs): For each visual presentation of a referrable entity, one VO is added.
- Gesture Objects (GOs): For each gesture performed either by the user or the system, a GO is added.

Each modality object is linked to a corresponding discourse object. The central layer of the discourse model is the discourse object layer. A Discourse Object (DO) represents a concept that can serve as a candidate for referring expressions, including objects, events, states and collections of objects. When a concept is newly introduced by a multimodal communicative act of the user or the system, a DO is created. For each concept introduced during a dialogue, there exists only one DO, regardless of how many modality objects mention this concept.

The compositional information for the particular DOs that represent collections of objects, is provided by partitions (Salmon-Alt, 2001). A partition provides information about possible decompositions of a domain object. Such partitions are based either on perceptual information (e.g., a set of movie titles visible on the screen) or discourse information (e.g., "Do you have more information about the first and the second movie?" in the context of a list of movie titles presented on the screen). Each element of a partition is a pointer to another DO, representing a member of the collection. The elements of a partition are distinguishable from one another by at least one differentiation criterion like their relative position on the screen, their size, or color. For instance, the TV listing shown in Fig. 10 is one DO that introduces 13 new DOs corresponding to particular broadcasts.

The domain object layer provides a mapping between a DO and instances of the domain model. The instances in the domain model are Ontological Objects (OO) that

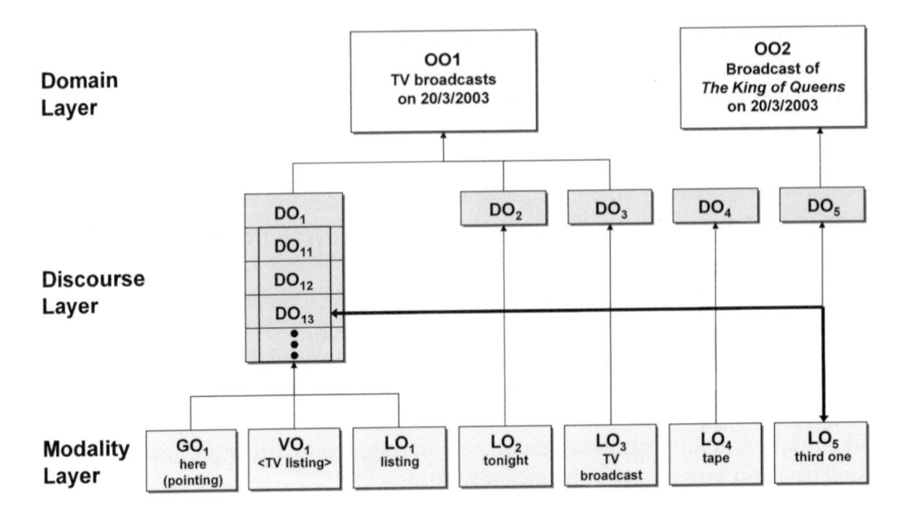

Fig. 11. An excerpt from SMARTKOM's multimodal discourse model

provide a semantic representation of actions, processes, and objects. SMARTKOM's domain model is described in the ontology language OIL (Fensel et al., 2001).

Let us discuss an example of SMARTKOM's methodology for multimodal discourse modeling. The combination of a gesture, an utterance, and a graphical display that is generated by SMARTKOM's presentation planner (see Fig. 10) creates the gestural object GO_1, the visual object VO_1 and the linguistic object LO_1 (see Fig. 11). These three objects at the modality layer are all linked to the same discourse object DO_1 that refers to the ontological object OO1 at the domain layer. Note that DO_1 is composed of 13 subobjects. One of these subobjects is DO_{13}, which refers to OO2, the broadcast of "The King of Queens" on 20 March 2003 on the ZDF channel. Although there is no linguistic antecedent for the one-anaphora "the third one", SMARTKOM can resolve the reference with the help of its multimodal discourse model. It exploits the information that the spatial layout component has rendered OO1 into a horizontal list, using the temporal order of the broadcasts as a sorting criterion. The third item in this list is DO_{13}, which refers to OO2. Thus, the cross-modal one-anaphora "the third one" is correctly resolved and linked to the broadcast of "The King of Queens" (see Fig. 11).

During the analysis of turn (3) in the dialogue excerpt above, the discourse modeler receives a set of hypotheses. These hypotheses are compared and enriched with previous discourse information, in this example stemming from (1). Although (3) has a different topic to (1) (it requests information about the cinema program, whereas (3) concerns the TV program), the temporal restriction (tonight) of the first request is propagated to the interpretation of the second request. In general, this propagation of information from one discourse state to another is obtained by comparing a current intention hypothesis with previous discourse states, and by enriching it (if possible) with consistent information. For each comparison, a score has to be computed reflecting how well this hypothesis fits in the current discourse state. For this purpose, the nonmonotonic OVERLAY operation (an extended probabilistic unification-like scheme, see Alexandersson and Becker (2003)) has been integrated into SMARTKOM as a central computational method for multimodal discourse processing.

10 Beyond Restricted Domains: From SmartKom to SmartWeb

Although SMARTKOM works in multiple domains (e.g., TV program guide, telecommunication assistant, travel guide), it supports only restricted-domain dialogue understanding. Our follow-up project SmartWeb (duration: 2004–2008) goes beyond SMARTKOM in supporting *open-domain question answering* using the entire Web as its knowledge base.

Recent progress in *mobile broadband communication* and *semantic Web technology* is enabling innovative mobile Internet information services, that offer much higher retrieval precision than current Web search engines like Google or Yahoo!. The goal of the SmartWeb project (Reithinger et al., 2005) is to lay the foundations for multimodal interfaces to wireless Internet terminals (e.g., smart phones, Web phones, PDAs) that offer flexible access to various Web services. The SmartWeb consortium brings together experts from various research communities: mobile Web services, intelligent user interfaces, multimodal dialogue systems, language and speech technology, information extraction, and semantic Web technologies (see http://www.smartweb-project.org).

SmartWeb is based on the fortunate confluence of three major efforts that have the potential to form the basis for the next generation of the Web. The first effort is the *Semantic Web* (Fensel et al., 2003) which provides the tools for the explicit markup of the content of webpages; the second effort is the development of *semantic Web services* which results in a Web where programs act as autonomous agents to become the producers and consumers of information and enable automation of transactions. The third important effort is information extraction from huge volumes of rich text corpora available on the Web. There has been substantial progress in extracting named entities (such as person names, dates, locations) and facts relating these entities, for example "Winner[World_Cup, Germany, 1990, Italy]", from arbitrary text.

The appeal of being able to ask a question to a mobile Internet terminal and receive an answer immediately has been renewed by the broad availability of always-on, always-available Web access, which allows users to carry the Internet in their pockets. Ideally, a multimodal dialogue system that uses the Web as its knowledge base would be able to answer a broad range of questions. Practically, the size and dynamic nature of the Web and the fact that the content of most webpages is encoded in natural language makes this an extremely difficult task. However, SmartWeb exploits the machine-understandable content of semantic webpages for intelligent question-answering as a next step beyond today's search engines. Since semantically annotated webpages are still very rare due to the time-consuming and costly manual markup, SmartWeb is using advanced language technology, information extraction

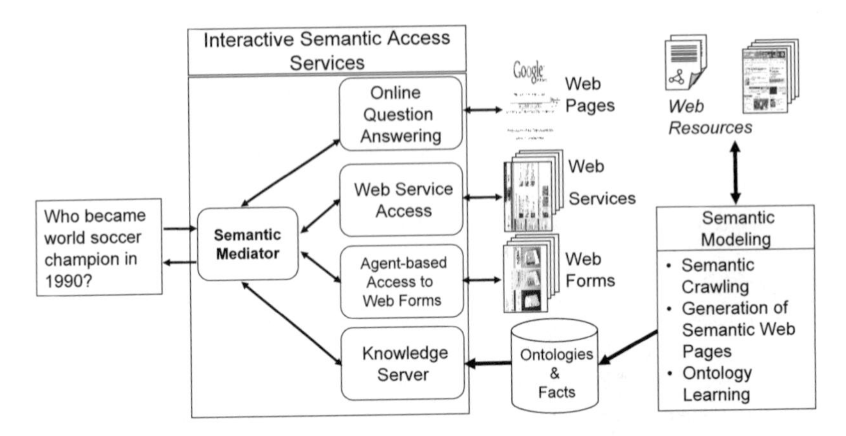

Fig. 12. The semantic mediator of SmartWeb

methods and machine learning for the automatic annotation of traditional webpages encoded in HTML or XML. SmartWeb generates such semantic webpages offline and stores the results in an ontology-based database of facts that can be accessed via a knowledge server (see Fig. 12). In addition, the semantic mediator of SmartWeb uses online question answering methods based on real-time extraction of relevant information from retrieved webpages.

But SmartWeb does not only deal with information-seeking dialogues but also with task-oriented dialogues, in which the user wants to perform a transaction via a Web service (e.g., program his navigation system to find the soccer stadium). Agent-based access to web forms allows the semantic mediator to explore the so-called Deep Web, including webbed databases, archives, dynamically created webpages and sites requiring login or registration.

SmartWeb provides a *context-aware user interface*, so that it can support the user in different roles, e.g., as a car driver, a motor biker, a pedestrian or a sports spectator. One of the demonstrators of SmartWeb is a personal guide for the 2006 FIFA World Cup in Germany, which provides mobile infotainment services to soccer fans, anywhere and anytime (see Fig. 13). The academic partners of SmartWeb are the research institutes DFKI (consortium leader), FhG FIRST, and ICSI together with university groups from Erlangen, Karlsruhe, Munich, Saarbrücken, and Stuttgart. The industrial partners of SmartWeb are BMW, DaimlerChrysler, Deutsche Telekom, and Siemens as large companies, as well as EML, Ontoprise, and Sympalog as small businesses. The German Federal Ministry of Education and Research (BMBF) is funding the SmartWeb consortium with grants totaling 13.7 million €.

11 The Roadmap for Multimodality

This book presents the foundations of multimodal dialogue systems using our fully fledged SmartKom system as an end-to-end working example. However, in this

Fig. 13. SmartWeb: open-domain and multimodal question-anwering

young research field many foundational questions are still open, so that intensive research on multimodality will be needed throughout the next decade. Our research roadmap for 2006–2010 shown in Fig. 14 outlines the research agenda for multimodality (Bunt et al., 2005).

Three "lanes" in the road identify three areas of research and development, including empirical and data-driven models of multimodality, advanced methods for multimodal communication and toolkits for multimodal systems. From 2006 to 2010, in the area of models of multimodality, we envision biologically inspired intersensory coordination models, test suites and benchmarks for comparing, evaluating and validating multimodal systems, and eventually computational models of the acquisition of multimodal communication skills, among other advancements. Advanced methods will include affective, collaborative, multiparty, and multicultural multimodal communication. Toolkits will advance from real-time localization and motion/eye tracking, to the incorporation of multimodality into virtual and augmented reality environments, and resource-bounded multimodality on mobile devices. The annual international conference on multimodal interfaces (ICMI) has become the premier venue for presenting the latest research results on multimodal dialogue systems (see, e.g., Oviatt et al. (2003)).

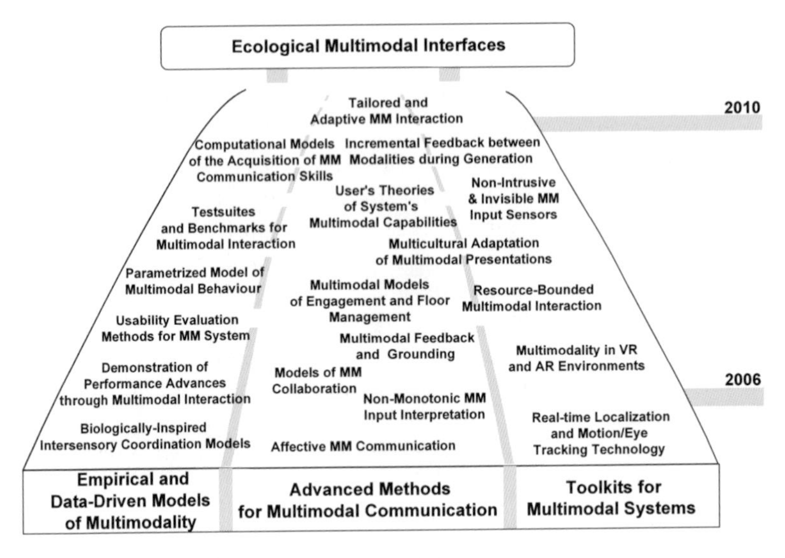

Fig. 14. Research roadmap for advanced multimodal systems

12 The Economic and Scientific Impact of SmartKom

The industrial and economic impact of the SMARTKOM project is remarkable. Up to now, 52 patents concerning SMARTKOM technologies have been filed by members of the SMARTKOM consortium, in areas such as speech recognition (13), dialogue management (10), biometrics (6), video-based interaction (3), multimodal analysis (2), and emotion recognition (2).

In the context of SMARTKOM, 59 new product releases and prototypes have surfaced during the project's life span. 29 spin-off products have been developed by the industrial partners of the SMARTKOM consortium at their own expense. The virtual mouse, which was invented by Siemens, is a typical example of such a technology transfer result. The virtual mouse has been installed in a cell phone with a camera. When the user holds a normal pen about 30 cm in front of the camera, the system recognizes the tip of the pen as a mouse pointer. A red point then appears at the tip on the display. For multimodal interaction, the user can move the pen and point to objects on the cell phone's display.

Former researchers from the SMARTKOM consortium have founded six start-up companies, including Sonicson, Eyeled, and Mineway.[1] The product spectrum of these companies includes multimodal systems for music retrieval, location-aware mobile systems, and multimodal personalization systems.

In addition to its economic impact, SMARTKOM has a broad scientific impact. The scientific results of SMARTKOM have been reported in 255 publications and 117 keynotes or invited lectures. During the project, six SMARTKOM researchers were

[1] http://www.sonicson.com, http://www.eyeled.com, and http://www.mineway.de

awarded tenured professorship. 66 young researchers have finished their master's or doctoral theses in the context of the SMARTKOM project.

SMARTKOM's MULTIPLATFORM software framework (see Sect. 6) is being used at more than 15 industrial and academic sites all over Europe and has been selected as the integration framework for the COMIC (COnversational Multimodal Interaction with Computers) project funded by the EU (Catizone et al., 2003). SMARTKOM's multimodal markup language M3L had an important impact on the definition of MMIL, which is now actively used in the ISO standardization effort towards a multimodal content representation scheme in ISO's Technical Committee 37, Subcommittee 4 "International Standards of Terminology and Language Resource Management". In addition, M3L had an obvious impact on the W3C effort towards a standard for a natural language semantics markup language (see http://www.w3.org/TR/nl-spec/).

The sharable multimodal resources collected and distributed during the SMARTKOM project will be useful beyond the project's life span, since these richly annotated corpora will be used for training, building, and evaluating components of multimodal dialogue systems in coming years. 448 multimodal Wizard-of-Oz sessions resulting in 1.6 terabytes of data have been processed and annotated (Schiel et al., 2002). The annotations contain audio transcriptions combined with gesture and emotion labeling.

Acknowledgments

The SMARTKOM project was made possible by funding from the German Federal Ministry of Education and Research (BMBF) under grant 01 IL 905. I would like to thank my SMARTKOM team at DFKI: Jan Alexandersson, Tilman Becker, Anselm Blocher (project management), Ralf Engel, Gerd Herzog (system integration), Heinz Kirchmann, Markus Löckelt, Stefan Merten, Jochen Müller, Alassane Ndiaye, Rainer Peukert, Norbert Pfleger, Peter Poller, Norbert Reithinger (module coordination), Michael Streit, Valentin Tschernomas, and our academic and industrial partners in the SMARTKOM project consortium: DaimlerChrysler AG, European Media Laboratory GmbH, Friedrich-Alexander University Erlangen-Nuremberg, International Computer Science Institute, Ludwig-Maximilians University Munich, MediaInterface GmbH, Philips GmbH, Siemens AG, Sony International (Europe) GmbH, and Stuttgart University for the excellent and very successful cooperation.

References

J. Alexandersson and T. Becker. The Formal Foundations Underlying Overlay. In: *Proc. 5th Int. Workshop on Computational Semantics (IWCS-5)*, pp. 22–36, Tilburg, The Netherlands, February 2003.

A. Batliner, R. Huber, H. Niemann, E. Nöth, J. Spilker, and K. Fischer. The Recognition of Emotion. In: W. Wahlster (ed.), *Verbmobil: Foundations of Speech-to-Speech Translation*, pp. 122–130, Berlin Heidelberg New York, 2000. Springer.

H. Bunt, M. Kipp, M. Maybury, and W. Wahlster. Fusion and Coordination for Multimodal Interactive Information Presentation. In: O. Stock and M. Zancanaro (eds.), *Multimodal Intelligent Information Presentation*, vol. 27 of *Text, Speech and Language Technology*, pp. 325–340, Berlin Heidelberg New York, 2005. Springer.

R. Catizone, A. Setzer, and Y. Wilks. Multimodal Dialogue Management in the CoMIC Project. In: *Proc. EACL-03 Workshop on "Dialogue Systems: Interaction, Adaptation and Styles of Management"*, Budapest, Hungary, April 2003. European Chapter of the Association for Computational Linguistics (EACL).

P.R. Cohen, M. Johnston, D. McGee, S.L. Oviatt, J.A. Pittman, I. Smith, L. Chen, and J. Clow. QuickSet: Multimodal Interaction for Distributed Applications. In: *Proc. 5th Int. Multimedia Conference (ACM Multimedia '97)*, pp. 31–40, Seattle, WA, 1997. ACM.

D. Fensel, J. Hendler, H. Lieberman, and W. Wahlster (eds.). *Spinning the Semantic Web. Bringing the World Wide Web to Its Full Potential.* MIT Press, Cambridge, MA, 2003.

D. Fensel, F. van Harmelen, I. Horrocks, D.L. McGuinness, and P.F. Patel-Schneider. OIL: An Ontology Infrastructure for the Semantic Web. *IEEE Intelligent Systems*, 16(2):38–45, 2001.

I. Gurevych, S. Merten, and R. Porzel. Automatic Creation of Interface Specifications from Ontologies. In: H. Cunningham and J. Patrick (eds.), *Proc. HLT-NAACL 2003 Workshop on Software Engineering and Architecture of Language Technology Systems (SEALTS)*, pp. 59–66, Edmonton, Canada, 2003. Association for Computational Linguistics.

G. Herzog, H. Kirchmann, S. Merten, A. Ndiaye, P. Poller, and T. Becker. MULTIPLATFORM Testbed: An Integration Platform for Multimodal Dialog Systems. In: H. Cunningham and J. Patrick (eds.), *Proc. HLT-NAACL 2003 Workshop on Software Engineering and Architecture of Language Technology Systems (SEALTS)*, pp. 75–82, Edmonton, Canada, 2003. Association for Computational Linguistics.

M. Johnston, S. Bangalore, G. Vasireddy, A. Stent, P. Ehlen, M. Walker, S. Whittaker, and P. Maloor. MATCH: An Architecture for Multimodal Dialogue Systems. In: *Proc. 10th ACM Int. Symposium on Advances in Geographic Information Systems*, pp. 376–383, Washington, DC, 2002.

M. Löckelt, T. Becker, N. Pfleger, and J. Alexandersson. Making Sense of Partial. In: C.M. Johan Bos, Mary Ellen Foster (ed.), *Proc. 6th Workshop on the Semantics and Pragmatics of Dialogue (EDILOG 2002)*, pp. 101–107, Edinburgh, UK, September 2002.

S. LuperFoy. *Discourse Pegs: A Computational Analysis of Context-Dependent Referring Expressions.* PhD thesis, University of Texas at Austin, December 1991.

D.L. Martin, A.J. Cheyer, and D.B. Moran. The Open Agent Architecture: A Framework for Building Distributed Software Systems. *Applied Artificial Intelligence*, 13(1–2):91–128, 1999.

M.T. Maybury and W. Wahlster. Intelligent User Interfaces: An Introduction. In: M.T. Maybury and W. Wahlster (eds.), *Readings in Intelligent User Interfaces*, pp. 1–13, San Francisco, CA, 1998. Morgan Kaufmann.

S. Oviatt. Multimodal Interfaces. In: J.A. Jacko and A. Sears (eds.), *The Human-Computer Interaction Handbook: Fundamentals, Evolving Technologies and Emerging Applications*, pp. 286–304, Mahwah, NJ, 2003. Lawrence Erlbaum.

S.L. Oviatt, T. Darrell, M.T. Maybury, and W. Wahlster (eds.). *Proc. Int. Conf. on Multimodal Interfaces (ICMI'03)*, Vancouver, Canada, November 5–7 2003. ACM.

N. Pfleger, J. Alexandersson, and T. Becker. Scoring Functions for Overlay and Their Application in Discourse Processing. In: *Proc. KONVENS 2002*, pp. 139–146, Saarbruecken, Germany, September–October 2002.

N. Reithinger, S. Bergweiler, R. Engel, G. Herzog, N. Pfleger, M. Romanelli, and S. Sonntag. A Look Under the Hood — Design and Development of the First SmartWeb System Demonstrator. In: *Proc. Int. Conf. on Multimodal Interfaces (ICMI'05)*, pp. 159–166, Trento, Italy, 2005.

S. Salmon-Alt. Reference Resolution Within the Framework of Cognitive Grammar. In: *Int. Colloquium on Cognitive Science*, pp. 1–15, San Sebastian, Spain, May 2001.

F. Schiel, S. Steininger, and U. Türk. The SmartKom Multimodal Corpus at BAS. In: *Proc. 3rd Int. Conf. on Language Resources and Evaluation (LREC 2002)*, pp. 35–41, Las Palmas, Spain, 2002.

S. Seneff, R. Lau, and J. Polifroni. Organization, Communication, and Control in the Galaxy-II Conversational System. In: *Proc. EUROSPEECH-99*, pp. 1271–1274, Budapest, Hungary, 1999.

W. Wahlster. User and Discourse Models for Multimodal Communication. In: J.W. Sullivan and S.W. Tyler (eds.), *Intelligent User Interfaces*, pp. 45–67, New York, 1991. ACM.

W. Wahlster. SmartKom: Fusion and Fission of Speech, Gestures, and Facial Expressions. In: *Proc. 1st Int. Workshop on Man-Machine Symbiotic Systems*, pp. 213–225, Kyoto, Japan, 2002.

W. Wahlster, E. André, W. Finkler, H.J. Profitlich, and T. Rist. Plan-Based Integration of Natural Language and Graphics Generation. *Artificial Intelligence*, 63:387–427, 1993.

W. Wahlster, N. Reithinger, and A. Blocher. SmartKom: Multimodal Communication with a Life-like Character. In: *Proc. EUROSPEECH-01*, vol. 3, pp. 1547–1550, Aalborg, Denmark, September 2001.

W. Wahlster and R. Wasinger. The Anthropomorphized Product Shelf: Symmetric Multimodal Interaction with Instrumented Environments. In: E. Aarts and J. Encarnação (eds.), *True Visions: The Emergence of Ambient Intelligence*, Berlin Heidelberg New York, 2006. Springer.

Facts and Figures About the SmartKom Project

Anselm Blocher

DFKI GmbH, Saarbrücken, Germany
blocher@dfki.de

Summary. The SMARTKOM project *Dialogue-Based Human Computer Interaction by Coordinated Analysis and Generation of Multiple Modalities* was one of six lead projects in the area of human computer interaction funded by the Federal Ministry of Education and Research and the Federal Ministry of Economics and Labour of Germany. We describe the intention of this initiative and summarize the organizational and funding structure of the SMARTKOM project. The final functionalities of the demonstrator system are compiled as well as an overview of the reception of the project in the research community and in the media.

1 Lead Projects in Human Computer Interaction

In 1998 the Federal Government of Germany initiated an ideas competition for lead projects on human computer interaction. Based on the main interaction and communication means of humans including speech, gestures, facial expressions, haptic signals and visualization and integrated in a multimodal approach these projects were to develop new assistance systems and agents for intelligent information and knowledge processing.

The projects were to be strongly interdisciplinary and to generate both scientific impact for further research and application-oriented prototypes as demonstrators addressing a broader public. Inviting research partners from science and industry to participate in this kind of large strategic collaborative project was aimed at developing attractive scientific solutions with a high market potential and it resulted in a total of 89 project proposals. High emphasis was put on the usability and the ergonomics of the new forms of interaction and their user acceptance. Design and development of systems and prototypes were based on conducting preliminary user studies as well as accompanying and final evaluations.

Finally emerging out of the project ideas proposed, six major lead projects were selected by a jury and recommended for funding. Starting in mid-1999, the Federal Ministry of Education and Research and the Federal Ministry of Economics and Labour launched the four-year projects with a funding volume of 82.6 million €. The total funding requirement for a total of 102 project partners yielded 152.2 million €.

Covering a large range in human computer interaction the lead projects funded addressed additional but disjunct areas focussing on the following highlights:

- ARVIKA:
  Mobile systems for action in mixed real/virtual working environments. Practical concepts for the design of augmented reality systems
- EMBASSI:
  Intelligent user interfaces for consumer electronics. Individually adaptable access to public terminal systems for disabled persons
- INVITE:
  Combining intuitive visualization and personalized navigation, semantic search, automated classification and implicit recording of knowledge
- MAP:
  New ways of using and integrating computers in mobile activities. A wireless, networked pocket-size personal assistant
- MORPHA:
  Service robots for use in private homes and in production. Quick intuitive programming by pointing and showing what is to be done
- SMARTKOM:
  Human computer interaction in dialogues. Situation-based understanding of vague, ambiguous or incomplete multimodal input at the semantic and pragmatic levels

In parallel to the start of the lead projects a scientific advisory board with international membership was installed. The advisory board continuously monitored and evaluated the projects during the funding period and advised the two ministries on the steering decisions to be taken. The members of the advisory board were:

- Professor R. Reichwald, TU Munich (chairman)
- U. Klotz, IG Metall, Frankfurt/M. (SMARTKOM reviewer)
- M. Bartels, IPmotion GmbH, Marburg
- Professor P. Cohen, Oregon Graduate Institute, Beaverton, USA
- Professor M. Gross, ETH Zurich, Switzerland
- Dr. G.B. Hantsch, Deutsches Handwerksinstitut e.V., Karlsruhe
- Professor R. Hoffmann, TU Dresden
- Professor M.K. Lang (Ordinarius i.R.), TU Munich
- Professor S. Maaß, University of Bremen (SMARTKOM reviewer)
- Dr. M. Maybury, MITRE Corporation, Bedford, USA
- Professor A. Ourmazd, Communicant AG, Frankfurt/Oder
- Dr. J. Redmer, Remshalden-Grunbach
- Professor G. Rigoll, TU Munich
- Professor J. Sauter, ART+COM AG, Berlin
- P. Zoche, Fraunhofer-ISI, Karlsruhe

For a comprehensive overview of the six lead projects on human computer interaction see Krahl and Günther (2003).

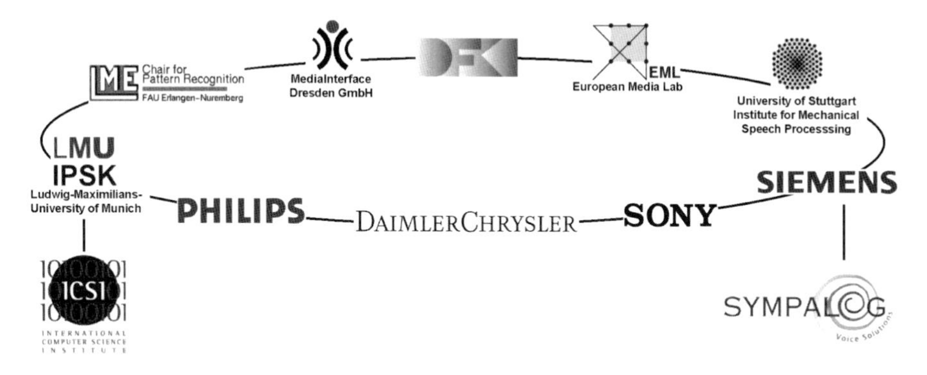

Fig. 1. The SMARTKOM consortium

2 SmartKom: A Lead Project in Human Computer Interaction

The SMARTKOM project was designed as a four-year initiative. The network plan was set up in spring 1999 defining nine subprojects with a total of 70 workpackages. Within the twelve research groups (see next section) a total of 218 person years were allocated. The project started September 1, 1999 and ended September 30, 2003.

2.1 The SmartKom Partners

The SMARTKOM consortium consisted of twelve partners, integrating research centers, large-scale industry, SME and universities (see Fig. 1).

- **Research centers:**
  - German Research Center for Artificial Intelligence GmbH, Kaiserslautern and Saarbrücken (DFKI, Main Contractor)
  - European Media Laboratory GmbH, Heidelberg (EML)
  - International Computer Science Institute, Berkeley/USA (ICSI)
- **Large-scale industry:**
  - DaimlerChrysler AG, Stuttgart (DCAG)
  - Philips Speech Processing, Aachen
  - Siemens Aktiengesellschaft, Munich
  - Sony International (Europe) GmbH, Stuttgart
- **SME:**
  - MediaInterface Dresden GmbH (MID)
  - Sympalog Voice Solutions GmbH, Erlangen
- **Universities:**
  - Chair for Pattern Recognition, Friedrich-Alexander-University Erlangen-Nuremberg (FAU)
  - Institute for Natural Language Processing, University of Stuttgart (IMS)
  - Institute of Phonetics and Speech Communication, Ludwig-Maximilians-University Munich (LMU)

2.2 Funding and Controlling

The SMARTKOM project was funded by the German Federal Ministry of Education and Research (BMBF, see Fig. 2, grant no. 01 IL 905) and additionally financed by the industrial and SME partners. Until the year 2003, 16.8 million € was allocated to the project. In addition the industrial and SME partners brought in 8.9 million € (see Table 1). Universities and research centers received a full 100% funding, while industrial partners contributed 60% of their costs.

Table 1. Funding

BMBF-Funding, 01.09.1999 – 30.09.2003	16.8 Mio. €
Industrial and SME investment	8.9 Mio. €
Overall volume	**25.7 Mio. €**

The project was controlled by the German Aerospace Research Establishment (DLR), Berlin (see Fig. 3).

SPONSORED BY THE
Federal Ministry
of Education
and Research

Fig. 2. German Federal Ministry of Education and Research

Fig. 3. German Aerospace Research Establishment

2.3 Management and Organization Structure

The SMARTKOM project was jointly managed by the Scientific Management, the SMARTKOM Steering Committee and the Group of Module Managers (see Fig. 4). The Scientific Management organised, coordinated and supervised the project and evaluated the progress of the implementation in close contact with the funding agency.

The assignment of the SMARTKOM Steering Committee was to support the Scientific Management in ensuring the scientific excellence of the work. The SMARTKOM Steering Committee consisted of ten project partner leaders to adjust the project

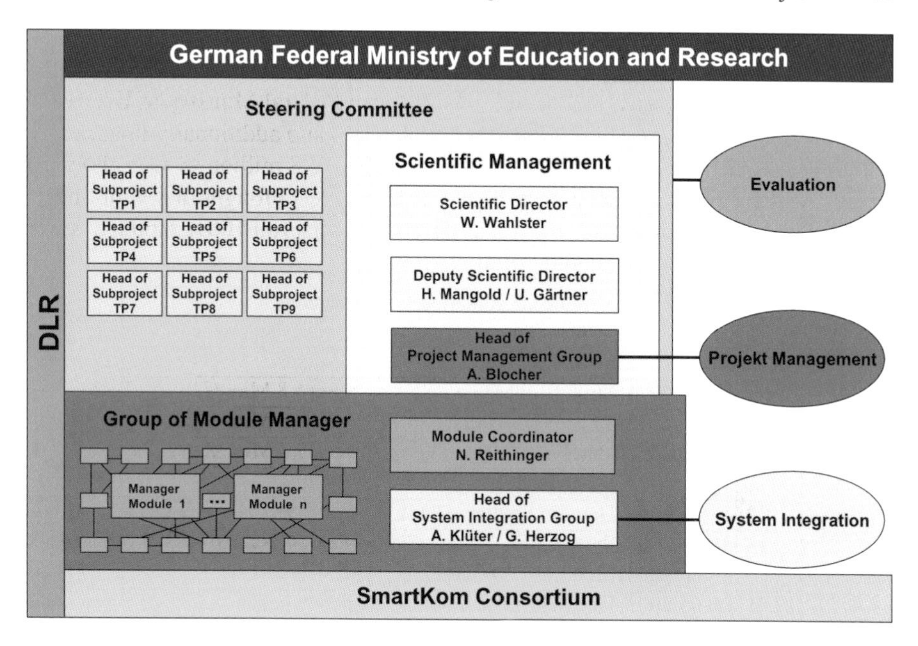

Fig. 4. The management structure of the SmartKom Project

plan to new scientific results or trends in HCI technology and to react immediately to problems that arose in the development and implementation process. A total of 15 meetings (including the constituent one) were held. Within the SmartKom Steering Committee there were assigned responsibilities for subprojects and working areas (WA) (see Table 2).

Table 2. Subprojects and working areas

Subproject/WA	Title	Lead partner
1	Modality specific analysis	DCAG
2	Multimodal interaction	FAU
3	Application interface	Sony
4	Genaration and multimodal media design	IMS
5	SmartKom-Public	Siemens
6	SmartKom-Mobile	EML
7	SmartKom-Home/Office	Philips
8	System integration	DFKI
9	Project management	DFKI
WA	Evaluation	LMU
WA	Technical integration	MID

The main task of the Group of Module Managers was to tie together and bundle the distributed development forces. In the meetings of the Group of Module Man-

Table 3. Large project workshops

Date	Topic	Place
11.-13.10.1999	Data collection	Munich
17.-18.11.1999	System integration	Kaiserslautern
29.-30.11.1999	Domains	Aachen
10.-11.04.2000	System integration	Kaiserslautern
11.-12.05.2000	Interpretation and ontology	Stuttgart
08.-09.08.2000	Module coordination	Stuttgart
14.12.2000	Module coordination	Munich
14.-15.03.2001	System integration	Kaiserslautern
18.-19.04.2001	Function modeling	Stuttgart
26.06.2001	Module coordination	Ulm
09.-10.10.2001	System integration	Kaiserslautern
13.11.2001	Module coordination	Saarbrücken
15.01.2002	Module coordination	Stuttgart
10.-11.04.2002	Domain ontology/XML	Saarbrücken/Heidelberg
18.-19.06.2002	System integration	Kaiserslautern
18.02.2003	Module coordination	Kaiserslautern
19.05.-20.05.2003	UserState processing	Saarbrücken

agers all important decisions concerning interface definitions, offline data flow, delivery schedules, and integration cycles were prepared, discussed and communicated. The implementation process was tightly monitored and supervised. It also provided contact points for the Scientific Management and gave feedback to the SmartKom Steering Committee.

Beside numerous bilateral and multilateral workshops, 17 large project meetings were held to discuss the scientific matters, to fine-tune details of the realization process, and to synchronize the inter-module communication (see Table 3).

Table 4. Project steering and review meetings

Project steering meetings	
10.05.2000	Stuttgart
22.05.2001	Heidelberg
10.05.2002	Heidelberg

Project review meetings	
13.12.2000	Munich
18.12.2002	Saarbrücken
05.09.2003	Stuttgart

2.4 Reviews and HCI Events

The SmartKom reviewers and the BMBF/DLR were informed in three milestone review meetings and in three steering meetings about the progress of the project (see Table 4). During these large project meetings the results achieved as well as the ongoing and planned work were presented in a total of 51 scientific talks with more than 1000 slides. In the accompanging demonstration sesssions the current SMARTKOM systems were shown and industrial SMARTKOM spin-offs of the partners were first presented.

In the framework of the six HCI lead projects two International Status Conferences were held to reach a high degree in visibility both in the international research community and in the media. In order to optimize the cooperation with and to strengthen the understanding of the HCI partner projects several workshops on topics of general interest were organized (see Table 5).

Table 5. HCI events

Date	HCI event
28.09.1999	HCI Start Meeting
29.09.2000	WS Multimodal Interaction and Modeling
29.01.2001	HCI Congress
22.02.2001	WS Domain Modeling and Ontologies
17.05.2001	WS Architecture
03.07.2001	WS Tracking
25.09.2001	WS Adaptivity
26.10.2001	Int. Status Conference Lead Projects "Human Computer Interaction" 2001
25.02.2002	WS Usability
07.03.2002	WS Benchmarking
03.-04.06.2003	Int. Status Conference Lead Projects "Human Computer Interaction" 2003

3 The SmartKom System

3.1 Milestones and Final System Functionalities

The first important milestone, the first fully integrated SMARTKOM demonstrator, was reviewed in December 2000 (13.12.2000). This first SMARTKOM demonstrator was unveiled to the public at the national conference Mensch&Computer 2001 (04.-08.03.2001). The first English SMARTKOM demonstrator was presented at the international conference Eurospeech (03.-07.09.2001, see Wahlster et al. (2001)). The first SMARTKOM demonstrator system covered basic functionalities in the SMARTKOM-Public scenario and worked in the cinema reservation domains: selection of cinemas, information on movies, and seat reservation.

The final SMARTKOM demonstrator was presented in June 2003 at the HCI Status Conference (03.-04.06.2003; scenarios Public and Home/Office) and during Mensch&Computer 2003 (07.-10.09.2003; scenario Mobile). At CeBIT'04 (18.-24.03.2004) all three scenarios were shown. The final SMARTKOM demonstrator allows the user symmetric multimodal interaction in 14 applications with 52 functionalities (see Table 6). An exemplary interaction with the SMARTKOM system can be found in Reithinger and Herzog (2006). For technical data of the SMARTKOM demonstrator see Herzog and Ndiaye (2006).

Table 6. Multimodal addressable functionalities in SMARTKOM

Home			
EPG (Electronic-Programming Guide)	General program Channel selection Channel information Selection based on genre	Information for one single broadcast Time-based operations Help functions for genres	7
TV	On/off	Channel selection	2
VCR control	On/off Record Play Pause	Wind/rewind Programming using EPG and the calendar	6
Lean-Forward/ Lean-Backward	Select Lean-Backward Deactivate Lean-Backward	Context-aware presentations	3
Total Home			**18**
Public			
Telephone	Manipulative key operations Telephony functions	Audio handling Address book	4
Hand contour biometry	Selection of biometry type Hand contour biometry	Presentation and camera control Address book (see above)	3
Voice biometry	Presentation and audio control Voice biometry	Address book (see above) Selection of biometry type (see above)	2
Signature biometry	Presentation and tablet control Signature biometry	Address book (see above) Selection of biometry type (see above)	2
Fax	Presentation and interaction Fax handling	Address book (see above) Camera control	3
E-mail	Presentation and interaction E-mail handling	Address book (see above) Camera control (see above)	2
Cinema	General program Movie information	Seat reservation Cinema location	4
Total Public			**20**
Mobile			
Car navigation	Selection of start and goal city Route type selection Car route computation	Selection of parking garage Information about parking garages	5
Pedestrian navigation	Selection of map type Selection of start and goal Route computation	Selection of points of interest Information for points of interest Integrated car and pedestrian route planning	6
Map manipulation	Resize Change viewpoint	Help functions for map interactions	3
Total Mobile			**14**
Total SMARTKOM			**52**

The economic and scientific impact of the SMARTKOM project is described in Wahlster (2006). A list of SMARTKOM-related scientific papers, keynotes, talks and presentations is available at the web site http://www.smartkom.org.

3.2 Public Relations

SMARTKOM has been shown at two press conferences and in various presentations to a large number of representatives of science, industry and politics (see Table 7).

Table 7. Press conferences and highlight presentations

Press conferences	
14.12.2000	Munich
05.09.2003	Stuttgart

Highlight presentations	
03.-07.09.2001	EUROSPEECH'01
22.05.2002	Dr. Angela Merkel (Party President CDU)
22.08.2002	Josef Brauner (Deutsche Telekom AG)
	Saarland Minister for Economic Affairs and Labour Dr. Hanspeter Georgi
26.08.2002	German Federal President Johannes Rau
	Minister President of the Saarland Peter Müller
	Saarland Minister for Education, Culture and Science Jürgen Schreier
	President of Saarland University Prof. Margret Wintermantel
17.-20.09.2002	ICSLP'02
07.01.2003	EU executive Mr. Horst Forster
24.09.2003	Indian State Secretary of the Department of Science and Technology Prof. V. S. Ramamurthy
03.12.2003	Information Day "Mensch-Technik-Interaktion: Impulse für den Maschinenbau" of the German Engineering Federation (VDMA)
18.-24.03.2004	CeBIT'04
11.05.2004	German Language Technology Congress "LT Summit" at DFKI
29.-30.06.2004	Congress and Exposition "Empower Deutschland"
15.04.2005	Jürgen Gallmann,
	General Manager, Microsoft Germany, Vice President, Microsoft EMEA
	Dr. Said Zahedani, Director Developer Platform & Strategy Group
	Pierre-Yves Saintoyant, Director European Microsoft Innovation Center
	Walter Seemayer, NTO Microsoft Germany

In addition the SMARTKOM system has been demonstrated — amongst others — to representatives of: Adam Opel AG, ALCATEL, ATR Spoken Language Communication Research Labs, Bentley Motors, BMW AG, CapInfo, FUJI RIC, Harvard University, Hitachi Central Research Lab, NTT DoCoMo and DoCoMo Euro-Labs, NTT (Nomura Research Institute), RICOH (Japan), Robert Bosch GmbH, Siemens AG, Siemens Medical Solutions, Volkswagen Autostadt and Volkswagen Research.

The project was covered in several scientific or news TV programs and far more than 100 relevant press articles were published about SMARTKOM (see Table 8).

Table 8. Selected TV programs and press articles

Selected TV programs	
11.03.2001	3sat “hitec”
21.04.2001	N-TV
29.11.2001	zdf “heute”
20.01.2002	RTL “SPIEGEL TV Magazin”
19.08.2002	RTL “future Trend”
29.09.2002	ZDF “heute”
05.09.2003	Saarländischer Rundfunk “Aktueller Bericht”
03.02.2004	ZDF “heute-journal”
08.10.2004	3sat “Nano - Die Welt von Morgen”
Selected press articles	
03.03.1999	Saarbrücker Zeitung *Bonn fördert Saarbrücker Forscher mit Millionen*
15.03.1999	c’t *Rechner ohne Tastatur und Maus*
22.03.1999	Computerworld *Nur wegbeamen kann man sich nicht*
06.10.1999	Technisch Weekblad *Spraakgestuurde databankprojecten*
30.01.2001	Die Welt *Eine Maschine versteht Gesten des Menschen*
19.02.2001	Handelsblatt *Sprechen und Zeigen statt Tippen*
23.03.2001	VDI nachrichten *Maschinen lesen Wünsche von den Augen ab*
01.06.2001	c’t-News *Maschine versteht Sprache, Gesten und Augenbewegungen*
07.06.2001	ComputerZeitung *Mimik und Augenbewegungen steuern Computer*
04.02.2002	CHIP *Er hört aufs Wort*
12.03.2002	Frankfurter Allgemeine Zeitung *Sprich mit mir*
14.03.2002	Financial Times Deutschland *Verständige Maschinen*
22.04.2002	Salzburger Nachrichten Online *Deutscher Zukunftspreisträger entwickelt einen Computer-Butler ohne Maus und Tastatur*
06/2002	Spektrum der Wissenschaft *Redselige Chips*
27.08.2002	Saarbrücker Zeitung *Mit dem Bus zum roten Teppich*
29.09.2002	Handelsblatt *Maschinen, die Menschen besser verstehen*
08.10.2002	Süddeutsche Zeitung *Grimassen für den Computer*
26/2002	c’t *Talkmaster*
05.03.2003	Hindustan Times *Talking computer in the making*
14.04.2003	i-com - Zeitschrift für interaktive und kooperative Medien *SmartKom - Multimodale Mensch-Technik-Interaktion*
10.06.2003	Berliner Zeitung *Verständnisvolle Computer*
13.06.2003	VDI Nachrichten *Computer und Roboter lernen Menschen zu verstehen*
16.06.2003	c’t *“Wenn wir schreiten Seit’ an Seit’ ...”*
07.07.2003	dpa *Die Technik soll den Menschen verstehen lernen*
15.09.2003	Computerzeitung *Intelligente Schnittstellen erhöhen Produktakzeptanz*
06.11.2003	Handelsblatt *Computer reagieren künftig auch auf Fingerzeig*
28.01.2004	Saarbrücker Zeitung *Der Computer gehorcht aufs Wort / Das Handy lernt sehen*
13.03.2004	Handelsblatt *Elektronische Sekretärinnen erkennen Gesten*
07.04.2004	Die Zeit *Und wie viele Chips hat Ihrer?*
13.05.2004	Saarbrücker Zeitung *Mit den Maschinen reden lernen*

To give a deeper insight into the final SMARTKOM system, its scenarios, applications and functionalities and into multimodal interaction in general terms, videos — in English and German — can be downloaded from the SMARTKOM web site http://www.smartkom.org:

- SMARTKOM Complete: Covering all three scenarios Public, Home and Mobile (MPEG, $\approx$ 170 MB)
- SMARTKOM Mobile: Special long version of the Mobile scenario (MPEG, $\approx$ 67 MB)

Fig. 5. The SMARTKOM logo

References

G. Herzog and A. Ndiaye. Building Multimodal Dialogue Applications: System Integration in SmartKom, 2006. In this volume.

R. Krahl and D. Günther (eds.). *Proc. Human Computer Interaction Status Conference 2003*, Berlin, Germany, June 2003. DLR.

N. Reithinger and G. Herzog. An Exemplary Interaction with SmartKom, 2006. In this volume.

W. Wahlster. Dialogue Systems Go Multimodal: The SmartKom Experience, 2006. In this volume.

W. Wahlster, N. Reithinger, and A. Blocher. SmartKom: Multimodal Communication with a Life-like Character. In: *Proc. EUROSPEECH-01*, vol. 3, pp. 1547–1550, Aalborg, Denmark, September 2001.

An Exemplary Interaction with SmartKom

Norbert Reithinger and Gerd Herzog

DFKI GmbH, Saarbrücken, Germany
{reithinger,herzog}@dfki.de

Summary. The different instantiations of the SmartKom demonstration system offer a broad range of application functions and sophisticated dialogue capabilities. We provide a first look at the final SmartKom prototype from the point of view of the end user. In particular, a typical interaction sequence is presented in order to illustrate the functionality of the integrated multimodal dialogue system.

1 Introduction

The three different usage scenarios of SmartKom and its various application functions allow for a wide range of possible interactions. Aiming at flexible and natural multimodal dialogues, we need to define the intended behaviour of the system as well as its general look and feel. In particular, we have to lay out basic dialogue steps, which the user can freely combine during his or her interaction with the system. This design task leads to an iterative process, which takes the initial project definition as a starting point. System developers, scenario experts and prospective users (e.g., through wizard-of-oz experiments) collaborate to design and refine the capabilities of the multimodal system to be build. Following the paradigm of scenario-based design (Carroll, 2000; Rosson and Carroll, 2003), the initial focus of the design activity is not a formal functional specification but a description of how people will use the system to accomplish work tasks and other activities.

In the next section, we first present the dialogue descriptions we use to define and document the basic interactions in SmartKom as well as the illustrative dialogue protocols that can be generated automatically from the extensive log data resulting from a system run. Then we walk through an original sample dialogue between a test user and the system prototype to provide an insight into the capabilities of the SmartKom demonstrator. A presentation on paper, of course, can only provide a rough sketch of a multimodal interaction. The SmartKom Web site, located at `http://www.smartkom.org`, provides a comprehensive video that complements the description of the integrated SmartKom prototype given here.

2 From Dialogue Drafts to Dialogue Protocols

For a large, distributed project it is necessary to coordinate the design and development efforts on various levels. One important task is to agree on those dialogue steps and discourse phenomena that the system should be able to process. At the beginning, the different scenario experts performed user studies (see, e.g., Horndasch et al. (2006)). The goal was to come up with preferred interaction metaphors and interaction sequences that should be realised. In addition, the wizard-of-oz data collection (Schiel and Türk, 2006) provided important insights concerning natural interaction sequences.

With this in mind, we designed a template for dialogue descriptions, which were the basis for the communication between interface designers and scenario experts. The idea was to collect all relevant information for turns, i.e., input/output sequences between end user and system, and turn sequences that were to be realised.

Figure 1 shows a typical excerpt from such a document and presents one turn for the English Mobile scenario. A turn description starts with possible inputs from the user. The list of utterances is not exhaustive and contains category types like names. The explanatory notes section below contains remarks about the processing of these input utterances. In the example, it contains all entries of the English base lexicon for sights in the town of Heidelberg. This is also the place to document limitations of the system. The turn description ends with possible verbal system reactions and example screenshots.

Initially, we took the interaction descriptions as defined by the scenario experts and wrote one document for each functionality. The documents have been made available via the SMARTKOM intranet for project-wide discussion and potential enhancements. During the realization phase, the information in the descriptions was further augmented with real processing results from the system. The scenario experts, who sometimes had no immediate access to the latest version of the development system, could then comment on the results and provide feedback. To ensure consistency, the head of the system integration group was in charge of all additions and changes.

Of course, it is difficult to argue about a dialogue-based interaction unless you have a video of the interaction or some other sort of protocol. The SMARTKOM testbed (Herzog et al., 2004) is able to trace all data communication between the various modules of the system. This option is very useful during system development to debug the system on various levels. As the log contains the results from speech recognition, the modality analysis components and the presentation modules, it provides all information necessary to automatically create a protocol of a particular interaction sequence. Even though it does not contain the animations, important changes in the screen display and the final screenshot are included in the log file and can be extracted together with the textual output of the system. The information in the trace file is encoded in M3L (Herzog and Reithinger, 2006; Herzog and Ndiaye, 2006), the XML-based language that is used for data exchange between SMARTKOM software components. XSLT style sheets (see, e.g., Gardner and Rendon (2001)) are employed to automatically extract and format the relevant information from an interaction log.

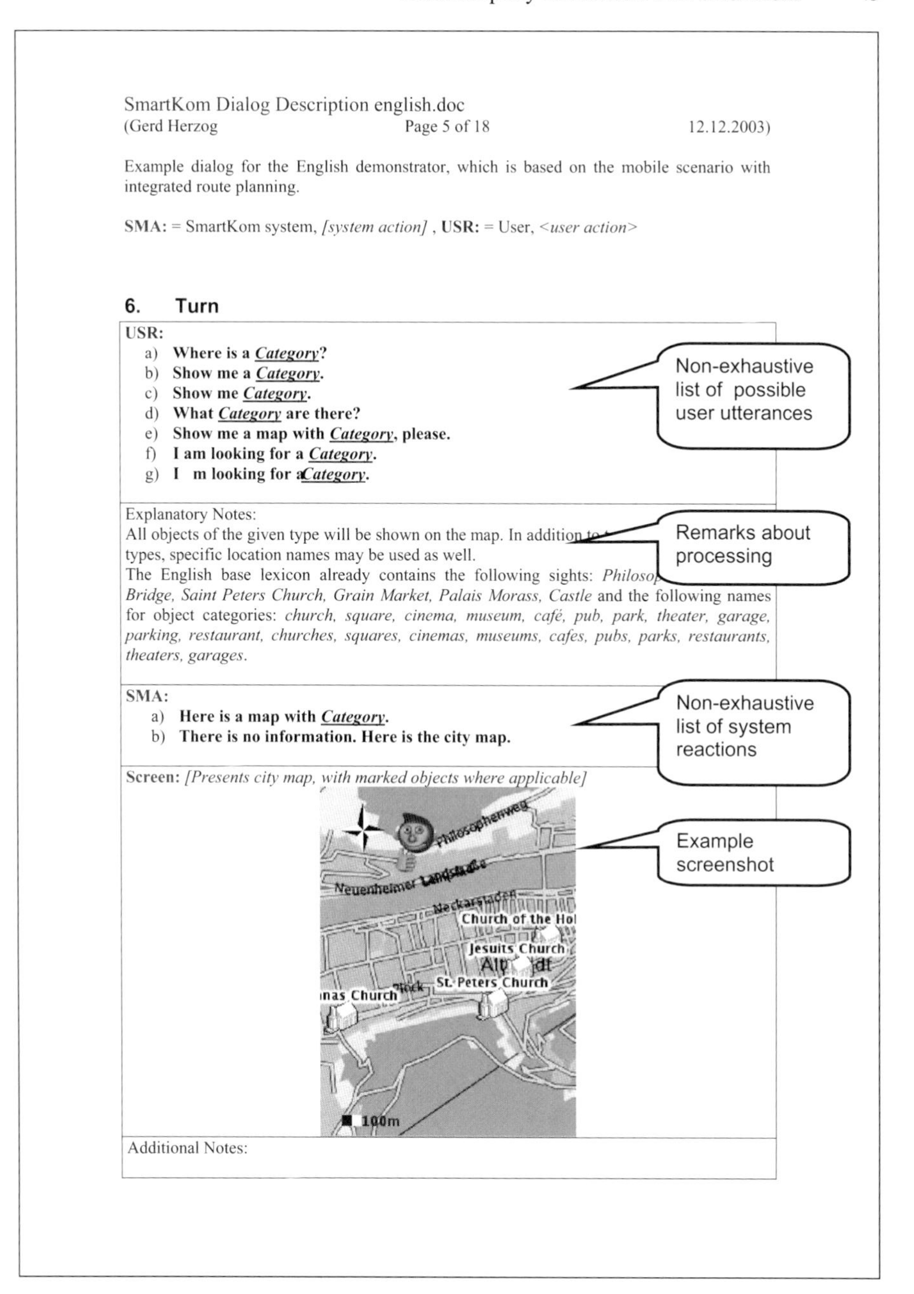

SmartKom Dialog Description english.doc
(Gerd Herzog Page 5 of 18 12.12.2003)

Example dialog for the English demonstrator, which is based on the mobile scenario with integrated route planning.

SMA: = SmartKom system, *[system action]* , **USR:** = User, *<user action>*

6. Turn

USR:
a) **Where is a *Category*?**
b) **Show me a *Category*.**
c) **Show me *Category*.**
d) **What *Category* are there?**
e) **Show me a map with *Category*, please.**
f) **I am looking for a *Category*.**
g) **I m looking for a *Category*.**

Explanatory Notes:
All objects of the given type will be shown on the map. In addition types, specific location names may be used as well.
The English base lexicon already contains the following sights: *Philoso Bridge, Saint Peters Church, Grain Market, Palais Morass, Castle* and the following names for object categories: *church, square, cinema, museum, café, pub, park, theater, garage, parking, restaurant, churches, squares, cinemas, museums, cafes, pubs, parks, restaurants, theaters, garages*.

SMA:
a) **Here is a map with *Category*.**
b) **There is no information. Here is the city map.**

Screen: *[Presents city map, with marked objects where applicable]*

Additional Notes:

Fig. 1. An example page of the English dialogue description

As a result, a browsable HTML page and a self-contained document in PDF format are created, which contain the condensed information for the specific interaction. This documentation of an interaction sequence is used to discuss the implemented dialogue functionality and may initiate a new development iteration, which can also lead to changes in the corresponding dialogue description.

Figure 2 shows a page of the dialogue protocol for the example dialogue presented in the next section. Each entry starts with the internal message number of the corresponding data exchange. On the top of the page, there are two screenshots which are the result of the previous user interaction, where the user asked for the cinema program. The next message in the protocol contains the best hypothesis of the speech recogniser, followed by the word chain that has been selected as the basis for input interpretation in the system. The chosen hypothesis can be different from the best chain since the speech analysis module is able to parse the complete word lattice in order to select the most appropriate semantic interpretation. The next entry represents the derived user intention, which is used by the action planning component to select a suitable system reaction. In this case, the style sheet transformations create a compact predicate-argument structure from the original M3L description, which condenses the sometimes rather lengthy XML markup. Finally, the system output is printed as a text string, followed by the final screenshot of the animated presentation.

In the end, the final dialogue descriptions and meaningful protocols of current interactions provide a comprehensive overview of the detailed capabilities of the multimodal dialogue system.

3 An Extended Example Dialogue

In this section, we will walk through an example dialogue to provide the reader with a feeling of how the actual system works. As the public information kiosk integrates most of the functionality and features of SMARTKOM (Berton et al., 2006; Horndasch et al., 2006; Malaka et al., 2006; Portele et al., 2006), the presentation will use the scenario SMARTKOM-Public instead of the English Mobile system (Gelbart et al., 2006). The dialogue interaction in SMARTKOM-Public is based on German, so we also provide English translations.

The assumed location of the system installation is in the main railway station in the city of Heidelberg, and the current date is September 6, 2003. Imagine a user who just arrived in Heidelberg. She accesses the system and is presented with the initial display (see Fig. 3, left). To activate the system, the user can, for example, place a hand in the focus of the gesture recognition camera.

The interaction agent "Smartakus" appears on the screen (see Fig. 3, right) and greets the user:

(1) SYS: Herzlich willkommen beim SMARTKOM-System. Ich bin Smartakus wie kann ich Ihnen helfen?
(Welcome to the SMARTKOM *information system. I am Smartakus. How may I help you?)*

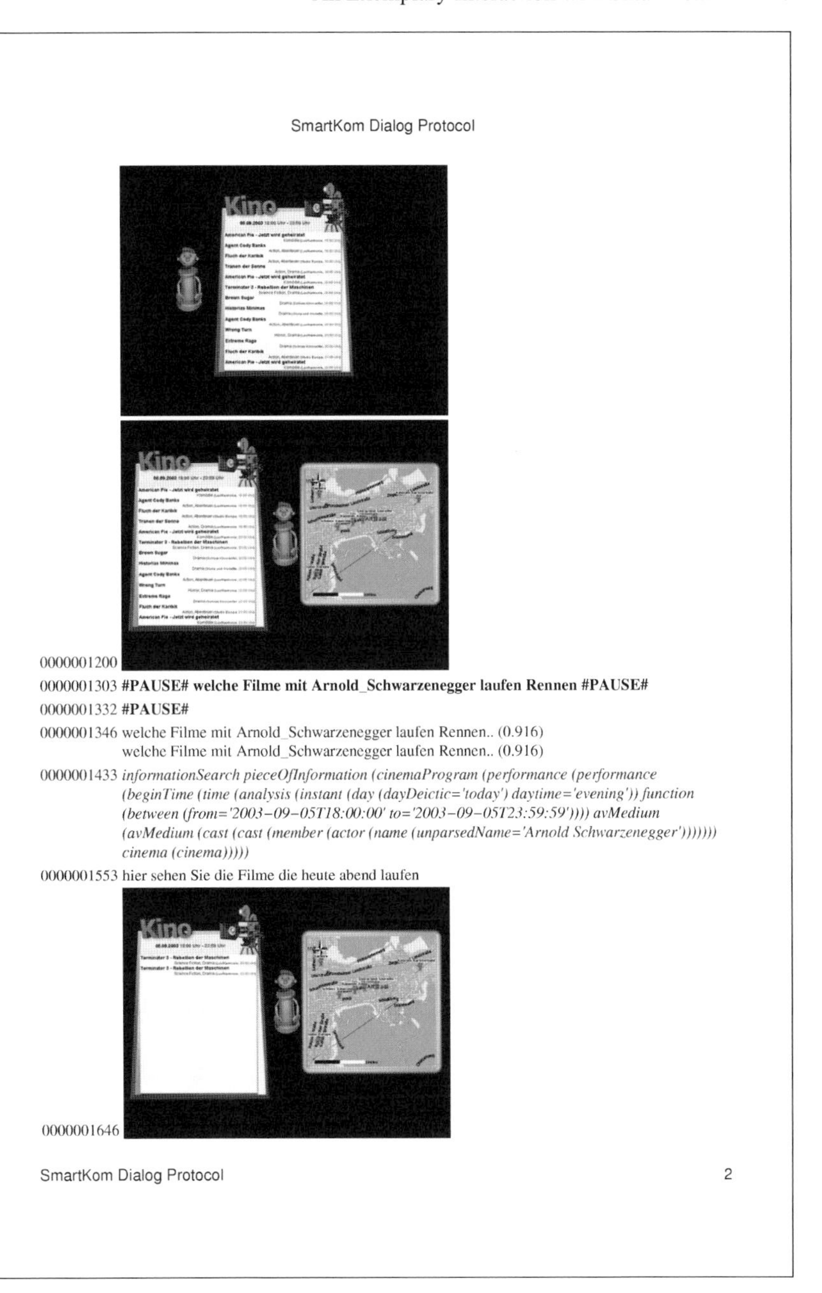

SmartKom Dialog Protocol

0000001200

0000001303 **#PAUSE# welche Filme mit Arnold_Schwarzenegger laufen Rennen #PAUSE#**

0000001332 **#PAUSE#**

0000001346 welche Filme mit Arnold_Schwarzenegger laufen Rennen.. (0.916)
welche Filme mit Arnold_Schwarzenegger laufen Rennen.. (0.916)

0000001433 *informationSearch pieceOfInformation (cinemaProgram (performance (performance (beginTime (time (analysis (instant (day (dayDeictic='today') daytime='evening')) function (between (from='2003-09-05T18:00:00' to='2003-09-05T23:59:59')))) avMedium (avMedium (cast (cast (member (actor (name (unparsedName='Arnold Schwarzenegger')))))) cinema (cinema)))))*

0000001553 hier sehen Sie die Filme die heute abend laufen

0000001646

SmartKom Dialog Protocol 2

Fig. 2. A page of the interaction protocol for the example dialogue

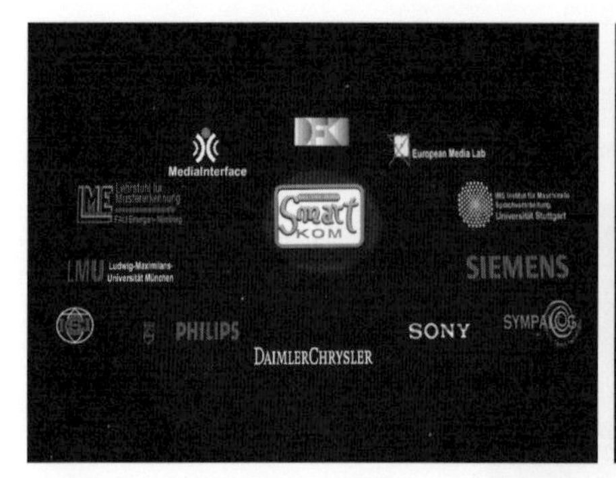

Fig. 3. Start screens of the SMARTKOM system

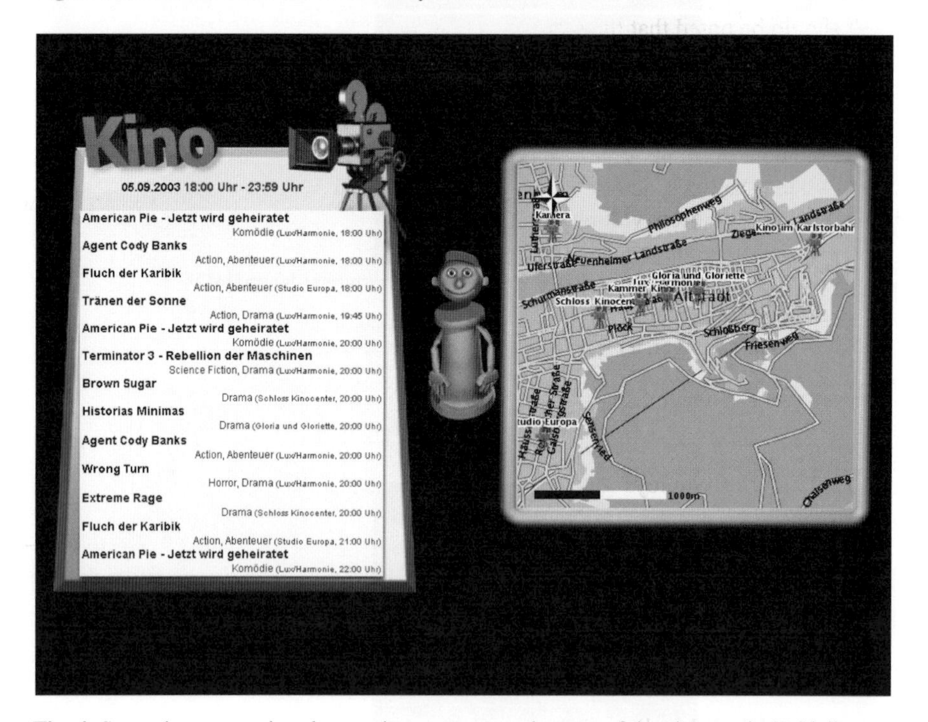

Fig. 4. Smartakus presenting the evening program and a map of the cinemas in Heidelberg

The user wants to go to the movies tonight and therefore asks:

(2) USR: Was kommt heute Abend im Kino?
(What is playing at the cinema tonight?)

Figure 4 shows the processing results: On the left hand-side is the cinema program, and on the right-hand side is a city map of Heidelberg with cinema locations. Note that SMARTKOM anticipated that the user might be interested to know where the cinemas are. This is particularly important, if, for example, a movie is playing at various theatres. Smartakus provides as spoken information:

(3) SYS: Hier sehen Sie die Filme die heute abend laufen. Auf dieser Karte sind die Kinos markiert.
(These are the movies playing tonight. The cinemas have been marked on the map.)

Our user recalls that there is a new movie with Arnold Schwarzenegger, so she asks:

(4) USR: Welche Filme mit Arnold Schwarzenegger laufen denn?
(Are there any movies featuring Arnold Schwarzenegger?)

It should be noted that the words "Arnold" and "Schwarzenegger" are not in the base lexicon of the system. The result of a database query—in this case movie titles and names of actors—may contain words that are not in the lexicon of the speech recognizer or the speech synthesis component. Hence, those modules have to be informed about possible words the user or the system might utter when presenting or addressing the titles, actors and other information. In SMARTKOM, the dedicated lexicon module (Goronzy et al., 2006) creates the phonetic transcriptions for new terms—since titles are often in foreign languages it has to provide pronunciation variants—and propagates the added lexicon entries to the speech modules.

The user query results in an information display that contains only movies with Arnold Schwarzenegger, as shown in Fig. 5 on the left side. The system utters:

(5) SYS: Hier sehen Sie die Filme die heute abend laufen.
(These are the movies playing tonight.)

Now the user employs multimodality and, while pointing to one of the movie tiles, says:

(6) USR: Gib mir Informationen zu diesem.
(Give me information about this one.)

SMARTKOM replaces the movie information frame and displays the description for this specific film (see the enlarged display in Fig. 5, on the right), telling the user:

(7) SYS: Ich habe diese Informationen zu "Terminator 3 – Rebellion der Maschinen".
(I have got this information about "Terminator 3: Rise of the Machines".)

The user wants to visit this cinema showing with her friend and tells Smartakus:

(8) USR: Bitte 2 Karten für diesen Film reservieren.
(Please reserve 2 seats for this film.)

In the context of the ongoing dialogue it is not ambiguous which performance she addresses, and the system presents the floor plan of the selected cinema (see Fig. 6, on the left), asking:

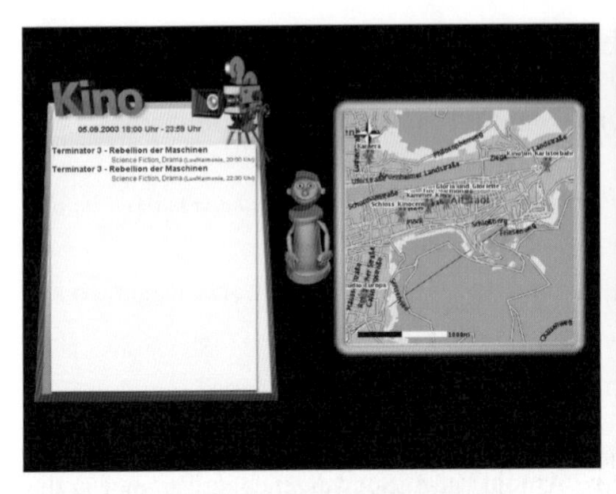

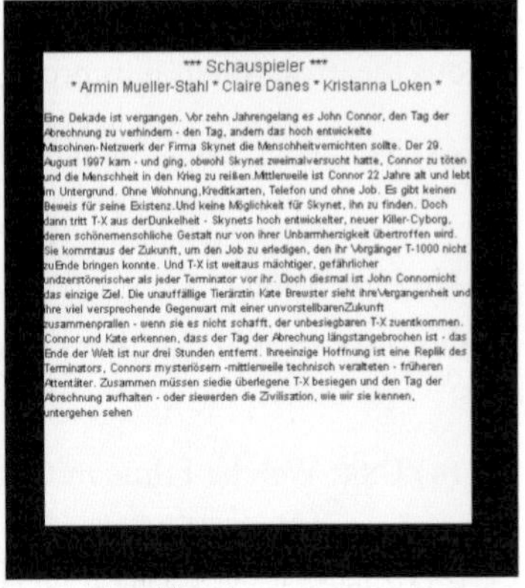

Fig. 5. Information about the film *Terminator 3*

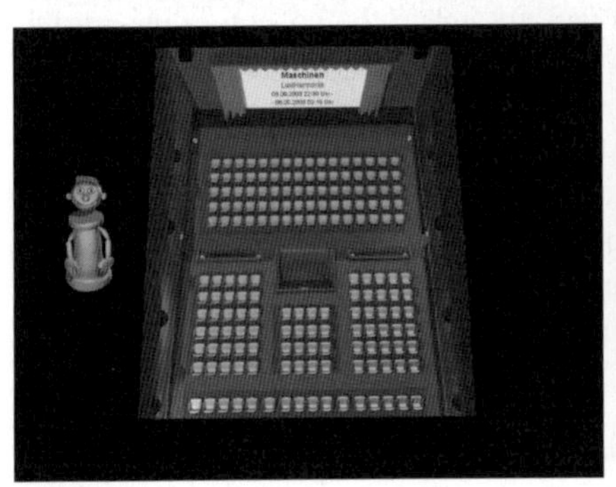

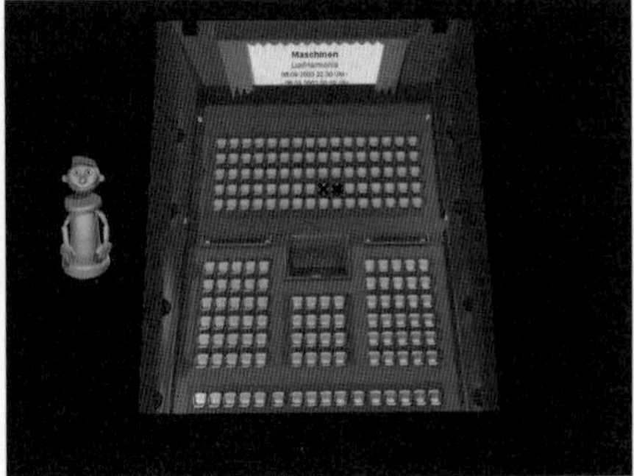

Fig. 6. Floor plan of the cinema

(9) SYS: Zeigen Sie mir wo Sie sitzen wollen.
(Show me where you want to sit.)

The user approximately encircles the desired area and says:

(10) USR: Dort möchte ich sitzen.
(I want to sit there.)

Again, the discourse context provides the information that she wants two seats and thus a proposal can be made. Smartakus marks two convenient seats (see Fig. 6, on the right) and asks:

(11) SYS: Ist das so richtig?
(Is this correct?)

The user is convinced, agrees and gets the reservation number from Smartakus.

(12) USR: Ja, genau.
(Yes, perfect.)

(13) SYS: Ich habe die gewünschte Reservierung für Sie durchgeführt. Ihre Reservierungsnummer ist 20. Sie können die Karten bis spätestens eine halbe Stunde vor Vorstellungsbeginn an der Kinokasse abholen.
(I reserved your seat. Your reservation number is 20. Please get your tickets not later than 30 minutes before the movie starts.)

Now she wants to know the route to the cinema and asks:

(14) USR: Wie komme ich von hier zum Kino “Studio Europa”?
(Can you show me the way from here to “Studio Europa” cinema?)

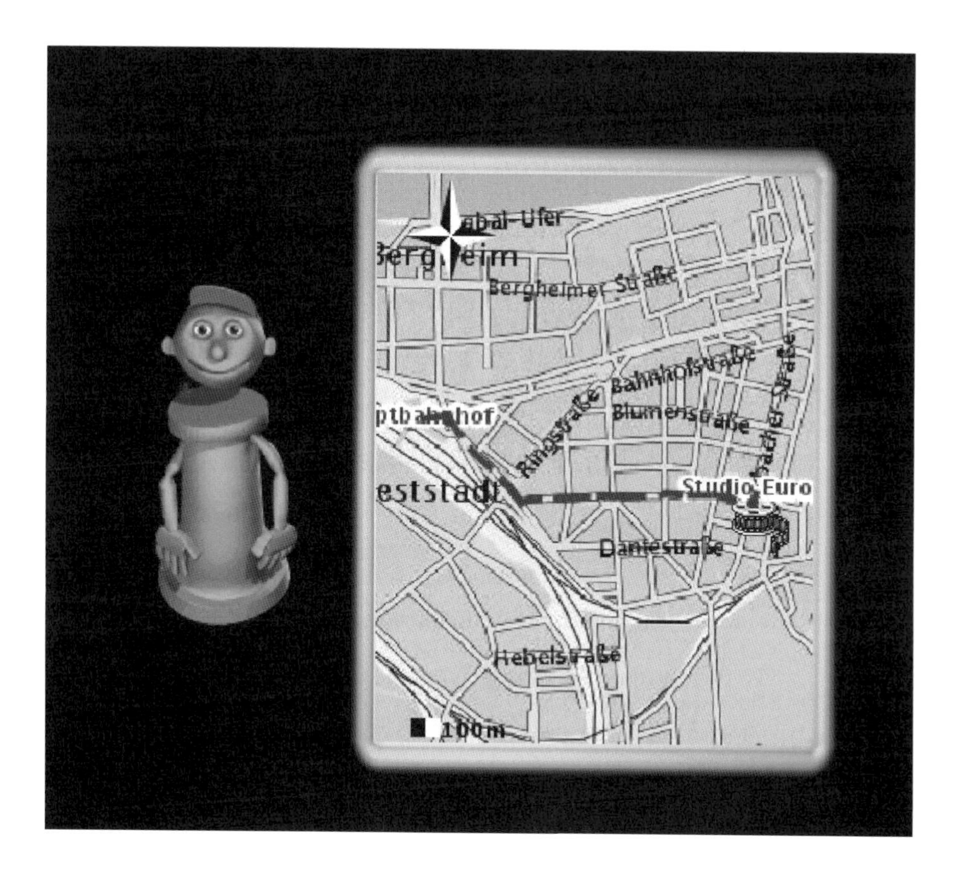

Fig. 7. Pedestrian route from main station to cinema

Originally, the functionality for pedestrian navigation was conceived for the scenario SMARTKOM-Mobile. The consistent, knowledge-based approach of SMARTKOM, however, facilitated the additional inclusion of this application in other scenarios. So, later on, the pretrip planning was simply added to SMARTKOM-Public. The system knows its own location from its general context knowledge source and is thus able to resolve the adverb “here”. The public information kiosk is located at the main station in Heidelberg, and SMARTKOM can compute the pedestrian route from there to the selected cinema. Figure 7 shows the map with the route inserted. Since the computation of the route takes some time, Smartakus also informs the user to be patient and says:

(15) SYS: Die Route wird berechnet. Hier sehen Sie die Karte mit der Route von Hauptbahnhof zu Kino “Studio Europa”.
(The route is being computed. Here you can see the map with the route from main station to cinema “Studio Europa”.)

After she has taken care of today’s evening entertainment, our user recalls that there is an important document that needs to be sent to her director. She tells Smartakus:

(16) USR: Ich will ein Dokument an Wolfgang Wahlster verschicken.
(I would like to send a document to Wolfgang Wahlster.)

Fig. 8. Document scanning area and address information

Sending a document requires the address of the recipient. SMARTKOM therefore must combine two different applications to satisfy a single user request. It looks up its address database for the coordinates of Wolfgang Wahlster. Being successful, Smartakus presents a scanning area and the address information (see Fig. 8, on the left), saying:

(17) SYS: Ich habe den Eintrag von Wolfgang Wahlster gefunden. Bitte legen Sie das Dokument auf die dafür vorgesehene Fläche.
(I found the address for Wolfgang Wahlster. Please put the document on the scanning area.)

Following the request of the system, the user places the document on the area. The document camera detects the object and scans it. At the end of the scan process, Smartakus tells the user:

(18) SYS: Bitte nehmen Sie es jetzt wieder weg.
(Please remove it now.)

The system displays the scanned document in the scanning area (see Fig. 8, on the left) and finally sends it via electronic mail.

(19) SYS: Das Dokument wurde aufgenommen. Das Dokument wurde abgeschickt.
(The document has been scanned. The document has been sent.)

The user is satisfied with SMARTKOM and leaves the system to meet her friend, saying:

(20) USR: Tschau.
(Good bye.)

(21) SYS: Auf Wiedersehen.
(See you again.)

4 Conclusion

In this contribution, we gave an impression of how the multimodal dialogue capabilities of SMARTKOM are defined, starting from dialogue description of scenario experts, resulting in detailed documentation of dialogues and their individual turns. The recorded example dialogue presented here provides an idea of one particular interaction sequence, which combines seven different applications—cinema program, city information, seat reservation, pedestrian navigation, document scanning, address book and email—in a coherent, seamless interaction. The subsequent chapters of this volume will provide a thorough discussion of the underlying methods and techniques that are required to achieve this kind of advanced multimodal dialogue functionality.

References

A. Berton, D. Bühler, and W. Minker. SmartKom-Mobile Car: User Interaction With Mobile Services in a Car Environment, 2006. In this volume.

J.M. Carroll. *Making Use: Scenario-Based Design of Human-Computer Interactions.* MIT Press, Cambridge, MA, 2000.

J.R. Gardner and Z.L. Rendon. *XSLT and XPATH: A Guide to XML Transformations.* Prentice Hall, 2001.

D. Gelbart, J. Bryant, A. Stolcke, R. Porzel, M. Baudis, and N. Morgan. SmartKom English: From Robust Recognition to Felicitous Interaction, 2006. In this volume.

S. Goronzy, S. Rapp, and M. Emele. The Dynamic Lexicon, 2006. In this volume.

G. Herzog and A. Ndiaye. Building Multimodal Dialogue Applications: System Integration in SmartKom, 2006. In this volume.

G. Herzog, A. Ndiaye, S. Merten, H. Kirchmann, T. Becker, and P. Poller. Large-Scale Software Integration for Spoken Language and Multimodal Dialog Systems. *Natural Language Engineering*, 10(3–4):283–305, 2004. Special Issue on Software Architecture for Language Engineering.

G. Herzog and N. Reithinger. The SmartKom Architecture: A Framework for Multimodal Dialogue Systems, 2006. In this volume.

A. Horndasch, H. Rapp, and H. Röttger. SmartKom-Public, 2006. In this volume.

R. Malaka, J. Häußler, H. Aras, M. Merdes, D. Pfisterer, M. Jöst, and R. Porzel. SmartKom-Mobile: Intelligent Interaction With a Mobile System, 2006. In this volume.

T. Portele, S. Goronzy, M. Emele, A. Kellner, S. Torge, and J. te Vrugt. SmartKom-Home: The Interface to Home Entertainment, 2006. In this volume.

M.B. Rosson and J.M. Carroll. Scenario-Based Design. In: J.A. Jacko and A. Sears (eds.), *The Human-Computer Interaction Handbook: Fundamentals, Evolving Technologies and Emerging Applications*, pp. 1032–1050, Mahwah, NJ, 2003. Lawrence Erlbaum.

F. Schiel and U. Türk. Wizard-of-Oz Recordings, 2006. In this volume.

Part II

Multimodal Input Analysis

The SmartKom Architecture: A Framework for Multimodal Dialogue Systems

Gerd Herzog and Norbert Reithinger

DFKI GmbH, Saarbrücken, Germany
{herzog,reithinger}@dfki.de

Summary. SMARTKOM provides an adaptive and reusable dialogue shell for multimodal interaction that has been employed successfully to realize fully-fledged prototype systems for various application scenarios. Taking the perspective of system architects, we review the overall design and specific architecture framework being applied within SMARTKOM. The basic design principles underlying our approach are described, and the different instantiations of the conceptual architecture are presented to illustrate the adaptibility and flexibility of the generic framework.

1 Introduction

The next generation of user interfaces aims at intelligent systems that are able to adapt to common forms of human dialogues and hence provide more intuitive and natural ways of interaction (Bernsen and Dybkjær, 1998; Blattner and Dannenberg, 1992; Maybury, 1993; Maybury and Wahlster, 1998b; Oviatt, 2003; Sullivan and Tyler, 1991). This ambitious goal, however, poses new challenges for the design and implemention of interactive software systems. Compared with traditional user interface software, these kinds of intelligent interfaces, which are capable of employing natural language as well as other modalities for communication, require more sophisticated architecture models.

The development of a complex multimodal system like SMARTKOM, which realizes three different application scenarios, requires an architecture that accomodates various, sometimes conflicting goals:

- Natural interaction behaviour: Facilitate timely, user adapted interaction.
- Flexibility: Handle a large number of modalities and applications.
- Openness: Allow for different processing approaches.
- Manageability: Support distributed development.

In this chapter, we elucidate the architectural design of SMARTKOM, which provides a practical framework for the successful realization of effective multimodal dialogue applications. The following section discusses essential requirements for multimodal dialogue systems and presents a more elaborate high-level architecture for intelligent interfaces. Based on these design foundations, a detailed description of the specific architecture framework that has been developed within the SMARTKOM project will be provided in Sect. 3. Following the introduction of the underlying framework, the different instantiations of the generic SMARTKOM architecture are presented in Sect. 4. There we highlight the flexibility and adaptibility of our approach.

2 Architectural Requirements for a Multimodal Interaction System

Conceptual architectures like the ones described in Allen et al. (2000), Bordegoni et al. (1997), Elting et al. (2003), Feiner (1991), Hill et al. (1992) and Maybury and Wahlster (1998a) contrast the system design of traditional handcrafted interfaces with intelligent user interfaces. Within the classical Seeheim reference model (Pfaff, 1985), the internal structure of the user interface component that mediates between the end user and the underlying application system is partitioned into three distinct conceptual tiers:

- The *presentation* level, or display management, determines the appeareance and low-level behaviour of the user interface.
- *Dialogue control* constitutes the intermediary processing stage which, manages the flow and order of interaction.
- The *application interface* as the third logical component provides the semantic model of the application, which is needed to access the functional core of the system.

This coarse modularization, however, is not sufficiently detailed to capture the basic structure of multimodal interfaces and to serve as a starting point for the development of the modular structure of a complex multimodal system.

Figure 1 provides a sketch of a more refined architecture at a high level of granularity. The conceptual architecture is based on the functional blocks of a generic multimodal dialogue system and decomposes the processing task into the following parts:

- Sensor-specific input processing: This layer of functionalities comprises all technical input devices like microphones and gesture recognition technologies. It provides a standardized, hardware-independent signal- or symbol-level interface for further processing.
- Modality-specific analysis: This processing level consists of the recognizers that transform sensor signals from the environment into symbolic information and analyzers that provide a meaning description for these signals.
- Modality fusion: This stage merges the meanings recorded in different modalities to one coherent, unified meaning representation. A core aspect of modality fusion is the mutual disambiguation of multimodal input on semantic and pragmatic levels.

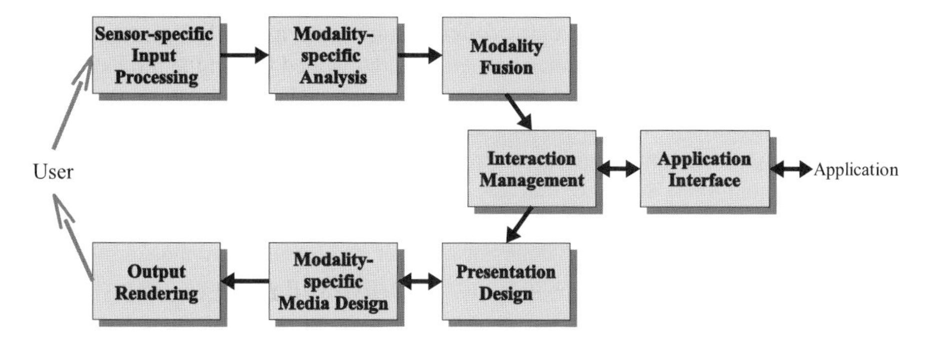

Fig. 1. A high-level architecture for multimodal interaction

- Interaction management: Here resides the "mind" of the machine. This part of the system identifies the intention of the user, determines the next steps to be taken by the system, and addresses the various application functions.
- Application interface: This layer defines an abstract interface for the interaction with the background applications. It hides application-specific technical details and thus provides a generic way to control the applications.
- Presentation design: Given a system intention, the information to be conveyed needs to be transformed into coordinated multimodal output. This functional block constitutes the first stage of modality fission and deals with the overall organization of the intended multimodal presentation. It includes in particular subtasks like content selection, media and modality allocation, layout design, and coordination.
- Modality-specific media design: In close cooperation with the presentation design stage, the processing on this layer is specifically concerned with the conversion of abstract content structures into presentable media objects.
- Output rendering: The final function block comprises the technical means to present the coordinated system reaction in the defined media channels like screens, loudspeakers, or force-feedback devices.

Of course, this breakdown of the architecture cannot and does not define sharp lines between the functional blocks. For example, one could argue about the output interface of sensor-specific input processing. For speech, it could either be a standardized audio stream after echo compensation or the output of the speech recognizer, if that component already incorporates all lower-level signal processing functionalities. In practice, the exact boundaries between the processing steps may depend on the constraints imposed by available components that are to be used for building a multimodal system.

In addition to the processing steps mentioned above, a rich variety of knowledge sources that are not shown in the diagram are also vital within the system and may require shared access from multiple components. Important knowledge sources for intelligent multimodal interaction include explicit models of the user, domain, context, task, and discourse as well as the different media and modalities. Consider, for example, reference resolution processes, which need access to data from discourse context storage. This information has to be provided by the output components and is delivered on demand to all interested parties.

The architecture sketch highlights the major flow of information through the system, leading to a linear structure that resembles a processing pipeline. However, this depicts only the main and principal succession of processing steps, including analysis, execution, and generation. It does not rule out additional interconnections. The individual components of the system could be regarded as communicating processes rather than just pipeline stages.

If the system designer now refines this high-level architecture, various additional aspects have to be considered (Bunt et al., 2001), including

- Declarative data models: The benefits of a flexible and configurable knowledge-based approach can only be exploited if system design strictly follows the principle that there should be no processing and presentation without explicit representation.
- Time-coordinated parallelism: Modality-specific analyzers and generators run in parallel. Therefore the analysis and generation of events must be coordinated, and the architecture must support this behaviour. Especially important are time constraints on processing as well as the provision of timing information within data formats.
- Standardised interfaces supporting the handling of probabilistic information: Modality recognizers usually provide alternative interpretation hypotheses augmented with information about the recognition quality. The design of the integration platform has to support the maintenance and handling of multiple, competing hypotheses.
- Incremental processing: It cannot be assumed that only complete representations are passed between the function blocks in the system. Increments of information, either originating in the user's input style or caused by processing methods, must be handled properly in the system.
- Iterative development: A complex multimodal system usually starts from a limited core of modalities and functionalities. Step-by-step, the coverage of the system is extended, with new input/output modalities being added. Therefore, the design must be extensible, not limiting additions by hardwired interaction and control paths. This requires accessible interfaces and extensive support for debugging and monitoring.
- Distributed development: Depending on specific organizational and technical constraints, software modules from various sites and with different backgrounds have to be integrated. For instance, it is very typical that off-the-shelf recognizers are to be used together with bespoke analysis components.

In the next sections, we will show how the SmartKom system realizes these requirements in three different application scenarios.

3 SmartKom as an Adaptive and Reusable Shell for Multimodal Dialogues

SmartKom represents a new generation of multimodal dialogue systems, dealing not only with basic modality integration and synchronization but addressing the full spectrum of dialogue phenomena arising in multimodal interaction (Reithinger et al., 2003b; Wahlster, 2002). With a clear focus on reusability, SmartKom has been designed as a transmutable system that can engage in many different types of tasks in various usage scenarios (Reithinger et al., 2003a; Wahlster, 2003).

3.1 Distinguishing Features of the SmartKom Approach

The SmartKom system relies on the so-called *situated, delegation-oriented dialogue paradigm* (Wahlster et al., 2001) to provide an anthropomorphic and affective user interface. This interaction metaphor is based upon the idea that the user delegates a task to the virtual communication assistant, which is visualized as a life-like character. The interface agent recognises the user's intentions and goals, asks the user for feedback if collaboration is necessary, accesses the various application services on behalf of the user, and presents the collated results in an appropriate form. The interaction style being employed within SmartKom targets in particular nondesktop scenarios, such as smart rooms, kiosks, or mobile application contexts.

An interesting property of the SmartKom dialogue shell is its ability to integrate multiple services, i.e., different application functions that have been developed independently from each other, into a coherent, value-added system. Another peculiarity of the system is a build-in self-monitoring capability which continuously determines the overall processing state using status information from the individual components. In the sense of a reflective architecture (Maes and Nardi, 1988), this kind of metalevel reasoning and control can be exploited to provide the user with appropriate feedback concerning system activity and potential error situations.

3.2 Anatomy of the SmartKom Core System

The SmartKom kernel shown in Fig. 2 provides the core building blocks for a multimodal interactive system. The diagram documents the software architecture using a component-and-connector viewtype as defined in Clements et al. (2003). Each component represents one of the principal processing units of the executing system, and the highlighted components are being reused in all application cases. Since SmartKom has been implemented as a distributed system (see Herzog et al. (2004) for details), every component corresponds to an independent process. The connectors illustrate the pathways of interaction. The underlying architectural style is based on communicating processes, more specifically on the publish–subscribe style. In this style, the interaction is event-driven using directed communication links between a message sender, who acts as data producer or event source, and a set of recipients, the data consumers.

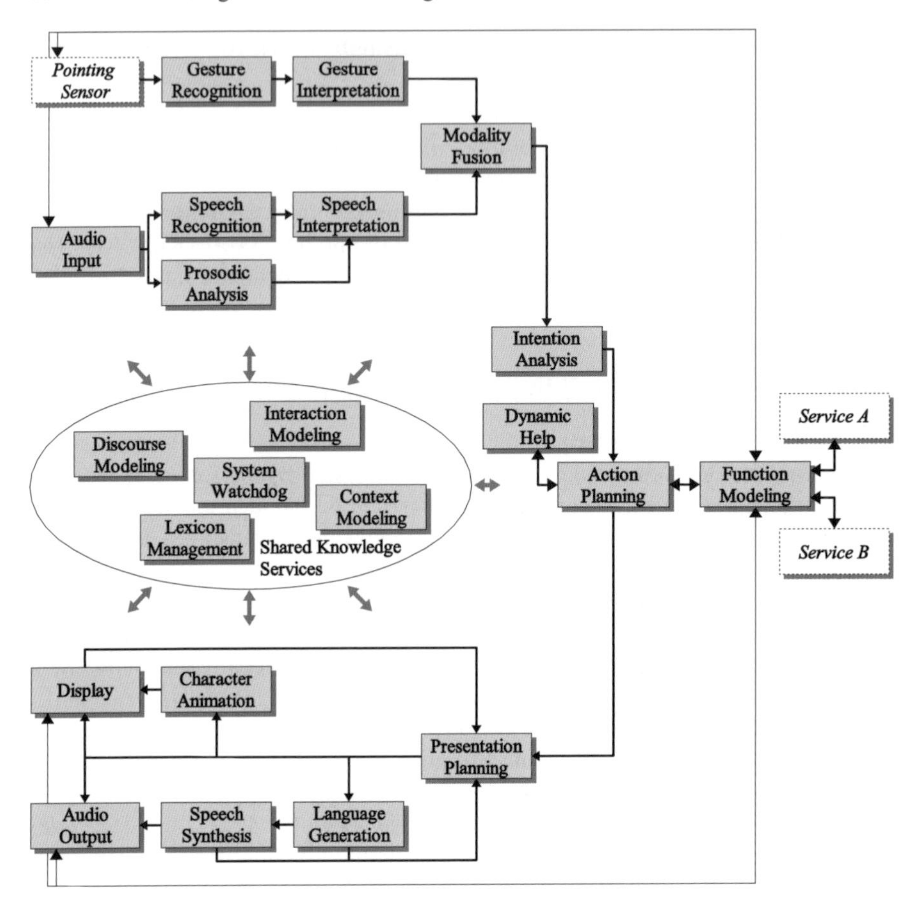

Fig. 2. SMARTKOM architecture blueprint

To provide a clearer outline of the system structure, the connectors chosen for Fig. 2 denote major information flows and may combine several publish–subscribe communication links. The various publish–subscribe connections between the shared knowledge sources and the other components of the dialogue backbone have only been sketched.

A basic SMARTKOM system supports multimodal input using speech plus gestures and integrates some application-specific services. The specific sensor type for gesture input depends on the technical setup of the given application scenario. Practical options that have been used for the SMARTKOM prototypes include either pen-based pointing or a vision system for hand tracking.

The separation of sensor-specific input processing into distinct components, which connect some technical device to the system and which encapsulate access to the underlying hardware, constitutes a first decomposition principle of the architecture framework. In order to provide maximum flexibility for system deployment, the same approach is followed with respect to the output devices that are needed for the rendering stage.

Modality-specific input analysis can be modularized into separate components, and it has proven useful to differentiate further between modality-specific recognition, i.e., processing of sensory data on the lexical and syntactical layer, and subsequent semantic interpretation of the derived symbolic representation.

Interaction management within SMARTKOM comprises several components. The main task of intention analysis is to select the most plausible input interpretation from the given set of hypotheses. Action planning constitutes the heart of dialogue control and is backed by a supplementary help component that is activated whenever difficulties occur during the interaction or if additional help is needed. The main processing stages are supported by different components that actively maintain shared knowledge sources:

- The multimodal discourse model is utilized for the semantic and pragmatic interpretation during input and output processing. It is dynamically updated as system output progresses and performs contextual reasoning and scoring.
- Contextual information, as it is needed to handle references to situative parameters like current place and time, is provided by the context model.
- Interaction modeling is concerned with different aspects like available modalities and user preferences for specific forms of communication as well as the affective state of the user. The interaction model allows one to dynamically adapt the communicative behaviour of the system.
- The lexicon is a dynamic knowledge source, which is updated with additional lexical entries depending on dynamic application data as it is received from the external information services. Lexicon updates are propagated to all components that process natural language input and output.
- The system watchdog monitors the processing status of all individual components to offer up-to-date information concerning the system state. This is used to provide immediate feedback or to initiate helpful reactions in case of processing problems.

The function modeling component realizes the application interface and in addition controls all input and output devices to coordinate access depending on an explicit state model. The application layer itself is modularized into multiple service components, each of which realizes some application-specific functionality.

The remaining elements of the component ensemble deal with modality fission. A second planning component is responsible for the presentation design. Part of the modality-specific output design is carried out by the character animation component, which realizes the perceivable behaviour of the embodied conversational agent.

In addition to the integration of application-specific service components, the architecture framework provides other extension points to expand the core dialogue components with even more modalities. A practical example for an optional input modality is the interpretation of facial expressions to obtain information about the affective state of the user. The SMARTKOM prototypes focus mainly on audiovisual sensor input and presentations. With respect to other human senses, it would be feasible to include haptic sensations as well, e.g., using a force-feedback device. A limited form of tactile interaction is already provided when using pen-based input.

3.3 Representing Data with Multimodal Markup Language

The interaction between the distributed components within the SMARTKOM architecture is based on the exchange of structured data through messages. The external data format for the representation of structured information employs XML notation. XML, the *eXtensible Markup Language*, provides a flexible standard for the definition of specific data exchange formats that ensures interoperability between heterogeneous software components.

In the context of the SMARTKOM project we have developed M3L (Multimodal Markup Language) to cover all data interfaces within the entire multimodal dialogue system in a declarative way. Instead of using several unrelated XML languages for the various information flows, we aimed at an integrated and coherent language specification, which includes all substructures that may occur on the different communication links.

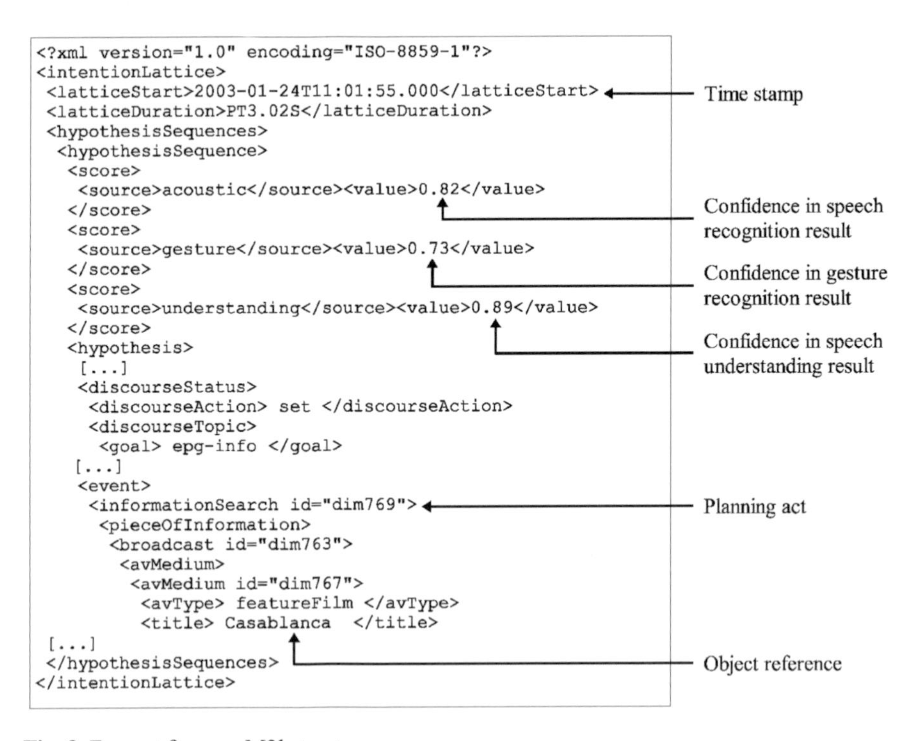

Fig. 3. Excerpt from an M3l structure

The basic information flow from user input to system output continuously adds further processing results so that the representational structure will be refined step-by-step. A typical example of such a complex M3L expression is shown in Fig. 3. The partial listing contains a so-called intention lattice that represents the interpretation result for a multimodal user input which combines a verbal information request ("Tell me more about this film") with a pointing gesture.

Conceptual taxonomies provide the foundation for the representation of domain knowledge as it is required within a dialogue system to enable a natural conversation in the given application scenario. In order to exchange instantiated knowledge structures between different system components they need to be encoded in M3L. The corresponding parts of the M3L definition are derived from an underlying ontology that captures the required terminological knowledge (Gurevych et al., 2003). For example, in Fig. 3 the representation of the event structure inside the intention lattice originates from the ontology. M3L can easily be updated as the ontology evolves while novel application domains are being modeled.

4 Instantiations of the Generic Architecture

The SMARTKOM architecture framework is designed to support a wide range of collaborative and multimodal dialogues that allow users to intuitively and efficiently access the functionalities needed for their task. The SMARTKOM prototypes address three different application scenarios that exhibit varying characteristics that influence the architectural design.

4.1 Public Information and Communication Kiosk

SMARTKOM-Public realizes an advanced multimodal information and communication kiosk for train stations, airports, or other public places. It supports users seeking information concerning movie programs, offers reservation facilities, and provides personalized communication services using telephone, fax, or electronic mail.

Figure 4 provides an overview concerning the specific software architecture of SMARTKOM-Public. Scenario-specific components are highlighted using gray boxes.

Gesture input is based on an infrared camera system that tracks the user's hand as it moves across a video projection of the graphical output. This kind of vision-based sensor does not require direct physical contact with the horizontal projection surface and hence allows for more natural hand gestures.

SMARTKOM-Public integrates facial expressions as an additional input modality. In combination with acoustic indicators derived from prosodic analysis, the interpretation of facial expressions is used to recognize the current affective state of the user.

The kiosk system is equipped with two application-specific input components for pen-based signature input and an additional camera system that takes digital images of objects placed on the projection panel. In the given application scenario, the following functions are made available through the main service components:

- Biometric authentication using either voice, signature, or hand contour

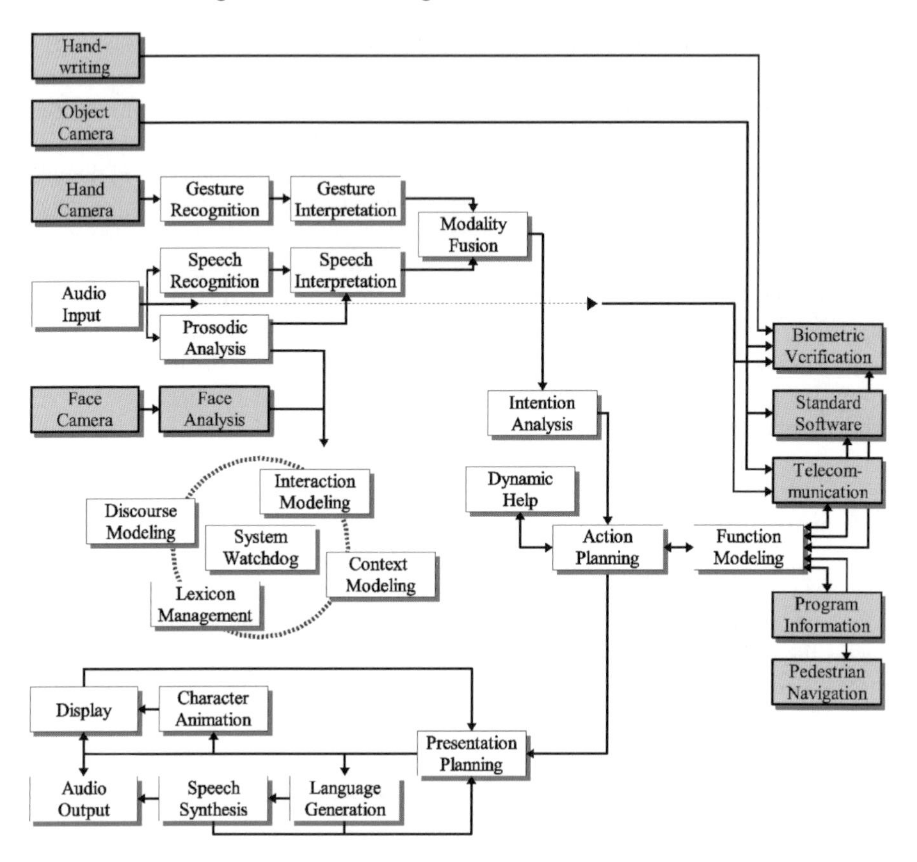

Fig. 4. Component architecture of SMARTKOM-Public

- Access to standard applications like address book and e-mail transfer
- Telephone connections and fax transmission
- Cinema program and reservation service
- Provision of maps with cinema locations and marked routes to the desired destination

The function model of the application interface coordinates the interplay of services to carry out specific tasks. The need for flexible application-specific use of input and output components is the reason why the basic control of all shared devices has been allocated to the function modeling component. The corresponding control lines have been omitted from Fig. 4.

Depending on the current application context, audio input can be rerouted to the biometrics component or to the telecommunication service. Input from the object camera is used to send documents (via e-mail or fax), and the input component can also record the hand contour for biometric verification.

4.2 Infotainment Assistant for the Living Room at Home

In its basic configuration, SMARTKOM-Home aims to provide a multimodal portal to home entertainment services. Using a portable tablet PC, the user is able to utilize the system as an electronic program guide or to easily control consumer electronics devices like a television set or a digital video casette recorder.

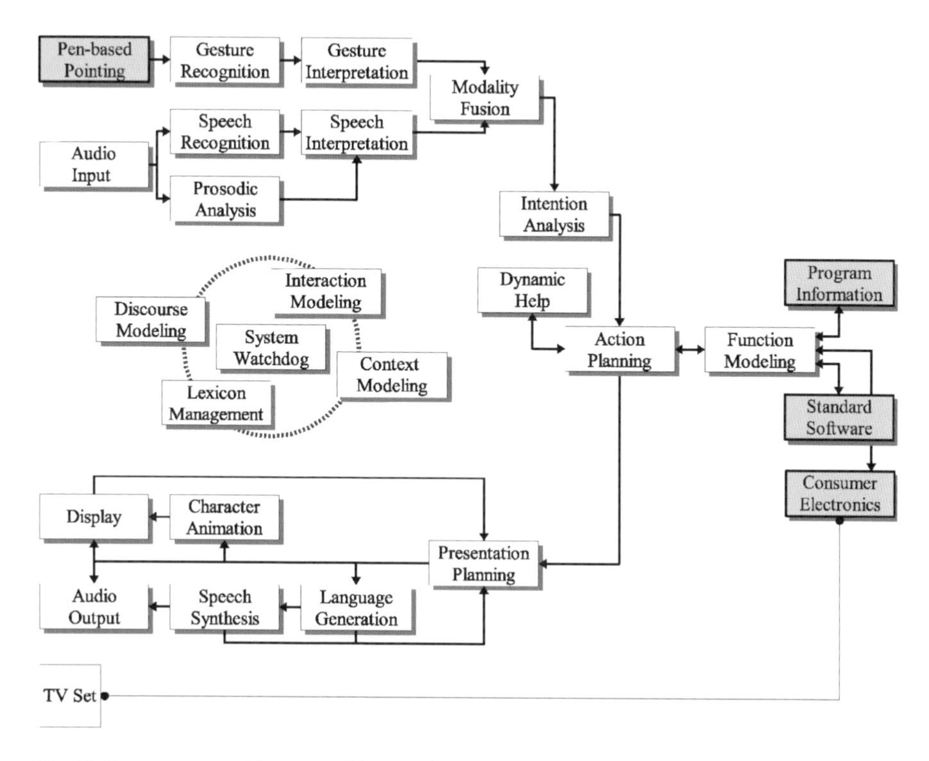

Fig. 5. Component architecture of SMARTKOM-Home

Figure 5 highlights the specific components of the SMARTKOM-Home prototype. The system supports the basic modalities that are associated with the core multimodal system and the user can employ a pen for stylus-drawn gestures on the tablet PC's screen. In the context of the Home scenario, however, two different styles of interaction modes are supported. In so-called *lean-forward* mode, coordinated speech and gesture input can be used for multimodal interaction with the system. Alternatively, the user is able to put away the tablet PC and change to so-called *lean-backward* mode. This interaction style is constrained to verbal communication. To compensate for the unavailable graphical view in this case, the presentation design needs to dynamically adapt its presentation strategies using more elaborate natural language output. The user can easily request the activation or deactivation of the display to switch between the two modes.

SMARTKOM-Home shares two of its service components with the Public scenario but employs different application functions from these services. The program database is primarily used to obtain detailed information concerning upcoming televison broadcasts and the calender management utility from the standard software component is utilized to schedule off-air recordings of selected programs. The third service component provides a high-level interface to various appliances. In particular, the consumer electronics service provides remote control of a TV set, a VCR, and a satellite receiver. The TV screen can be regarded as an application-specific output device, which is not directly available for system output that is routed through the interface components.

SMARTKOM offers a very flexible and open approach for application integration. Services constitute connector components that provide a well-defined link to some application-specific functionality and often encapsulate complete and complex application-specific subsystems. A typical case is the integration of commercial off-the-shelf software. The standard software component in SMARTKOM, for example, constitutes a bridge to external standard applications provided by a separate installation of IBM Lotus software. The service component for consumer electronics reuses already available software modules from the related Embassi project (Elting et al., 2003). It is basically a wrapper for a complex multiagent system, which in fact relies on a different kind of middleware solution than the publish–subscribe style used to interconnect the SMARTKOM components.

4.3 Mobile Travel Companion

SMARTKOM-Mobile provides a travel companion that can be used inside the car and while walking around. This application scenario comprises typical services like integrated trip planning and incremental route guidance through a city via GPS. Outside the car, the user operates the system using a handheld computer as a front end. Inside the instrumented car, the onboard equipment of the vehicle is used for displaying visual information and for audio input/output. A pecularity of the mobile travel companion is the seamless transition from one device to the other without losing the current dialogue context.

Figure 6 shows the basic components of SMARTKOM Mobile. Multimodal interaction is possible using pen-based input on the handheld device. The in-car display, however, is not touch-sensitive.

The handover between alternative devices constitutes one of the central technical challenges imposed by this specific application scenario. The component named *mobile device* provides a generic solution to manage different remote front-end devices. In order to avoid time-consuming component restarts and shutdowns during system runtime, audio input and output as well as the display component have been enhanced with streaming capabilities for input and output redirection. RTP, the Internet-standard protocol for the transport of real-time data, is used for bidirectional audio connections and VNC (Richardson et al., 1998) enables remote access to the display component. The mobile device component serves as a special kind of input device.

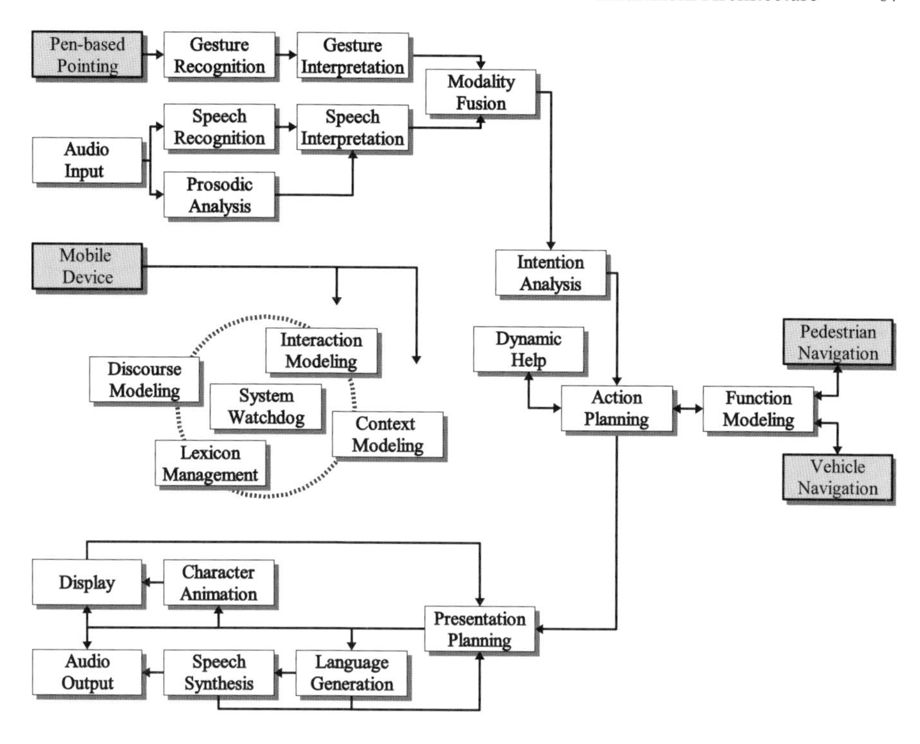

Fig. 6. Component architecture of SMARTKOM-Mobile

After pressing the push-to-activate button on the handheld or in the car, the characteristics of the new front-end device—also including address parameters for video and audio redirection—are propagated by the mobile device component to dynamically update the interaction model. The significant difference in screen format of the handheld computer and the in-car display enforces the presentation design to automatically adapt the visual output accordingly.

The vehicle navigation component includes commercial off-the-shelf navigation software to generate map displays with marked vehicle routes. It also provides information concerning parking locations. The pedestrian navigation service offers feature-rich functions for map generation, touristic information, pretrip planning, and incremental route guidance in Heidelberg. The component encapsulates a complex subsystem that is realized in a modular fashion, using a multiagent framework.

SMARTKOM-Mobile introduces an important new aspect for the dialogue management. In addition to user inputs, the system also needs to handle external events since navigation instructions are generated on the fly by the navigation service to provide incremental route guidance.

4.4 Other System Configurations

The SMARTKOM prototypes for the three sample scenarios use German for speech input and output. A special English system is based on a subset of SMARTKOM-Mobile, excluding incremental route guidance. SMARTKOM-English uses adapted versions of language-specific recognition, analysis, generation, and synthesis components.

In addition to the basic prototypes, many other plug-and-play variations of system setups with extended or restricted capabilities can be defined easily, using a flexible configuration mechanism for system instantiations. For example, for demonstration purposes, the SMARTKOM reference installation is often used as an extended Public system that integrates all available service components.

5 Conclusion

Multimodal dialogue systems hold the promise of providing more intuitive and natural ways of human–computer interaction. In order to construct such advanced user system interfaces, an effective architectural design is of utmost importance. SMARTKOM provides an adaptive and reusable dialogue shell that has been employed successfully to realize fully-fledged prototype systems for various application scenarios.

SMARTKOM is based on an open architecture for multimodal dialogue systems that is flexible and adaptive. All available components can be reused directly, and modular organization into different service components simplifies the integration of application-specific functions. With respect to the dialogue backbone, the realization of new application scenarios primarily becomes a knowledge engineering task instead of an extensive programming exercise. In order to extend the domain-specific dialogue behaviour, mainly the component-specific declarative knowledge sources have to be adapted. The generic SMARTKOM architecture with its core components and their interconnections promises to provide a solid base structure, even in cases where a customized multimodal dialogue system may require more advanced adaptations and extensions.

References

J. Allen, D. Byron, M. Dzikovska, G. Ferguson, L. Galescu, and A. Stent. An Architecture for a Generic Dialogue Shell. *Natural Language Engineering*, 6(3):1–16, 2000.

N.O. Bernsen and L. Dybkjær. Is Speech the Right Thing for Your Application? In: *Proc. ICSLP-98*, pp. 3209–3212, Sydney, Australia, 1998.

M.M. Blattner and R.B. Dannenberg (eds.). *Multimedia Interface Design*. ACM, New York, 1992.

M. Bordegoni, G. Faconti, S. Feiner, M.T. Maybury, T. Rist, S. Ruggieri, P. Trahanias, and M. Wilson. A Standard Reference Model for Intelligent Multimedia Presentation Systems. *Computer Standards & Interfaces*, 18(6–7):477–496, 1997.

H. Bunt, M.T. Maybury, and W. Wahlster. Coordination and Fusion in Multimodal Interaction. Dagstuhl Seminar Report 325, IBFI, Wadern, Germany, 2001.

P. Clements, F. Bachmann, L. Bass, D. Garlan, J. Ivers, R. Little, R. Nord, and J. Stafford. *Documenting Software Architectures: Views and Beyond*. Addison-Wesley, Boston, MA, 2003.

C. Elting, S. Rapp, G. Möhler, and M. Strube. Architecture and Implementation of Multimodal Plug and Play. In: *Proc. 5th Int. Conf. on Multimodal Interfaces*, pp. 93–100, Vancouver, Canada, 2003. ACM.

S. Feiner. An Architecture for Knowledge-Based Graphical Interfaces. In: J.W. Sullivan and S.W. Tyler (eds.), *Intelligent User Interfaces*, pp. 259–279, New York, 1991. ACM.

I. Gurevych, S. Merten, and R. Porzel. Automatic Creation of Interface Specifications from Ontologies. In: H. Cunningham and J. Patrick (eds.), *Proc. HLT-NAACL 2003 Workshop on Software Engineering and Architecture of Language Technology Systems (SEALTS)*, pp. 59–66, Edmonton, Canada, 2003. Association for Computational Linguistics.

G. Herzog, A. Ndiaye, S. Merten, H. Kirchmann, T. Becker, and P. Poller. Large-Scale Software Integration for Spoken Language and Multimodal Dialog Systems. *Natural Language Engineering*, 10(3–4):283–305, 2004. Special Issue on Software Architecture for Language Engineering.

W. Hill, D. Wroblewski, T. McCandless, and R. Cohen. Architectural Qualities and Principles for Multimodal and Multimedia Interfaces. In: Blattner and Dannenberg (1992), pp. 311–318.

P. Maes and D. Nardi (eds.). *Meta-Level Architecture and Reflection*. North-Holland, Amsterdam, The Netherlands, 1988.

M.T. Maybury (ed.). *Intelligent Multi-Media Interfaces*. AAAI/MIT Press, Menlo Park, CA, 1993.

M.T. Maybury and W. Wahlster. Intelligent User Interfaces: An Introduction. In: M.T. Maybury and W. Wahlster (eds.), *Readings in Intelligent User Interfaces*, pp. 1–13, San Francisco, CA, 1998a. Morgan Kaufmann.

M.T. Maybury and W. Wahlster (eds.). *Readings in Intelligent User Interfaces*. Morgan Kaufmann, San Francisco, CA, 1998b.

S. Oviatt. Multimodal Interfaces. In: J.A. Jacko and A. Sears (eds.), *The Human-Computer Interaction Handbook: Fundamentals, Evolving Technologies and Emerging Applications*, pp. 286–304, Mahwah, NJ, 2003. Lawrence Erlbaum.

G.E. Pfaff (ed.). *User Interface Management Systems: Proc. the Seeheim Workshop*. Springer, Berlin Heidelberg New York, 1985.

N. Reithinger, J. Alexandersson, T. Becker, A. Blocher, R. Engel, M. Löckelt, J. Müller, N. Pfleger, P. Poller, M. Streit, and V. Tschernomas. SmartKom: Adaptive and Flexible Multimodal Access to Multiple Applications. In: *Proc. 5th Int. Conf. on Multimodal Interfaces*, pp. 101–108, Vancouver, Canada, 2003a. ACM.

N. Reithinger, G. Herzog, and A. Ndiaye. Situated Multimodal Interaction in SmartKom. *Computers & Graphics*, 27(6):899–903, 2003b.

T. Richardson, Q. Stafford-Fraser, K.R. Wood, and A. Hopper. Virtual Network Computing. *IEEE Internet Computing*, 2(1):33–38, 1998.

J.W. Sullivan and S.W. Tyler (eds.). *Intelligent User Interfaces.* ACM, New York, 1991.

W. Wahlster. SmartKom: Fusion and Fission of Speech, Gestures, and Facial Expressions. In: *Proc. 1st Int. Workshop on Man-Machine Symbiotic Systems*, pp. 213–225, Kyoto, Japan, 2002.

W. Wahlster. SmartKom: Symmetric Multimodality in an Adaptive and Reusable Dialogue Shell. In: R. Krahl and D. Günther (eds.), *Proc. Human Computer Interaction Status Conference 2003*, pp. 47–62, Berlin, Germany, June 2003. DLR.

W. Wahlster, N. Reithinger, and A. Blocher. SmartKom: Multimodal Communication with a Life-like Character. In: *Proc. EUROSPEECH-01*, vol. 3, pp. 1547–1550, Aalborg, Denmark, September 2001.

Modeling Domain Knowledge: Know-How and Know-What

Iryna Gurevych*, Robert Porzel, and Rainer Malaka

European Media Laboratory GmbH, Heidelberg, Germany
{robert.porzel,rainer.malaka}@eml-d.villa-bosch.de
iryna.gurevych@eml-r.villa-bosch.de
* Current affiliation EML Research gGmbH

Summary. The approach to knowledge representation taken in the multimodal, multidomain, and multiscenario dialogue system — SMARTKOM — is presented. We focus on the ontological and representational issues and choices helping to construct an ontology, which is shared by multiple components of the system and can be reused in different projects and applied to various tasks. Finally, two applications of the ontology that highlight the usefulness of our approach are described.

1 Introduction

The ways in which knowledge has been represented in multimodal dialogue systems (MMDSs) show that individual representations with different semantics and heterogeneously structured content can be found in various formats within single natural language processing (NLP) systems and applications. For example, a typical NLP system, such as TRAINS (Allen et al., 1996) employs different knowledge representations for parsing, action planning, and generation, despite the fact that what is being represented is common to all those representations, e.g., the parser representation for *going from A to B* has no similarity to the action planner's representation thereof (Ferguson et al., 1996). Also central concepts, for example, *city*, are represented in multiple ways throughout the system.

The origin for this state of affairs is that the respective knowledge stores are handcrafted individually for each task. Sometimes they are compiled into code and cease to be externally available. Where an explicit knowledge representation is used, we find a multitude of formats and inference engines, which often cause both performance and tractability problems. In this paper we introduce the results of an effort to employ a single knowledge representation, i.e., an *ontology*, throughout a complete multimodal dialogue system. Therefore, we will describe the underlying modeling principles and the benefits of such a rigorously crafted knowledge store for the actual and future MMDS.

The SMARTKOM system comprises a large set of input and output modalities which the most advanced current systems feature, together with an efficient fusion and fission pipeline. SMARTKOM features speech input with prosodic analysis, gesture input via infrared camera and recognition of facial expressions and their emotional states. On the output side, the system features a gesturing and speaking life-like character together with displayed generated text and multimedia graphical output. It currently comprises a set of modules running on a parallel virtual machine-based integration software called *Multiplatform* (Herzog et al., 2003).[1]

As mentioned in the introduction, complex MMDSs such as SMARTKOM can benefit from a homogeneous world model. This model serves as a common knowledge representation for various modules throughout the system. It represents a general conceptualization of the world (top-level or generic ontology) as well as of particular domains (domain-specific ontologies). This way, the ontology represents language-independent knowledge. The language-specific knowledge is stored elsewhere, e.g., in the lexicon containing lexical items together with their meanings defined in terms of ontology concepts.

The initial version of the ontology described herein was designed as a general-purpose component for knowledge-based NLP. It includes a top-level developed following the procedure outlined by Russell and Norvig (1995) and originally only covered the tourism domain encoding knowledge about sights, historical persons, and buildings. Then, the ontology was successfully extended and modified until it covered SMARTKOM domains, e.g., new media and program guides. The top-level ontology was reused with some slight extensions. Further developments were motivated by the need of a *process hierarchy*. This hierarchy models processes which are domain-independent in the sense that they can be relevant for many domains, e.g., *InformationSearchProcess* (see Sect. 3.3 in this chapter for more details).

Currently, the ontology employed by the system has about 730 concepts and 200 relations. The acquisition of the ontology went in two directions: top-down to create a top level of the ontology and bottom-up to satisfy the need of mapping lexical items to concepts. The purpose of the top level ontology is to provide a basic structure of the world, i.e., abstract classes to divide the universe in distinct parts as resulting from the ontological analysis (Guarino and Poli, 1995). The domain concepts emerged through a comprehensive corpus analysis. The most important modeling decisions will be discussed in Sect. 3. Once available, the ontology was augmented with comments containing definitions, assumptions and examples that facilitate its appropriate use in a multicomponent system such as SMARTKOM and its possible reuse in other systems. Such descriptions of ontology classes are particularly important as the meanings associated with them may vary considerably from one ontology to another.

In Sect. 2 we will introduce the representational formats pertinent to our ontology. In Sect. 3 we discuss the modeling principles underlying the ontology. Sect. 4 presents a few examples of the various ways in which the common ontology is employed throughout the system. Concluding remarks are given in Sect. 5.

---

[1] The abbreviation stands for "MUltiple Language/Target Integration PLATform FOR Modules".

2 The Representational Formalism Used

We give a brief outline of the formalism pertinent to the following description of the ontology. Various standardization efforts originating, e.g., in W3C and Semantic Web projects came up with a number of several knowledge modeling standards: Resource Description Framework (RDF), DARPA Agent Mark-up Language (DAML), Ontology Interchange Language (OIL) and Web Ontology Language (OWL). [2] The XML-based languages OIL/DAML+OIL are of particular interest for domain and discourse knowledge modeling. The advantages of the XML standard, availability of tools and their wide use let us choose the OIL-RDFS syntax. A detailed characterization of the formal properties of the OIL language has been given by Fensel et al. (2001).

We used the OIL language in particular as a whole range of software is freely available to support the ontology construction as mentioned above. In addition, the usage of the ontology in Semantic Web applications would be simplified. The FACT [3] system can be used as a reasoning engine for OIL ontologies, providing some automated reasoning capabilities, such as class consistency or subsumption checking. Graphical ontology engineering front ends and visualization tools are available for editing, maintaining, and visualizing the ontology. [4]

The semantics of OIL is based on description logic extended with concrete datatypes. The language employs a combination of frame and description logic. It provides most of the modeling primitives commonly used in the frame-based knowledge representation systems. Frames are used to represent concepts. These frames consist of a collection of classes along with a list of slots and attributes. Under the term *class* or *class expression*, a class name, or an enumeration, or a property-restriction, or a Boolean combination of class expressions is to be understood. Slots are interpreted as a collection of properties. They are divided into those that relate classes to other classes (so-called *object properties*) and those that relate classes to datatype values (so-called *datatype properties*). Slots can be filled by: class names, names of the atomic elements, collection of the above (conjunctive sets *and*, disjunctive sets *or*, or negation *not*) or concrete datatypes (*integers* and *strings*).

Then, domain and range restrictions of the slots can be defined. A domain restriction asserts that the property only applies to the instances of particular class expressions. A range restriction specifies that the property only assumes values that are instances of the respective class expressions. Slot fillers can have several types of further constraints, also called *facets*. These include *value-type* restrictions (all fillers must be of a particular class) and *has-value* restrictions (there must be at least one filler of a particular class). The *value-type* restriction corresponds to the universal quantifier of the predicate logic. The *has-value* restriction is analogous to the

---

[2] See http://www.w3c.org/RDF, http://www.ontoknowledge.org/oil, http://www.daml.org and http://www.w3.org/2001/sw/WebOnt/ for the individual specifications.

[3] See also http://www.cs.man.ac.uk/~horroks/FaCT/.

[4] See OilEd (http://oiled.man.ac.uk) for editing and FrodoRDFSViz (http://www.dfki.uni-kl.de/frodo/RDFSViz) for visualization.

existential quantifier. Another constraint on the slot fillers is *cardinality*, which limits the number of possible fillers of the given class. Atomic elements or individuals can also be associated with a class definition via slot constraints. The decision to restrict ourselves to schemes based on the description logic that allows us to represent enough knowledge for the effective operation of envisaged NLP applications, e.g., those described in Sect. 4.

3 Our Approach to Knowledge Representation

3.1 Top-Level Ontology

Following the distinctions made by Guarino and Welty (2000), we initially defined a collection of concepts that have *primary* ontological status. The guiding principle was to differentiate between the basic ontological entities and the roles taken by them in particular situations, events, or processes. For example, a building can be a hospital, a railway station, a school, etc. But while taking all these roles, it does not cease to be a building. Another example is a person who can take the role of a school teacher, a mother, etc., but it still remains a person for their entire life.

Here the question arises of how deep the differentiation should go. Consider the example of a person: We give a concept *Person* a primary ontological status, but what about the concepts *Man* and *Woman*? Should they be given the same status? Our answer is positive and is based, on one hand, on the assumption that sex is the primary property that defines a person as a man or a woman, and on the other hand, a functional approach shows that relations of these two classes to other classes and their other attributes can be determined by this property. In this way, the basic top-level ontological categorization in our system divides all concepts into two classes *Type* and *Role* (Fig. 1). As the class *Type* includes concepts with primary ontological status independent of the particular application, every system using the ontology for its specific purposes deals with the class *Role*.

Role is the most general class in the ontology representing concrete roles that any entity or process can perform in a specific domain. It is divided into *Event* and *AbstractEvent*. Along with concrete events, i.e., freestanding entities existing essentially in space or in time, our model includes abstract objects, e.g., numbers or abstract properties, such as spatial relations, and abstract processes or rather abstracted states every real process can go through, such as *Start*, *Done*, *Interrupt*, etc. These are modeled separately thereby allowing a uniform description of the processes throughout the ontology.

On the *Role* level we distinguish between *Event* and *AbstractEvent*. *Event* is used to describe a kind of role any entity or process may have in a real situation or process, e.g., a school or an information search. It is contrasted with *AbstractEvent*, which is abstracted from a set of situations and processes. It reflects no reality and is used for the general categorization and description, e.g., *Number*, *Set*, *SpatialRelation*. *AbstractEvent* has subclasses *AbstractObject* and *AbstractProcess*. *Event*s are further classified in *PhysicalObject* and *Process*. In contrast to abstract objects, they have a

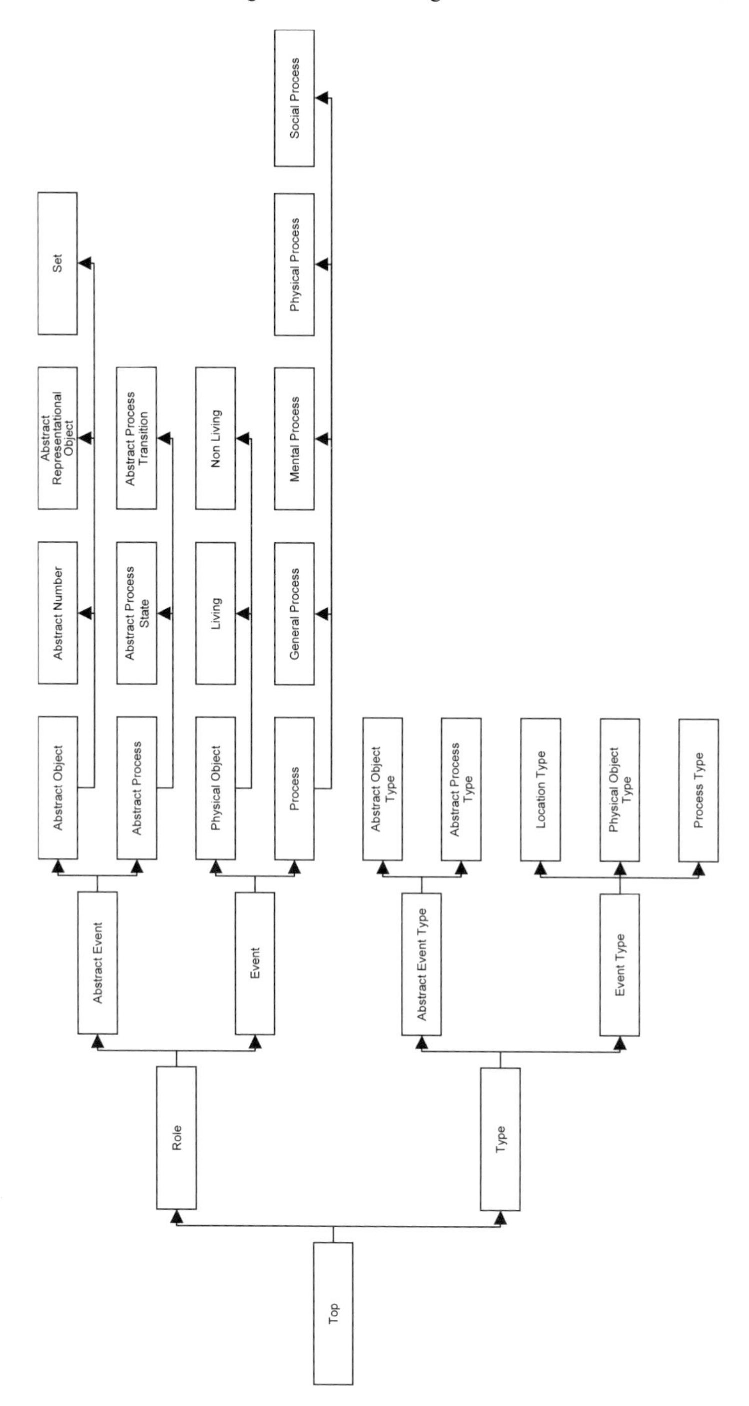

Fig. 1. Top-level part of the ontology

location in space and time. The class *PhysicalObject* describes any kind of objects we come in contact with — living as well as nonliving. These objects refer to different domains, such as *Sight* and *Route* in the tourism domain, *AvMedium* and *Actor* in the TV and cinema domain, etc., and can be associated with certain relations in the processes via slot constraint definitions.

The modeling of *Process* as a kind of event that is continuous and homogeneous in nature follows the frame semantic analysis used for generating the FRAMENET data (Baker et al., 1998). Based on the analysis of our dialogue data, we developed the following classification of processes (Fig. 2):

- *GeneralProcess*, a set of the most general abstract processes such as duplication, imitation, or repetition processes
- *MentalProcess*, a set of processes such as cognitive, emotional, or perceptual processes
- *PhysicalProcess*, a set of processes such as motion, transaction, or controlling processes
- *SocialProcess*, a set of processes such as communication or instruction processes

The *MentalProcess* subtree includes *CognitiveProcess*, *EmotionProcess*, and *PerceptualProcess*. Under *CognitiveProcess* we understand a group of processes that aim at acquiring information or making plans about the future. The further division of *EmotionProcess* into the following subclasses — *EmotionExperiencerObjectProcess* and *EmotionExperiencerSubjectProcess* — is due to the fact that an emotion can be either provoked by an object (e.g., The cry scared me) or can be experienced by an agent towards some object (e.g., I want to go home).

The *PhysicalProcess* has the following subclasses: The semantics of *ControllingProcess* presupposes the controlling of a number of artifacts, e.g., devices; *MotionProcess* models different types of agent's movement regarding some object or point in space; *PresentationProcess* describes a process of displaying some information by an agent, e.g., a TV program by an artificial character embedding the SMARTKOM system; *StaticSpatialProcess* consists in the agent's dwelling in some point in space; *TransactionProcess* presupposes an exchange of entities or services among different participants of the process. Another subclass of the *Process* – *SocialProcess* includes *CommunicativeProcess*, which consists in communicating by the agent a message to the addressee by different means, and *InstructiveProcess* which describes an interaction between an agent and a trainee.

3.2 Slot Hierarchy

The slot structure also reflects the general intention to keep abstract and concrete elements apart. A set of most general properties has been defined with regard to the role an object can play in a process: *agent*, *theme*, *experiencer*, *instrument* (or *means*), *location*, *source*, *target*, *path*. These general roles applied to concrete processes may also have subslots. Thus an agent in a process of buying (*TransactionProcess*) is a *buyer*; the one in the process of cognition is a *cognizer*. This way, slots can also build

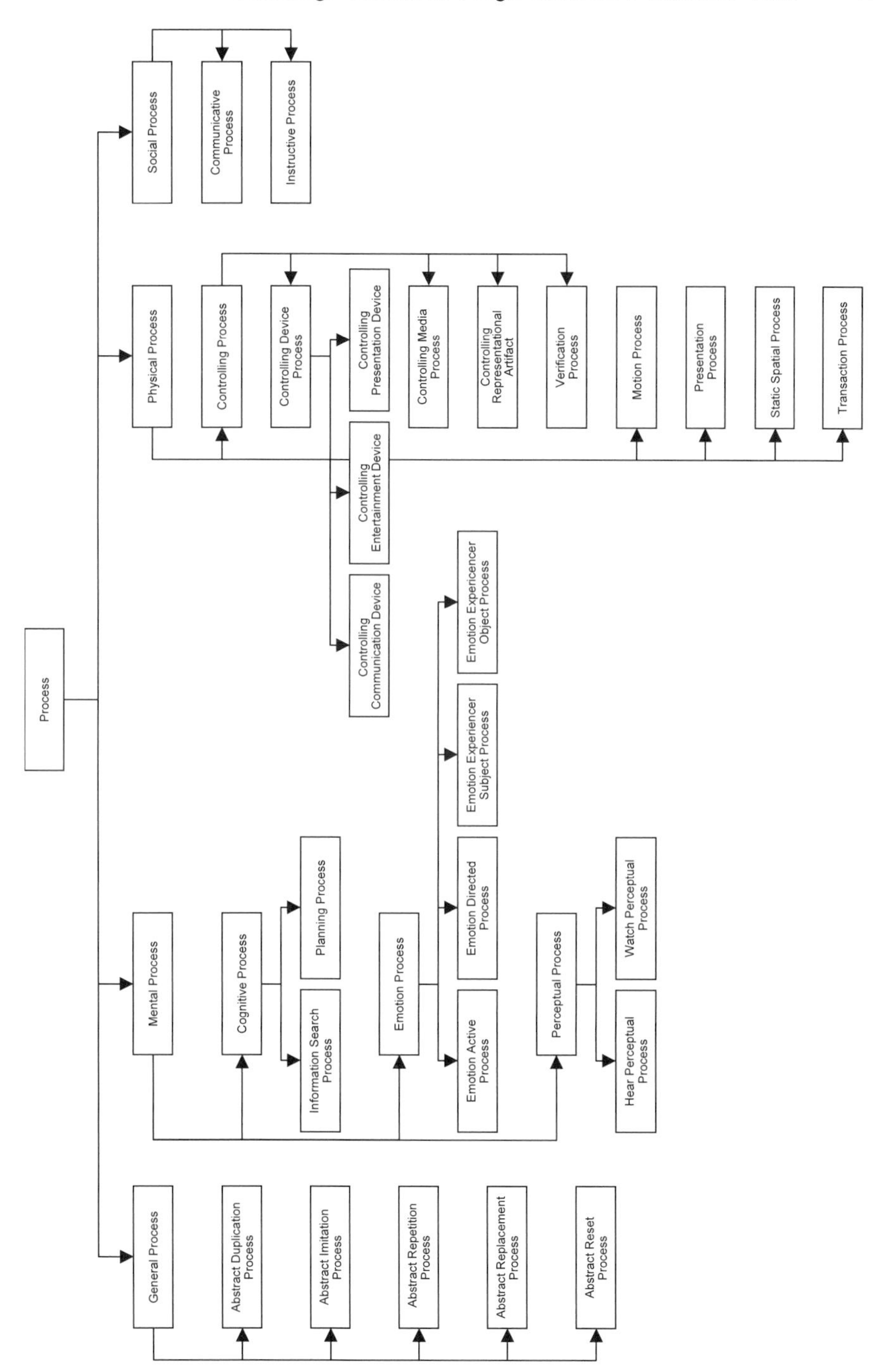

Fig. 2. Process hierarchy

hierarchical trees. The property *theme* in the process of information search is a required *piece-of-information*; in a presentation process it is a *presentable-object*, i.e., the item that is to be presented, etc.

Consider the class *Process*. It has the following slots: *begin-time*, a time expression indicating the starting point; *end-time*, a time expression indicating the time point when the process is complete; *state*, one of the abstract process states. These slots describe properties that are common to all processes, and as such they are inherited by all subclasses of the *Process* class.

An *EmotionExperiencerSubjectProcess* inherits the slots of the *Process* class, among them the slot *theme* that can be filled with any process or object (the basic idea is that any physical entity or the performance of any process can become an object of someone's emotion). It also has several additional properties such as *experiencer* to denote the one who undergoes the process, and *preference* to define the attitude an experiencer has to the object of its emotion.

3.3 Ontology Instances

Consider the definition of the *InformationSearchProcess* in the ontology. It is modeled as a subclass of the *CognitiveProcess*, which is a subclass of the *MentalProcess*. Therefore it inherits the following slot constraints: *begin-time*, a time expression indicating the starting time point; *end-time*, a time expression indicating the time point when the process is complete; *state*, one of the abstract process states, e.g., start, continue, interrupt, etc.; and *cognizer*, filled with a class *Person* including its subclasses.

The *InformationSearchProcess* features one additional slot constraint, *piece-of-information*. The possible slot fillers are a range of domain objects, e.g., *Sight*, *Performance*, or whole *sets* of those, e.g., *TvProgram*, but also processes, e.g., *ControllingTvDeviceProcess*. This way, an utterance such as:[5]

(1) *I hätte gerne Informationen zum Schloss*
I would like information about the castle

can be mapped onto the *InformationSearchProcess*, which has an agent of the type *User*, and a piece of information of the type *Sight*. *Sight* has a name of the type *Castle*. Analogously, the utterance:

(2) *Wie kann ich den Fernseher steuern*
How can I the TV control

can also be mapped onto the *InformationSearchProcess*, which has an agent of the type *User*, and has a piece of information of the type *ControllingTvDeviceProcess*.

Another example demonstrating how slot structures can be shared between some super- and subclasses: the subclass *AvEntertainment* inherits from its superclass *Entertainment* the following slots: *duration*, *end-time*, and *begin-time*, filled by the

[5] All examples are displayed with the Germano riginal on top and a glossed translation below.

TimeDuration and *TimeExpression*, respectively. The class *AvEntertainment* features two additional slots: *language*, whose filler is an individual *Language*, and *av-medium*, whose filler is a class *AvMedium*. The class *AvEntertainment* has further subclasses — *Broadcast*, which represents an individual entry in a TV program, and *Performance*, which models an entry in a cinema program. Both of them inherit the slots of the superclasses *Entertainment* and *AvEntertainment*, while also featuring their own additional slots, e.g., *channel* and *showview* for the *Broadcast*, *cinema* and *seat* for the *Performance*. This feature can be effectively utilized by a specific dialogue interpretation algorithm called OVERLAY(Pfleger et al., 2002).

4 Applications of the Ontology

There is no agreed methodology for ontology evaluation. In our opinion, the usefulness of an ontology can be evaluated by examining the ways it is employed within the system, allowing us to draw tentative conclusions as for the reusability of the ontology and its portability with respect to new applications and NLP tasks. The ontology described here is used by the complete core of the system, i.e., as a knowledge repository and model for language interpretation, discourse modeling, intention recognition and action planning (Engel, 2002; Löckelt et al., 2002; Alexandersson and Becker, 2003; Gurevych et al., 2003b; Porzel et al., 2003b). In the next sections we describe some examples of the usage within the project.

4.1 Semantic Coherence Scoring

One of the major challenges in making an MMDS reliable enough to be deployed in more complex real world applications is an accurate recognition of the users' input. In many cases both correct and incorrect representations of the users' utterances are contained in the automatic speech recognizer's n-best lists. Facing multiple representations of a single utterance poses the question, which of the different hypotheses most likely corresponds to the user's utterance. Different methods have been proposed to solve this problem. Frequently, the scores provided by the recognition system itself are used. More recently, also scores provided by the parsing system have been employed, e.g., Engel (2002). In this application, we propose a new ontology-based method and show that knowledge-based scores can be successfully employed to re-score the speech recognition output.

We introduced the notion of *semantic coherence* as a special measurement that can be applied to estimate how well a given speech recognition hypothesis (SRH) fits with respect to the existing knowledge representation (Gurevych et al., 2003a).

The software for scoring the SRHs and classifying them in terms of their semantic coherence employs the ontology described herein. This means that the ontology crafted as a general knowledge representation for various processing modules of the system is additionally used as the basis for evaluating the semantic coherence of sets of concepts.

The scoring software performs a number of processing steps:

- Converting each SRH into a concept representation. For this purpose, each entry of the system's lexicon was augmented with zero, one, or multiple ontology concepts.
- Converting the domain model, i.e., an ontology, into a directed graph with concepts as nodes and relations as edges.
- Scoring concept representations using a scoring metric based on the shortest path between concepts.

For example, in our data (Gurevych et al., 2002) a user expressed the wish to get more information about a specific church, as:

(3) *Kann ich bitte Informationen zur Heiliggeistkirche*
May I please information about the Church of Holy Spirit
bekommen
get

Looking at two SRHs from the ensuing n-best list we found that Example 5 constituted a suitable representation of the utterance, whereas Example 4 constituted a less adequate representation thereof, labeled accordingly by the human annotators:

(4) *Kann ich Information zur Heiliggeistkirche kommen*
May I information about the Church of Holy Spirit come

(5) *Kann ich Information zur Heiliggeistkirche bekommen*
May I information about the Church of Holy Spirit get

According to the lexicon entries, the SRHs are transformed into two alternative concept representations:

CR_1:{Person; Information Search Process; Church; Motion Directed Transliterated Process};
CR_2:{Person; Information Search Process; Church; Transaction Process}.

The scores are normalized as numbers on a scale from 0 to 1 with higher scores indicating better semantic coherence. Then, the resulting score assigned to Example 4 is 0.6, and the score of Example 5 is 0.75. The evaluation of the method against the hand-annotated corpus has shown that it successfully classifies 73.2 in a German corpus of 2.284 speech recognition hypotheses as either coherent or incoherent, given a baseline 54.55 from the annotation experiments (the majority class).

Semantic coherence scoring method has been successfully applied to the task of identifying the best SRH in the output of the automatic speech recognizer (Porzel and Gurevych, 2003). Its additional application is the calculation of a semantic coherence score for SRHs taking into account their conceptual context (Porzel et al., 2003a). Also, ontological concepts can be used to perform ontology-based automatic domain recognition and domain change detection.

4.2 Generating Interface Specifications

In this ontology application, we proposed to use the knowledge modeled in the ontology as the basis for defining the semantics and the content of information exchanged between various modules of the system (Gurevych et al., 2003b). In NLP systems, modules typically exchange messages, e.g., a parser might get word lattices as input and produce corresponding semantic representations for later processing modules, such as a discourse manager. The increasing employment of XML-based interfaces for agent-based or other multiblackboard communication systems sets a de facto standard for syntax and expressive capabilities of the information that is exchanged amongst modules. The content and structure of the information to be represented are typically defined in corresponding XML schemata (XMLS) or document type definitions (DTD).

As discussed above, ontologies are a suitable means for knowledge representation, e.g., for the definition of an explicit and detailed model of a system's domains. That way, they provide a shared domain theory, which can be used for communication. Additionally, they can be employed for deductive reasoning and manipulations of models. The meaning of ontology constructs relies on a translation to some logic. This way, the inference implications of statements, e.g., whether a class can be related to another class via a subclass or some other relation, can be determined from the formal specification of the semantics of the ontology language. However, this does not make any claims about the syntactic appearance of the representations exchanged, e.g., an ordering of the properties of a class.

An interface specification framework, such as XMLS or DTD, constitutes a suitable means for defining constraints on the syntax and structure of XML documents. Ideally, the definition of the content communicated between the components of a complex dialogue system should relate both the syntax and the semantics of the XML documents exchanged. Those can then be seen as instances of the ontology represented as XMLS-based XML documents. However, this requires that the knowledge, originally encoded in the ontology, is represented in the XMLS syntax.

The solution proposed states that the knowledge representations to be expressed in XMLS are first modeled in OIL-RDFS or DAML+OIL as *ontology proper*, using the advantages of ontology engineering systems available, and then transformed into a communication interface automatically with the help of the software developed for that purpose.[6]

Employing this approach, XMLS and DTDs are created such that they:

- stay logically consistent
- are easy to manage
- enable a straightforward mapping back to the respective knowledge representation for inference

[6] This is a free software project. The package and respective documentation can be obtained from http://savannah.nongnu.org/projects/oil2xsd.

- allow the handling of a range of NLP tasks immediately on the basis of XMLS, e.g., the discourse module operates on the XML schema obtained via ontology transformation(Alexandersson and Becker, 2003).

The resulting schemata capture the hierarchical structure and a significant part of the semantics of the ontology. We therefore provide a standard mechanism for defining XMLS-based interface specifications, which are *knowledge rich*, and thus can be used as a suitable representation of domain and discourse knowledge by NLP components. Since the software that has been developed completely automates the transformation process, the resulting XMLS are congruent with the XML schema specifications. Furthermore, the ontology can be reused in multiple systems as a single ontology can be used to generate application-specific communication interfaces.

However, the main advantage of our approach is that it combines the power of ontological knowledge representation with the strengths of XMLS as an interface specification framework in a single and consistent representation. Our experience shows, this would not have been possible for a complex dialogue system if XML schemata were defined from scratch or handcrafted, and constitutes a step toward building robust and reusable NLP components.

5 Conclusion

An ontology-based approach to modern MMDS uses a single knowledge representation. We described the major modeling principles and the design choices made. Furthermore, we sketched some examples of the ontology application within the system.

Together these examples suffice to demonstrate the benefits of using a single knowledge representation throughout a dialogue system as opposed to using multiple knowledge representations and formats. An additional advantage of such a homogeneous world model that defines the processing interfaces as well as the system's world knowledge is that no costly mappings between them are no longer necessary. This means that modules receive only messages whose content is congruent to the terminological and structural distinctions defined in the ontology.

Our additional concern while designing the ontology was the reusability of this component within our MMDS as well as other NLP systems. So far, the top-level ontology proved stable. We found the extensions on the lower levels of the ontology to be comparatively cheap. This single knowledge base was successfully tested and applied to multiple NLP problems, e.g., resolving bridging expressions in texts as well as for the resolution of metonymical and polysemous utterances next to defining communication interfaces for NLP components of the system and scoring of speech recognition hypotheses described above. Consequently, the approach to knowledge representation and the ontology itself can be reused in the future design of MMDS.

Acknowledgments

This work was partially funded by the German Federal Ministry of Education, Science, Research, and Technology (BMBF) in the framework of the SMARTKOM project under Grant 01 IL 905 K7 and by the Klaus Tschira Foundation. We would also like to thank Tatjana Medvedeva, Elena Slinko, Lutz Wind, and Iryna Zhmaka for their valuable contributions.

References

J. Alexandersson and T. Becker. The Formal Foundations Underlying Overlay. In: *Proc. 5th Int. Workshop on Computational Semantics (IWCS-5)*, pp. 22–36, Tilburg, The Netherlands, February 2003.

J.F. Allen, B. Miller, E. Ringger, and T. Sikorski. A Robust System for Natural Spoken Dialogue. In: *Proc. 34th Annual Meeting of the Association for Computational Linguistics*, pp. 62–70, Santa Cruz, CA , June 1996.

C.F. Baker, C.J. Fillmore, and J.B. Lowe. The Berkeley FrameNet Project. In: *Proc. COLING-ACL'98*, pp. 86–90, Montreal, Canada, 1998.

R. Engel. SPIN: Language Understanding for Spoken Dialogue Systems Using a Production System Approach. In: *Proc. ICSLP-2002*, pp. 2717–2720, Denver, CO, 2002.

D. Fensel, F. van Harmelen, I. Horrocks, D.L. McGuinness, and P.F. Patel-Schneider. OIL: An Ontology Infrastructure for the Semantic Web. *IEEE Intelligent Systems*, 16(2):38–45, 2001.

G. Ferguson, J.F. Allen, B. Miller, and E. Ringger. The Desgin and Implementation of the TRAINS-96 System. Technical Report 96-5, University of Rochester, New York, 1996.

N. Guarino and R. Poli. Formal Ontology in Conceptual Analysis and Knowledge Representation. *Special issue of the Int. Journal of Human and Computer Studies*, 43, 1995.

N. Guarino and C. Welty. A Formal Ontology of Properties. In: R. Dieng and O. Corby (eds.), *Proc. EKAW-2000: The 12th Int. Conf. On Knowledge Engineering and Knowledge Management*, vol. 1937, pp. 97–112, Berlin Heidelberg New York, 2000. Springer.

I. Gurevych, R. Malaka, R. Porzel, and H.P. Zorn. Semantic Coherence Scoring Using an Ontology. In: *Proc. Human Language Technology Conference / North American Chapter of the Association for Computational Linguistics Annual Meeting 2003*, pp. 88–95, Edmonton, Canada, 2003a.

I. Gurevych, R. Porzel, E. Slinko, N. Pfleger, J. Alexandersson, and S. Merten. Less Is More: Using a Single Knowledge Representation in Dialogue Systems. In: *Proc. HLT-NAACL'03 Workshop on Text Meaning*, pp. 14–21, Edmonton, Canada, May 2003b.

I. Gurevych, R. Porzel, and M. Strube. Annotating the Semantic Consistency of Speech Recognition Hypotheses. In: *Proc. 3rd SIGdial Workshop on Discourse and Dialogue*, pp. 46–49, Philadelphia, PA, July 2002.

G. Herzog, H. Kirchmann, S. Merten, A. Ndiaye, and P. Poller. MULTIPLATFORM Testbed: An Integration Platform for Multimodal Dialog Systems. In: *Proc. HLT-NAACL 2003 Workshop on Software Engineering and Architecture of Language Technology Systems (SEALTS)*, pp. 76–83, Edmonton, Canada, May 2003.

M. Löckelt, T. Becker, N. Pfleger, and J. Alexandersson. Making Sense of Partial. In: C.M. Johan Bos, Mary Ellen Foster (ed.), *Proc. 6th Workshop on the Semantics and Pragmatics of Dialogue (EDILOG 2002)*, pp. 101–107, Edinburgh, UK, September 2002.

N. Pfleger, J. Alexandersson, and T. Becker. Scoring Functions for Overlay and Their Application in Discourse Processing. In: *Proc. KONVENS 2002*, pp. 139–146, Saarbruecken, Germany, September–October 2002.

R. Porzel and I. Gurevych. Contextual Coherence in Natural Language Processing. In: P. Blackburn, C. Ghidini, R.M. Turner, and F. Giunchiglia (eds.), *Proc. of Modeling and Using Context – 4th International and Interdisciplinary Conference, CONTEXT 2003*, LNCS 2680, pp. 272–285, Berlin Heidelberg New York, 2003. Springer.

R. Porzel, I. Gurevych, and C. Müller. Ontology-Based Contextual Coherence Scoring. In: *Proc. 4th SIGdial Workshop on Discourse and Dialogue*, Sapporo, Japan, July 2003a.

R. Porzel, N. Pfleger, S. Merten, M. Loeckelt, I. Gurevych, R. Engel, and J. Alexandersson. More on Less: Further Applications of Ontologies in Multi-Modal Dialogue Systems. In: *Proc. 3rd IJCAI 2003 Workshop on Knowledge and Reasoning in Practical Dialogue Systems*, Acapulco, Mexico, 2003b.

S.J. Russell and P. Norvig. *Artificial Intelligence. A Modern Approach.* Prentice Hall, Englewood Cliffs, NJ, 1995.

Speech Recognition

André Berton, Alfred Kaltenmeier, Udo Haiber, and Olaf Schreiner

DaimlerChrysler AG, Research and Technology, Ulm, Germany
{andre.berton,alfred.kaltenmeier,udo.haiber,
olaf.schreiner}@daimlerchrysler.com

Summary. The human machine interaction of SmartKom is a very complex task, defined by natural, spontaneous language, speaker independence, large vocabularies, and background noises. Speech recognition is an integral part of the multimodal dialogue system. It transforms the acoustic input signal into an orthographic transcription representing the utterance of the speaker. This contribution discusses how to enhance and customize the speech recognizer for the SmartKom applications. Significant improvements were achieved by adapting the speech recognizer to the environment, to the speaker, and to the task. Speech recognition confidence measures were investigated to reject unreliable user input and to detect user input containing unknown words, i.e., words that are not contained in the vocabulary of the speech recognizer. Finally, we present new ideas for future work.

1 Introduction

Speech recognition as part of multimodal dialogue systems transforms the acoustic input signal into an orthographic transcription representing the utterance of the speaker. This contribution discusses how to optimize the speech recognizer for the SmartKom applications. It starts with a general description of the speech recognition engine, which explains the signal processing, vector quantization, and decoding of the recognizer. The decoding that relies on acoustic Hidden Markov Models (HMMs) and statistical language models is described, particularly the training procedure.

After describing the speech recognition basics, we focus on the four main research interests for speech recognition in multimodal environments, namely how to adapt to the environment, to the speaker, to the task, and how to estimate confidences for the correctness of the words hypothesized by the recognizer. The system adapts itself to the environment by enabling Barge-In using an echo cancellation technique in the acoustic device, by reducing environmental noises using spectral substraction, and by adjusting the input signal level using automatic gain control. Several supervised and unsupervised speaker adaptation techniques are presented: vocal tract normalization, maximum likelihood linear regression, codebook mixture, and speaker-dependent training. The SmartKom system adapts itself to the task or

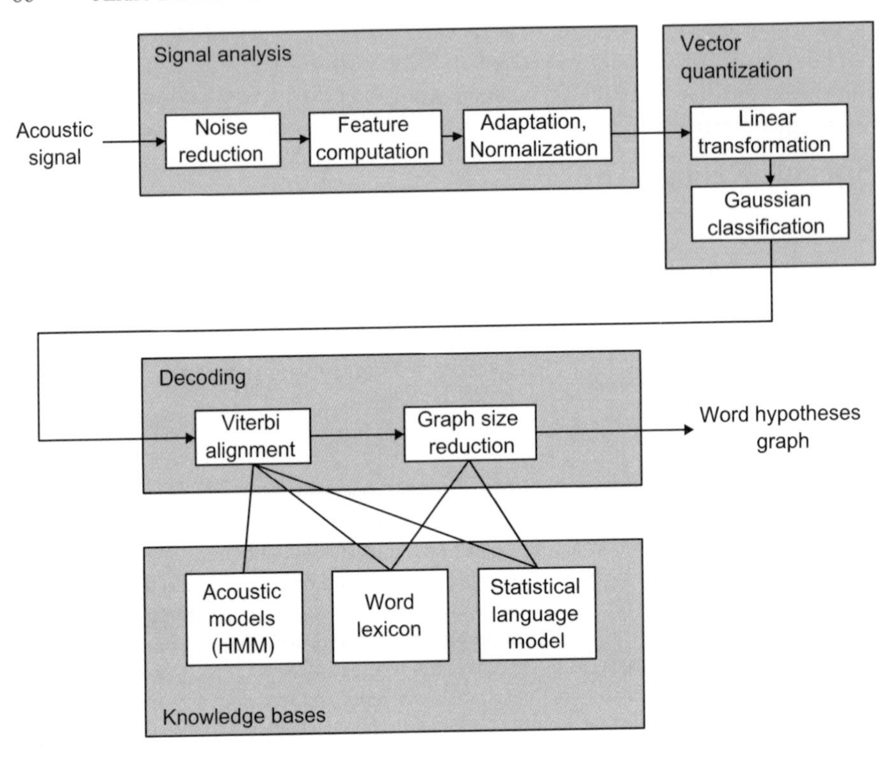

Fig. 1. Setup of the DaimlerChrysler speech recognition engine

application by dynamically updating the vocabulary and the members of the categories in the statistical language model. The last part of the research interests introduces two methods for word confidence estimation: The first method directly calculates the confidence from very large word graphs, while the second method collects features that correlate with the word's correctness and estimates confidences using known classification techniques. Furthermore, we discuss the problem of unknown words. These are words that the speaker may utter but that are not included in the recognizer vocabulary. We introduce an approach how to model such words in order to detect them. Finally, conclusions about the results achieved in the research areas are drawn, and ideas for future work are presented.

2 Speech Recognition Engine

An automatic speech recognizer transforms the acoustic signal into a sequence of hypothesized words in orthographic representation. The speech recognizer implemented at DaimlerChrysler follows the semicontinuous Hidden Markov Model paradigm (SCHMM), which consists of three main processing stages: signal analysis, vector quantization, and decoding (Fig. 1).

Signal processing transforms the speech signal into a vector of features that represent the phonetic contents of the signal. Details about signal analysis can be found in the sections that describe the adaptation to the environment and to the speaker. Vector quantization and decoding are explained in detail in the following sections. The decoding process needs certain knowledge bases to find the best matching word sequences. We distinguish three knowledge bases:

- acoustic models: HMMs
- word lexicon: vocabulary
- language model: statistical trigram model

2.1 Feature Transform and Classification (Vector Quantization)

The second block of the speech recognition engine (Fig. 1) includes feature transformation and classification. After signal analysis 9 adjacent mel-frequency cepstrum (MFC) feature vectors (covering 90 ms of the input speech signal) were combined to form a 117-dimensional super-vector. Linear transforms such as linear discriminant analysis (LDA) or principal component analysis (PCA) can be applied to reduce the vector dimension, from 117 to 32 in our case. Transformed vectors have an inherent component ordering; the first component is the most important one for separation or reconstruction, resp. Transform matrices are computed during forward/backward training of acoustic HMMs. Both transforms yield very similar recognition results with LDA slightly superior to PCA.

Transformed feature vectors are vector-quantized (VQ) with a Gaussian classifier consisting of 1024 classes with a full covariance matrix (full matrix inverse) for each class. Due to the symmetry of covariance matrices and inverses, 33*34/2=561 parameters have to be stored for each class, i.e., 561*1024 parameters for the whole classifier. Taking advantage of component ordering, classification can be performed in a hierarchical structure of classifier levels. Each level takes into account eight components of a transformed input vector: The first level uses the eight most important components and estimates likelihoods for all classes. The second level then adds another eight components and reestimates likelihoods only for classes considered important by the first level. This procedure is iterated for all four levels. Nonhierarchical classification requires 561*1024*100 scalar product operations (multiply/add) per second; our approach reduces computational requirements by more than a factor of 5 ($\approx$10,000,000 operations per second) without *any loss* in recognizer performance.

The classifier output is finally limited to a given likelihood threshold ($\varepsilon = 10^{-3}$) and ten classes and associated likelihoods at maximum. The average number of classes is about 8.0 for our training and test databases.

2.2 Acoustic Models and Training

Recognition is based on statistical models — HMMs — of phonetic units. Many recognition systems described in the literature use generalized triphone models. Our approach uses a more extended set:

- Several models for background noise
- Models for nonverbal utterances (laughter, coughing, clearing one's throat, smacking, breathing, ...)
- Whole-word models for important content words (yes, no, cardinal and ordinal numbers, navigation or entertainment applications, ...)
- 30 models for spelling (a, b, ..., z, ä, ö, ü, and ß)
- Context-dependent models of frequent initial or final consonant clusters of syllables, e.g., |Str...| as in "Straße" or |...nst| as in "Dienst"
- Context-dependent models of diphthongs and vowel–r combinations, e.g., |...aU...| as in "Baum" or |...O6...| as in "dort"
- Context-dependent triphone or generalized triphone models, biphone and monophone models as usual

Model parameters are trained with the well-known forward/backward algorithm in an iterative manner, with four training steps in our case:

1. Silence and background noise models, keep all other models fixed
2. + Whole word, alphabet, nonverbal utterances, monophone models
3. + Initialize triphone and syllable group models with monophone parameters and train the complete model set together
4. Generalized consonant and vowel clusters for out-of-vocabulary words (OOV)

After training, model parameters (weights of classes or emissions) are clipped to a certain threshold ($\varepsilon = 10^{-6}$). Hence our recognizer has the following setup:

- 1750 models with 7300 states
- 715,000 weights (instead of $\approx$7,400,000 weights for full storage)

Neglecting the ε-weights significantly reduces storage and computational requirements without *any loss* in performance.

The LDA approach for vector quantization requires some class information that is defined by states or state-clusters. Basically, we have 7300 states (or classes) that have to be reduced to 1024 for VQ. Class reduction is performed by bottom-up entropy clustering; most similar states are automatically grouped together in an iterative clustering procedure taking the HMM type into account:

- States of noise models may only be clustered with states of other noise or silence models (noise/silence forms a separate superclass).
- States of alphabet models may be clustered only with states of other alphabet models (alphabets forms another superclass).
- All other model states may be clustered together.

2.3 Decoding and Word Hypotheses Graph Processing

Speech recognition is based on a frame-synchronous Viterbi algorithm and uses following knowledge bases (Fig. 1):

- HMMs as described above
- A lexicon tree constituted from HMMs and the recognizer vocabulary, which may be updated dynamically
- A class-dependent language model (LM) for the active vocabulary, which may be updated dynamically, too

The *nodes* of the tree represent *states* of HMMs; the *arcs* represent *transitions* to other states with associated state-transition probabilities and approximated LM unigram and bigram values.

Decoding is performed on frame-by-frame basis:

- For each frame vector represented by VQ classes and likelihoods, observation probabilities are computed for all states. This requires approximately 7300 * 8.0 * 100 $\approx$ 6,000,000 operations (multiply/add) per second.
- Using these acoustic probabilities, all active paths through the lexicon tree are extended for one frame. At word ends (leaf nodes of the tree) we not only store the best path and its associated word sequence, but we also store all promising word candidates forming the so-called word hypotheses graph (WHG). The nodes of this graph represent time stamps, and the arcs represent words with associated acoustic probabilities.
- If words end at a time frame, the lexicon tree is "restarted" again.

The WHG is finally reduced in the last block of the decoder module; i.e., words and paths may be eliminated from the graph. Reduction is performed on a node-by-node basis. At each "anchor" node, WHG size is reduced considering acoustic and LM probabilties. Anchor nodes are normally constituted by speech pauses or well-recognized words, i.e., words with no competing alternatives.

For the SMARTKOM lexicon with about 5000 words, our recognizer requires about *0.4-times* real-time on a IBM T30 laptop computer. Recognition was evaluated with three test sets. Table 1 gives word error rates (WER) and sentence error rates (SER).

Table 1. Recognition results for the final SMARTKOM recognizer and three test sets "Public", "Mobil", "EML"

Test set name	No. of sentences	No. of words	WER	SER
Public	1144	7411	11.2%	36.1%
Mobil	1136	4874	7.9%	19.3%
EML	1508	8788	21.1%	49.2%

3 Adaptation to the Environment

The SMARTKOM system must be able to adapt itself to the continuously changing environmental conditions, like noises and overlapping speech. A great variety of instationary noises make the complex task of speech recognition even harder. This section presents three techniques, namely echo cancellation, noise reduction, and automatic gain control, which are applied in the SMARTKOM system in order to achieve better system behavior due to environmental adaptation.

3.1 Echo Cancellation and Barge-In

One of the common expectations that the majority of state-of-the-art dialogue systems do not meet is that the user can interrupt the system anytime, even while the system speaks to the user. This functionality that allows the user to speak his request while the system is speaking is called *Barge-In*. The user expects an appropriate system response to the request on interpretation level. This kind of interaction relates to all modules of the dialogue system. Barge-In can be applied in the following scenarios:

1. Shortcut: e.g., early selection of a list item (before the end of the list)
2. Cancel: e.g., interruption due to misunderstandings or irrelevant system output
3. Back-channel: maintainance of the communication by the user during long system outputs (e.g., user utterance of "Uhm")
4. Off-talk: utterances not directed to the SMARTKOM system

The following technical requirements can be defined for handling Barge-In intelligently:

1. Two input channels (system output and user input)
2. Speech recognition in parallel with speech synthesis output
3. Echo cancellation of the synthesis prompt
4. Back-channel models (user signals attention, but does not intend to interrupt): uhm, yeah, right, okay, I see
5. Semantic constructions for back-channel models
6. Contextual interpretation
7. Dialogue control models for different types of Barge-In
8. Time alignment of synthesis, generation, and audio output (duration estimation for utterances and segments of utterances)
9. Optimization of recognition control by empirical tuning

The most important component that needs to be added to the dialogue system to allow Barge-In is *echo cancellation*. The microphone of the dialogue system needs to be open almost anytime, so that the system responses from the loudspeakers will be included in the signal received by the microphone additionally to the utterances of the user. The sound of the speakers reflected by the walls of the room arrives at slightly different times and volumes at the microphone. Larger rooms lead to longer signal response times. The speech recognizer requires a speech signal that is distorted as little as possible to achieve high recognition accuracy. Therefore, the echo cancellation needs to substract the system output from the input signal on the microphone in order to ensure that the input signal of the recognizer contains only the

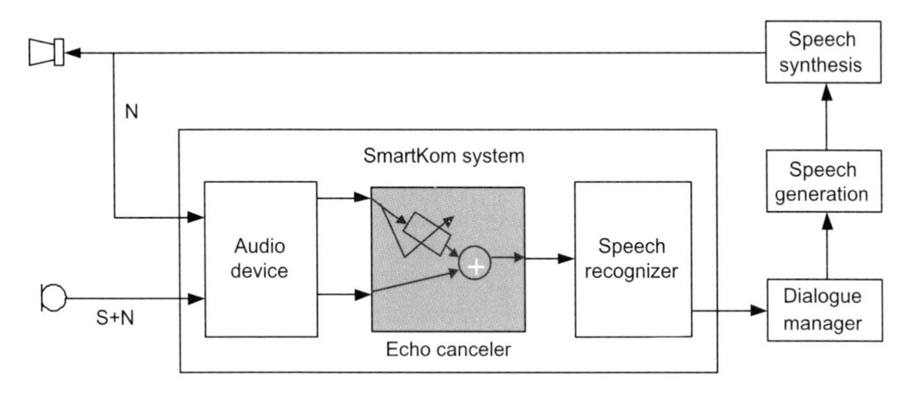

Fig. 2. Technical setup of the echo cancellation in the SMARTKOM system

speech uttered by the user. The noise signal N (system response) is simultaneously recorded with the input signal S from the microphone on a different channel of the audio device (Fig. 2), since only synchronous computation allows for canceling the system response.

The following three tasks of an echo canceler can be distinguished:

1. Estimation of the signal impulse response of the system output (variation of the signal between loudspeaker and microphone)
2. Convolution of the speaker output signal with the system impulse response
3. Substraction of the microphone signal after convolution

The impulse response consists of the room impulse response and the impulse response of the loudspeaker and microphone. It is iteratively estimated by the language model score (LMS) algorithm.

The echo cancellation has been implemented in cooperation with MediaInterface Dresden. DaimlerChrysler provided the echo cancellation algorithm, and MediaInterface integrated it in the audio device and optimized it for the particular environments.

The echo cancellation algorithm adapts itself to environmental changes and is a pure software solution. Experiments have shown that the recognizer's word error rate decreases from 100% to about 30% for Barge-In and continuous utterances if echo cancellation is applied. That means that Barge-In can only be allowed if a good echo canceler is at hand. The echo canceler has still to be optimized for every new room because of varying sound characteristics (length of filter stages and adaptation coefficients). The cancellation leads to an average signal gain of about 40 dB at very low computational costs and very little delay caused by the incremental processing. However, the estimation of the impulse responses needs a certain onset time, because it follows the principle of iterative error minimization. It is still a scientific challenge to shorten the onset time.

3.2 Noise Reduction

Another aspect of environmental adaptation is a more precise modeling of speech and noises to be able to separate one from another. Noises can either be substracted from the speech signal or additionally modeled.

The technique commonly applied to reduce stationary noises is *spectral substraction* (SPS). An adaptive version of that method has been developed at DaimlerChrysler. This technique reduces noises in the spectral representation of the speech signal. The stationary noises that will be substracted from the signal can be estimated in speech pauses. The SPS requires a well-working voice activity detector, which segments the signal. The automatic gain control that has been implemented to improve voice activity detection will be described in Sect. 3.3.

We also investigated a new technique for robust speech recognition, namely the *missing data method.* This method assumes that not all frequency bands are distorted by noise. The aim of our investigations was to recognize speech only in the nondistorted frequency bands. The selection of nondistorted bands proved to be very difficult. The method performed better on noisy speech than a speech recognizer that was trained only on clean speech, but worse than a recognizer also trained on noisy speech.

So far, we described methods to reduce noise in the signal. The following methods, however, will try to model noise as precisely as possible and to separate it from speech. That allows us to hypothesize noises on the HMM level and to reject them if needed. We trained various HMMs for technical and human noises, so that such speech segments will not be mixed up with words.

We also considered an approach called *parallel model combination* (PMC) to additively model noises by a particular mean estimation and parallel noise models. This approach led to recognition improvements only for recognizers trained on clean speech. Since the DaimlerChrysler recognition engine is usually trained on noisy speech, there has been no need to apply PMC so far.

Finally, we investigated a promising new hybrid method: *multistream recognition*, since such an approach combines the advantages of neural networks and HMMs. That method employs a neural network (multilayer perceptron) to estimate the posterior probabilities of each HMM state for a given speech signal on frame basis. These posterior probabilities are considered as emission probabilities in the HMMs and can be used directly in Viterbi decoding. Various acoustic feature streams and their combinations have been investigated in the acoustic front end:

1. MFC: mel-frequency cepstrum (the classical short-term feature stream)
2. PLP: perceptual linear prediction (oriented toward human perception)
3. MSG: modulation spectogram (covers long-term signal characteristics)

The Performance of the new method, which includes the different feature streams, was superior to our standard recognizer on a speech database covering digit sequences with artificially added noises (Aurora-2). In particular, the combination of PLP and MSG feature streams led to a significant improvement in recognition accuracy of up to 60%. That improvement can be explained by the complementary properties of the feature streams: the PLP features cover the short-term properties of the signal (about 20 ms), while the MSG features represent the long-term characteristics (about 2 s), like reverberation. However, the new methods did not prove successful for large vocabulary continuous speech recognition, because the estimation of posterior probabilities was not discriminative enough for large sets of HMMs. Therefore, the multistream approach has not been further considered for the SMARTKOM project.

3.3 Automatic Gain Control

Adapting the speech recognizer to the environment also means coping with various signal levels. Some users speak louder than others, and there are different microphone setups and various distances between speaker and microphone. The speech recognizer however performs best if the signal level is well-adjusted to about half the maximum level. Such an adjustment is performed by an *automatic gain control* (AGC), improving the speech-pause segmentation of the voice activity detector and thus the recognition accuracy. In contrast to the standard approach, the method developed at DaimlerChrysler adjusts the signal only in speech segments, while non-speech segments remain unchanged. This has a very positive effect on segmentation accuracy, particularly at low signal-to-noise ratios. The AGC makes sure the input signal is not overamplified, and thus distorted, by converting the signal for the speech recognizer to the range of 60% of the maximum amplitude. Figure 3 shows an example of the effect of AGC by amplifying the low signal in the upper part to a level well-suited for speech recognition.

4 Adaptation to the Speaker

Speaker adaptation plays a great role in SMARTKOM, since the multimodal dialogue system should be able to handle the requests of any user. Different utterances of each speaker vary greatly, but utterances of different speakers vary even more. Speaker adaptation is intended to improve recognition accuracy, so that it contributes largely to user acceptance. Adaptation can be realized by supervised and unsupervised methods. Unsupervised adaptation is performed while the user interacts with the system without precisely knowing the transcription of the user utterance. The transcription of user utterance is known in supervised adaptation, where the user is requested to speak some predefined sentences (typically 10 to 100) in a training phase. In this section we present three unsupervised methods: vocal tract normalization (VTN), maximum likelihood linear regression (MLLR), and codebook mixture (CBM), and one supervised adaptation method: speaker dependent training (SDT).

A particular motivation for performing speaker adaptation is to better cope with problem speakers. Problem speakers are characterized by one or more of the following phenomena: dialect, unclear articulation or strong coarticulation, and syntactical mistakes. These phenomena decrease recognition accuracy significantly, so that each of the following methods should contribute to improving adaptation.

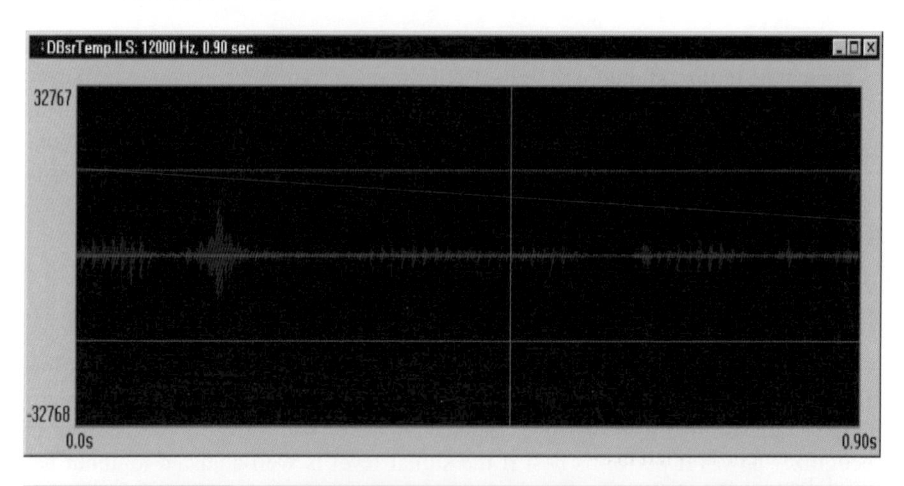

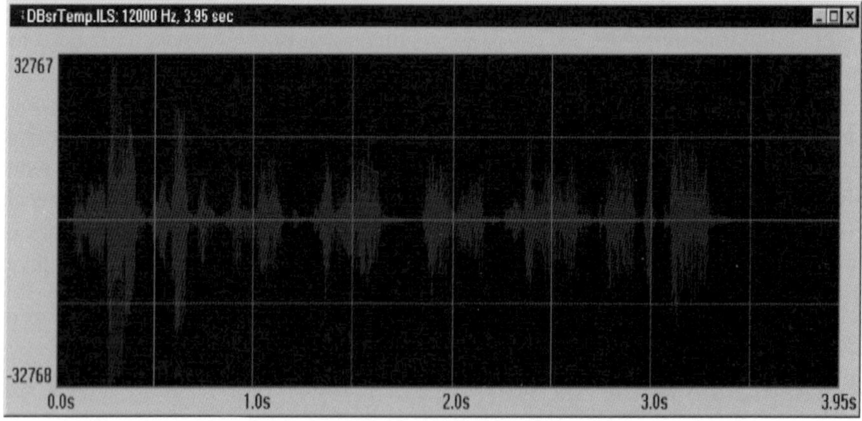

Fig. 3. The effect of automatic gain control on a lowlevel speech signal

4.1 Vocal Tract Normalization and Maximum Likelihood Linear Regression

The well-known speaker adaptation methods: Vocal tract normalization and maximum likelihood linear regression had already been integrated in the DaimlerChrysler speech recognizer before the beginning of the SmartKom project. Both methods were optimized for the SmartKom applications. VTN tries to normalize the vocal tract length of all speakers using a pitch estimation and a spectral adjustment (shift). The dynamic estimation of the shift factor was optimized. Normalization on phrase level proved superior to normalization on frame level.

Speaker adaptation by MLLR is based on shifting the classes of the vector quantization codebook in order to better represent the speaker properties. The HMM state sequence that is needed for adaptation can be directly extracted from Viterbi decoding or in an additional forced alignment of the hypothesized words. Codebook classes may also be marked invariant in order to prevent them from transformation.

Some HMM states, like pauses and noises, are excluded from MLLR, so that they do not contribute to the estimation of the transformation.

Each of the two adaptation methods, VTN as well as MLLR, leads to an average improvement in recognition accuracy of between 1% and 2% absolute. Both methods are applied in the SMARTKOM system since the improvements can be provided at low computational cost.

4.2 Speaker-Dependent Training

The SDT method explicitely adapts the vector quantization codebooks to a particular speaker. Therefore, the speaker has to utter 20 to 100 sentences that are given by the system. This supervised training method should be considered only if the voice of the speaker is recognized with low accuracy and if the speaker frequently uses the system. SDT can be applied in cars, but not while driving since such a training procedure would distract the driver significantly. The Baum–Welch algorithm is incorporated for SDT, because it is easier to implement than the maximum a posteriori (MAP) method, and also allows us to configure the recognizer user-specifically simultaneously with the speaker getting used to the system itself. Experiments performed on test sets of city names, spelling units, and digits sequences showed a maximum relative improvement in recognition accuracy of 33% for digit sequences and an average improvement of about 20% for the entire test set (see Table 2). The SDT method proved to be particularly suitable for problem speakers since for these speakers it achieved an average improvement in recognition accuracy of almost 25%. Empirical studies indicated that the more sentences the speaker enters for SDT, the better the adaptation result. However, uttering 100 sentences is quite a burdensome task and should be considered the maximum amount for training.

Table 2. Recognition accuracy and its relative improvement for speaker-dependent training for all speakers and the problem speakers using 20 to 100 utterances

	All speakers	Problem speakers
Baseline	22.5%	28.2%
SDT-20	20.2%	24.3%
SDT-50	19.0%	22.5%
SDT-100	18.0%	21.2%
Improvement (rel.)	20.0%	24.8%

4.3 Codebook Mixture (CBM)

The mixture of the vector quantization codebooks can be considered as an extension of the SDT method. SDT is incorporated to compute a vector quantization codebook for every single speaker given all his or her utterances in the training set. Principal component analysis is then used to reduce the number of codebooks to 10 to 100 prototypic mixtures. These given prototype codebooks can be linearly combined to give a whole new range of codebook means. The prototype codebooks are loaded during the initialization phase of the recognizer, while the weights for the linear combination are estimated online, so the method belongs to the category of unsupervised adaptations. The experimental results in Table 3 prove that CBM technique outperforms the MLLR and VTN techniques, particularly for small adaptation sets. It achieves about 60% of the improvement of the SDT method, but has the clear advantage of an unsupervised method. That is, the user does not need to undergo a special training procedure. Therefore, CBM is used in SMARTKOM, while SDT is not.

Table 3. Recognition accuracy and its relative improvement for codebook mixture training for all speakers and the problem speakers using 20 to 100 utterances

	All speakers	Problem speakers
Baseline	22.5%	28.2%
CBM-10	20.7%	25.7%
CBM-100	19.6%	23.6%
Improvement (rel.)	12.5%	16.2%

5 Confidence Measures

Confidence measures have become common practice in state-of-the-art speech recognition systems. In this contribution we focus on confidence estimations on the word level, i.e., the confidence of each word is estimated as the posterior probability for the word being correct. The theoretical framework for confidence estimation is introduced in Young (1994) and Eide et al. (1995).

Confidence estimation methods strongly depend on the decoding algorithm of the recognizer. N-Best list decoders (Weintraub, 1995), word graph decoders (Schaaf and Kemp, 1997), and hybrid HMM/ANN (Artificial Neural Networks) decoders (Williams and Renals, 1997) all require different confidence estimators. This report compares two different estimation methods for word graph decoders. The first method computes the confidence from a large word graph (Wessel et al., 1998), and the second collects a set of features that correlate with the correctness of a word (Eide et al., 1995). Both methods are modified and extended in this contribution.

Our aim is to find the best-suited confidence estimation method for in-car recognition tasks. So far, command speech is the common way to operate phone, radio, CD player, and navigation systems in a car. The Linguatronic II system, which is available as a supplementary feature in Mercedes-Benz cars, is one such advanced product. Since customers find remembering a set of predefined commands a burdensome task, they demand more natural ways to input speech. SMARTKOM is a first step toward natural speech interaction covering complex navigation, tourist information, and parking place information tasks.

Many applications of confidence measures have already been presented: rejection, word error minimization (Fetter et al., 1996), speaker adaptation (Pitz et al., 2000), and language identification (Metze et al., 2000). In car applications, we use confidence measures in the framework of dialogue systems, where the dialogue manager decides what action to take next based on the given confidence level. A low confidence is interpreted as rejection and leads to a repetition prompt. A medium confidence triggers a confirmation step, and a highly confident recognition result causes the user input to be executed.

The following sections describe two confidence estimation methods explored at DaimlerChrysler. We also present our evaluation framework for confidence estimation, the experiments and results, and finally the conclusions which can be drawn.

5.1 Confidence Estimation Methods

This section presents two different methods for confidence estimation. The first method computes confidences directly from a large word graph, whereas the second method is based on a typical two-stage classification framework. That is, features that correlate with the word's correctness are collected and fed into a classifier which estimates posterior probabilities for the word being correct.

5.2 Direct Estimation from Word Graphs

In this method the large hypothesis graph, which results from the first decoding stage and which has been pruned, acts as the search space and is the sole input for confidence estimation. The probability mass lost due to pruning may be neglected if the graph is large enough.

How can confidences be estimated from the word graph only? Wessel et al. (1998) present a method that computes for each edge i in the word graph the probability $\gamma(i)$ that it is passed. The calculation is based on the well-known forward/backward algorithm, where acoustic probabilities replace the emission probabilities and language model probabilities replace the transition probabilities. Given the forward variable $\alpha(i)$ and the backward variable $\beta(i)$, the probability $\gamma(i)$ can be computed as follows:

$$\gamma(i) = \frac{\alpha(i)\beta(i)}{\sum\limits_{j:e(j)=T} \alpha_T(j)} \qquad (1)$$

Since the nodes are the connection points between successive edges, any efficient algorithm must consider them. T is the final node in the word graph. The sum in the denominator in Eq. (1) is the total probability that the utterance is contained in the word graph.

Since word graphs usually contain overlapping instances of the same word in competing paths, the probabilities of these instances should be combined in some way as they represent the same word. This can be achieved by a homogeneity measure HAP(i) (*homogeneity of all paths*), which is calculated on a frame-by-frame basis:

$$\mathrm{HAP}(i) = \frac{\sum_{t = t_a(i)}^{t_e(i)} \sum_{j \,:\, t_a(j) \le t \le t_e(j)} \gamma(j)\,\delta(w_i, w_j)}{\sum_{t = t_a(i)}^{t_e(i)} \sum_{j \,:\, t_a(j) \le t \le t_e(j)} \gamma(j)} \tag{2}$$

where $\delta(w_i, w_j)$ is the Kronecker function.

5.3 Feature Collection and Classification

The second method for confidence estimation is based on statistical decision theory (Eide et al., 1995). A set of confidence features m, each correlating with the correctness of the hypothesized word, is mapped into a binary class k with values 0 for "erroneous" and 1 for "correct" word hypothesis. The posterior probability can then be calculated as follows:

$$P(k=1|m) = \frac{P(m,k=1)}{P(m,k=1)+P(m,k=0)} \tag{3}$$

A set of utterances must be excluded from recognizer training in order to train the confidence classifier. The target vector for classification training is computed by a recognition run on the heldout set and an alignment of the word hypothesis on the reference transcription. Correctly recognized words receive the label 1, and recognition errors receive the label 0 in the target vector. This is because errors should ideally lead to a very low posterior probability and correct results to a high probability. Because we used Levenshtein alignment for the experiments in this study, only insertions and substitutions, but not deletions, are covered. A schematic diagram for this two-stage method can be found in Fig. 4.

Which of the many possible features should be used must be decided by a human expert. Features were selected from all recognition stages mainly heuristically. The most important features are listed below:

1. Signal analysis:
   a) SNR: signal-to-noise ratio (Eide et al., 1995)
   b) Fo: fundamental frequency
2. Vector quantization:
   a) VQMaxProb: maximum probability per frame
   b) VQNumClass: number of competing codebook classes
3. Decoding:
   a) NormScore: length-normalized acoustic score (Schaaf and Kemp, 1997)
   b) LMScore: average language model score (Chase, 1997)

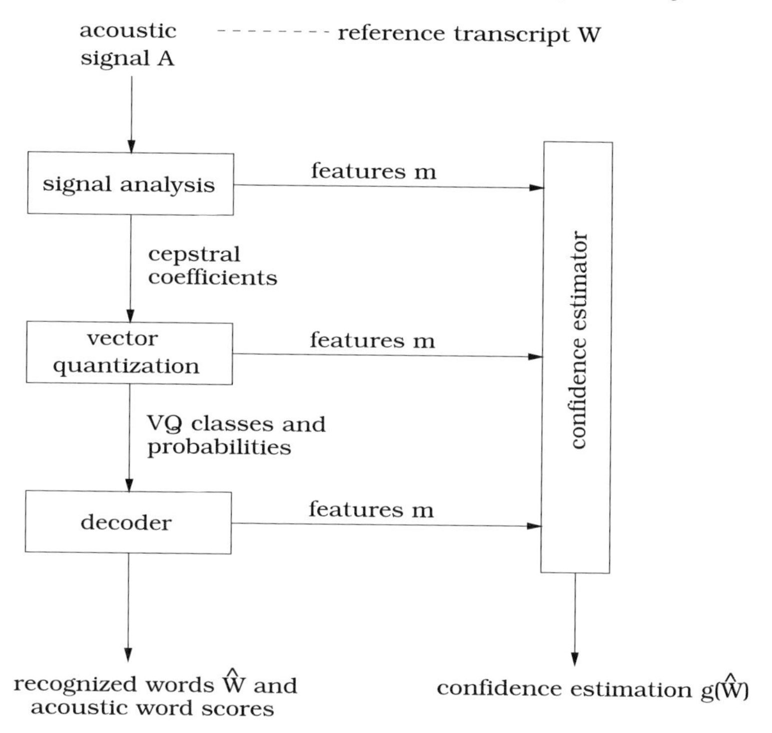

Fig. 4. Schematic diagram of two-stage confidence classification

   c) LMScoreBestPath: language model score on the best path
   d) LRWordPhonFree: likelihood ratio of scores of word decoding to scores of free phoneme decoding (Young, 1994)
   e) LREmiBestAvgFree: likelihood ratio of best and average emission probabilities per frame (for active model states)
   f) GraphHomo: graph homogeneity HAP in Sect. 5.2 (a local confusability measure)
   g) Astabil: acoustic stability (Finke et al., 1996)
   h) ActWordEnds: number of competing word ends
   i) SpeechRate: simple speech rate estimation as frames per state
   j) AvgLossProbFrm: average probability loss per frame
4. Word-specific properties from recognizer training:
   a) WordCorRate: word-correct rate of each word in the training data (Schaaf and Kemp, 1997)
   b) WordFreqTrain: frequency of the word in the training data (Chase, 1997)

A large number of additional features have also been evaluated, but they are neglected here since they do not improve classification performance.

Many different classification techniques can be employed for confidence estimation. We decided to investigate four different approaches, each with its own particular advantages:

1. Gaussian mixture models (GMM) — easy and fast to train; most features fulfill the normal distribution assumption
2. Polynomial classifiers (PC) — better functional approximation than GMM; practical solution by a system of linear equations
3. Multilayer perceptron (MPL) — complex and time-consuming training procedure; robust against outliers; suggested if no knowledge about feature distributions exists
4. Support vector machine (SVM) — prevents overadaptation by minimizing the empirical risk; focuses on modeling the class boundaries

5.4 Evaluation Metrics

The confidence measures in this study are applied to word rejection. This means a binary output of "accept" or "reject" is sufficient. Since the introduced confidence measures are estimations of posterior probabilities, a threshold operation allows the continuous values to be transformed into binary values. In order to compare the performance of different confidence estimators, we are looking for a scalar measure that represents the relative amount of errors. One such well-known measure is the classification error rate (CER), the ratio of the number of erroneous classifications to the overall number of classifications. Classification errors are rejections of correctly recognized words (type I errors) and acceptances of recognition errors (type II errors).

Type I errors increase with decreasing type II errors, and vice versa. For some applications it makes sense to allow more rejections in order to be very sure about the accepted samples. The optimal operating point for each application can be found by investigating the detection error tradeoff (DET) curves that plots the two types of errors against each other. The DET diagrams also help us gain insight into the performance of the estimators by comparing curves.

5.5 Experimental Setup

A semicontinuous HMM-based speech recognition system (Class et al., 1993) was used for our experiments. This system was described in detail in Sect. 2. For our studies two different recognizers are compared. The first one was trained on the command speech database CARCOMM of DaimlerChrysler, which contains commands spoken in the noisy car environment. The second one was trained on the spontaneous speech database Verbmobil II, which consists of human-to-human travel-planning dialogues. Word transitions were modeled by a restricted command grammar for CARCOMM, and by a statistical language model for Verbmobil. Spectral subtraction (SPS) is used in signal preprocessing for CARCOMM only. Table 4 summarizes some important statistical data of these databases.

Table 4. Training and test databases for word recognition and confidence classification

Database	No. of HMMs	No. of words in train. set	Size of vocabulary	WER (%)
CARCOMM	1500	24,336	406	9.7
Verbmobil II	1500	35,943	10,166	37.6

5.6 Results and Discussion

This section shows the performance of the direct confidence estimation method (HAP). Results are presented in terms of relative reduction in CER over the baseline system. The baseline CER assumes a naive classifier that always chooses the most frequent class, in our case the class "correct". It is basically WER excluding deletion rate.

Table 5 shows results for CARCOMM and Verbmobil II. CER can be significantly decreased (about 30%) for both databases.

Table 5. Results for direct estimation of acoustic confidences (HAP)

Database	No. of words in rec. set	Size of vocabulary	WER (%)	CER %
CARCOMM	8825	406	9.7	30.5
Verbmobil II	10,914	10,166	37.6	30.9

The classifier for the two-stage method is initialized with just the best-performing feature. Then all other features are added one by one in the order of their performance. Tables 6 and 7 show the results using GMM classifiers for command and spontaneous speech respectively.

Table 6. Results for the GMM classifier on command speech

Database	Feature	CER (%)
CARCOMM	WrdCorRate	47.3
	+ GraphHomo	55.0
	+ LMProbBestPath	57.9
	+ Astabil	58.2
	+ ActWrdEnds	60.4

The *WrdCorRate* feature strongly influences the confidence performance in command speech due to the smaller sized vocabulary. *GraphHomo* also plays an important role in both kinds of speech. It proved to be the best feature for spontaneous speech. Some other decoder features reduced CER further, but these features differ for command and spontaneous speech.

The best GMM classifier for command speech performs almost twice as well as the best GMM classifier for spontaneous speech. That can be explained by the

Table 7. Results for the GMM classifier on spontaneous speech

Database	Feature	CER (%)
Verbmobil II	GraphHomo	30.9
	+ WrdCorRate	34.0
	+ SpeechRate	34.1
	+ LREmiBestAvgFree	34.2
	+ ActWrdEnds	34.7

fact that command speech is more restricted, so that the distributions of confidence features are better separated.

The assumption made by the GMM approach that feature distributions can be approximated by a mixture of Gaussians does not always hold. Therefore, we examined some other classification techniques. These techniques lead to further CER reductions (Table 8), particularly for the SVM classifier. Precise modeling of class boundaries and minimizing empirical risk instead of classification error on the training database pays off here.

Table 8. Comparison of classification methods

Database	GMM	PC	MLP	SVM
CARCOMM	60.4	63.0	64.2	66.1
Verbmobil II	34.7	36.7	34.8	36.2

Finally, the estimation methods for both speech databases are compared in a DET curve (Fig. 5). The two-stage method clearly outperforms the direct estimation method for command speech (two lower curves). The performance difference between the two methods is not so significant for spontaneous speech (two upper curves).

6 Modeling Unknown Words

In natural speech input the user is not restricted in the way how to put his or her words. Thus, naturally, a very large vocabulary is required to cover all the utterances a user might enter into the system. Now, at this point, two aspects have to be regarded: first, the recognition performance of a speech recognition system degrades with growing vocabulary size due to decreasing separability. Second, the gain in recognition performance obtained by adding more ever-less-frequent words to the vocabulary will constantly shrink and approach zero. Hence it makes sense to limit the vocabulary size, which is inevitable in real-world applications, anyway.

This leaves us with a certain (unknown) number of words that the recognizer cannot detect. Since these are the least frequent words, we can assume that they will not bear any meaning critical to the application. Thus, in a given utterance, we

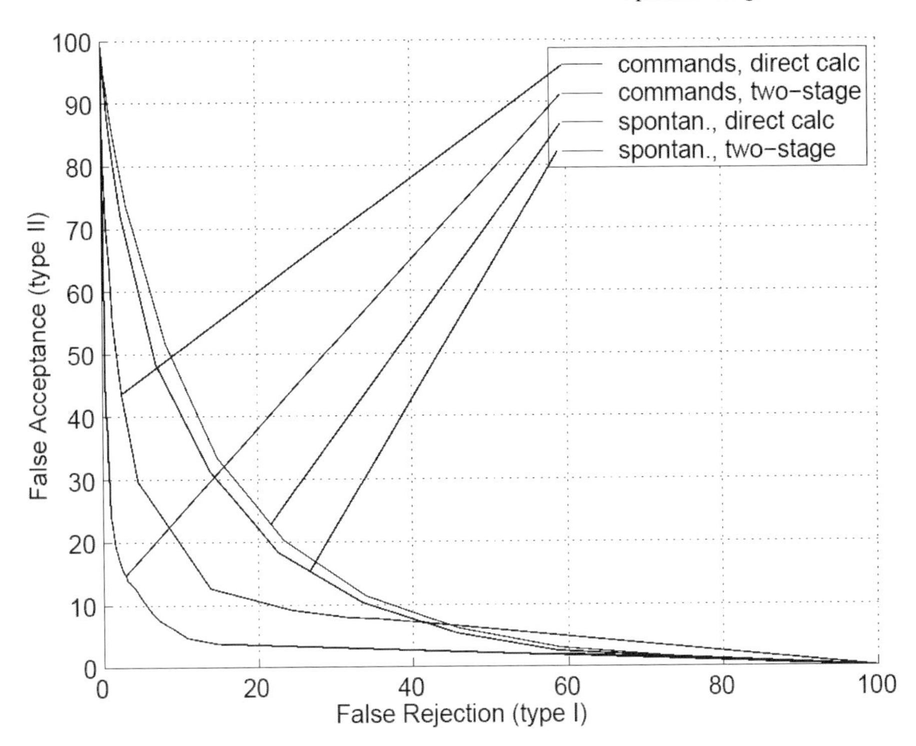

Fig. 5. DET curves for direct estimation and two-stage approaches on command and spontaneous speech

will be happy to ignore these words, but we want the rest of the utterance perfectly preserved. Therefore, the word boundaries need to be preserved, whereas a typical recognizer would do anything, including shifting the word boundaries, in order to match the utterance with the unknown words against words that it does have in its vocabulary.

To fill the holes left by unknown words, acoustic models are needed that match better against the unknown word than a wrong alignment of vocabulary words would, including the fact that a wrong alignment might produce lower scores for wrong surrounding words. Such acoustic out-of-vocabulary (OOV) models have been suggested as one single garbage model (Class et al., 1993) and as a set of 15 length-dependent models (Fetter, 1998). In the case of a single model either all nonapplication words or all words of a dictionary are trained to that model. It is quite obvious that a single model cannot be a reasonable acoustic match to many acoustically different words. Therefore, Fetter divided the training vocabulary into 15 sets of words by number of phonemes (SAMPA characters) and trained them to HMMs with 15 topologies, respectively.

In the present approach, a further step is made toward capturing the acoustic properties of unknown words more closely. Every training word is regarded as consisting of syllables. Every syllable in turn consists of a number of start consonants, a core of vowels, and a number of end consonants. A certain number of phoneme classes are introduced for each, the start consonants, vowels, and end consonants, respectively. To obtain these classes, gradually phonemes with the greatest similarity (see Sect. 2.2) are merged together to form a class. Each class is trained to a separate HMM. In the experiments, five sets of phoneme classes with different stages of granularity are regarded (see Tabel 9).

Table 9. OOV sets/HMMs and lexicon entries

Set No.	Start cons. classes	Vowel classes	End cons. classes	No. of HMMs	Lexicon entries
1	1	1	1	3	33
2	2	1	2	5	399
3	2	2	2	6	567
4	2	2	3	7	4509
5	3	2	3	8	4936

For each set, the phonemes of each training word (60,000 German city names) are mapped to the generalized phoneme classes, and the resulting different generalized words are introduced into the recognizer vocabulary as OOV words. The 33 lexicon entries for set 1 roughly correspond to Fetter's length-dependent models.

In the test setup, a statistical trigram language model with backoff is used, where all OOV words were in a single garbage word class. For the test purpose, the OOV word rate in the test material was artificially raised to 23.2% by choosing the 1000 sentences of the LM training material that contained the most unknown words. In a first recognition run the overall performance was evaluated (see Tabel 10). After the first run, the timestamps of the recognizer output were used to cut out the signal parts that were marked OOV. These signal parts should bear the unknown words that were really uttered. In a second recognition run these signal portions were recognized with a dictionary that contained those words instead of the OOVs with a single word recognizer. This second run measures how well the word boundaries are preserved by the OOV models.

Table 10. OOV/recognition results

	1st Run			2nd Run
Set No.	WER (%)	RCL (%)	PRC (%)	WER (%)
1	33.9	86.1	78.2	83.6
2	31.0	89.7	76.3	77.3
3	28.9	92.7	76.3	69.7
4	18.8	97.1	80.4	40.0
5	18.7	97.0	80.9	39.2

The recognition results clearly show a great increase in performance regarding word error rate (WER), OOV recall rate (RCL) and OOV precision (PRC). All figures of the first run improve significantly until set 4. Sets 4 and 5 are almost equal. Thus, set 4 is proposed for use in the application. The second run, which tested if the signal portions marked OOV by the first run really contain the uttered word, shows dramatic improvement by increasing the number of OOV words. This means that the preservation of word boundaries could also be greatly improved. Again, in this test, sets 4 and 5 are very close.

In a natural speech input environment the recognizer dictionary can never hold all words that may possibly be uttered. Acoustic OOV models are a means to fill in the holes left by unknown words. Literature shows that length-dependent models outperform the single model approach. Experiments prove that we were able to improve the system performance substantially by replacing the length-dependent approach by a greater number of generalized word models.

6.1 Conclusion

Significant improvements in speech recognition have been achieved in all four main research areas investigated in SmartKom. The research demonstrators for the different SmartKom scenarios are able to cope with natural, spontaneous language input, speaker independence, large vocabularies, and background noises. A speech recognizer based on semi continuous Hidden Markov Models and statistical language models was trained for the SmartKom application on a large set of training data.

The system adapts itself to the environment by enabling Barge-In, reducing environmental noises, and adjusting the input signal level using automatic gain control. Barge-In allows for user input while the system speaks to the user. This functionality is enabled by an echo cancellation technique integrated in the acoustic device. Several noise reduction and noise modeling methods were investigated, but only spectral substraction proved to be useful in the SmartKom applications.

A number of speaker adaptation methods have been presented in this contribution, each improving recognition accuracy by approximately 10% to 20% relative. The SmartKom system incorporates several unsupervised speaker adaptation techniques: Vocal tract normalization is mandatory, while maximum likelihood linear regression and codebook mixture can be used optionally.

The SmartKom system adapts itself to the task or application by dynamically updating the vocabulary and the members of the categories in the statistical language model. Vocabulary sizes range from 3000 words to 12,000 words, depending on the application. Typical language model categories are city names, street names, movie titles, actor names, genres, numbers, etc. Adaptation to the task has a major influence on recognition accuracy.

Two methods for word confidence estimation were introduced. The first method calculates word confidences from very large word graphs directly, while the second method collects features that correlate with the word's correctness and estimates confidences using known classification techniques. The direct calculation and the two-stage methods both lead to significant improvement of confidence estimations. A

36.7% relative reduction in confidence error rate is achieved for spontaneous speech, and a 66.1% reduction is obtained for command speech. Confidence estimation is more reliable on command speech, since vocabulary size and perplexity are lower than in spontaneous speech. The feature-based two-stage approach is superior to the direct calculation method, but also requires more computation time and memory.

For SMARTKOM, however, two-stage calculation has some drawbacks:

1. Two different recognition algorithms (for German and English, resp.). Hence a *feature-based* confidence classifier strongly depends on specific algorithms implemented.
2. Any changes in the recognizer setup (e.g., signal analysis, vector quantization, HMM modeling, vocabulary size, and even parameters to speed up computation) may degrade classifier performance. In order to overcome these disadvantages of two-stage calculation, classifiers should be strictly adapted to a given setup.

Considering these drawbacks, we decided for direct estimation of confidences. This is because we have a well-defined interface between the Viterbi decoding module and the confidence estimation module: *the word hypotheses graph.*

Finally, we investigated how to model words that are not covered by the vocabulary of the speech recognizer. Using an approach of generalized word models we were able to detect a large number of unknown words and improve the system performance substantially.

References

L.L. Chase. *Error-Responsive Feedback Mechanisms for Speech Recognizers.* PhD thesis, Carnegie Mellon University, Pittsburgh, PA, 1997.

F. Class, A. Kaltenmeier, and P. Regel-Brietzmann. Optimization of an HMM-based Continuous Speech Recognizer. In: *Proc. EUROSPEECH-93*, pp. 803–806, Berlin, Germany, 1993.

E. Eide, H. Gish, P. Jeanrenaud, and A. Mielke. Understanding and Improving Speech Recognition Performance Through the Use of Diagnostic Tools. In: *Proc. Int. Conf. on Acoustics, Speech, and Signal Processing (ICASSP-95)*, pp. 221–224, Detroit, MI, 1995.

P. Fetter. *Detection and Transcription of Out-Of-Vocabulary Words in Continuous Speech Recognition.* PhD thesis, Technical University of Berlin, 1998.

P. Fetter, F. Dandurand, and P. Regel-Brietzmann. Word Graph Rescoring Using Confidence Measures. In: *Proc. ICSLP-96*, pp. 10–13, Philadelphia, PA, 1996.

M. Finke, T. Zeppenfeld, M. Maier, L. Mayfield, K. Ries, P. Zhan, and A. Waibel. Switchboard April 1996. Technical report, DARPA, 1996.

F. Metze, T. Kemp, T. Schultz, and H. Soltau. Confidence Measure Based Language Identification. In: *Proc. Int. Conf. on Acoustics, Speech, and Signal Processing (ICASSP-2000)*, Istanbul, Turkey, 2000.

M. Pitz, F. Wessel, and H. Ney. Improved MLLR Speaker Adaptation Using Confidence Measures for Conversational Speech Recognition. In: *Proc. ICSLP-2000*, Beijing, China, 2000.

T. Schaaf and T. Kemp. Confidence Measure for Spontaneous Speech Recognition. In: *Proc. Int. Conf. on Acoustics, Speech, and Signal Processing (ICASSP-97)*, pp. 875–878, Munich, Germany, 1997.

M. Weintraub. LVCSR Log-Likelihood Ratio Scoring for Keyword Spotting. In: *Proc. Int. Conf. on Acoustics, Speech, and Signal Processing (ICASSP-95)*, pp. 887–890, Detroit, MI, 1995.

F. Wessel, K. Macherey, and R. Schlüter. Using Word Probabilities as Confidence Measures. In: *Proc. Int. Conf. on Acoustics, Speech, and Signal Processing (ICASSP-98)*, pp. 225–228, Budapest, Hungary, 1998.

G. Williams and S. Renals. Confidence Measures for Hybrid HMM/ANN Speech Recognition. In: *Proc. EUROSPEECH-97*, pp. 1955–1958, Rhodes, Greece, 1997.

S. Young. Detection of Misrecognitions and Out-Of-Vocabulary Words. In: *Proc. Int. Conf. on Acoustics, Speech, and Signal Processing (ICASSP-94)*, pp. 21–24, Adelaide, Australia, 1994.

Class-Based Language Model Adaptation

Martin C. Emele, Zica Valsan, Yin Hay Lam, and Silke Goronzy

Sony International (Europe] GmbH
Sony Corporate Laboratories Europe, Advanced Software Laboratory, Stuttgart, Germany
{emele,valsan,lam}@sony.de, silke@goronzy-privat.net

Summary. In this paper we introduce and evaluate two class-based language model adaptation techniques for adapting general n-gram–based background language models to a specific spoken dialogue task. The required background language models are derived from available newspaper corpora and Internet newsgroup collections. We followed a standard mixture-based approach for language model adaptation by generating several clusters of topic-specific language models and combined them into a specific target language model using different weights depending on the chosen application domain. In addition, we developed a novel word n-gram pruning technique for domain adaptation and proposed a new approach for thematic text clustering. This method relies on a new discriminative n-gram–based key term selection process for document clustering. These key terms are then used to automatically cluster the whole document collection. By selecting only relevant text clusters for language model training, we addressed the problem of generating task-specific language models. Different key term selection methods are investigated using perplexity as the evaluation measure. Automatically computed clusters are compared with manually labeled genre clusters, and the results provide a significant performance improvement depending on the chosen key term selection method.

1 Introduction

Statistical language modeling has been successfully applied to many domains such as automatic speech recognition (ASR (Jelinek, 1997; Niesler and Woodland, 1997), spoken language understanding (Wang, 2001), information retrieval (Miller et al., 1999) and handwriting recognition (Srihari and Baltus, 1992)). In particular, trigram language models (LMs) have been demonstrated to be highly effective for these domains. The need to go beyond trigram models is advocated in Jelinek and Lafferty (1991). Long-distance models, which try to capture statistical dependencies that go beyond trigrams, include models such as caching (Clarkson and Robinson, 1997; Kuhn and De Mori, 1990), higher order n-grams (Goodman, 2001), sentence mixture (Iyer and Ostendorf, 1999), and class-based models (Kneser and Ney, 1993; Bellegarda et al., 1996), etc.

An important issue for any ASR system is the adaptivity to a specific domain or task for which it is intended. Models used in such systems should be specific to the style of language used in a particular domain. As such, when confronted with utterances from another domain, performance will degrade substantially. Therefore, the adaptation of the language model to new domains is required. Techniques such as caching or mixture-based language models are commonly used to tackle this problem. For the latter approach, which we have followed here, one needs to generate several LMs specific to a particular topic or style of language (Iyer and Ostendorf, 1999; Dempster et al., 1977; Seymore and Rosenfeld, 1997).

Language modeling attempts to capture, represent, and exploit the regularities and characteristics of natural language. In large-vocabulary continuous speech recognition (LVCSR), the language model provides a prior estimation of the probability of a spoken word sequence. It helps to disambiguate acoustically ambiguous utterances. It also facilitates the reduction of the search space to word sequences that occur more frequently in natural language. State-of-the-art statistical n-gram modeling (Bahl et al., 1983) models the frequency of the occurrence of word sequences. It assumes that the occurrence of a word depends only on the preceding n-1 words. Neither words with a distance more than n nor the semantic contexts are taken into consideration. The probability of a word sequence $W = w_1, w_2, w_3, \ldots, w_N$ is estimated as:

$$P(W) = \prod_{i=1}^{N} P(w_i | w_{i-n+1}, \ldots, w_{i-1}). \tag{1}$$

All the probability components on the right-hand side of Eq. (1) are usually estimated by taking statistics on a large collection of written texts such as newspaper articles.

In application-specific spoken dialogue systems, in order to recognize spontaneous task-specific utterances correctly, domain-specific language models, which are able to capture spontaneous speech phenomena, are essential for the recognition engine. Nevertheless, n-gram language models trained on newspaper articles do not really reflect the characteristics of the spontaneous speech as used in real dialogue systems. Therefore labor- and cost-intensive Wizard-of-Oz (WOZ) experiments (Schiel and Türk, 2006) have to be conducted to collect tas-specific dialogue data. Compared to the size of written texts, the manually collected data is usually small and hence is not a reliable basis for constructing adequate language models. On the other hand, Internet resources such as newsgroup posts, mails etc. provide abundant amount of texts with characteristics and styles similar to spontaneous speech.

In Sect. 2, we first describe a newsgroup-based text corpus development and preparation system, which automatically collects newsgroups, filters junk material, and converts them to a high-quality general corpus that can be used for language model training. In Sect. 3, we describe the generalization of standard n-gram LMs to class-based LMs where the contents of classes can be dynamically changed at runtime. In Sect. 4, we introduce two complementary language model adaptation techniques, which adapt the general corpora to application- and task-specific language models. In Sect. 5, we discuss the clustering performance of three different term selection methods for the thematic text-clustering approach and evaluate the adapted LM in terms of word error rates using different test sets from the SmartKom multimodal dialogue collections. Finally, we present our conclusions in Sect. 6.

2 Copora Preparation

Since we are dealing with German, we had to collect and prepare suitable training corpora for the language model training. Besides using newspaper text from the Süddeutsche Zeitung (SZ), we set up an automatic collection of domain-specific texts from newsgroups and collected dialogue turns which were recorded during the WOZ experiments. Since the corpora are not directly in a format that can be used for language model training, a set of filtering and formatting tools was developed. The module consists of two main components:

1. **Removal of junk content.** For the corpora, the filtering script removes the junk material, e.g., the binary streams such as audio and video files in the newsgroups. Then, it also removes all non-German news articles by using a simple language guesser based on a wide-coverage dictionary for German, i.e., CELEX (Baayen et al., 1993), containing 359,611 words. It also removes nonalphabetic words, like emoticons etc.
2. **Text reformatting.** This component does tokenization and detection of sentence boundaries by trying to disambiguate interpunctuations, like period, whether it is part of an abbreviation or whether it marks the end of a sentence. Since many news articles make use of a special abbreviation language to speed up communication, e.g., "cu" expands to "see you," etc., we try to expand those abbreviations by using a dictionary of common abbreviations. Next, we try to clean up the text by normalizing German umlauts from ASCII notation into their corresponding ISO 8859-1 encoding scheme. Finally, we run a simple spell checker across the text by checking the spelling of words against the CELEX dictionary.

After the preprocessing phase of the SMARTKOM corpora, we yield 3,297,970 sentences for training. The SZ corpus mainly serves as the background corpus to capture the general characteristics of German utterances. The dialogue turns that partly result from the WOZ experiments and partly from manual construction based on the analysis of the particular tasks are supposed to reflect the specific scenarios for which SMARTKOM is intended. The collection of newsgroup texts in addition to the dialogue data is necessary, since the amount of dialogue-specific data is not sufficient to construct a reliable LM. Furthermore the nature of writing in newsgroups is somehow closer to spontaneous speech. All these corpora are used as the basis for the language model adaptation techniques, which are described in Sect. 4.

3 Class-Based Language Models

The LM used in this project is a statistical, class-based, 3-gram LM. The LM is tightly coupled to the lexicon (Goronzy et al., 2006) and also has to reflect the dynamic nature of the vocabulary. To be able to do so, frequently changing words, such as actor names, movie titles, streets in different cities, etc., are categorized and put into word classes. The 3-grams are then based on these classes, rather than on the words themselves. The advantage of the class-based LM is that it is no longer necessary to have all words in all contexts in the training data. For all words in one class the histories of all the single words in that class can automatically be associated with all words in that class. For example, consider two sample sentences: "I liked James Bond very much," and "I hated Star Trek". Using the class MOVIETITLE the sentences will be mapped to "I liked MOVIETITLE very much," and "I hated MOVIETITLE". The resulting LM then covers the original sentences as well as all possible combinations, e.g., "I liked Star Trek very much" etc., since both "Star Trek" and "James Bond" are in the MOVIETITLE class.

In SMARTKOM the content of the word classes can dynamically change. New words can be added/removed to/from a class. We just need to recompute the in-class probability for all words in the classes, which is trivial if all words in a class are equally weighted. However, it would be possible to increase the weight of those words in a class that are currently displayed on the screen.This would account for the fact, that if, e.g., a list of 20 movie titles is the result of a user request and only 10 can be displayed on the screen (to view the others you would have to scroll down), then the user is more likely to say one of those titles that he can currently see on the screen. Therefore their probability could be increased.

By updating the contents of the classes, we can achieve the adaptation of the LM to a specific domain to a certain extent, without the need to completely change and reload the LM as is done by many other dialogue systems, e.g., those based on VoiceXML. Such an approach is very time-consuming and often not feasible in dialogue systems of this coverage. Furthermore, by not restricting the LM to subparts for certain kinds of expected dialogue turns (which is an often used strategy), the capability of being able to speak words from all subdomains, no matter if they are within the range of expected utterances at this stage of the dialogue, remains.

Currently, 31 word classes are defined. However, only the following classes will be dynamically updated, because their content changes frequently depending on the current dialogue state: ACTOR, CITY, GARAGE, GENRE, HOTEL, LOCATION, MOVIETHEATER, and MOVIETITLE.

4 Language Model Adaptation

Within the SMARTKOM project, two different approaches are adopted to tackle the problem of LM adaptation: language model adaptation using word *n*-gram pruning (Lam and Emele, 2001) and language model adaptation using thematic text clustering (Valsan and Emele, 2003).

4.1 LM Adaptation Using Word *N*-Gram Pruning

The vocabulary size of the background corpora, i.e., SZ and newsgroups, is large. On the other hand, for spoken dialogue systems, it is a common practice to define the vocabulary beforehand based on the underlying tasks, e.g., as those in the SMARTKOM system. Such a task-specific system usually employs a small or medium size vocabulary. Therefore, the probability distribution of n-grams generated from such general corpora usually does not reflect those that occur in the spoken dialogue system due to the presence of large amount of out-of-vocabulary (OOV) words. Therefore, in the n-gram pruning approach described below, n-grams with OOV words are pruned.

N-gram pruning using count cutoff has been used in the literature to reduce the size of language models. Any word n-grams with counts below a predefined threshold are pruned. Instead of using such a count cutoff, a novel word n-gram pruning technique based on OOVs is introduced to adapt a language model on the n-gram statistical level. It exploits the basic assumption of n-gram language models that the probability of a given word in a word sequence is only influenced by its n-1 preceding words. In word n-gram pruning, all word n-gram frequency counts are computed. Then, any word n-grams that contain OOVs are pruned. N-gram probabilities are directly estimated from the pruned word n-gram frequency counts. As OOVs are completely eliminated in the pruning step, the probability of unknowns or the likelihood of seeing an OOV in the test set can be set to values that are appropriate for the application if an open vocabulary is required. This approach also allows rapid adaptation to new vocabularies as the new vocabulary is only required during the pruning stage and recounting of word n-grams from the corpora is not necessary. The pruning step can also be interpreted as a corpus adaptation technique in which the adapted corpora are formed of sentence segments, which consist of at least n continuous non-OOVs. In addition to word n-gram pruning, which is an adaptation on the level of n-grams, we developed another adaptation technique on the level of selecting only relevant documents from a general background corpus by applying an automatic text-clustering approach.

4.2 Language Model Adaptation Using Thematic Text Clustering

In general, an application domain can be characterized by the subtasks that comprise the overall dialogue flow. Each task can be characterized by a topic or a set of topics to which the sample dialogues belong. Similarly, a given document collection like newspaper texts can be categorized into specific text clusters according to a given set of topics. Usually, the newspaper/newsgroup articles are already manually classified into different genres like news, sports, movies, etc. Together with the title of the article or words that appear in the beginning of the documents, this input can be used to derive the main topic of the documents. Based on this information, topic-specific text clusters can be derived and further, for each cluster, a topic-specific LM can be generated. The probability estimates from these component LMs are interpolated to produce an overall probability, where the interpolation weights are chosen to reflect the topic or style of language currently being recognized.

But, in general, such a naive approach, which uses only manually assigned topic labels and limited context information like title words and abstracts, is not sufficient to cover the dynamic topic shifts that occur even within one single document. The result might be a false labeling of documents, leading finally to a cluster that covers many different topics. Hence, the topic-generated LM is actually a more general LM rather than a specific one.

Therefore, in SMARTKOM, we proposed another automatic text-clustering approach for topic assignment based on three different key term selections that are used to define the topics of the documents. These three term selections are compared with the naive baseline reference, which rely on manually clustered documents according to the genre information.

The process of term selection is based on standard information retrieval (IR) techniques (Rasmussen, 1992), but in our approach the importance of a term is evaluated by averaging over the whole document collection. The accuracy of the classifier is therefore subject to the selection of key terms. The semantic classification of the document context is done by using a combination of standard techniques like text preprocessing, random mapping projection (RMP, see Kohonen et al. (2000)) to reduce the size of the term vectors, and vector quantization (VQ) techniques like the split LBG algorithm (Linde et al., 1980) for identifying the clusters in the reduced vector space.

4.2.1 Term Selection and Document Representation

Text-clustering algorithms partition a set of documents into groups or clusters. In this respect a document has to be transformed into a representation which is suitable for the classification task. We used the vector space model (VSM) approach (Salton and McGill, 1983) for coding the documents, where the documents are represented as an m-dimensional stream of indexing *terms*. In general, terms can be n-grams, instead of simple words ($n > 1$):

$d = (w^{(1)}_{n-gram}, w^{(2)}_{n-gram}, w^{(3)}_{n-gram}, \ldots, w^{(m)}_{n-gram})$.

Defining and selecting the most representative terms in a corpus using this technique is based on the assumption that the terms in a document are conditionally independent of each other (Robertson and Spärck, 1976). Due to this independence assumption some information about the document is lost. Additional information about the context in terms of n-grams ($n > 1$) can be an effective way to alleviate the limitation of the simple bag-of-words model. On the other hand, using, for instance, all bigrams or trigrams, the size of the coded document is increased now to around m^2 or to around m^3, respectively. Therefore, in order to not increase the dimension of the term-category matrix too much, one can apply the term selection process only to those n-grams where **all** n-constituent words belong to a predefined vocabulary. This selection is efficient for most practical dialogue systems, where a fixed medium-sized vocabulary is already defined by the domain. The advantage of using such an approach is twofold: First, it decreases the computational cost by discarding all irrelevant documents (those whose corresponding n-grams are missing). Second, it improves the classifier performances by providing clusters, which cover documents with the same or at least related topics.

After removing words that occur in the stop word list, and after applying a standard stemming algorithm, the most relevant terms in the collection need to be selected. This is done by defining a *term-weighting* function which supplies a weight for each term (w_i) given a document d_j in the collection.

Since we are interested in defining the terms that are the most representative for the entire collection and not in a particular document (as is the case for standard IR systems) the formula below was used:

$$weight(w_{n-gram}) = \left(\frac{1}{N_{n-gram}} \sum_{k=1}^{N_{n-gram}} tf_{n-gram,k} \right) \cdot \log \frac{N}{N_{n-gram}} \tag{2}$$

where each term is an n-gram where $n \geq 1$, N_{n-gram} is the number of documents that contain the term w_{n-gram}; $tf_{n-gram,k}$ is the number of concurrencies of term w_{n-gram} in the document k.

Thus, the n-gram terms weights are computed according to their "average" importance in the corpus. To select a subset of m terms in all documents, which represent the key terms, the m n-grams with the highest weight exceeding an empirical threshold are chosen. The value of the vector elements $w_{n-gram}^{(i)}$ for a document is a function of the frequency counts of the n-grams in the corresponding document. The documents are coded based on these values, represented by vectors in the vector space, and then normalized to the unit vector. In this way the direction of the vector will reflect the content of the document.

4.2.2 Methods for Dimensionality Reduction

The main problem of the vector-space model is its large vocabulary for any sizable collection of free-text documents. Such term count vectors are typically very sparse and very high dimensional; therefore it is almost impossible to use them as such. A computationally simple method of dimensionality reduction that does not introduce a significant distortion in the data is thus desirable. Random Mapping Projection is a solution that provides a computationally feasible method for reducing the dimensionality of the data so that the mutual similarities between the data vectors are approximately preserved (Kaski, 1998).

4.2.3 Estimation and LM generation

Once the documents have been encoded and mapped to a lower dimension, so that similarity of the texts is reflected in the similarity of the respective document vectors, the remaining task is to create text clusters out of these documents.

We resort to a VQ technique such as split LBG (Linde et al., 1980), which uses for comparison purposes 32 dimensions (classes) for the code vectors, exactly the same number as the manually generated topics in the corpus. In the next section we evaluate various term selection variants in terms of perplexity and present the word error evaluation results using the optimised LM.

5 Experimental Results

In this work 100,117 documents were used for training. The whole document collection was manually categorized into 32 clusters, using genre information, with each cluster representing a single topic. This categorization serves only as a reference for evaluating the automatic clustering algorithm described above. Furthermore, we will refer to these clusters as genre-clusters. For each cluster a 3-gram–based LM was generated using the CMU SLM toolkit (Rosenfeld and Clarkson, 1997).

5.1 Comparative Study

Three sets of comparative experiments for the key term selection process were performed, which are defined as follows:

1. **Word Frequency Counts** (WFC): Select as key terms in the corpus all words with counts higher than a predefined threshold (as one of the standard procedures).
2. **Word Score Counts** (WSC): Select as key terms in the corpus all words whose weight computed by Eq. (2) exceeds a predefined threshold.
3. **Bigram Score Counts** (BSC): Select as key terms in the corpus all bigrams whose weight computed by Eq. (2) exceeds a predefined threshold.

Some preprocessing steps are required for each of the three above approaches in order to find a "good" set of features. A combination of several methods is used as follows: All nontextual information, all numbers, and all words in the stop list are removed; words are converted to their base form using a stemmer; and finally, infrequent and high-frequency words are pruned (see Valsan and Emele (2003) for details).

The resulting set of features defines the vector of key terms, and all documents in the corpus are coded according to this vector. Because the dimensionality of this document vector is huge (around 180,000 for the first two approaches and 450,000 for the third one), we used the RMP method to reduce the size of each of the document vectors to a reasonable value. The mapped size must satisfy two contradictory requests. On one hand, it must be large enough to not distort the mutual similarities of the data vectors too much, and on the other hand it must be small enough to be feasible by the classifier. Therefore, we have chosen a value of 100, which is a good trade-off between precision and feasibility.

The reduced documents matrix was used as input data for the split LBG algorithm. The number of codevectors was 32, the same as the number of manually derived clusters for the corpus. After applying the split LBG algorithm, each document was assigned to a codevector. All documents corresponding to the same codevector were grouped together and define a text cluster. There are 32 such clusters, which in the following will be referred to as smart-clusters. The next section describes the evaluation of the clustering performance based on perplexity results.

5.1.1 Clustering Evaluation Based on Perplexity Results

Several experiments were conducted to investigate how well the automatic clustering algorithm picked out particular topics. Therefore, we have evaluated the performances of the manual clustering against the automatic clustering for all three approaches. Two steps are required for this evaluation:

Step A: manually clustering. Thirty-two Genre-clusters were produced out of all training documents in the corpus using the a priori labeling. Accordingly, 32 LMs were generated for each topic, and we refer to them as LM_{G-i}, i=1,...,32. By using some additional test documents from the same corpus, which have been correctly assigned to their topics, another 32 test sets were produced (each test set corresponds to a topic; the topics are the same as those used for genre-clusters). Perplexity of each of the $LM_{G-i}$was evaluated on each of these test sets. Each time when a LM_{G-i} computes the smallest perplexity value on the corresponding test set i, we assumed that the cluster contains documents related to the topic of the test set i. In this case, we counted the manually assigned topic label as a correct one and we called such a cluster a topic-winner.

Step B: automatic clustering. One LM was also generated for each of the 32 smart-clusters in order to evaluate the separability between topics and at the same time to give the smart-clusters a proper label. For convenience we refer to them as LM_{S-i}. Perplexities of all LM_{S-i} were evaluated on the same 32 test sets as used in step A. The outcome is a matrix of 32×32 perplexity values. For each test set, whenever any of the 32 LM_{S-i}, which covers an unknown topic, outputs a perplexity value smaller than the smallest perplexity value corresponding to a $LM_{G-i,}i$=1,. . . ,32, the smart-cluster is considered as a smart-cluster winner. In this way we considered the smart-cluster winner as covering the topic of the test set and it was labeled accordingly. The results for all experiments are summarized in Table 1.

Table 1. Clustering performance evaluation

Clustering approach	Manual	WFC	WSC	BSC
No. of winner clusters/32 (%)	12/32 (37.5%)	19/32 (59.37%)	26/32 (81.25%)	29/32 (90.62%)
Average perplexity	70	55	38	25

The results proved that the automatic text clustering based on any of these approaches outperforms the manual clustering. In case of the WFC-based term selection, the discriminative power of terms is measured by their number of occurrences. The clustering performance is about 59.37%. Using instead the proposed new selection criterion according to Eq. (2) and/or additional context information by using bigrams as terms, a better separability of different topic areas is achieved. In case of WSC a clustering performance of 81.25% is achieved and, in case of BSC the clustering performance is 90.62%, leading to an improvement of 31.25% compared to the standard WFC baseline approach. Because the ultimate goal of text clustering is language model adaptation, we found perplexity to be a more appropriate metric to measure the performance of the classifier for clustering the documents. Moreover, we have used perplexity to assign a label to each cluster, whenever we considered it as being representative for a certain topic. A huge improvement in term of perplexity is achieved whenever any of three methods (WFC, WSC, and BSC) is used to provide text clusters instead of manually clustering.

5.1.2 Language Model Evaluation using WER

We also evaluated the system in terms of word error rate (WER) using the SMARTKOM recognizer that was developed by DaimlerChrysler (see Berton et al., this volume). We did so by using different speech corpora that were recorded by different partners. The results are listed in Table 2. "Pub" again reflects the public scenario (recorded by DaimlerChrysler) and "MobI" are dialogues from the mobile scenario (car navigation, recorded by DaimlerChrysler). "MobII" are also dialogues from the mobile scenario (pedestrian navigation, recorded by EML); "MTI" are dialogues for the home scenario that were recorded during the MTI exhibition in October 2001; SKI and SKII were dialogues recorded in the SMARTKOM WOZ recordings.

The results for the WOZ recordings are rather bad. The reason is that these recordings include huge amounts of off-talk that was completely out of domain (e.g., users starting to read movie descriptions that are displayed in the home scenario when the task was to select a movie, etc.). For the more scenario-specific dialogues, which did not include that much off-talk, the results are much better, ranging from 8.8% WER to 29.9%. Remember that these experiments included the dynamic update of the word classes and the automatic generation of pronunciations for these words. Please also note that SMARTKOM is a multimodal system. Thus by combining speech with gestures a lot of sentences can be interpreted correctly, even if the speech recognition rate is not perfect.

Table 2. Speech recognition evaluation using different corpora

Corpus	Pub	MobI	MobII	MTI	SKI	SKII
No. speakers	11	11	29	2	3	50
No. sent	100	100	1507	23	26	1234
Perplexity	7.2	14.6	12.6	16.7	19.7	42.2
WER	10.7	8.8	27.6	25.5	29.9	65.4

6 Conclusion

In this paper, we have shown that newspaper and Internet resources, such as newsgroups, are a valuable resource for language model adaptation. Based on these background corpora we have compared and evaluated two LM adaptation techniques and applied them to dialogue-specific language model adaptation. Besides developing a novel word n-gram pruning for domain adaptation, we proposed a new approach for thematic text clustering. This method relies on a new discriminative n-gram–based term selection process, which reduces the influence of the corpus inhomogeneity, and outputs only semantically focused n-grams as being the most representative key terms in the corpus. These key terms are then used to automatically cluster the whole document collection. By selecting only relevant text clusters for LM training, we addressed the problem of generating task-specific LMs. Different key term selection methods are investigated using perplexity as the evaluation measure. Automatically computed clusters are compared with manually labeled genre clusters. Compared to the manual clustering, a significant performance improvement between 21.87% and 53.12% is observed, depending on the chosen key term selection method.

We also described the class-based language model that allows dynamic updates of classes during runtime, depending on the current dialogue state. Furthermore, we evaluated the adapted language model within the SMARTKOM dialogue system in terms of word error rates using the SMARTKOM recognizer developed by DaimlerChrysler. For the scenario-specific dialogues the word error rates ranged from 8.8% to 29.9%. Compared to the much more restricted first version of the dialogue system, the results using the adapted class-based LMs show a big improvement because the final system demonstrator covers a huge variety of natural interaction dialogues ranging from device control, movie selection, document sending, and route planning to seat reservations, etc.

Acknowledgments

We would like to thank Ralf Kompe for the helpful discussions at the early stage of the project. Many thanks also to Philipp Cimiano and Marion Freese, who worked on the tools for the corpus preparations and helped in evaluating the language models. We would also like to thank our SMARTKOM colleagues from DaimlerChrysler, Andre Berton, Alfred Kaltenmeier, Klaus Mecklenburg, and Peter Regel-Brietzmann, for the very good cooperation and for providing us test data and relevant parameter tunings for their speech recognizer.

References

H. Baayen, R. Piepenbrock, and H. van Rijn. *The CELEX Lexical Database (CD-ROM).* Linguistic Data Consortium, University of Pennsylvania, 1993.

L.R. Bahl, F. Jelinek, and R.L. Mercer. A Maximum Likelihood Approach to Continuous Speech Recognition. *IEEE Transactions on Pattern Analysis and Machine Intelligence (PAMI)*, 5(2), 1983.

J.R. Bellegarda, J. Butzberger, W. Chow, N. Coccarao, and D. Naik. A Novel Word Clustering Algorithm Based on Latent Semantic Analysis. In: *Proc. Int. Conf. on Acoustics, Speech, and Signal Processing (ICASSP-96)*, vol. 1, pp. 172–175, Atlanta, GA, 1996.

P.R. Clarkson and A.J. Robinson. Language Model Adaptation Using Mixtures and an Exponentially Decaying Cache. In: *Proc. Int. Conf. on Acoustics, Speech, and Signal Processing (ICASSP-97)*, Munich, Germany, 1997.

A. Dempster, N. Laird, and D. Rubin. Maximum Likelihood From Incomplete Data Using the EM Algorithm. In: *Annals of the Royal Statistical Society*, vol. 39, pp. 1–38, London, UK, 1977. Royal Statistical Society.

T. Goodman. A Bit of Progress in Language Modeling. *Computer Speech and Language*, 15(403-434), 2001.

S. Goronzy, S. Rapp, and M. Emele. The Dynamic Lexicon, 2006. In this volume.

R. Iyer and M. Ostendorf. Modeling Long Distance Dependence in Language: Topic Mixtures Versus Dynamic Cache Models. *IEEE Transactions on Speech and Audio Processing*, 7(1):30–39, January 1999.

F. Jelinek. *Statistical Methods for Speech Recognition*. MIT Press, Cambridge, MA, 1997.

F. Jelinek and D. Lafferty. Computation of the Probability of Initial Substring Generation by Stochastic Context Free Grammars. *Computational Linguistics*, 17(3): 315–323, 1991.

S. Kaski. Dimensionality Reduction by Random Mapping Fast Similarity Computation for Clustering. In: *Proc. IJCNN'98* , vol. 1, pp. 413–418, Piscataway, NJ, 1998.

R. Kneser and H. Ney. Improved Clustering Techniques for Class-Based Statistical Language Modeling. In: *Proc. EUROSPEECH-93*, pp. 973–976, Berlin, Germany, 1993.

T. Kohonen, S. Kaski, K. Lagus, J. Salojärvi, J. Honkela, V. Paatero, and A. Saarela. Self Organization of a Massive Document Collection. *IEEE Transactions on Neural Networks*, 11(3):574–585, 2000.

R. Kuhn and R. De Mori. A Cache-Based Natural Language Model for Speech Reproduction. *IEEE Transactions on Pattern Analysis and Machine Intelligence (PAMI)*, 12(6):570–583, 1990.

Y.H. Lam and M.C. Emele. Application-Specific Language Model Adaptation Using Internet Resources. In: *Proc. 11th Sony Research Forum '01*, 2001.

Y. Linde, A. Buzo, and R.M. Gray. An Algorithm for Vector Quantizer Design. *IEEE Transactions on Communications*, pp. 702–710, 1980.

D.R.H. Miller, T. Leek, and R.M. Schwartz. A Hidden Markov Model Information Retrieval System. In: *Proc. 22nd Int. Conf. on Research and Development in Information Retrieval*, pp. 214–221, Berkley, CA, 1999.

T.R. Niesler and P.C. Woodland. Modeling Word-Pair Relations in a Category-Based Language Model. In: *Proc. Int. Conf. on Acoustics, Speech, and Signal Processing (ICASSP-97)*, Munich, Germany, 1997.

E. Rasmussen. Clustering Algorithms. In: W.B. Frakes and R. Baeza-Yates (eds.), *Information Retrieval: Data Structures and Algorithms*, pp. 419–442, Englewood Cliffs, NJ, 1992. Prentice Hall.

S.E. Robertson and J.K. Spärck. Relevance Weighting of Search Terms. *Journal of American Society for Information Science*, 27:129–146, 1976.

R. Rosenfeld and P. Clarkson. Statistical Language Modeling Using the CMU-Cambridge Toolkit. In: *Proc. EUROSPEECH-97*, Rhodes, Greece, 1997.

G. Salton and M.J. McGill. *Introduction to Modern Information Retrieval*. McGraw-Hill, New York, 1983.

F. Schiel and U. Türk. Wizard-of-Oz Recordings, 2006. In this volume.

K. Seymore and R. Rosenfeld. Large-Scale Topic Detection and Language Model Adaptation. Technical report, School of Computer Science, Carnegie Mellon University, June 1997.

R. Srihari and C. Baltus. Combining Statistical and Syntactic Methods in Recognizing Handwritten Sentences. In: *Proc. AAAI Symposium: Probabilistic Approaches to Natural Language*, pp. 121–127, Cambridge, MA, 1992.

Z. Valsan and M. Emele. Thematic Text Clustering for Domain Specific Language Model Adaptation. In: *Proc. IEEE Automatic Speech Recognition and Understanding ASRU 2003*, St. Thomas Island, USA, 2003.

Y.Y. Wang. Robust Spoken Language Understanding. In: *Proc. EUROSPEECH-01*, Aalborg, Denmark, 2001.

The Dynamic Lexicon

Silke Goronzy, Stefan Rapp, and Martin Emele

Sony International (Europe) GmbH
Sony Corporate Laboratories Europe, Advanced Software Laboratory, Stuttgart, Germany
silke@goronzy-privat.net, {rapp,emele}@sony.de

Summary. The dynamic lexicon is one of the central knowledge sources in SMARTKOM that provides the whole system with the capabability to dynamically update the vocabulary. The corresponding multilingual pronunciations, which are needed by all speech-related components, are automatically generated.

Furthermore, a novel approach for generating nonnative pronunciation variants and therefore for adapting the speech recognizer to nonnative accents is presented. The capability to deal with nonnative accents is crucial for dialogue systems dealing with multimedia and therefore often with multilingual content.

1 Introduction

This chapter describes the dynamic lexicon. This module is distinguished from comparable components in other dialogue systems by its capability to dynamically update its vocabulary with new words and to automatically generate multilingual pronunciations for these words using grapheme-to-phoneme conversion. It therefore serves as a central knowledge source for several SMARTKOM modules that deal with the recognition, analysis and synthesis of speech. For dialogue systems like SMARTKOM that are characterised by their highly variable application content, this is an essential feature. In applications such as city information systems, cinema or TV program guides, or multimedia databases, it is not possible to foresee the vocabulary for the speech-related modules at the time of the design of the system as they are determined by accessing dynamically changing content (Rapp et al., 2000). Online updates of the vocabulary that are consistently propagated to all modules involved are therefore absolutely necessary to guarantee a smoothly running system. We describe such a lexicon along with all involved technologies as well as design issues that are of concern for the whole dialogue system.

It is also part of this chapter to describe a novel technique to automatically generate nonnative pronunciation variants. Such pronunciations are necessary in systems where the user is using words or sentences in a language that is different from his/her mother tongue. In these cases the words will be pronounced in an accented way, which causes some difficulties for speech recognisers that are usually trained on native speech only. Generating alternative pronunciations that reflect accented pronunciations and keeping them in the dictionary can help alleviate this problem to a certain extent.

The remainder of the chapter is organised as follows: Sect. 2 illustrates the need for a dynamic lexicon in SMARTKOM before Sect. 2.1 explains how the lexicon is integrated in the SMARTKOM architecture. The updateable content is described in Sect. 2.2 before Sect. 2.3 explains how dynamic changes of the vocabulary take place. The automatic generation of multilingual pronunciations via grapheme-to-phoneme conversion is described in Sect. 3. Finally, Sect. 4 outlines the novel approach for the automatic generation of accented pronunciation variants.

2 The Dynamic, Multilingual Lexicon

Information systems like SMARTKOM often need to access databases, whose content might change frequently. In all three scenarios — SMARTKOM-Home (Portele et al., 2006), SMARTKOM-Public (Horndasch et al., 2006) and SMARTKOM-Mobile (Berton et al., 2006a; Malaka et al., 2006) — this is the case for the content of the information the user might want to access.

One example is a TV program that is accessed via an Internet-based electronic program guide (EPG) in SMARTKOM-Home. In this case it is obvious that the content changes daily, and sometimes even more often. One might argue that these EPGs are often available several days in advance, but generating a static dictionary every five days and loading it to the system is too expensive and cumbersome. In addition, sudden program changes might occur that need to be reflected in the system.

When retrieving the TV program for a certain time, channel etc. specified by the user, the results will typically be a list of TV shows that are then adequately presented to the user via the presentation manager (Poller and Tschernomas, 2006). Even though SMARTKOM offers the possibility of using pointing gestures for selecting icons, etc., on the screen, the system should not be limited to this modality for the selection of show titles. Speech input should also be possible, which in this case means that the user can speak the TV show titles resulting from the user query. Currently dialogue systems usually cannot automatically handle this kind of dynamic content. As a result, they need to keep as many words as possible in the system dictionary right from the beginning, which increases the dictionary size and thus the speech recognition error rates. It also causes the subsequent components (e.g., speech analysis) to become more complex. Furthermore, it is often not possible to determine the necessary words in advance (like in the TV program example), and thus such systems risk working with an incomplete dictionary that restricts the user to use only a certain set of words.

As a result, a system like SMARTKOM requires dynamic updates of the recogniser dictionary to reflect the actual titles. Thus, there is the need for a lexicon module that is able to dynamically update the vocabulary of the system and inform all other modules that might depend on this information. In the case of the lexicon this means that new words can be added (and removed) to (from) the dictionary and the pronunciation of these new words will be provided automatically. Especially in the context of multimedia content and information such as movies or music titles, the original language of the content needs to be taken into account when the pronunciation is generated.

While it seems to be an obvious step to allow words to be added to the lexicon, it is not so clear how exactly this can happen. There are many open questions that need to be answered. First of all, an identification of words that do actually change frequently is necessary. Simply allowing everything for the update-able content is not suggested since this might lead to unnecessary lexicon updates. Second, it is not a trivial point to determine which module in the SMARTKOM architecture is best aware about the current status of the other modules and can thus deliver the information to the lexicon, which words are needed for the next dialogue steps and which are to be added to the lexicon. It then needs to be clarified what is the appropriate timing for lexicon updates and how this can be synchronised with other modules to allow for a more responsive system. Last but not least, it needs to be clarified which kind of pronunciations are to be generated, taking into account the requirements of the different scenarios.

All these questions played an important role for the design of the dynamic lexicon. How these problems were solved is described in the following sections.

2.1 The Lexicon as a Central Knowledge Source

In the above EPG example, the speech recogniser was named one of the components that needs to access dynamic vocabulary information. However, besides the speech recogniser (Berton et al., 2006b), other modules also need to do so. More concretely, the prosody recogniser (Zeißler et al., 2006) also needs to be able to recognise new words. Furthermore the speech analysis (Engel, 2006) needs to be able to interpret a user's input, possibly referring to the new content. Finally the speech synthesis (Schweitzer et al., 2006) needs to be able to generate output that might also contain new words.

Thus the lexicon can be considered as one of the central knowledge sources of SMARTKOM. As such, all dynamic information is managed in only one place so that it can be ensured that all above-mentioned modules use the same information at any time. This increases consistency, which is mandatory to show consistent system behaviour during the interaction with the user. The placement of the lexicon in the SMARTKOM system architecture is shown in Fig. 1. This figure does not show the SMARTKOM system in detail but focuses on those components that are relevant for describing the functionality of the lexicon. The normal arrows indicate the direct links between the modules, while the bold arrows are meant to show those modules in particular that depend on the lexicon.

2.2 "Updateable" Content

It is to be expected that not all words that are used in a dialogue system change frequently. There will always be a certain set of words that the system should always

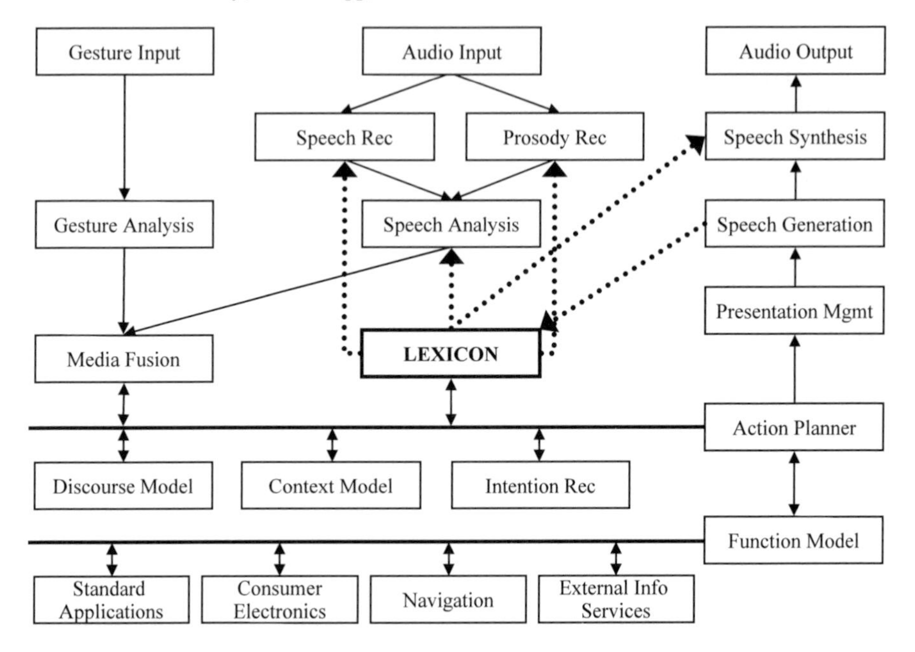

Fig. 1. SMARTKOM architecture (part of the system)

be able to recognise and process, whereas there are others that are only important in certain dialogue turns. "General" words like "please, show, me, ...", etc., are contained in a base lexicon whereas, e.g., movie titles, which are only important when the user is talking about movies, are dynamically loaded. We thus defined a set of certain word classes that can be updated dynamically. In the EPG example such a class would be called MOVIETITLE. Other word classes that were used are ACTOR, CITY, GARAGE, GENRE, HOTEL, LOCATION and MOVIETHEATER. These are closely coupled with the word classes that are used by the language model as described in Goronzy et al. (2003). A detailed explanation can be found in Emele et al. (2006).

2.3 Dynamic Updates

2.3.1 Adding Words to the Lexicon

The dynamic updates of the lexicon are initiated by the action planner (Löckelt, 2006). The action planner is the module that also initiates requests to external services such as the EPG. It is thus aware that a request for "movies on TV for tonight" will result in a list of movie/TV show titles. Whenever the query result is received by the action planner, it forwards the relevant information to the lexicon. Since the SMARTKOM system is using the pool architecture as described in Herzog and Reithinger (2006), the lexicon extracts all relevant words after having found a vocabulary update in the output pool of the action planner. It then tests whether these words are already contained in the current lexicon. If so, it is checked whether a pronunciation in the requested language, specified by a language tag in the update request, exists. If everything is there already, no action is taken. If the word is not yet contained in the lexicon, an entry for it is added. Additionally, the pronunciation for the word is generated by automatic grapheme-to-phoneme conversion (G2P), as described in Sect. 3. An update is written to the output pool as described in Sect. 2.4.

2.3.2 Replacing Words in the Lexicon

A situation that often occurs when a user is searching for some content is the following: he/she will ask the system for something and will receive an answer. However, often a refinement of the user's request to reach the final decision will happen. In the EPG example, users might ask for "movies that are on TV tonight". They might get a lot of results, so they refine this query to "only news shows for tonight." Such multiple requests will result in repeated lexicon updates of the same word classes (in this case MOVIETITLE). But since it is not necessary to always keep all words in the lexicon, even the results from the very first request, the contents of the word classes are replaced by the result of the new request. This is the default behaviour of the lexicon, but other behaviour like adding/removing words can always be requested explicitely.

2.3.3 Removing Words from the Lexicon

Sometimes a user request might result in a very long result list, e.g., when the request is not very specific. All these words are then added to the lexicon. Also, when the system is used for a longer time words might be repeatedly added to the lexicon. This can increase the lexicon size such that it might negatively affect the speech recogniser performance. It thus makes sense to remove words again that are no longer needed. This can, e.g., be done if the topic of the interaction changed. Once the user selected a movie to be recorded at home on his VCR and then requests some information about Heidelberg, all the previously loaded movie titles are considered to be no longer needed and are thus removed from the lexicon.

2.4 Update Mechanisms

After the update request was processed as described above, the new information needs to be propagated to the modules that need the lexical information. For this two different possibilities were implemented, the full update and the incremental update. At the startup of the lexicon it can be specified which of the two updates should be performed, as it depends on the other module's capabilities which updates can best be handled. In both cases the updates are written to a pool, called modelled.lexicon, that the speech recogniser, the speech analyser and the prosody recogniser have subscribed to. The speech synthesis is treated a little differently in this respect, as will be explained in Sect. 2.4.3.

2.4.1 Full Update

In this mode, whenever there is a change in lexicon initiated by the action planner, the full lexicon is written to the corresponding pool and read by all modules that subscribed to it. This solution is rather impractical for frequent dynamic updates and was more intended to serve as a "lexicon synchronisation" help during system testing. In real system use, this update is of minor importance.

2.4.2 Incremental Update

The update mode that is most frequently used in SMARTKOM is the incremental update. In this case the lexicon writes only the differential lexicon to the pool, meaning those parts that were changed. That means if a list of words is added to the lexicon, only these words together with the generated pronunciations are written to the pool and read by the other components.

2.4.3 Special Requests

In Sect. 2.1 the speech synthesis module was also named one of the modules that needs to have access to the lexicon. However, this module is treated differently from the others. While the other modules always need the complete lexicon for their tasks, they need to be informed about all updates. The synthesis, on the other hand, needs the pronunciations for the words it is supposed to synthesize. As a consequence, there is a dedicated pool that can be used to explicitly request pronunciations for certain words and sentences. These are looked up or, if necessary, generated and then used to synthesize a system answer. In SMARTKOM the speech generation generates the sentence that is to be synthesized as the system response to the user. Thus the speech generation sends a request to the pool modelled.lexicon.request containing the words to be synthesized. The lexicon then adds the pronunciation for each word if it is contained in the dictionary; if not, the word is added as usual to the lexicon and the pronunciation is automatically generated. The words together with all pronunciations are then written to the pool modelled.lexicon.response, where they are read by the synthesis module and used for generating the speech output.

2.5 Intelligent Handling of Movie Titles

In the Home scenario the problem arose that, e.g., the movie titles that are contained in the EPG database are often rather long and furthermore often also contain subtitles, like "James Bond, 007: *Tomorrow never dies*". If this title were added to the lexicon as described above, users would have to speak the complete title if they want to select this movie. In reality, however, users tend to say "record James Bond" or "record 007", rather than the complete title. This is accounted for by the lexicon by generating multiple pronunciations that are assigned to the complete title. The complete title needs to be kept as the orthography for the shortened pronunciations to ensure consistency for further database requests concerning this title. The various pronunciations then reflect the shortened versions of the titles as the users usually say them. The shortened title versions are determined by splitting the titles at special symbols such as "-", "!", ",", etc., that we found to often separate titles and subtitles. This feature allows a more natural use of SMARTKOM while interacting with the EPG application.

3 Automatic Generation of Pronunciations

After discussing the need for a dynamic lexicon and how it is integrated into the SMARTKOM system, we now describe how we actually derive the pronunciations for new words. Automatic grapheme-to-phoneme conversion (G2P), the task of mapping a written word (that is, a sequence of letters) to its pronunciation (that is, a sequence of phonemes) can be achieved in various ways. First, in the knowledge-based approach, a set of manually written rules can be deployed that define how a given letter or group of letters is pronounced in a given graphemic, phonemic, morphological or other context (such as a specific foreign language origin). Such rules can frequently be found printed at the beginning or end of pronunciation dictionaries to advise speakers on the pronunciation of words not found in the dictionary, that is, for a task very similar to the one the dynamic lexicon has in the SMARTKOM system. In practice, however, manually written rules for G2P are often not very precise. For instance, we implemented a subset of the rules from a standard German dictionary (those that were reasonably easy to implement) and found that our rules achieve about 15% errors on the phoneme level, so that only less than every second word is correct.

Second, it was proposed to use corpus-based (or data-driven) methods to learn the mapping by some machine learning (ML) program such as neural networks (Sejnowski and Rosenberg, 1987). Besides neural networks, various other ML techniques can be used. In our work, we rely on the induction of decision trees (Black et al., 1998; Rapp and Raddino, 1999). Obviously, whereas manually written rules must be implemented again for every new language, the corpus-based approach has the advantage that it is largely independent of the language in question and only presupposes the existence of a rather large pronunciation dictionary as a training corpus. In our work, we found that the automatically trained methods achieve much higher precision than our manually written rules (Rapp and Raddino, 1999).

The contribution of our G2P work is twofold. First, we propose a hybrid two-stage architecture for multilingual G2P. In the first stage, phonemic pronunciations are generated using a corpus-based approach, and in the second stage, manually written rules map from phonemic to phonetic pronunciations, including pronunciation variants. Second, we introduce boosting, a meta–learning technique to the decision tree induction of the first stage and find that this considerably improves accuracy at the expense of extra memory. We examine the influence of training parameters on size and accuracy of decision tree classifiers to optimise space in the first stage. The parameters we vary are the amount of pruning applied during construction of the classifiers, the number of classifiers and the size of training set.

3.1 Architecture

We propose a two-stage process for G2P, with the orthography as input, an abstract phonemic representation as an intermediate step and the phonetic representation as output. The intermediate phonemic representation is also applicable for automatic speech recognition, but performance is expected to be better for the phonetic representations.

The two-stage architecture offers flexibility in creating lexicons. The phonemic representation output from the first stage gives a canonical ("standard") pronunciation. The second stage can be altered to derive pronunciation variants, such as those needed to represent different dialects, foreign accents or recognizer phone sets. For SMARTKOM, the two-stage approach simplifies development and contributes to the recognition performance.

For one, we could take advantage of the rather large coverage German pronunciation lexicon CELEX (Baayen et al., 1993) to achieve good prediction accuracy. CELEX has listed about 360,000 pronunciation entries, much more than what would have been available from the SMARTKOM and Verbmobil data collections. On the other hand, the pronunciations found in CELEX are more phonemic (or canonical) oriented than what is typically deployed in state-of-the-art speech recognizers. This difference is accounted for in the second stage. By comparing the output of the first stage's classifier with the transcription on which the recogniser is trained, rules were developed by phonetic experts that map from the learned phonemic representation to the more phonetic phone set of the recogniser. In this process either a single or multiple pronunciations can be generated. Thus, the recogniser is supplied with acoustically reasonable pronunciations while the corpus-based learning can be kept simpler, as the data for the learning task is more consistent. Finally, by exchanging the standard rules of the second stage by speaker-dependent ones—either by modifying the rules or by using a data-driven approach as described in Sect. 4, the system can be adapted to dialectal or nonnative variations of speech.

For the first step, the prediction of the phonemic representation, we deploy decision tree classifiers that are induced from language-specific training data. The induction itself is language-independent. As described in Rapp and Raddino (1999), we use the C5.0 tool,[1] which first builds tree classifiers and then optionally transforms them into more memory-efficient rule sets of comparable accuracy. Preparation of the data for C5.0 requires alignment of the dictionary to relate each grapheme symbol to one or more phoneme symbols, and vice versa. One-to-one relations between graphemes and phonemes are established by inserting additional "null" graphemes or phonemes using Viterbi decoding: If there was a 1:1 relation between grapheme and phoneme, the inserted null phoneme will correspond to the inserted null grapheme. Otherwise, either a grapheme or a phoneme will correspond to a null symbol, see Rapp and Raddino (1999).

Pronunciations are predicted moving grapheme by grapheme from right to left, mapping each grapheme symbol to a phoneme symbol. For every grapheme, two contexts are considered: (1) three (non-null) graphemes to the right and to the left, and

[1] http://www.rulequest.com

(2) the already predicted phonemes corresponding to the three (non-null) graphemes to the right. To allow for prediction of word stress and syllable boundaries as well, the number of stressed syllables to the right of the current syllable and the number of consonants up to the next vowel to the right are also taken into account.

For the second step, the generation of one or more phonetic representations, we use rules written by phonetic experts. Currently, we have implemented such rules for American English and German. For American English, they generate pronunciation variants. For German, they currently only constitute some postlexical rules (Sect. 4).

3.1.1 Building Classifiers

Like most machine learning algorithms, C5.0 can be adapted to different learning tasks by setting some training parameters. These parameters have an influence on the degree of generalisation of the resulting classifiers. There has been speculation that generalisation is often harmful in natural language tasks and that the best results are achieved by highly specialised classifiers (Daelemans et al., 1999).

The parameter with the strongest influence on generalisation in C5.0 is the prepruning parameter m, which specifies the minimum number of cases that must be covered by each branch in the tree. Small values for m yield highly specialised, large trees, while large values give smaller classifiers that generalise more strongly.

Boosting is a machine learning technique that is known to significantly improve prediction accuracy for many tasks by combining several classifiers (Bauer and Kohavi, 1999). During training, each classifier emphasizes the cases misclassified by the previous one. For prediction, each classifier is applied, and the class predicted by most classifiers will be chosen. Boosting requires more memory since several trees have to be stored. An interesting question is whether the accuracy improvements of boosting can be maintained even if the individual trees are pruned strongly to save space.

3.1.2 Experiments

We investigated the relation between size and word accuracy rate of the tree classifiers for English and German. We varied the prepruning parameter m (2 to 256) and the number of trees used for boosting (1, 5, 10, 15) on training data of 20K and 100K words sampled from Baayen et al. (1993). The classifiers were evaluated ignoring errors in the assignment of word stress and syllable boundaries.

Figure 2 shows the results for German. We are comparing the size (number of nodes in the trees, in logarithmic scale, on the x-axis) and the word accuracy rate (in %, on the y-axis) of the classifiers.

The best word accuracy rates can be achieved for larger sets of training data when using boosting. However, both factors result in larger trees, as seen in Fig. 2. Results for the 100K training set are marked by gray symbols, and results for the 20K training set are indicated by white symbols. In general, the gray 100K series in the diagram show better accuracies with larger numbers of nodes than the white 20K series. We used boosting with 5, 10 and 15 classifiers, represented by the triangle,

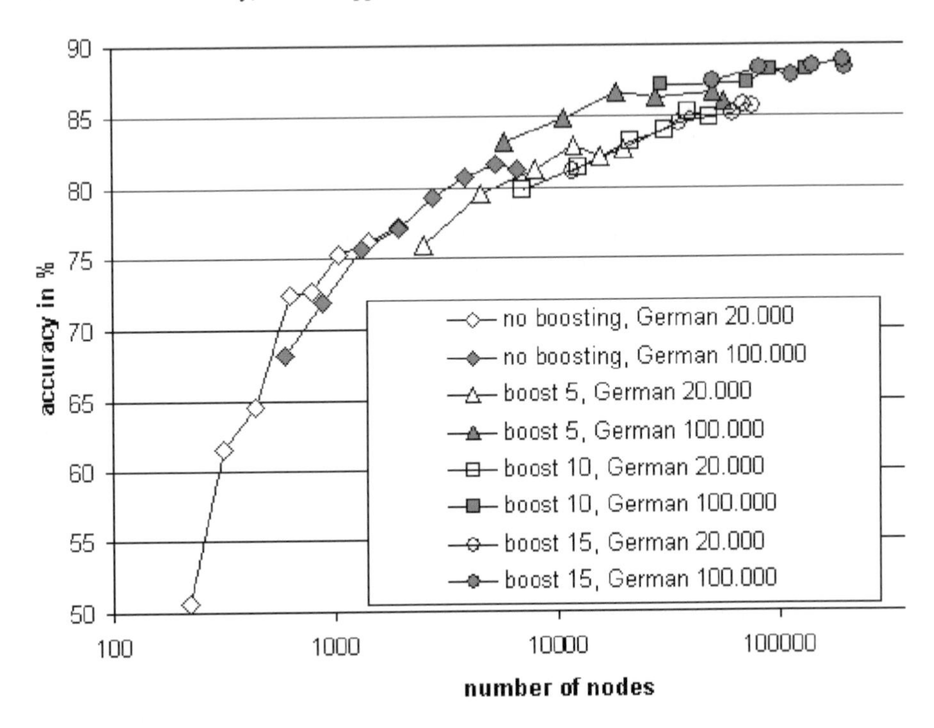

Fig. 2. Word accuracy rates for classifiers trained on german data

square and circle series, respectively. Boosting gave higher rates than did the series without boosting, as indicated by the diamond symbols, again at the expense of larger classifiers.

An interesting comparison is between the strongly pruned trees trained on the 100K training set and the less strongly pruned trees built from the smaller 20K set. The trees were comparable in size, but the more specialised 20K trees outperformed the 100k trees (cf. the series with the white and gray diamond symbols in the diagram). This means that the generalisation originating from stronger pruning is more harmful than the one originating from a smaller training set.

As for the question whether the high accuracy rates achieved for boosting can still be maintained if the individual trees are strongly pruned, the answer is yes. Moreover, when reducing the pruning factor to apply less pruning, word accuracy rates only improve down to a pruning factor of $m = 32$, then stagnate or deteriorate for $m < 32$, as the trees get bigger and bigger. (For British English, the threshold was found to be $m = 16$). Also, boosting with 15 classifiers did not yield significantly better accuracy rates (the circle series and the square series overlap in Fig. 2).

For British English, the same tendencies can be observed. Word accuracy rates figure at about 9% below the ones for German. The best result for British English, 80.4%, was achieved using boosting with ten classifiers and a pruning factor of

$m = 16$ on training data from a 100K dictionary, with a total of approx. 195K nodes. This accuracy compares favourably to the results of others. The same parameters on 20K training data yielded 69.5% accuracy. For American English, first experiments on training data of 53K entries show lower accuracies, namely 63%. The difference between American and British English is attributed to the different lexical resources that were deployed. On the other hand, the considerably improved accuracy of German over English is consistent with the findings of others and can be explained with the conservative writing system of English.

3.2 Discussion of G2P results

We presented a language-independent procedure to obtain classifiers for G2P and evaluated it on German, British English and American English data. It can be said that the optimal training parameters are heavily dependent on the language, the size of the training data and the desired size of the trees. Optimal word accuracy rates can be reached with large amounts of training data using boosting in combination with careful pruning of the individual classifiers. However, optimal classifiers come at the cost of huge memory consumption. For many applications, a reasonable compromise between size and accuracy has to be found, as it is not always possible to afford arbitrary memory for every system component. In its totality we could not confirm the position of Dealeman et al. that generalization is harmful for the G2P task. If also memory consumption is considered as an additional dimension to optimize for, it becomes obvious that generalization is to be pursued. In our data we also can see clearly the advantage of using boosting for the G2P task. For the 20K lexicon, boosting improves word accuracy by about 9%, and for the 100K lexicon by about 7%.

3.3 Multilingual Pronunciations

For the base lexicon a default language is defined in the M3L document containing all words of the base vocabulary. If new words are added to the lexicon, they are generated in this default language unless it is explicitly requested to generate them in another language. The languages supported by the lexicon are German and British English. These two languages can be provided by using two decision trees for G2P, one trained on German, the other trained on English. Again, movie titles are treated differently from other words. Especially for movie and music titles, it is difficult to determine a default, since they are often mixed. Sometimes English subtitles are translated to German, while the original title remains unchanged, sometimes everything is translated, sometimes the title is completely English. Since we do not know what was translated to German and what was not, both an English and a German pronunciation is always generated for movie and music titles to allow the greatest flexibility in terms of pronunciation for the user. Additionally, there is the problem of a user speaking, e.g., an English title, with a German-accented pronunciation. This case needs special treatment for the generation of appropriate pronunciations, and an approach for this is described in Sect. 4.

4 Pronunciation Adaptation

The methods implemented for lexicon updates allow to add multiple pronunciations for words. In Sect. 3.3 we already discovered the problem of nonnative pronunciations. If we want to reflect such pronunciations to increase recognizer robustness w.r.t. nonnative speech, the question is how to derive such accented or nonnative pronunciations that users typically use for speaking, e.g., English titles with a German accent. A novel method was developed within SMARTKOM and will be outlined in this section. A detailed description can be found in Goronzy (2002) and Goronzy et al. (2004). There are several approaches to pronunciation modeling. They can be categorized into two broad classes, which are knowledge-based and data-driven approaches.

Knowledge-based approaches make use of existing knowledge that was derived by experts (e.g., dictionaries). Such knowledge is often used to generate rules that are able to generate typical pronunciation variants from canonical pronunciations or from the orthography of a word. Data-driven approaches try to derive the pronunciation variants directly from the speech signal, e.g., using manual transcriptions or automatic transcription using phoneme recognisers. In the latter case, in order to obtain a certain generalization capability to new words, the generated variants are often not used as they are. Rules can be generated therefrom, or decision trees can be grown that are able to predict the pronunciation variants.

In general, the pronunciation variants that are generated can then be added to the dictionary. A detailed overview on pronunciation modeling can be found in Strik and Cucchiarini (1999).

While knowledge-based approaches are completely task-independent, they tend to overgenerate. In contrast, data-driven approaches are rather database dependent. For our problem we would need to explicitly record accented speech databases, which is a cumbersome and expensive task.

4.1 Automatic Generation of Nonnative Pronunciation Variants

Since we are interested in pronunciations for various accents and languages, both approaches are too expensive. We need an approach that can flexibly be applied to various combinations of source (the native language of a speaker) and target languages (the language they are trying to speak).

The automatic method that was developed exploits the fact that many native databases exist. It is a data-driven approach but solely uses native speech to derive nonnative pronunciations. This is achieved by simulating one of the many different possibilities of humans to learn a new language, namely to listen to native speakers of the target language and try to reproduce what was just heard. If we assume that the learner is completely new to the target language two things are likely to happen:

- He/she will use phonemes known from the source language.
- He/she will try to use phonotactic constraints inherent to the mother tongue.

The first item simply means that phonemes that exist in both languages will be pronounced as in the source language rather than as in the target language. This can be handled easily by speaker adaptation techniques that adapt the Hidden Markov Models (HMMs) to the current speaker. The second point means that speakers will tend to reproduce certain phonotactic structures characteristic to their source language when speaking the target language, thus resulting in wrong phoneme sequences. The consonant–vowel structure of Japanese might serve as an example here. Beacause Japanese speakers often experience difficulty in pronouncing subsequent consonants, they often insert vowels, like in "table": /t eI b l/ becomes /t eI b u r u/. In addition to the pronunciation of the /l/ as an /r/, two phonemes (/u/) are inserted before and after the /r/. Standard acoustic speaker adaptation algorithms are often used to improve performance for nonnative speakers. While taking into account that certain phonemes are pronounced differently, they totally neglect potentially different phoneme sequences. In the above example, acoustic adaptation would try to adapt the HMM for /l/ to /u r u/ since it uses the native canonical pronunciation as a reference for adapting the HMMs. However, in reality three instead of one phonemes have been produced, which would cause a misadaptation that might cause the overall recognition rates to decrease.

In order to generate these accented pronunciations, we simulate the learning process by using a phoneme recognizer that is trained on the source language and uses a source language phoneme-bigram to reflect the phonotactic constraints of the source language. This source language phoneme recognizer simulates the learner trying to reproduce a word. We thus let this recognizer recognize native speech of the target language. For example, we use a British English phoneme recognizer to recognize German speech, and the result is an English-accented pronunciation for German words. Since phoneme recognition usually is rather error-prone, some kind of generalisation is necessary to filter out pronunciations that contain a lot of errors. We achieve this by training a decision tree that learns a mapping from the German canonical to an English-accented pronunciation. Adding these generated variants to the lexicon, recognition rates on nonnative speech could be improved by up to 5.2%, as described in Goronzy et al. (2004). The method is particularly effective when combined with speaker adaptation techniques, as was also shown in Goronzy et al. (2004). Then recognition rates can be improved by up to 18%. The processes of generating the variants (1) and then employing these later in the recognition phase (2) are shown in Fig. 3.

5 Conclusion

In this chapter the dynamic lexicon was described in detail. A module like this allows dialogue systems to flexibly react to user requests that refer to dynamic content such as music databases or electronic program guides for TV. State-of-the-art dialogue systems usually are not able to handle such dynamic content. As a consequence, users cannot speak, e.g., movie titles, which renders the dialogue very inflexible. In SMARTKOM a very flexible behaviour concerning this aspect was achieved by using

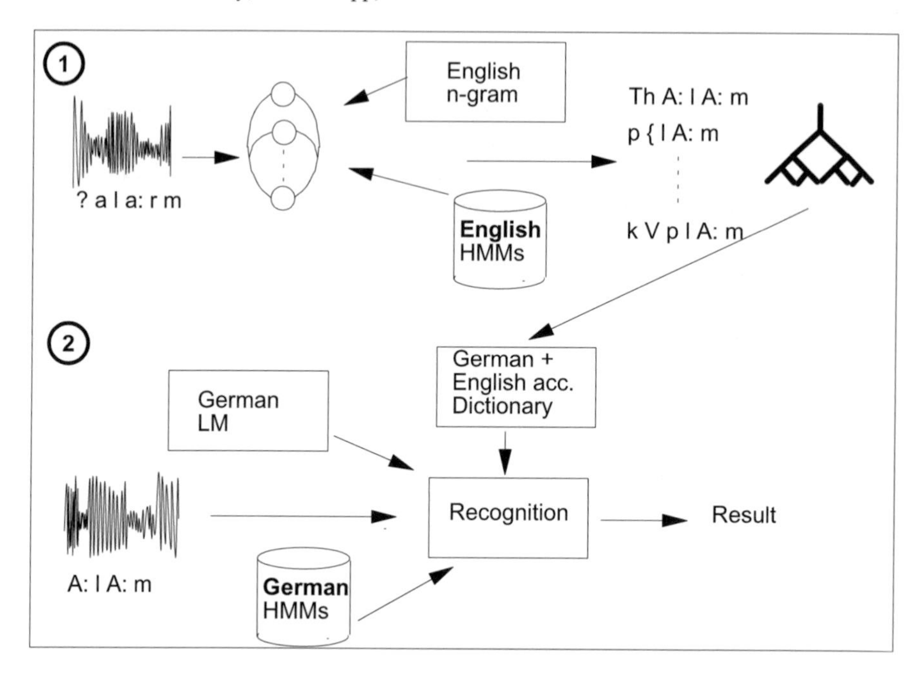

Fig. 3. *1*: Method to generate nonnative pronunciations. *2*: Final recognition system with extended dictionary

a lexicon that can accept such content for inclusion and that is well synchronized with all necessary dialogue modules. We described the integration of the lexicon into the SMARTKOM architecture and the methods to perform the dynamic updates.

The automatic generation of multilingual pronunciations using a grapheme-to-phoneme conversion approach was also explained. Furthermore, a novel approach to automatically generate accented pronunciations for inclusion into the lexicon that is very flexible w.r.t. different accents and does not need any accented speech data was outlined.

Acknowledgments

We would like to thank Ralf Kompe for the helpful discussions, which helped give not only the lexicon but also SMARTKOM its direction. Our colleague Jürgen Schimanowski was an enormous help in the implementation and tuning phase — thanks! We are also very grateful to Antje Schweitzer, who made a crucial contribution to the works concerning G2P. Anton Batliner shared his expertise on all phonetic issues that arose and manually checked a vast number of base lexicon pronunciations. Last but not least, we want to thank all other SMARTKOM colleagues involved in the lexicon activities for the very good cooperation.

References

H. Baayen, R. Piepenbrock, and H. van Rijn. *The CELEX Lexical Database (CD-ROM)*. Linguistic Data Consortium, University of Pennsylvania, 1993.

E. Bauer and R. Kohavi. An Empirical Comparison of Voting Classification Algorithms: Bagging, Boosting, and Variants. *Machine Learning*, 36(1–2):105–139, 1999.

A. Berton, D. Bühler, and W. Minker. SmartKom-Mobile Car: User Interaction With Mobile Services in a Car Environment, 2006a. In this volume.

A. Berton, A. Kaltenmeier, U. Haiber, and O. Schreiner. Speech Recognition, 2006b. In this volume.

A. Black, K. Lenzo, and V. Pagel. Issues in Building General Letter to Sound Rules. In: *Proc. 3rd ESCA Workshop on Speech Synthesis*, pp. 77–80, Jenolan Caves, Australia, 1998.

W. Daelemans, A. van den Bosch, and J. Zavrel. Forgetting Exceptions Is Harmful in Language Learning. *Machine Learning*, 34:11–41, 1999.

M. Emele, Z. Valsan, Y.H. Lam, and S. Goronzy. Class-Based Language Model Adaptation, 2006. In this volume.

R. Engel. Natural Language Understanding, 2006. In this volume.

S. Goronzy. *Robust Adaptation to Non-Native Accents in Automatic Speech Recognition*. LNAI 2560. Springer, Berlin Heidelberg New York, 2002.

S. Goronzy, S. Rapp, and R. Kompe. Generating Non-Native Pronunciation Variants for Lexicon Adaptation. *Speech Communication*, 42(1):109–123, 2004.

S. Goronzy, Z. Valsan, J. Schimanowski, and M. Emele. The Dynamic, Multi-Lingual Lexicon in SmartKom. In: *Proc. EUROSPEECH-03*, vol. 3, pp. 1937–1940, Geneva, Switzerland, 2003.

G. Herzog and N. Reithinger. The SmartKom Architecture: A Framework for Multimodal Dialogue Systems, 2006. In this volume.

A. Horndasch, H. Rapp, and H. Röttger. SmartKom-Public, 2006. In this volume.

M. Löckelt. Plan-Based Dialogue Management for Multiple Cooperating Applications, 2006. In this volume.

R. Malaka, J. Häußler, H. Aras, M. Merdes, D. Pfisterer, M. Jöst, and R. Porzel. SmartKom-Mobile: Intelligent Interaction With a Mobile System, 2006. In this volume.

P. Poller and V. Tschernomas. Multimodal Fission and Media Design, 2006. In this volume.

T. Portele, S. Goronzy, M. Emele, A. Kellner, S. Torge, and J. te Vrugt. SmartKom-Home: The Interface to Home Entertainment, 2006. In this volume.

S. Rapp and D. Raddino. Trained Automatic Grapheme to Phoneme Conversion. In: *Proc. 9th Sony Research Forum (SRF'99)*, Tokyo, Japan, 1999. Sony Corporation Corporate Technology Dept.

S. Rapp, S. Torge, S. Goronzy, and R. Kompe. Dynamic Speech Interfaces. In: *Proc. Workshop Artificial Intelligence in Mobile Systems at the 14th ECAI (AIMS 2000)*, Berlin, Germany, 2000.

A. Schweitzer, N. Braunschweiler, G. Dogil, T. Klankert, B. Möbius, G. Möhler, E. Morais, B. Säuberlich, and M. Thomae. Multimodal Speech Synthesis, 2006. In this volume.

T. Sejnowski and C. Rosenberg. Parallel networks that learn to pronounce English text. *Complex Systems*, 1:145–168, 1987.

H. Strik and C. Cucchiarini. Modeling Pronunciation Vartiation for ASR: A Survey of the Literature. *Speech Communication*, 29:225–246, 1999.

V. Zeißler, J. Adelhardt, A. Batliner, C. Frank, E. Nöth, R.P. Shi, and H. Niemann. The Prosody Module, 2006. In this volume.

The Prosody Module

Viktor Zeißler, Johann Adelhardt, Anton Batliner, Carmen Frank, Elmar Nöth, Rui Ping Shi and Heinrich Niemann

Friedrich-Alexander Universität Erlangen-Nürnberg, Germany
{zeissler,adelhardt,batliner,frank,noeth,shi,
niemann}@informatik.uni-erlangen.de,

Summary. In multimodal dialogue systems, several input and output modalities are used for user interaction. The most important modality for *human computer interaction* is speech. Similar to human human interaction, it is necessary in the human computer interaction that the machine recognizes the spoken word chain in the user's utterance. For better communication with the user it is advantageous to recognize his internal emotional state because it is then possible to adapt the dialogue strategy to the situation in order to reduce, for example, anger or uncertainty of the user.

In the following sections we describe first the state of the art in emotion and user state recognition with the help of prosody. The next section describes the *prosody module*. After that we present the experiments and results for recognition of user states. We summarize our results in the last section.

1 The State of the Art

Prosody refers to the segments of speech larger than phonemes, e.g., syllables, words, phrases, and whole utterances. These segments are characterized with properties like pitch, loudness, duration, speaking rate, and pauses. The machine can analyze these properties and detect prosodic events, such as accents and phrase boundaries, as well as decide the mood or emotion in which a human expresses a certain utterance (Adelhardt et al., 2003). With the help of prosody it is consequently possible to get more knowledge about the user who is "talking" with the system, as has been, for instance, shown in some studies (Dellaert et al., 1996; Amir and Ron, 1998; Li and Zhao, 1998; Petrushin, 2000).

User states are an extension of the well-known term of emotion with some internal states of a human like, e.g., *"hesitant"*, that are important in the context of human computer interaction (HCI). This extension of emotion refers to the interaction of users with the system, for instance, if the user shows hesitance or uncertaincy because he does not know how the machine can help him. For details, see also Streit et al. (2006).

One problem with the recognition of user states is the difficulty of data collection. In most cases actors "create" emotions according to some certain scenario, but recognizers trained with these *actor data* are not applicable for emotion detection with naive speakers. An alternative method to collect data for training an emotional recognizer is the so-called Wizard-of-Oz experiment (WOZ, see Schiel and Türk (2006)).

In the research of emotion recognition through prosody, generally three base features are used: fundamental frequency (F0 or pitch), duration, and energy. Furthermore, these base features can be combined with several other features. In Dellaert et al. (1996) actor data of five speakers are used to detect four emotions — joy, sadness, rage, and fear. The authors use several F0 features, speaking rate and statistics about the whole utterance. In Amir and Ron (1998) actor data of more than 20 persons are used to detect joy, anger, grief, fear, and disgust. The authors use, e.g., F0, energy and derived features based mainly on the whole utterance. In Li and Zhao (1998) joy, anger, fear, surprise, sadness, and neutral are detected with actor data of five speakers. The authors use, e.g., formants, F0, energy, and derived short-term features as well as several derived long-term features. In Petrushin (2000) data of 30 persons showing 5 emotions, joy, rage, sorrow, fear, and neutral, in their speech are classified. The authors use features like F0, energy, speaking rate, formants, and bandwidth of formants. In addition, minima, maxima, average, and regression of the features above are used.

Different classification techniques can be applied to emotion classification. In Dellaert et al. (1996) maximum likelihood Bayes classification, kernel regression, and k-nearest neighbor are used for classification. In Amir and Ron (1998) the authors use two methods for classification. The first method computes wordwise emotion scores averaged over the utterance, which are compared against each other to determine the emotion. The second method suggests framewise classification followed by the final decision based on majority voting of the frames in each emotional class. In Li and Zhao (1998) the choice of the features for classification is based on principal component analysis (PCA), while classification results from vector quantization, Gaussian mixture models, and *artificial neural networks* (ANN) are also used in Petrushin (2000).

Another contribution that is very important to our work stems from Huber (2002). The author uses wordwise as well as turnwise prosodic features and linguistic information for the classification of emotion (for the features, see Sect. 2.4 and Nöth et al. (2000); Kießling (1997)).

There are some other methods for emotion detection based on evaluation of linguistic information. One possibility is keyword spotting, where the utterance is checked against certain words (Lee et al., 2002; Arunachalam et al., 2001). Another method is the use of *part of speech* features (POS) introduced in Batliner et al. (1999).

In the SMARTKOM project, speech, gesture, and facial expressions are used for emotion recognition. In the further context of emotion recognition, there are several studies in the area of the term "affective computing," which has been established mainly by R. Picard. Affective computing combines several information channels to get the emotion of the user. An interesting introduction to this field is given in

Picard (1997). It covers, e.g., emotion in speech, facial expression, and physiological signals.

2 Module Description

The prosody module used in the SMARTKOM demonstrator is based on the Verbmobil prosody module described in detail in Batliner et al. (2000a) and Nöth et al. (2000). Compared to the Verbmobil version, several major changes have been made concerning both implementation and classification models. Since the SMARTKOM system provides a different approach for module communication (Herzog and Ndiaye, 2006), the module interface has been fully reimplemented. The classification core remains essentially the same, except for some minor changes, which increase the stability and performance of the module. All existing classification models for the recognition of prominent words, phrase boundaries, and questions have been retrained on the actual SMARTKOM WOZ dataset (Schiel and Türk, 2006). This makes it possible to achieve much better recognition results than those obtained with the old models on the current dataset (Sect. 3.1). Additionally, the user state classifier has been trained and integrated into the module.

In the following sections we first give a brief overview of the overall module structure and the data flow in the module (Sect. 2.1). The issues concerning the execution of the module in the SMARTKOM system and synchronization of input data streams are covered in Sect. 2.2 and 2.3, respectively. Afterwards the features used for classification are presented in Sect. 2.4, followed by the description of the prosodic classifiers in Sect. 2.5.

2.1 Architecture

The goal of the prosody module in SMARTKOM is to analyze the speech as a modality of the user input in order to detect the prosodic events as well as the most likely emotional state of the user (Streit et al., 2006). As shown in Fig. 1, the module has two main inputs: the speech signals from the *audio module* and the word lattices (*word hypothesis graphs*, WHGs) from the *speech recognizer*. After running through feature extraction and classification steps, the detected prosodic events are added to the original input WHG and the user state lattice is generated. In more detail, the subsequent processing steps are described below:

- *XML parser:* According to the SMARTKOM communication paradigm all data exchanges are done in the XML format. The incoming XML packets have to be parsed and filtered. Thus, we can check the data consistency, drop the irrelevant information, and convert the useful data to a compact internal representation.
- *Stream synchronization:* This component compares the time stamps of incoming packets to find the correct mapping between the WHGs and the speech data. It also ensures the amount of data is enough to trigger the feature extraction for a given WHG. For the detailed description see Sect. 2.3.

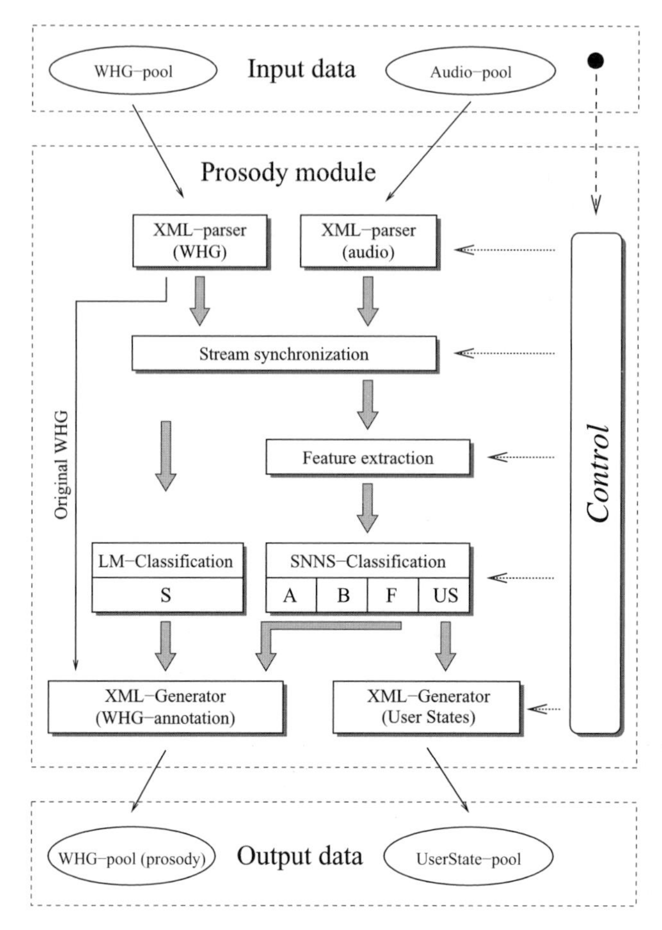

Fig. 1. Architecture of the prosody module

- *Feature extraction:* After the synchronization of the input data, a speech signal segment corresponding to each word in the WHG is located. The various F0, energy, and durational features of one segment are combined into a single feature vector to be used at the classification step.
- *ANN classification:* The ANN classification block consists of four independent classifiers used for the detection of:
  - phrase *A*ccents (A-labels),
  - phonetic phrase *B*oundaries (B-labels),
  - rising intonation for *Q*ueries (Q-labels),
  - *U*ser *S*tates (US-labels).

  For each word and label the classifiers generate a likelihood score representing a confidence of the recognition. The scores of each classifier are normalized to resemble the real probabilities (summed up to 1), though their distribution might be quite different.

- *LM classification:* The recognized words taken from the WHG are used here to detect *S*yntactic phrase boundaries (S-labels) with a *Language Model classifier* described in Batliner et al. (2000a); Schukat-Talamazzini (1995).
- *XML generators:* The output of classifiers is used to annotate the original WHG stored at the parsing step. Technically, we generate the XML structure only for the annotation part and paste it to the original lattice. This improves the performance of the module because disassembling of the complex WHG structure and reassembling of a new WHG from scratch can be avoided. Additionally, the user state labels are used to generate the user state lattice for the current turn. After generation, both lattices are sent to the output pools.
- *Control:* The control component determines the order of execution of all other parts and also handles the data transfer.

Apart from the main input pools shown in Fig. 1 the prosody module also gets data from the lexicon and the configuration pool. The processing of lexicon data makes it possible to update the internal static lexicon with new entries. The configuration data are used to set the internal module parameters from the GUI.

2.2 Execution

After being started by the SMARTKOM system the prosody module goes through several initialization steps including reading the configuration files, setting up the classifiers and loading statistical models. To be able to interact with the rest of the system, the module then subscribes the needed communication resources and installs the pool handlers.

The SMARTKOM communication system is based on the so-called *Pool Communication Architecture* (PCA) described in Herzog and Ndiaye (2006). There are several I/O FIFOs called communication pools which run in an asynchronous manner. If any module puts data into a pool, all modules that have subscribed this pool will be notified and can access the data. There are two possibilities to get notification events. The module can wait for these events by calling a special function or it can install handlers to be called when the notification event arrives. The prosody module handles the controlling events, such as exit or restart commands in the former and the pool notification events in the latter way.

When there are new data in one of the input pools subscribed by the module, the installed pool handler is called and the control component becomes active. First of all, the appropriate XML parser is called to get the data from the pool. When a new WHG arrives, the control component tries then to find the matching speech signal region (synchronization component). If it succeeds, the module registers itself as *processing* to the SMARTKOM system and proceeds to the further steps: feature extraction, classification and the data output.

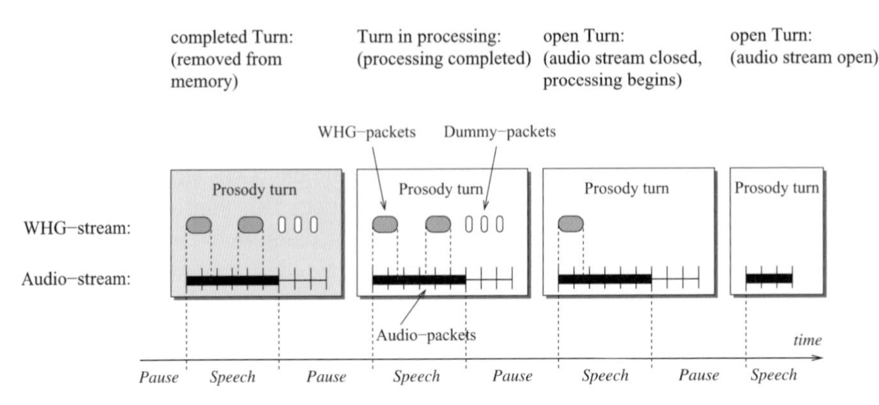

Fig. 2. Synchronization of audio and WHG streams in the prosody module

2.3 Data Synchronization

The essential part of the module managing the internal data flow is the stream synchronization component. Several problems must be solved to keep the data processing running smoothly from the technical point of view:

- memory allocation for the subsequently arriving data packets
- building the data structures needed for the following steps
- memory disallocation for the already processed data

Furthermore, some special features of the modules providing the input data should be considered. It concerns, for instance, the *silence detection* scheme. The audio module records the speech until silence is detected. The first and the last packet of the audio sequence between silences are labeled, so that such sequences can easily be identified. During silence dummy audio packets are generated in regular intervals. They are handled by the speech recognizer and can be ignored by the prosody module. The silence detection from the audio module regards only the signal energy and therefore only a robust detection of long pauses is possible. The speech recognizer has its own silence detection that also works for shorter and less distinct pauses. For every speech segment between such *short silences*, the speech recognizer generates a WHG and sends it to the output pool. Thus, there can be more than one WHG for a single audio sequence that needs to be properly aligned. Another factor to be considered is the processing of the dummy packets. The dummy packets from the audio module are transformed by the speech recognizer to dummy WHGs and sent to the WHG pool. These packets are needed for the robust detection of the user turns and dialogue management of SMARTKOM, therefore they should be passed through prosody module to the output pool "as is."

To reflect the SMARTKOM dialogue turns, the stream synchronization component works also with turn-based structures. Correspondingly, the *prosodic turn* is defined as a continuous speech sequence between the silences detected by the audio module. The speech data are collected from the incoming audio packets and stored in an array

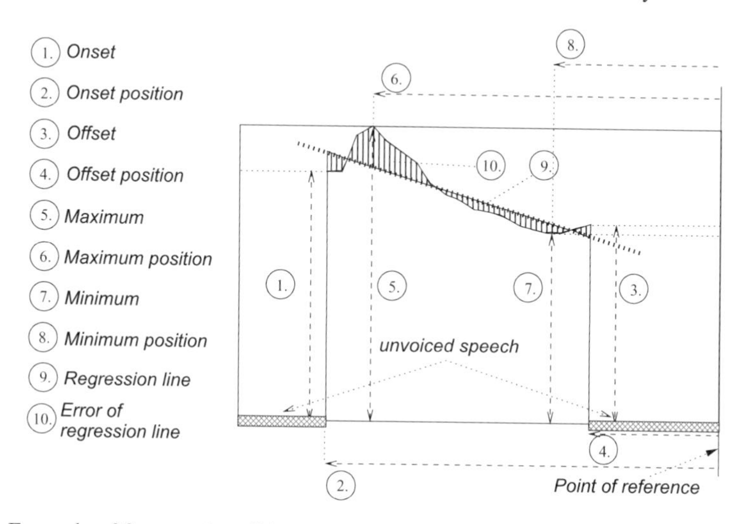

Fig. 3. Example of features describing the pitch contour (Batliner et al., 2000a)

in the turn structure. With the begin of a new audio sequence, the next turn structure is created to collect the incoming audio and WHG streams. When a new WHG packet arrives, the time stamps are compared to decide which turn the WHG belongs to. The processing scheme is illustrated in Fig. 2. Although this makes it necessary to hold open and manage several turns at a time, it has obviously the following advantages:

- The speech data belonging to one turn are stored in a single array and do not need to be pasted together from different packets. It also means smaller overhead at the feature extraction step.
- The memory management is done in the following manner: After the last WHG in a turn is processed, the complete turn, including all WHG, speech, and all intermediate data, is deleted and there is no need for costly garbage collection.

2.4 Feature Extraction

To recognize the prosodic characteristics of a speech signal we use the features describing such prosodic cues as energy, pitch, and duration, but also some linguistic information, in particularly POS features. The feature extraction is performed in two steps. First, we compute two *basic prosodic features*: integrated short time energy and pitch values framewise with a step size of 10 ms (Kießling, 1997). At the second step we take the WHG to determine the word boundaries and compute for each word a set of *structured prosodic features* (Nöth et al., 2000; Kießling, 1997). Every feature of the set has the following three configuration parameters:

- *Extraction interval:* To incorporate the word context information, the features can be computed on the subsequent words within a five-word interval beginning two words before and ending two words after the actual word. For example, we can take a two-word interval beginning one after the word in focus and ending two words after it. We use no more than a two-word range for such intervals. The larger intervals in our experiments brought no further improvement.

- *Basic prosodic cue:* Every feature can be computed using one of the following information cues: energy and pitch (computed at the first step), duration (taken from the WHG according to the chosen extraction interval), and linguistic information (taken from the dictionary).
- *Feature type:* For every prosodic cue there are a number of different features that can be subdivided into two main groups: *contour describing* and *integrated* features. The features of the first group including onset/offset/maximum/minimum values and their positions, regression coefficient, and regression error are shown in Fig. 3. They can be applied to the pitch and some of them to the energy contour. The second group includes mean and normalized values of the energy and duration cues. To eliminate the influence of microprosodic and global factors, such as the intrinsic word duration and the speech rate (when applied to duration cue), we use twofold normalization, described in Zeißler et al. (2002).

To sum up, we compute a total set of 91 prosodic and 30 POS features, described in detail in Batliner et al. (2000b). We use different subsets for classification of different prosodic events. The total set of 121 features is used for the detection of phrase accents and phrase boundaries. With a subset of 25 F0 features we classify the rising intonation (queries). For the classification of the user states, we use only 91 prosodic features.

2.5 Classification

For the prosodic classification we use a *multilayer perceptron*, a special kind of artificial neural network (ANN). We normalize all the input features to the mean of zero and the variance of one. The normalization coefficients are estimated in advance on the whole training dataset. To find an optimal training configuration, we need to know the following parameters: network topology, training weight for the RPROP training algorithm (Riedmiller and Braun, 1993), and random seed for the initialization. In preliminary tests we found out that complex topologies with two or more hidden layers do not improve the results for our dataset in comparison to a simple perceptron with one hidden layer. Hence, we restrict the number of hidden layers to one and look only for the optimal number of nodes in the hidden layer. We then evaluate different combinations of these parameters and choose the configuration with the best result on the validation set.

As primary classification method we use the wordwise classification. For each word ω_i we compute a probability $P(s \mid \omega_i)$ to belong to one of the given user states s. The probability maximum then determines the classification result. Further, we use these probabilities to classify the whole utterance, assuming the conditional independence between word classification events (Huber et al., 1998). The utterance probabilities are computed according to the following equation:

Table 1. The classwise averaged (Cl) and total (Rr) recognition rates (in %) yielded for a two-class problem on phrase accents, phrase boundaries, and rising intonation

Classifiers	SMARTKOM				Verbmobil			
Tested on	SK-train		SK-test		SK-test		Verbmobil	
	CL	RR	CL	RR	CL	RR	CL	RR
Prominent words	81.2	81.0	77.1	77.0	72.8	72.8	79.2	78.4
Phrase boundaries	90.7	89.8	88.5	88.6	82.2	82.7	81.5	85.8
Rising intonation	75.6	72.0	79.2	66.4	57.4	81.1	60.1	89.7

$$P(s \mid \omega_1, \omega_2, \ldots, \omega_n) \approx \prod_{i=1}^{n} P(s \mid \omega_i) . \tag{1}$$

3 Experiments

3.1 Detection of Prosodic Events

Like in Verbmobil, in SMARTKOM we have three different prosodic events to detect: phrase accents, phrase boundaries, and rising intonation. We use the WOZ dataset, since it contains the labeling we need (Schiel and Türk, 2006). For each event we use a two-class mapping of the existing labels described in Batliner et al. (2000a): one class for the strong occurrence of the event and the other for its absence.

In our experiments we use a part of the WOZ dataset consisting of 85 public wizard sessions. It includes 67 minutes of speech data (the total session length is about 2.5 hours) collected from 45 speakers (20 m/25 w). We divide the data into training and test sets built from 770 and 265 sentences, respectively. The test set contains only speakers unseen in the training set. The recognition rates on both sets are given in columns 1 to 4 of Table 1. To ensure the newly trained networks have a better performance than those of the old Verbmobil classifiers, we also tested the Verbmobil networks on the test data. The results are given in columns 5 and 6 of Table 1 (For comparison, the results of the Verbmobil networks on the Verbmobil dataset are given in columns 7 and 8).

Comparing the results on test and training sets we see small (but significant) differences. Nonetheless, we conclude that the trained classifier has rather good generalization capability for the recognition of prosodic events on the SMARTKOM data. On the other hand, we observe rather big differences if comparing it with the results of Verbmobil classifiers, especially in the detection of rising intonation. It illustrates the necessity to retrain the old classifiers on the actual dataset.

3.2 Detection of User States on the WOZ Data

For the detection of user states on the WOZ data we use the same subset of 85 sessions as described above. There was no special *prosodic* labeling of user states, thus we used a so-called *holistic* labeling based on the overall impression of the labeler,

Table 2. Original labels, used class mappings and the number of words in each class

<table>
<tr><th colspan="2" rowspan="2">Original Labels</th><th colspan="5">Mappings</th></tr>
<tr><th>7 Classes</th><th>5 Classes</th><th>4 Classes</th><th>3 Classes</th><th>2 Classes</th></tr>
<tr><td>Joyful (strong)</td><td>113</td><td>113</td><td rowspan="2">805</td><td colspan="2" rowspan="3">884</td><td rowspan="6">11,491</td></tr>
<tr><td>Joyful (weak)</td><td>692</td><td>692</td></tr>
<tr><td>Surprised</td><td>79</td><td colspan="2">79</td></tr>
<tr><td>Neutral</td><td>9236</td><td colspan="4">9236</td></tr>
<tr><td>Pondering</td><td>848</td><td rowspan="2">1371</td><td colspan="2" rowspan="2">1371</td><td rowspan="4">2030</td></tr>
<tr><td>Hesitant</td><td>523</td></tr>
<tr><td>Angry (weak)</td><td>483</td><td>483</td><td colspan="2" rowspan="2">659</td><td rowspan="2">659</td></tr>
<tr><td>Angry (strong)</td><td>176</td><td>176</td></tr>
</table>

taking both speech and facial expression into account: During the annotation the labeler could observe the test person and hear the recorded utterances simultaneously. Then he was supposed to detect the changes in the user state of the test person and assign the corresponding label to the actual time interval (Schiel and Türk, 2006). The disadvantage of this method is obvious: If the observed user state change is judged only after the user's facial expression, gesture, or/and the situational context, there will be no prosodic equivalent of the assigned label, and this fact will in turn have negative influence on the classifier training and recognition results.

After the preprocessing of the annotations, we have one of the different user state labels assigned to each word. Thus, we apply only a word-based classifier as described in Sect. 2.5. In our experiments we used different mappings to get 2-, 3-, 4-, 5-, and 7-class problems. The original labels, the used class mappings, and the number of words in each class are shown in Table 2. The main problem for the automatic classification is a strong unequal distribution of the different classes. In case we want to train a classifier for a two-class problem, we have 659 words labeled as angry vs. 11,491 words with other labels. To get stable recognition results under such unfavorable conditions we conduct only *leave-one-speaker-out* (LOSO) tests on the whole dataset with neural networks (ANN) and *leave-one-out* (LOO) tests with *linear discriminant analysis* (LDA) from the SPSS software (Norusis, 1998). The results are shown in two last columns of Table 3.

Above we pointed out that there was only a holistic labeling available, taking into account both speech and facial expression at the same time: *u*ser *s*tates, *h*olistic (USH). In a second pass, other annotators labeled all nonneutral user states based purely on facial expressions: *u*ser *s*tates *f*acial (USF); details can be found in Batliner et al. (2003). For further classification experiments, we divided the whole database into two disjunct subsets, based on the agreement of USH and USF for the four "basic" user states *positive*, *neutral*, *hesitant*, and *angry*: *agreeing* (USH = USF) and *not agreeing* (USH $\neq$ USF). LOO classification was this time done with LDA and decision trees (J48) (Witten and Frank, 2000), and all 91 prosodic and 30 POS features. Recognition rates for *not agreeing* cases were lower, and for *agreeing* cases higher than for all cases taken together. For example, for seven classes, LDA yields a

Table 3. Classwise averaged recognition rates (in %) yielded in Lo(S)O tests on the WOZ data for five different class mappings. The results of two different classifiers, ANN and LDA, are given in last two columns

No. of classes	User states							Results	
								ANN	LDA
7	Joyful (strong)	Joyful (weak)	Surprised	Neutral	Hesitant	Angry (strong)	Angry (weak)	30.8	26.8
5	Joyful		Surprised	Neutral	Hesitant	Angry		36.3	34.2
4	Positive			Neutral	Hesitant	Angry		34.5	39.1
3	Positive			Neutral	Problem			42.7	45.5
2	Not angry					Angry		66.8	61.8

classwise averaged recognition rate of 29.7%, J48 47.5; for the two classes *not angry* vs. *angry*, the figures are 68.9% for LDA and 76.5% for J48. These results indicate a mutual reinforcment of the two modalities, which, in turn, leads to a more stable annotation if holistic and facial impressions converge.

3.3 Detection of user states on the Multimodal Emogram (MMEG) data

In experiments described below we use the data from the MMEG collection introduced in Adelhardt et al. (2006). In this dataset we have sentences labeled with one of four user states: *joyful*, *neutral*, *helpless*, and *angry* (Streit et al., 2006). From all collected speech data with good quality we choose randomly 4292 sentences for the training set and 556 for the test set (*test1*). Notice that we have here approximately 60% of sentences with the same syntactic structure in both training and test sets that were definitely dependent on the used user state. For example, all sentences built after the pattern *"I don't like <some TV genre>"* belong to the user state *angry*. To ensure that we really recognize user states and not the different syntactic structures of the sentences, we additionally select 1817 sentences for another test set (*test2*). The second test set contains only sentences with a syntactic structure independent of the labeled user state, for instance, sentences consisting of an isolated name of a TV genre or special expressions (Adelhardt et al., 2006). Thus, for this set, the syntactic structure of the sentence could not be used as a key to a proper recognition.

To train the classifier we had first to find out the optimal feature set. We tried different subset combinations of F0-based features, all prosodic features (91 set), and linguistic POS features in both context-dependent and independent form. In context-independent feature sets we used only the features computed for the word in question. For all configurations we trained the neural networks and tested them on the test sets (see the results of *test1* vs. *test2* in Table 4). The classwise averaged recognition rates for the four class problems (in percent) are shown in Table 4. We computed both word based and sentence-based recognition rates as indicated in the second column.

In Table 4 we notice that the POS features yield remarkable improvement only on the *test1* set; the results on the *test2* set get worse (see columns 3 and 5). That means they reflect to a great extent the sentence structure and therefore cannot be properly

Table 4. Classwise averaged recognition rates (in %) for four user states classified with ANN using five different feature sets. For each test set the wordwise and the sentencewise recognition rates are given

Test set	Type	Without context			With context	
		F0 feat. 12 feat.	All pros. 29 feat.	Pros.+POS 35 feat.	All pros. 91 feat.	Pros.+POS 121 feat.
Test1	Word	44.8	61.0	65.7	**72.1**	86.6
	Sentence	54.0	64.5	72.1	**75.4**	81.4
Test2	Word	36.9	46.8	46.5	**54.5**	52.7
	Sentence	39.8	47.5	48.1	**55.1**	54.2

Table 5. Confusion matrix of user state recognition with the ANN on the best feature set (91 features) using LOSO. Both the wordwise and sentencewise results are computed (in %)

Reference	Wordwise				Sentencewise			
User state	Neutral	Joyful	Angry	Hesitant	Neutral	Joyful	Angry	Hesitant
Neutral	**68.3**	12.5	12.6	6.6	**67.7**	12.0	16.5	3.8
Joyful	13.8	**65.8**	10.6	9.8	14.3	**66.4**	13.8	5.5
Angry	14.5	11.3	**64.7**	9.5	13.9	9.2	**70.8**	6.1
Hesitant	10.0	10.8	9.9	**69.3**	10.0	6.5	15.3	**68.2**

applied for the user state recognition in our case due to the construction of the corpus. The best results were achieved with the 91 prosodic feature set (75.4% *test1*, 55.1% *test2*, sentencewise). To verify these results with the speaker-independent tests, we additionally conducted a LOSO test using the 91 feature set. Here we achieved an average recognition rate of 67.0% wordwise and 68.3% sentencewise. The confusion matrix of this test is given in Table 5.

4 Conclusion

The prosody module used in the SmartKom demonstrator is based on the Verbmobil prosody module described in detail in Batliner et al. (2000a) and is extended with several new features. The module detects phrase accents, phrase boundaries, and rising intonation (prosodic marked queries) and includes new features concerning module communication, data synchronization, and a classifier for user state recognition. User state classification is done in two steps. In the first step we use word-based classification to compute a probability to assign one of several user states to each word. In the second step we process the probability of the whole utterance to decide one of the several classes.

For classification of prosodic events with the test set, we obtain a classwise averaged recognition rate of 77.1% for phrase accents, 88.5% for phrase boundaries, and 79.2% for rising intonation. For user state classification we collected our own data, due to the lack of training samples in WOZ data. Regarding the recognition of the user states, we noticed that POS features yield a remarkable improvement only for the *test1* set containing sentences with the same syntactic structure as the training set. For the *test2* set, which contains only one-word sentences and special expressions, POS features have worsened rather than improved the results. Because of the construction of the MMEG database, the POS features reflect to a great extent the structure of sentence and thus cannot be properly applied to the user state recognition in this case. In a speaker-independent test for user state classification without POS features we achieved an average recognition rate of 67.0% wordwise and 68.3% sentencewise.

References

J. Adelhardt, C. Frank, E. Nöth, R.P. Shi, V. Zeißler, and H. Niemann. Multimodal Emogram, Data Collection and Presentation, 2006. In this volume.

J. Adelhardt, R.P. Shi, C. Frank, V. Zeißler, A. Batliner, E. Nöth, and H. Niemann. Multimodal User State Recognition in a Modern Dialogue System. In: *Proc. 26th German Conference on Artificial Intelligence (KI 03)*, pp. 591–605, Berlin Heidelberg New York, 2003. Springer.

N. Amir and S. Ron. Towards an Automatic Classification of Emotions in Speech. In: *Proc. ICSLP-98*, vol. 3, pp. 555–558, Sydney, Australia, 1998.

S. Arunachalam, D. Gould, E. Andersen, D. Byrd, and S. Narayanan. Politeness and Frustration Language in Child-Machine Interactions. In: *Proc. EUROSPEECH-01*, pp. 2675–2678, Aalborg, Denmark, September 2001.

A. Batliner, A. Buckow, H. Niemann, E. Nöth, and V. Warnke. The Prosody Module. In: W. Wahlster (ed.), *Verbmobil: Foundations of Speech-to-Speech Translation*, pp. 106–121, Berlin Heidelberg New York, 2000a. Springer.

A. Batliner, R. Huber, H. Niemann, E. Nöth, J. Spilker, and K. Fischer. The Recognition of Emotion. In: W. Wahlster (ed.), *Verbmobil: Foundations of Speech-to-Speech Translation*, pp. 122–130, Berlin Heidelberg New York, 2000b. Springer.

A. Batliner, M. Nutt, V. Warnke, E. Nöth, J. Buckow, R. Huber, and H. Niemann. Automatic Annotation and Classification of Phrase Accents in Spontaneous Speech. In: *Proc. EUROSPEECH-99*, vol. 1, pp. 519–522, Budapest, Hungary, 1999.

A. Batliner, V. Zeißler, C. Frank, J. Adelhardt, R.P. Shi, E. Nöth, and H. Niemann. We Are Not Amused – But How Do You Know? User States in a Multi-Modal Dialogue System. In: *Proc. EUROSPEECH-03*, vol. 1, pp. 733–736, Geneva, Switzerland, 2003.

F. Dellaert, T. Polzin, and A. Waibel. Recognizing Emotion in Speech. In: *Proc. ICSLP-96*, vol. 3, pp. 1970–1973, Philadelphia, PA, 1996.

G. Herzog and A. Ndiaye. Building Multimodal Dialogue Applications: System Integration in SmartKom, 2006. In this volume.

R. Huber. *Prosodisch-linguistische Klassifikation von Emotion*, vol. 8 of *Studien zur Mustererkennung*. Logos, Berlin, Germany, 2002.

R. Huber, E. Nöth, A. Batliner, A. Buckow, V. Warnke, and H. Niemann. You BEEP Machine – Emotion in Automatic Speech Understanding Systems. In: *TSD98*, pp. 223–228, Brno, Czech Republic, 1998.

A. Kießling. *Extraktion und Klassifikation prosodischer Merkmale in der automatischen Sprachverarbeitung*. Shaker, Aachen, Germany, 1997.

C.M. Lee, S.S. Narayanan, and R. Pieraccini. Combining Acoustic And Language Information For Emotion Recognition. In: *Proc. ICSLP-2002*, pp. 873–876, Denver, CO, 2002.

Y. Li and Y. Zhao. Recognizing Emotions in Speech Using Short-Term and Long-Term Features. In: *Proc. ICSLP-98*, vol. 6, pp. 2255–2258, Sydney, Australia, 1998.

M.J. Norusis. *SPSS 8.0 Guide to Data Analysis*. Prentice Hall, Upper Saddle River, NJ, 1998.

E. Nöth, A. Batliner, A. Kießling, R. Kompe, and H. Niemann. Verbmobil: The Use of Prosody in the Linguistic Components of a Speech Understanding System. *IEEE Transactions on Speech and Audio Processing*, 8(5):519–532, 2000.

V.A. Petrushin. Emotion Recognition in Speech Signal: Experimental Study, Development, and Application. In: *Proc. ICSLP-2000*, vol. IV, pp. 222–225, Beijing, China, 2000.

R.W. Picard (ed.). *Affective Computing*. MIT Press, Cambridge, MA, 1997.

M. Riedmiller and H. Braun. A Direct Adaptive Method for Faster Backpropagation Learning: The RPROP Algorithm. In: *Proc. IEEE Intl. Conf. on Neural Networks*, pp. 586–591, San Francisco, CA, 1993.

F. Schiel and U. Türk. Wizard-of-Oz Recordings, 2006. In this volume.

E.G. Schukat-Talamazzini. *Automatische Spracherkennung – Grundlagen, statistische Modelle und effiziente Algorithmen*. Vieweg, Braunschweig, Germany, 1995.

M. Streit, A. Batliner, and T. Portele. Emotion Analysis and Emotion Handling Subdialogs, 2006. In this volume.

I.H. Witten and E. Frank. *Data Mining – Practical Machine Learning Tools and Techniques With Java Implementations*. Morgan Kaufmann, San Francisco, CA, 2000.

V. Zeißler, E. Nöth, and G. Stemmer. Parametrische Modellierung von Dauer und Energie prosodischer Einheiten. In: *Proc. KONVENS 2002*, pp. 177–183, Saarbruecken, Germany, 2002.

The Sense of Vision: Gestures and Real Objects

Jens Racky, Michael Lützeler, and Hans Röttger

Siemens AG, Human Computer Interaction, München, Germany
{Jens.Racky,Michael.Luetzeler,Hans.Roettger}@siemens.com

Summary. Natural human communication is based on visual and acoustical signals. Thus a technical system needs the same senses to allow intuitive interaction. The acoustical interface has been described in earlier chapters. In this chapter we concentrate on the perception of conscious visual utterances, for example, hand gestures and handling of dedicated (real) objects, e.g., paper documents.

1 Gesture Recognition

The gesture recognition modules supply the SMARTKOM system with video-based interaction capabilities. The users pointing gestures are captured by a downward looking video camera. Based on the well-known Siemens SiViT technology (Maggioni and Röttger, 1999) the position of the user's fingertips is determined and transformed to screen coordinates. In a second step the individual measurements are aggregated to gesture strokes that provide deeper insight into the user's intentions.

1.1 Introduction

Gestures play an important role in face-to-face interaction between humans. In conjunction with speech it forms the major part of information transfer. Gestures are an easy to use and natural way of interaction. Additionally, pointing gestures and speech disambiguate each other. A speech phrase "I want this" and an appropriate pointing gesture provide a very efficient, multimodal approach to information transfer.

Capturing pointing gestures by a video-based interaction system can provide an intuitive and toolsfree input modality. To meet this goal several requirements have to be met:

- **Real time:** The evaluation of the video input must be performed in real time to give the user direct feedback of the recognized intention.
- **Robust:** Recognition has to be be robust, especially under video projection, e.g., when the graphical user interface is projected by a beamer.

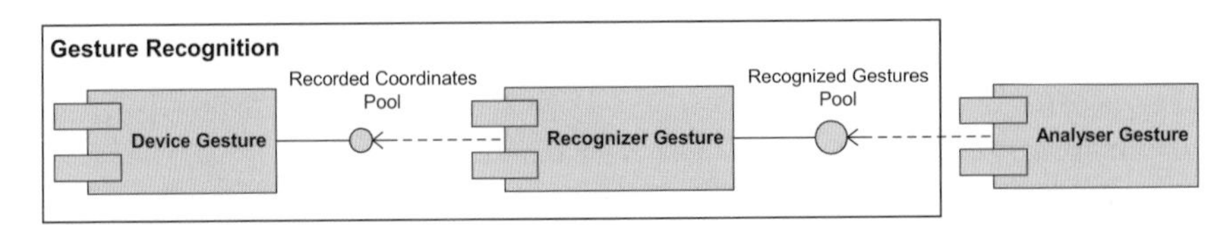

Fig. 1. Gesture recognition modules

- **Intuitive:** The system has to be adequate to the natural gesture vocabulary of the users, e.g., the users should be able to use their natural hand position and not be required to learn any specific hand position.

1.2 System Integration

A module for capturing pointing gestures can be seen as an abstract pointing device, similar to a computer mouse, a pen or a touch screen. These devices supply streams of single-coordinate readings. Individual measurements are then aggregated to gesture stokes to provide higher levels information. To treat different input devices, used in the SMARTKOM scenarios (`device.pen`, `device.gesture`, `device.mouse`) alike, an abstract representation for pointing devices was defined. Each input device supplies coordinates to the pool `recorded.coordinates` (see Fig. 1). The gesture recognition module reads these coordinates and identifies gesture strokes within a given vocabulary. Recognized gesture strokes are stored in the pool `recognized.gestures`, which provides the interface to the gesture analysis module for media fusion. This module correlates the recognized gestures with the current content on the display (Engel and Pfleger, 2006).

1.3 Gesture Vocabulary Definition

Classical gesture recognition, e.g., as provided by the SiViT technology, determines static gestures like a selection, when the user points at some location for a certain time span. As a substantial enhancement in this contribution the notion of *dynamic gestures* is introduced. Dynamic gestures result from motions the user performs while expressing a pointing gesture. A typical example for a dynamic gesture is encircling, a group of objects in order to select them.

Besides *tarrying* over a displayed object for selection and *encircling* the Wizard-of-Oz experiments accompanying the SMARTKOM project show that users tend to strike through an object to select it. A *linear stroke* is also used as supporting gesture while reading or searching in lists.

Thus a gesture vocabulary meeting the requirements of proper adaption to the natural gestures performed by users and taking into account the capabilities of the input device can be established on the basis of three classes (*tarrying*, *linear stroke* and *encircling*) and a class of undefined gestures. This vocabulary can be modeled by these geometric primitives: point, line and ellipse.

1.4 Interfaces

To support the above-defined abstraction of pointing input devices and the introduced vocabulary proper XML structures have been defined.

A pointing device supplies `coordinateData` to the pool `recorded.coordinates`. In Fig. 2 an example of a coordinate lattice is shown. Multiple single-coordinate recordings (`coordinateEdge`) are joined into one `coordinateData` structure. The information of the position reading is augmented by the event type, if applicable the pressure supplied (e.g., for pen input) and the confidence, modeled as probability, that a pointing event has been successfully detected.

```
<coordinateData>
  <source>         gesture                    </source>
  <sampleRate>     25                     </sampleRate>
  <startTime>      2001-09-25T15:25:33.002  </startTime>
  <endTime>        2001-09-25T15:25:33.953    </endTime>
  <numberOfSamples> 25               </numberOfSamples>
  <coordinateEdge>
    <recordedTime> 2001-09-25T15:25:33.082 </recordedTime>
    <screenPosition>
      <x>                    0.604022                <x>
      <y>                    0.770749                </y>
    </screenPosition>
    <pressure>               0               </pressure>
    <type>                   pointing            </type>
    <probabilityPointing>    1     </probabilityPointing>
    <probabilityUndefined>   0    </probabilityUndefined>
    <buttonDown>             false         </buttonDown>
  </coordinateEdge>
  . . .
</coordinateData>
```

Fig. 2. Example recorded coordinates

The `recognizer.gesture` takes measurements from any of the sources given in Sect. 1.2 and assembles them to meaningful gesture strokes. The data structure for a gesture stoke consists of a number of coordinate readings and the gesture primitive from the vocabulary defined in Sect. 1.3. The primitive is coded in the <type> structure. For encircling gestures the elliptical model is approximated by a bounding box, which proved to be sufficient information for the media fusion in the `analyzer.Gesture`. Figure 3 shows an example of a gesture stoke lattice. The lattice may contain multiple elements of the `gestureStroke` structure.

1.5 Recognition Process

The recognition process is dominated by the sequential nature of the input data. For a timely response, a geometric gesture primitive should be reported as early

```
<gestureStrokeLattice>
  <gestureStrokeCount>  1                 </gestureStrokeCount>
  <gestureStroke>
    <startTime>         2001-09-25T15:25:33.372 </startTime>
    <endTime>           2001-09-25T15:25:33.953   </endTime>
    <positionCount>     16                     </positionCount>
    <position>
      <x>               0.593003                          </x>
      <y>               0.354864                          </y>
    </position>
   . . .
    <type>              encircling                     </type>
    <probability>       0.8                     </probability>
    <geometricPrimitive>
      <probability>     0.8                     </probability>
      <boundingBox>
        <topLeft>
          <x>           0.578558                          </x>
          <y>           0.289846                          </y>
        </topLeft>
        <bottomRight>
          <x>           0.899767                          </x>
          <y>           0.692392                          </y>
        </bottomRight>
        <variance>      1.56025e-007              </variance>
      </boundingBox>
    </geometricPrimitive>
  </gestureStroke>
</gestureStrokeLattice>
```

Fig. 3. Example gesture lattice

as possible and be extended if more measurements arrive that fit into the current geometric model. Two distinct modes of operation can be separated.

In the first mode no gesture stroke has been identified yet. Measurements are added to the hypothesis for the gesture primitives in parallel. A number of heuristics are applied to verify each hypothesis. They can be described for the three classes by:

- **Tarrying:** Are the singular measurement points spatially and temporally sufficiently close to each other and is there a sufficient number of data readings?
- **Linear-stroke:** Does a straight line approximation fit the given data well?
- **Encircling:** Can the stroke be approximated by fitting an ellipse to the data-points?

If one of these hypotheses has been verified, the gesture stroke is passed to the output pool. Afterwards the second mode is entered. There the question is: Does the new measurement extend the current gesture stroke or does it initiate a new stroke?

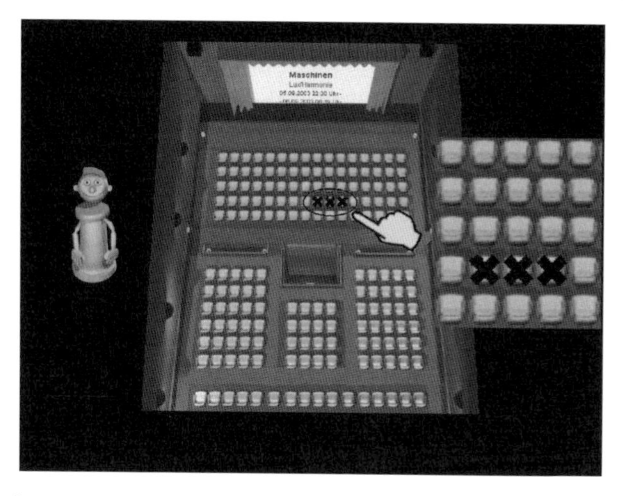

Fig. 4. Encircling gesture for cinema reservation

Experimental results within the SMARTKOM framework have demonstrated the robustness and suitability of the chosen approach. Gesture recognition is performed at 50 Hz on 320- by 240- resolution video images. Users accepted gesture input and used it intuitively. The gesture recognition modules are used in all three SMARTKOM scenarios (Public, Home, Mobil). Figure 4 shows the user's pointing gestures in the "cinema reservation" dialogue, when indicating which seats should be reserved. Another example for easy data input by pointing gestures is entering a phone number on a projected keypad for the public scenario.

2 Handling Real Objects

Besides video-based gesture recognition, yet another video-based approch is used to integrate real objects into the SMARTKOM world. By real objects we denote the class of those physical items by which a user may interact with SMARTKOM. Within the context of the SMARTKOM-Public case, these are objects whose visual appearance (or informational contents) shall be transmitted, e.g., *paper documents* (to be shown on the display, to be transmitted by e-mail or by fax) and also *hands* (for biometric authentication). In this section we give an overview of the functionalities of the module *device.realObjects* as well as implementation issues.

2.1 Interface to the SmartKom Testbed

The module provides three complementary interfaces:

1. The general XML-based interface by which it communicates with the other modules in the system

2. A private graphical user interface (GUI) for adapting the module to the current setup, e.g.,
   - Set input video amplifier parameters (brightness, contrast, saturation, hue)
   - Set threshold for contour detection
   - Set acquisition region for documents
   - Set *pan, tile, zoom* range for panoramic image acquisition
   - Have the recorded images rotated to allow for camera mounting flexibility
3. A GUI controlled by the testbed to set the current working parameters, e.g., activate (automatic) document detection, chose between simple and panoramic imaging, select binarization method for fax document generation

2.2 XML Interface

The XML interface[1] consists of three parts:

- The *control/command* interface
- The *State* interface (Sect. 2.2.2)
- The *Data* interface (Sect. 2.2.3)

An element called *documentID* is included in most data structures, as a reference to recorded real objects. This is necessary as recorded object may be referenced several times (e.g., send as e-mail, then send by fax) and the request–response mechanism of the SmartKom testbed is not suitable for this purpose. Implementation was done in C++ by creating a direct mapping of the XML types to C++ objects. In particular, XML derivation could be mapped to C++ derivation.

2.2.1 Command Interface

All commands to the module are derived from `RecognizerRealObjectsCommand`. This allows us to submit a sequence of commands to the module within one package. Furthermore, implementation of a command queue is straightforward when using C++ with STL containers.

- `RecognizerRealObjectsCommands` =
  recognizerRealObjectsCommand+ (`RecognizerRealObjectsCommand`)
  Abstract container for a sequence of control commands.
- `RecognizerRealObjectsCommand`
  Abstract control command for the device.realObjects recognizer.
- `GetRecognizerRealObjectsStatus`⟶`RecognizerRealObjectsCommand`
  Request state information.
- `ScanDocument`⟶`RecognizerRealObjectsCommand` =
  documentID (`NonNegativeInteger`),
  Scan real document into internal format.

[1] In the following section we use "⟶" to denote derivation, "+" to denote "one or more" and "{" "}" to denote choice of one element.

- `TranscribeImage`⟶`RecognizerRealObjectsCommand` =
  documentID (`NonNegativeInteger`),
  imageCodingType (`ImageCodingType`),
  imageSize (`ImageSize`),
  imageColorType (`ImageColorType`)

  Transcribe image with id *documentID* to representation of type *imageCodingType*. Optionally, the image's size and color type are converted.
- `ImageSize` =
  width (`NonNegativeInteger`),
  height (`NonNegativeInteger`)

  Size of the image in number of pixels.
- `ImageColorType`⟶`String` = { "Gray", "Color" }
  Color type of the image, either 24B-bit RGB or 8-bit gray.
- `ImageCodingType`⟶`String` = { "SFF", "PNG", "HCT" }
  Coding type of the image data, either Portable Network Graphic (PNG) or Structured Fax File (SFF) format. HCT is a proprietary format for transmitting the contour of a hand for biometric hand contour authentication (Grashey and Schuster, 2006).

2.2.2 State Interface

This interface is used to report the current module state, which is not only the state of the internal state machine (Sect. 2.3), but also information about the command received last.

- `TranscribeImage`⟶`RecognizerRealObjectsCommand` =
  documentID (`NonNegativeInteger`),
  recognizerState (`RecognizerState`),
  commandID (`NonNegativeInteger`),
  message (`RealObjectsMessage`),
  imageCodingType (`ImageCodingType`),
  imageSize (`ImageSize`),
  imageColorType (`ImageColorType`),
- `RecognizerState`⟶`String` = { "init", "scanning", "scanningHand", "scanned", "transcribing", "transcribed" }

  (State machine) state of the module (see Sect. 2.3).

2.2.3 Data Interface

This is the container for transmitting (binary) coded object data within the SmartKom system.

- `RealObjectRepresentation` =
  documentID (`NonNegativeInteger`),

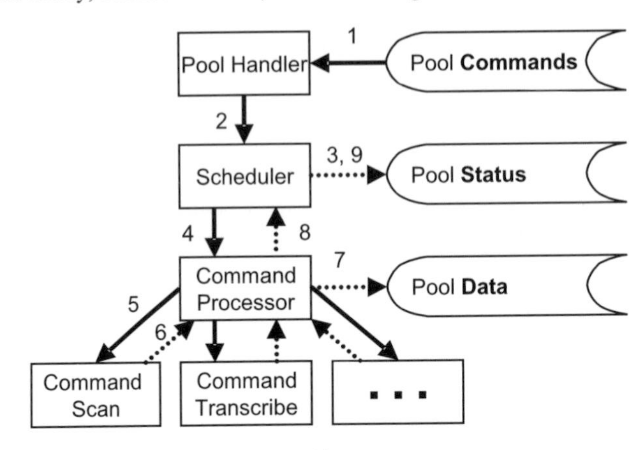

Fig. 5. General structure of Recognizer.Realdocument

This a base class for object representations. Currently only imagelike representations are used.

- `RealObjectImage` =

  imageCodingType (`ImageCodingType`),
  imageSize (`ImageSize`),
  imageColorType (`ImageColorType`),
  codedImageData (`Binary`),

  Binary (i.e., base-64 coded) image representation of the real object (according to *imageCodingType*). As all current coding formats are (originally) file based, the `Binary` data type was implemented with a stream interface.

2.3 General Module Structure

The general structure of the module is shown in Fig. 5. It communicates with the SMARTKOM system via three pools, i.e., a pool for commands, a pool for results (data) and a pool for telling its state. The latter ones are kept separately as not all of the module's subscribers need both kinds of information. The process flow on executing a command consists of the following steps. The step numbers correspond to those in Fig. 5, demonstrating a typical data flow:

1. The command is received from the pool by the pool handler.
2. The decoded XML information is a ready-to-use command object, which is added to the command execution queue of the command scheduler.
3. The scheduler sends a status message indicating the start of command processing to the status pool.
4. The scheduler computes the next internal command (i.e., target state) resulting from the external command and the current state, according to Fig. 6 (e.g., if a `ScanDocument` command has been received, the first internal command will be a `WaitForDocument`) and sends it to the processor for execution.

5. The processor then calls the concrete command implementation, which processes the data, e.g., converts a previously recorded image to the desired output format.
6. The result is written by the processor to the data pool, and
7. command completion is signaled to the scheduler,
8. which generates an appropriate status message for the status pool.
9. If the new state (now reached) is not stable, the next state is computed, continuing with step 4.

2.4 Object Detection

Detection of the presence of the real object to be acquired is a fundamental feature of the module. The implementation utilizes the same set up as gesture recognition, i.e., an IR camera–based approach resulting in a robust contour-based image segmentation.

A *real object* is considered to be present, if there is only one object, and it meets the following criteria:

- Contour and area exceed a given size/length.
- The contour does not hit the recording area's boundary.
- The objects dimensions do not vary too much for a given time.

2.5 Image Acquisition

Real objects are recoded via a PAL pan-tilt camera (Sony EVI-D30) connected by an S-VHS link and controlled through RS232. If not recording in panoramic mode (see next section) the resulting image to be encoded is just the "raw" camera image, optionally rotated to compensate a camera mounting angle.

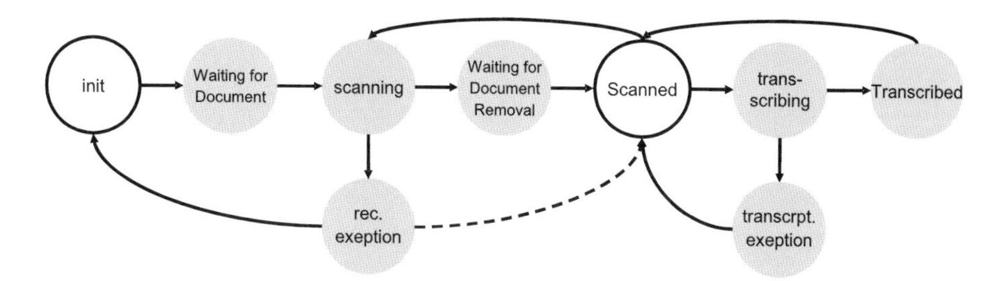

Fig. 6. Moore-type state machine of Recognizer.Realdocument. Stable states are denoted by a *nonfilled circle*; transient states by a *filled circle*

2.5.1 Panoramic Imaging

When operating in panoramic mode, the image to be transmitted is assembled from multiple images, recorded with a given overlap, as shown in Fig. 7.

The panoramic image recording consists of the following steps:

1. Recording of the image sequence. This is done by moving the camera to a given number of intermediate/interpolated positions between the limiting positions.
2. Estimation of the (camera) transformation between consecutive images of the sequence. We investigated two different approaches:
   - Tracking of a set of feature points followed by either computation of a translation vector or a linear transformation, as described, e.g., in Bhat et al. (2000): Let $\mathbf{x} = (x, y)$ be a pixel in an image I that translates to $\mathbf{x}' = (x + t_x, y + t_y)$ in the subsequent image J. The affine transformation between I and J is described by

     $$\mathbf{x}' = \begin{bmatrix} x' \\ y' \end{bmatrix} = \begin{bmatrix} ax + by + f \\ cx + dy + h \end{bmatrix},$$

     where $\mathbf{x}'$ is the coordinate of one (feature) point in image J, and $\mathbf{x}$ is its corresponding feature point in image I. The feature points are selected by the

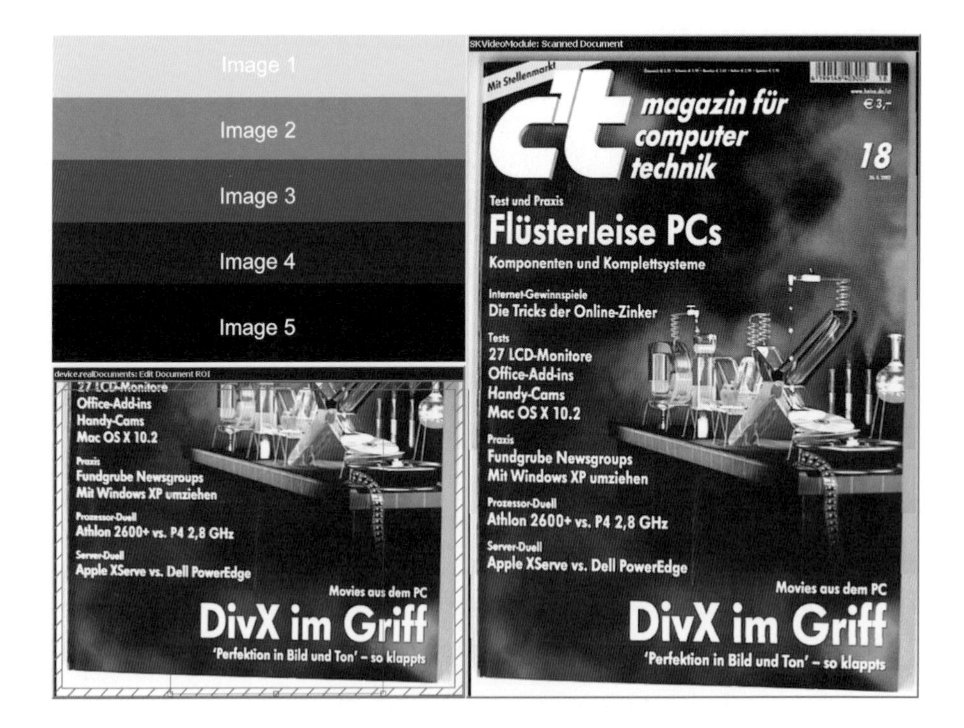

Fig. 7. Example of a (vertical) panoramic image creation. Recording of five overlapping images (scheme, *top left*) last (bottom) image (*bottom left*), resulting panoramic image (*right*)

algorithm from Shi and Tomasi (1994). Tracking is done by the algorithm described in Lucas and Kanade (1981) and Bouguet (1999), followed by an outlier removal based on variance and median filtering.
The approach showed good results and could be computed in real time if motion blur was low. Unfortunately the angular speed of the EVI-D30 depends on the current zoom, which leads to total motion blur at the required zoom level. The linear mapping was adequate for the given setup, but was very sensitive against erroneously detected feature points.
- Translation only: As the angular range within the SMARTKOM-Public setup is low, it was possible to use a correlation-based approach to robustly and sufficiently compute the translation. Figure 6 shows an exampled for this approach.

3. Stitching, i.e., assembling pieces of the recorded images to form the panoramic image. We distinguish two concepts: *Patching* uses exclusively the pixel information of one image to set a pixel value within the panoramic image and *blending* uses the pixel information from (some or all) overlapping pixels to compute the final value for the panoramic pixel. It has been found that for the SMARTKOM-Public setup (which utilizes an artificial illumination via the video projector) the first approach is sufficient, as can be seen from the example.

2.6 Preparation for Image Transmission

Image data can be transmitted via to ways: as an email attachment or as an ISDN fax. The main difference is that the chosen ISDN/fax data format provides only a black-and-white mode and requires the specific image width of exactly 1728 pixels.

2.6.1 Transmission by E-Mail

The Portable Network Graphics (PNG) format was chosen as a container to transport image data via e-mail. The system uses the free PNG library written in C (http://www.libpng.org/pub/png/) to implement the image coding. To enable data transport via the XML interface, a C++ stream wrapper was implemented. Images are always encoded as recorded, i.e., besides the optional camera rotation compensation no further processing takes place.

2.6.2 Transmission by ISDN–Fax

In SFF format images are run-length encoded black-and-white (b&w) bitmaps with a line width of 1728 (the format allows arbitrary line lengths, but it has been found that at least some G3 fax machines will not output the image if the line length differs from 1728). Our implementation features a stream-based interface, thus the implementation of the XML `Binary` type can be interfaced directly. As the recorded image does not meet these requirements, it is first converted to 8-bit gray level. Afterwards it is scaled up by an interpolating algorithm described in Rieg (1994), which has been stripped down to the gray-scale case. Finally, the color depth is reduced to b&w.

Color to B&W Conversion

First, some well-known dithering algorithms, like Floyd–Steinberg error diffusion were implemented. While the results are good as expected, the run-length encoding is not very efficient on the dithered data. Thus the resulting SFF file is very large, leading to very long transmission times via ISDN, which even causes some G3 fax devices to produce errors. In particular, the following algorithms were implemented, from Ulichney (1987):

- Floyd–Steinberg error diffusion
- Bayer's ordered dither with a 16×16 matrix
- Three sizes of 45-degree clustered-dot dither (3×3, 4×4, 8×8)

As a consequence, we investigated segmentation techniques, e.g., Otsu's algorithm, to perform the binarization. In particular, the following algorithms were implemented:

- Bipartite histogram partitioning minimizing intra- and interclass variance, according to Otsu (1979).
- Thresholding surface from a potential surface passing through the image at local maxima of the gradient image, according to Yanowitz and Bruckstein (1989).

Because of the present hardware setup (defined illumination), Otsu's algorithm shows good results, as long as the real object does not contain too much shading (e.g., a sheet of white paper with black text will be segmented well). As this is not always the case, we decided to use Yanowitz and Bruckstein's method by default. This proved to be a good compromise between the amount of data and the image quality. The methods described here have been applied successfully in the SMARTKOM-Public scenario to capture and transfer images of real objects and to enable hand-contour verification.

References

K.S. Bhat, S. Mahesh, and K. Pradeep. Motion Detection and Segmentation With Pan-Tilt Cameras Using Image Mosaics. *IEEE Int. Conf. on Multimedia and Expo*, 3:1577–1580, 2000.

J.Y. Bouguet. *Pyramidal Implementation of the Lucas Kanade Feature Tracker – Description of the Algorithm*. OpenCV Documentation, Intel Corporation, Microprocessor Research Lab, 1999.

R. Engel and N. Pfleger. Modality Fusion, 2006. In this volume.

S. Grashey and M. Schuster. Multiple Biometrics, 2006. In this volume.

B.D. Lucas and T. Kanade. An Iterative Image Registration Technique With an Application to Stereo Vision. In: *Proc. 7th IJCAI*, Vancouver, Canada, 1981.

C. Maggioni and H. Röttger. Virtual Touchscreen – A Novel User Interface Made of Light — Principles, Metaphors and Experiences. In: *Proc. 8th Intl. Conf. on Human-Computer Interaction*, pp. 301–305, Munich, Germany, 1999.

N. Otsu. A threshold selection method from gray level histograms. *IEEE Transactions Systems, Man and Cybernetics*, 9(1):62–66, 1979.

M. Rieg. Superelastisch verlustarmes Skalieren von Bitmap-Grafiken. *c't 11/94*, p. 302, 1994.

J. Shi and C. Tomasi. Good Features to Track. In: *Proc. Computer Vision and Pattern Recognition (CVPR'94)*, pp. 593–600, Seattle, WA, June 1994. IEEE Computer Society.

R. Ulichney. *Digital Halftoning*. MIT Press, Cambridge, MA, 1987.

S.D. Yanowitz and A.M. Bruckstein. A New Method for Image Segmentation. *Computer Vision, Graphics and Image Processing*, 46(1):82–95, April 1989.

The Facial Expression Module

Carmen Frank, Johann Adelhardt, Anton Batliner, Elmar Nöth, Rui Ping Shi, Viktor Zeißler, Heinrich Niemann

Friedrich-Alexander Universität Erlangen-Nürnberg, Germany
{frank,adelhardt,batliner,noeth,shi,zeissler,
niemann}@informatik.uni-erlangen.de,

Summary. In current dialogue systems the use of speech as an input modality is common. But this modality is only one of those human beings use. In human–human interaction people use gestures to point or facial expressions to show their moods as well. To give modern systems a chance to read information from all modalities used by humans, these systems must have multimodal user interfaces. The SMARTKOM system has such a multimodal interface that analyzes facial expression, speech and gesture simultaneously. Here we present the module that fulfills the task of facial expression analysis in order to identify the internal state of a user.

In the following we first describe the state of the art in emotion and user state recognition by analyzing facial expressions. Next, we describe the *facial expression recognition module*. After that we present the experiments and results for recognition of user states. We summarize our results in the last section.

1 State of the Art

The knowledge about a user's facial expression can be used, e.g., to recognize that a helpless user needs some help or to adapt presented information in case of confusion. The systems dealing with classification of facial images, which were developed over the past years, can be differentiated according to the following characteristics:

- sequence or single image
- recognizable classes
- used features
- used classifier

For the field of **single images**, the decision for a class is done with respect to one single image. Methods using single images can be found in Wiskott (1997); Chen and Huang (2002), Kumar and Poggio (2000), and Thomaz et al. (2001). **Images sequences** are used in Kaiser et al. (1998), Essa and Pentland (1997),and Müller et al. (2001). These methods and similar ones use the variation over time, for instance the optical flow.

Some systems cannot be associated with single-image or image-sequence methods. Samples of such hybrid systems can be found in Schwerdt et al. (2000), Lien et al. (1998a), and Tian et al. (2001). The first system classifies each frame of a sequence, and the results form a trajectory of changes in the face, which is classified using an eigenspace. Both other systems use a *neutral* image to find relative changes to the current one. This comparison allows users to determine the intensity of the facial expression.

Depending on the field of application, a facial expression recognition system must have a special **category set**, to which the faces should be classified. Very common is the concept of six basic emotions (happiness, sadness, anger, fear, disgust and surprise) used by Ekman (Lyons et al., 1999; Thomaz et al., 2001; Otsuka and Ohya, 1997). Subsets of these emotions are also used. Often there is one additional expression, "raised eyebrows" (Müller et al., 2001; Schwerdt et al., 2000; Essa and Pentland, 1995).

Besides these complex categories, there are several methods which use action units (AUs) (Ekman and Friesen, 1978) as a category set: Lien et al. (1998a), Tian et al. (2000), Tian et al. (2001), Otsuka and Ohya (1999), and Braathen et al. (2001). AUs are the smallest units in facial movements similar to phonemes for speech. The major interest to the AUs arises from their relevance in field of psychology. Because AUs are a standard technique for coding facial expressions, automatic coding would provide a much larger set of data for further experiments in psychology.

There are only a few papers concerning classification of user states that are not part of the basic emotions. Most systems that detect states necessary for a human computer interaction (attention, point of interest, stress, fatigue, etc.) use touch sensors. One example for the recognition of fatigue from an image sequence is shown in Li et al. (1999).

AUs are not only used as a category system, but also as **features**. Some systems using AUs as features are explained in Lien et al. (1998b), Otsuka and Ohya (1999), and Cohn et al. (1998).

For feature extraction to expression classification, one can distinguish between template-based and analytic methods. Template-based methods match a face model to a face. Then positions of facial features are extracted from this model, or muscle movements are derived. Template-based methods use graphs (Wiskott, 1997), models of facial features, e.g., the mouth (Oliver et al., 1997; Tian et al., 2000), or work with hand-segmented facial features. Essa explains (Essa and Pentland, 1997) a two-step system by matching a triangle mesh, using optical flow (Simoncelli, 1993) to the face and matching the result to a muscle model of the face.

In feature-based methods one can often find systems using wavelets. Two common types of wavelets are Haar wavelets (Kumar and Poggio, 2000) and Gabor wavelets (Lyons et al., 1999; Hu et al., 2002).

The systems for facial expression **classification** found in literature differ not only in the used features but also in the used classifier. They can be subdivided into model-, rule- and neural network–based methods. As model-based methods, principal component analysis, PCA (Schwerdt et al., 2000; Thomaz et al., 2001; Kirby and Sirovich, 1990) or support vector machines SVM (Braathen et al., 2001) are used. If the classification is done using image sequences and not only single images, the information of the single frames can be combined. For this task Hidden

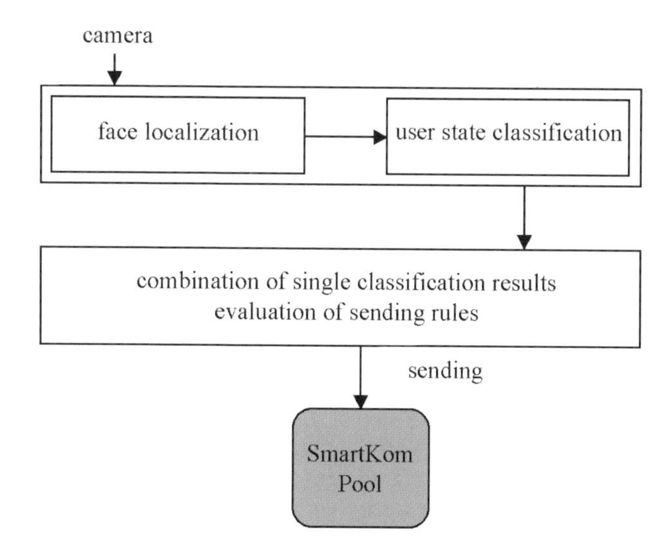

Fig. 1. The general architecture of the facial expression module

Markov Models are well known (Otsuka and Ohya, 1997; Müller et al., 2001; Oliver et al., 1997; Lien et al., 1998a). Rule-based methods include rules to map features to classes. In Chen and Huang (2002) the results from different classification stages are fused using handmade rules. An expert system using AUs for classifying facial expressions is explained in Pantic (2001). Methods based on neural nets can be found in Tian et al. (2000), Hu et al. (2002), and Zhang et al. (1998).

Furthermore, methods and systems for facial expression classification can be differentiated using facts like

- the use of handmade markers in the face
- the initialization of a sequence by hand
- the selection of features by hand
- person dependence or independence

The presented facial expression module uses a localization technique based on color because this is a feature common to all faces. As classifiers, eigenspaces PCA and SVM are used. Both allow the use of pixel values as features instead of segmented facial features such as eyebrows and lip corners. The motivation for not hand selecting features is the success of probabilistic, segmentation-free methods for spontaneous speech recognition. Facial expressions during a multimodal dialogue are equivalent to spontaneous utterances.

2 Module Description

The facial expression module consists of three main parts: **localization**, **classification**, and **combination and sending** of data (Fig. 1) in order to fulfill the task of

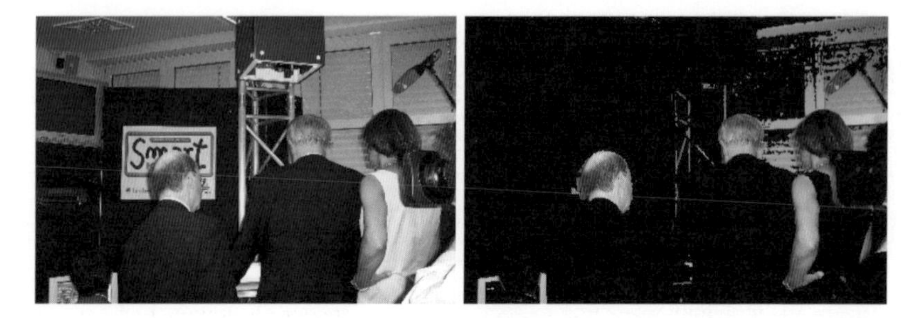

Fig. 2. Original image (*left*); result when skin color segmentation is done (*right*). The image parts with blue and green color shades disappear, but white shades and natural colors stay

facial expression recognition as part of a multimodal dialogue system. They are explained in more detail in the following sections.

The facial expression recognition does not try to detect the basic emotions like *disgust*, etc., but tries to detect user states. User states are all these internal states of a user that influence his or her interaction with the automatic system. For an information system similar to SmartKom , user states like *hesitant* or *angry* are suitable.

2.1 Localization

The localization of a face in the SmartKom environment has several constraints. There is at most one person using SmartKom , but this person is unknown to the system. The background is dynamic and is also unknown. It is not possible to use motion information to locate a person. For this reason we decided to use facial color as a first step in person localization.

Because skin color is not very clearly defined in the RGB color space, we use the YUV color space. In Fig. 2 one can see the results of a skin color segmentation. A nontypical image is used for better illustration: nonfacial skin color can be found as well. It is possible to eliminate "cold" colors. The resulting pixels are used as hypotheses for the further calculations.

To find the real position of the face, the facial expression module uses a holistic classifier and a filter. The holistic classifier can be a SVM, see Osuna et al. (1996) or PCA, see Turk and Pentland (1991a).

A problem when separating faces from "nonfaces" is that a nonface is not clearly defined. It is not possible to collect a set of all kinds of nonfaces. Therefore, the holistic classifiers sometimes select an area in an image as a face which is, for a human observer, clearly not a face. This problem can be reduced by a filter. We use a bank of circle frequency (CF) filters (Kawato and Tetsutani, 2002). The CF filter tries to find an area with alternation of light and dark values, which can be found near the between-eyes point (dark eyes, light forehead and nose). The effects of this filter are illustrated in Fig. 3.

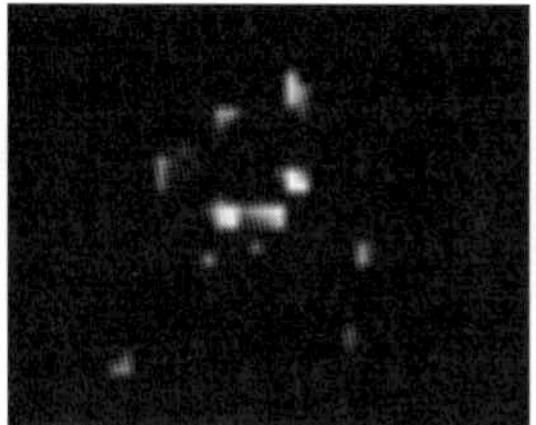
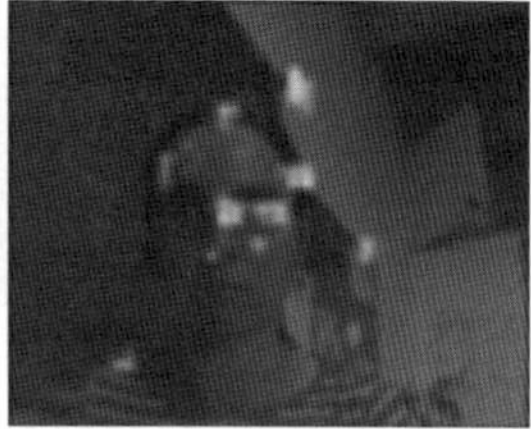

Fig. 3. The *left* face is the original face used to calculate the CF filter. The *middle* image shows the pixels with high weighting in light values. For a better illustration, both images are combined (right)

2.2 Classification

After the position of the face/the person is localized, the user state of this person can be classified. The classification is done by holistic classifiers as well. The method proposed by us for the recognition of facial expressions is a modification of a standard eigenspace classification for user identification. Eigenspace methods are well known in the field of face recognition (Turk and Pentland, 1991b; Yambor et al., 2000; Moghaddam and Pentland, 1994). In a standard face recognition system, one eigenspace for each person is created using different face images. The set of face images for each person is used to create a probability distribution or a representative for this person in the face space. Later, when classifying an image of an unknown person, this image is projected onto the face spaces. The probability distribution or representative that best matches the new image is chosen as the class in question.

To create an eigenspace with training images, a partial Karhunen–Loéve transformation, also called principal component analysis (PCA), is used. This transformation is a dimensionality reduction scheme that maximizes the scatter of all projected samples, using N sample images of a person $\{\mathbf{x}_1, \mathbf{x}_2, \ldots, \mathbf{x}_N\}$ with values in an n-dimensional feature space. Let μ be the mean image of all feature vectors. The total scatter matrix is then defined as

$$S_T = \sum_{k=1}^{N} (\mathbf{x}_k - \mu)(\mathbf{x}_k - \mu)^T \; . \tag{1}$$

In PCA, the optimal projection W_{opt} to a lower-dimensional subspace is chosen to maximize the determinant of the total scatter matrix of the projected samples

$$W_{opt} = \arg\max_{W} |W^T S_T W| = [w_1, w_2, \ldots, w_m] \; , \tag{2}$$

where $\{w_i | i = 1, 2, \ldots m\}$ is the set of n-dimensional eigenvectors of S_T corresponding to the set of decreasing eigenvalues. These eigenvectors have the same dimension as the input vectors and are referred to as “eigenfaces”. In Fig. 4 the first four eigenfaces of the *angry* eigenspace are shown. In the following we assume that high-order eigenvectors correspond to high eigenvalues. Therefore, high-order eigenvectors contain more relevant information.

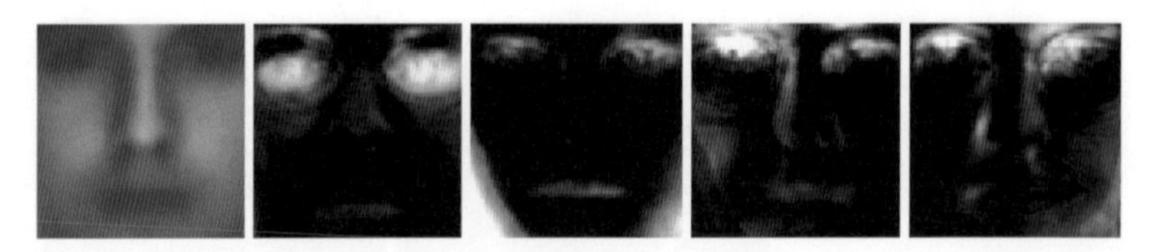

Fig. 4. The *left* image is the average image; the following images are the first four eigenvectors of the *angry* eigenspace (Frank and Nöth, 2001)

For face classification a new image is projected to each eigenspace and the eigenspace which best describes the input image is selected. This is accomplished by calculating the residual description error.

Imagine we have training sets F_κ of l samples $\mathbf{y}_i$ with similar characteristics for each class Ω_κ, $\kappa \in 1,\ldots,k$. There is different illumination, different face shape, etc., in each set F_κ. Reconstructing one image $\mathbf{y}_i$ with each of our eigenspaces results in k different samples $\mathbf{y}^\kappa$. The reconstructed images do not differ in characteristics like illumination, because this is modeled by each eigenspace. But they differ in the facial expression of specific regions, such as the mouth or the eyes area.

With a set of eigenspaces for each class Ω_κ we receive distances ν_κ of a test image $\mathbf{y}_i$ to each class

$$\nu_\kappa = ||\mathbf{y}_i - \mathbf{y}^\kappa||^2 \tag{3}$$

$$.k = \arg \min_{j \in 0\ldots k} \nu_\kappa . \tag{4}$$

An image is attributed to a class k with minimum distance as criterion.

In addition to the modeling of separate models for each user state, we use only a subset of the available pixels, i.e., only "significant" pixels; the extraction of this subset is described in Frank and Nöth (2003). This method creates a map of significancy of pixels and allows us to delete less significant ones. In Fig. 5 one can see such a map (leftmost image). Only pixels significant for facial expression (mouth and eye areas) are selected as features for the holistic classifier.

2.3 Combination and sending

The classification step produces results for every captured frame, but a facial expression has a longer duration than a single frame. Normally, it starts with a neutral expression, then the expression of the user state evolves and goes back to the neutral expression. Therefore, the single classification results are combined in order to send fewer messages to the SmartKom pool.

There are some rules to decide whether a message should be sent or not:

1. A new user state is detected.
2. No new user state has been detected for a long time.
3. The intensity of the recent user state changed significantly.

Fig. 5. The *leftmost* image shows the map of significance for a *smile–anger* classification. Different thresholds are used to create the other images (white pixels represent significant ones)

The second rule is necessary because the system needs to know that nothing happened, but the facial expression module is still "alive" and keeps watching the user. The values for *long time* and *significantly* used in rules 2 and 3 can be set using the GUI of the module. This is necessary because the amount of data processed by the facial expression module depends on the computational resources needed by other modules.

The sent data are a combination of single classification resulting from the last analyzed frames. The older a single classification result is, the lower is its influence to the overall result.

3 Experiments

For examination of the proposed classification methods we need a special dataset, including video sequences and corresponding user state labels. In the WOZ dataset of the SmartKom project (Schiel and Türk, 2006), one can only find a very small subset showing expressive user state. Therefore, we collected our own multimodal dataset with acted user states (Adelhardt et al., 2006). For the following experiments we used sequences of ten persons with, for a human being, recognizable expressions, in a leave-one-out manner. The whole face is used to create four eigenspaces for four facial expression classes. When a naive user tried to express an utterance in an angry manner, he/she often displayed neutral facial expressions before and after speaking, thus not every image of a sequence shows the same user state. The same is true for the other user states. The labeling of each single image is not possible due to the complexity. A consequence of that is a low recognition rate for neutral, because each class-specific dataset includes neutral faces. The classification of faces with internal movements, produced due to speaking, is very difficult. We achieve a low recognition rate of 32%. The confusion matrix is shown in Table 1. A problem is the user state *angry*. Anger a facial expression that is shown in many different ways by different users, whereas joy is consistently marked by risen lip corners.

The same procedure applied to a dataset (presented in Martinez and Benavente (1998)) of mugshots, showing four different facial expressions, yields 59%. The high recognition rate of the still images are due to reasons given above.

Table 1. Confusion matrix of user state recognition with facial expression data (%)

Reference user state	Results			
	Neutral	Joyful	Angry	Hesitant
Neutral	**7**	23	36	33
Joyful	5	**54**	22	20
Angry	4	62	**17**	16
Hesitant	6	12	35	**48**

One idea to solve the latter problem is the elimination of face regions influenced by speaking. To find out which part of a face is able to identify which facial expression, we performed experiments using only parts of a face for classification.

For this task it is necessary to know the position of eyes, nose and mouth and to have full-blown facial expression. Therefore, we use the database presented in Martinez and Benavente (1998). So the first-mentioned problem (not every frame shows an expression) does not influence our results and we can concentrate on the face regions.

To classify face regions, the faces are split into three parts (eye, nose, and moutharea). Samples of these three regions can be seen in Fig. 6. When only using one facial region for expression classification, the region centered at the mouth gives the best results (79%). The eye area achieves 63%, and the nose area 65%. The confusion matrices for all three face regions are shown in Table 2. It is noticeable that the different regions have different capabilities in indentifying different facial expressions. The mouth area gives really good results identifying *smile* and *shout* expressions, but *neutral* and *anger* are mixed up more often. However, these two classes can be told apart by the eye region. Astonishingly, the eye area had a good classification rate for identifying shouting. The reason for this is obvious when viewing images with the facial expression *shout* (Fig. 7). Normally, a human closes or pinches his/her eyes while shouting. This is one feature to separate *shout* from *open the mouth* (which is not part of the dataset). All other expressions are classified with nearly equal recognition rates.

So if one of the face regions is occluded or for other reasons is not available for classification, it is possible to recognize facial expressions. But this is not the case in all situations. While using an automatic dialogue system, a user has to speak. Speaking changes the mouth area but does not affect the user state. This makes the mouth

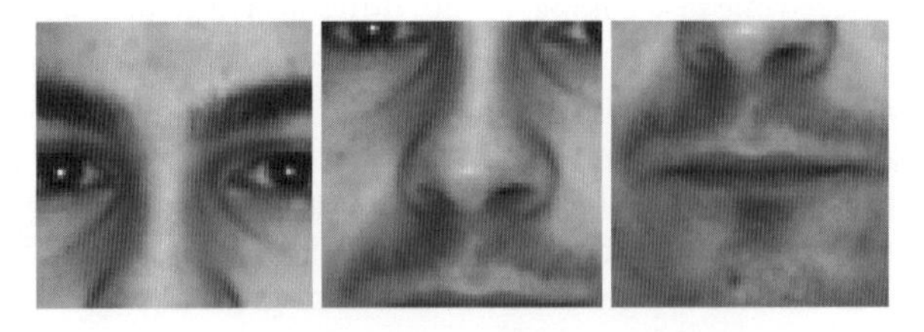

Fig. 6. The three parts of a face used for facial expression classification

Table 2. Confusion matrix (values in %) of facial expression recognition for the face areas. The best result in each line is in **boldface**: the eye area (**top**, overall recognition rate: 63%), the nose area (**middle**, overall recognition rate: 65%), the mouth area (**bottom**, overall recognition rate: 79%)

Eye area				
Reference facial expression	Results			
	smile	*neutral*	*anger*	*shout*
smile	**60**	13	19	8
neutral	14	**52**	29	5
anger	9	26	**59**	6
shout	5	8	4	**83**
Nose area				
Reference facial expression	Results			
	smile	*neutral*	*anger*	*shout*
smile	**77**	6	7	10
neutral	1	46	**47**	6
anger	2	25	**71**	2
shout	22	7	5	**66**
Mouth area				
Reference facial expression	Results			
	smile	*neutral*	*anger*	*shout*
smile	**97**	2	1	0
neutral	2	**55**	43	0
anger	0	33	**67**	0
shout	2	2	0	**96**

area classifier less trustworthy during speaking. A fusion of the classification results of other region classifiers may improve the overall result. Therefore, we examine the fusion of classification results.

3.1 Fusion

Each facial expression may be accentuated in another face region. We developed rules to combine results of the classifier for each face region dependent on user states. For the class *shout* it is clear (96%) that the mouth area is a trustworthy indicator.

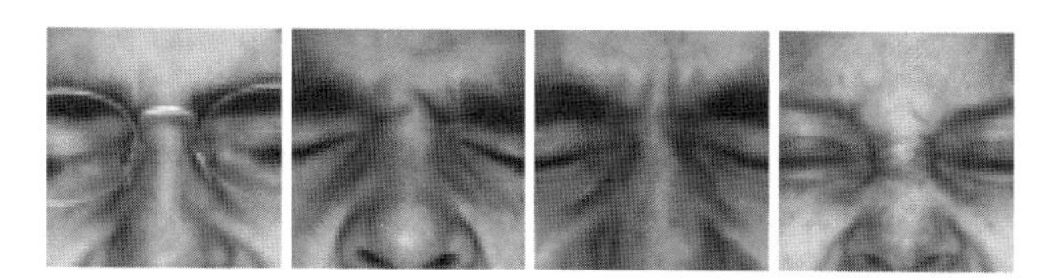

Fig. 7. Some samples of eye regions where the mouth concerning facial expression *shout* is clearly visible

The last section of Table 2 also shows that *smile* is classified correctly (with 97%). So the first two rules are:

Rule 1

IF	mouth area classifier classifies shout
THEN	overall result is shout

Rule 1

IF	mouth area classifier classifies smile
THEN	overall result is smile

These two rules are based on the low false alarm rate of the mouth area classifier for the classes *smile* and *shout* (4% for *smile* and 0% for *shout*). The classification rate is 96% and 97% for *shout* and *smile*, respectively. The high recognition rates combined with low false alarm rate makes the mouth area a trustworthy face region for classification of *smile* and *shout*.

During speaking it is not possible to use the mouth area for classification, because this face region has considerable changes due to moving lips. In this case the other regions of the face should be used for facial expression classification. The decision whether the person is speaking or not at the moment can be obtained from a speech recognition system.

The following experiments use only test images with classes not equal to *shout*. The eye area classifier achieves 57%, and the nose area classifier 64%. We introduce the following rules:

Rule 3

IF	both classifiers have same result
THEN	overall result is that result

Rule 4

IF	(one classifier says shout) AND (one classifier says smile)
THEN	overall result is smile

Rule 5

IF	(eyes classifier says neutral) AND (nose classifier says anger)
THEN	overall result is neutral

Rule 6

IF	not classified yet
THEN	overall result is that of nose classifier

The rules were developed using a subset of the total data. This subset was deleted from the test set in order to not influence recognition rates. Using these rules gives an improvement of the recognition rate of 10% (eye area), resp. 3% (nose area).

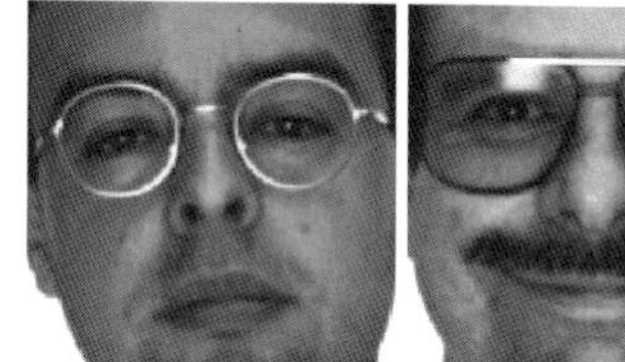

Fig. 8. Neither the eye nor the nose area classifier was able to classify these images. The displayed facial expressions are (from *left* to *right*): *neutral*, *smile*, *anger*. The first one was classified as *smile* (eye area), *shout* (nose area); The second as *anger* (both), the third as *smile* (eye area) and *neutral* (nose area)

The recognition rate of combining the eye area classifier and the nose area classifier achieves 67%. All classifiers missed the right decision for 23% of the images. It is not possible to solve the classification for these images. Some of these misclassified images are shown in Fig. 8. The confusion matrix for the fusion of results from eye and nose area classifiers is given in Table 3.

4 Conclusion

Facial expression recognition in an automatic dialogue system context has two additional problems compared to facial expression recognition known from the literature (Lien et al., 1998b; Essa and Pentland, 1995; Müller et al., 2001): the expressions are weak because users try to hide them, and users are speaking. The latter, a normal behavior for a dialogue, results in facial changes near the mouth not related to internal user states.

In cases where the mouth area could not be used for classification, a classification and fusion of local area classifiers allows the detection of the current user state. A disadvantage is that the combination results depend on the facial expression. Each face region displays each facial expression with different weights. The combination rules shown above obtain an improvement of 3% (resp., 10%) absolute to 67%.

Table 3. Confusion matrix (values in %) of fusion of results from eye and nose area classifiers; overall recognition rate is 67%

Reference	Results			
facial expression	*smile*	*neutral*	*anger*	*shout*
smile	**76**	8	11	5
neutral	9	**66**	19	6
anger	3	36	**58**	3

References

J. Adelhardt, C. Frank, E. Nöth, R.P. Shi, V. Zeißler, and H. Niemann. Multimodal Emogram, Data Collection and Presentation, 2006. In this volume.

B. Braathen, M.S. Bartlett, G. Littlewort, and J.R. Movellan. First Steps Towards Automatic Recognition of Spontaneous Facial Action Units. In: *Proc. ACM Conf. on Perceptual User Interfaces (PUI'03)*, pp. 319–242, Vancouver, Canada, 2001.

X.W. Chen and T. Huang. Facial Expression Recognition: A Clustering-Based Approach. *Pattern Recognition Letters*, 24:1295–1302, 2002.

J.F. Cohn, A.J. Zlochower, J.J. Lien, and T. Kanade. Feature-Point Tracking by Optical Flow Discriminates Subtle Differences in Facial Expressions. In: *Proc. Int. Conf. on Automatic Face and Gesture Recognition*, pp. 390–395, 1998.

P. Ekman and W.V. Friesen. *The Facial Action Coding System — A Technique for the Measurement of Facial Movement*. Consulting Psychologists Press, Palo Alto, CA, 1978.

I.A. Essa and A.P. Pentland. Facial Expression Recognition Using a Dynamic Model and Motion Energy. In: *Proc. 5th Int. Conf. on Computer Vision*, pp. 360–367, Cambridge, MA, 1995.

I.A. Essa and A.P. Pentland. Coding, Analysis, Interpretation, and Recognition of Facial Expressions. *IEEE Transactions on Pattern Analysis and Machine Intelligence (PAMI)*, 19(7):757–763, 1997.

C. Frank and E. Nöth. Automatic Pixel Selection for Optimizing Facial Expression Recognition Using Eigenfaces. In: *Pattern Recognition, Proc. 25rd DAGM Symposium*, pp. 378–385, Magdeburg, Germany, 2001.

C. Frank and E. Nöth. Optimizing Eigenfaces by Face Masks for Facial Expression Recognition. In: *Computer Analysis of Images and Patterns – CAIP 2003*, LNCS, pp. 1–13, Berlin Heidelberg New York, 2003. Springer.

T. Hu, L.C.D. Silva, and K. Sengupta. A Hybrid Approach of NN and HMM for Facial Emotion Classification. *Pattern Recognition Letters*, 23:1303–1310, 2002.

S. Kaiser, T. Wehrle, and S. Schmidt. Emotional Episodes, Facial Expressions, and Reported Feelings in Human-Computer Interactions. In: *Proc. 10th Conf. of the Int. Society for Research on Emotion*, pp. 82–86, Würzburg, Germany, 1998. ISRE Publications.

S. Kawato and N. Tetsutani. Real-Time Detection of Between-the-Eyes With a Circle Frequency Filter. In: *Proc. 5th Asian Conference on Computer Vision (ACCV)*, pp. 442–447, Melbourne, Australia, 2002.

M. Kirby and L. Sirovich. Application of the Karhunen–Loève Procedure for the Characterization of Human Faces. *TPAMI*, 12(1):103–108, 1990.

V.P. Kumar and T. Poggio. Learning-Based Approach to Real Time Tracking and Analysis of Faces. In: *Automatic Face and Gesture Recognition 2000*, pp. 96–101, 2000.

H. Li, A. Lundmark, and R. Forchheimer. Video Based Human Emotion Estimation. In: *Int. Workshop on Synthetic-Natural Hybrid Coding and Three Dimensional Imaging*, Sept. 1999.

J.J. Lien, T. Kanade, J.F. Cohn, and C.C. Li. A Multi-Method Approach for Discriminating Between Similar Facial Expressions, Including Expression Intensity Estimation. In: *Proc. Computer Vision and Pattern Recognition (CVPR'98)*, pp. 853–859, 1998a.

J.J. Lien, T. Kanade, J.F. Cohn, and C.C. Li. Automated Facial Expression Recognition Based on FACS Action Units. In: *Proc. Int. Conf. on Automatic Face and Gesture Recognition*, pp. 390–395, 1998b.

M.J. Lyons, L. Budynek, and S. Akamatsu. Automatic Classification of Single Facial Images. *IEEE Transactions on Pattern Analysis and Machine Intelligence (PAMI)*, 21(12):1357–1362, 1999.

A.M. Martinez and R. Benavente. The AR FAce DAtabase. Technical Report 24, Computer Vision Center, Purdue University, West Lafayette, IN, 1998.

B. Moghaddam and A.P. Pentland. Face Recognition Using View–Based and Modular Eigenspaces. In: *Automatic Systems for the Identification and Inspection of Humans, SPIE'94*, vol. 2257, 1994.

M. Müller, F. Wallhoff, F. Hülsken, and G. Rigoll. Facial Expression Recognition Using Pseudo 3-D Hidden Markov Models. In: *Pattern Recognition, Proc. 23rd DAGM Symposium*, pp. 291–297, Munich, Germany, 2001.

N. Oliver, A.P. Pentland, and F. Berard. LAFTER: Lips and Face Real Time Tracker. In: *Proc. Computer Vision and Pattern Recognition (CVPR'97)*, pp. 123–129, Puerto Rico, 1997.

E.E. Osuna, R. Freund, and F. Girosi. Support Vector Machines: Training and Application. Technical Report A. I. Memo No. 1602, Massachusetts Institute of Technology, Cambridge, MA, 1996.

T. Otsuka and J. Ohya. Recognizing Multiple Persons' Facial Expressions Using HMM Based on Automatic Extraction of Significant Frames From Image Sequences. In: *Int. Conf. on Image Processing (ICIP 1997)*, pp. 546–549, Oct 1997.

T. Otsuka and J. Ohya. Extracting Facial Motion Parameters by Tracking Feature Points. In: *Proc. 1st Int. Conf. AMCP'98*, LNCS 1554, pp. 433–444, Berlin Heidelberg New York, 1999. Springer.

M. Pantic. *Facial Expression Analysis by Computational Intelligence Techniques*. PhD thesis, Faculteit der Informatietechnology en Systemen, TU Delft, The Netherlands, 2001.

F. Schiel and U. Türk. Wizard-of-Oz Recordings, 2006. In this volume.

K. Schwerdt, D. Hall, and J.L. Crowley. Visual Recognition of Emotional States. In: *Proc. Int. Conf. on Multimodal Interfaces (ICMI'00)*, pp. 41–48, Beijing, China, 2000.

E.P. Simoncelli. *Distributed Analysis and Representation of Visual Motion*. PhD thesis, Massachusetts Institute of Technology, Department of Electrical Engineering and Computer Science, Cambridge, MA, 1993.

C.E. Thomaz, D.F. Gillies, and R.Q. Feitosa. Using Mixture Covariance Matrices to Improve Face and Facial Expression Recognitions. In: *Proc. 3rd Int. Conf. of Audio- and Video-Based Biometric Person Authentication AVBPA R01*, LNCS 2091, pp. 71–77, Berlin Heidelberg New York, June 2001. Springer.

Y. Tian, T. Kanade, and J.F. Cohn. Recognizing Action Units for Facial Expression Analysis. *IEEE Transactions on Pattern Analysis and Machine Intelligence (PAMI)*, 23(2):97–115, 2001.

Y.L. Tian, T. Kanade, and J.F. Cohn. Recognizing Lower Face Action Units for Facial Expression Analysis. In: *Proc. Int. Conf. on Automatic Face and Gesture Recognition*, pp. 484–490, 2000.

M. Turk and A.P. Pentland. Eigenfaces for Recognition. *Journal of Cognitive Neuroscience*, 3(1):71–86, 1991a.

M. Turk and A.P. Pentland. Face Recognition Using Eigenfaces. In: *Proc. Computer Vision and Pattern Recognition (CVPR'98)*, pp. 586–591, 1991b.

L. Wiskott. Phantom Faces for Face Analysis. *Proc. 7th Intern. Conf. on Computer Analysis of Images and Patterns, CAIP'97, Kiel*, 1296:480–487, 1997.

W.S. Yambor, B.A. Draper, and J.R. Beveridge. Analyzing PCA-Based Face Recognition Algorithms: Eigenvector Selection and Distance Measures. In: *2nd Workshop on Empirical Evaluation Methods in Computer Vision*, Dublin, Irland, 2000.

Z. Zhang, M. Lyon, M. Schuster, and S. Akamatsu. Comparison Between Geometry-Based and Gabor-Wavelets-Based Facial Expression Recognition Using Multi-Layer Perceptron. In: *Proc. 3rd Int. Conf. on Automatic Face and Gesture Recognition*, pp. 454–459, Nara, Japan, 1998.

Multiple Biometrics

Stephan Grashey and Matthias Schuster

Siemens AG, CT IC 5, München, Germany
{Stephan.Grashey,Matthias.Schuster}@siemens.com

Summary. Authentication is undoubtedly an important task for all systems providing a component responsible for the interaction between humans and computers. Traditional authentication techniques like PINs, passwords or ID cards show significant drawbacks: they might be forgotten, misplaced or lost, or even stolen, copied or forged. Biometrics use physical or behavioral characteristics to verify the identity of a person and thus overcome these problems. In this respect, biometric technology may be easier and more comfortable to use.

First, a short overview of biometric technology in general is given. Then, the main part of this chapter explains the biometric technologies integrated in the SMARTKOM system. In particular, a new approach to combine several biometrics in a multimodal device is presented. The performance of this proposed combination method is shown to be superior to that of the single biometric subsystems.

1 Biometrics

In the mobile and networked society, the use of many devices and services requires a preceding authentication. Today, authentication is mostly performed by checking possession (e.g., physical access using a key or ID card) or knowledge (e.g., access to a computer system using a password). For high-security transactions like banking, a combination of both (e.g., access to an ATM by credit card and PIN) is applied. The problem with possessions is that they can be lost, stolen, forgotten or misplaced. Knowledge bears the problem that it is difficult to remember the increasing number of passwords and PINs required. Many people write their PINs down or choose simple passwords that can easily be guessed by an adversary. As a result, the system cannot distinguish between an authorized person and an impostor who has acquired the necessary knowledge or possession.

Biometrics, on the other hand, is a method to provide genuine person identification. Biometric technology measures physical or behavioral characteristics to determine the identity of a person. The technology acts as a front end to a system that requires authentication before it can be accessed or used. While a few years ago biometrics were mainly used in science fiction, the technology is now mature and ready for application.

The assignment of an identity can be an identification or the verification of a claimed identity. In case of identification only the biometric feature is presented to the system and the system assigns the identity by comparing the feature to all references in a database (one-to-many search). For verification, the user first claims an identity and the system then checks the biometric feature against the reference assigned to this identity (one-to-one comparison).

Obviously, identification is very comfortable, since no user interaction besides the biometric feature acquisition is requested, but identification requires a precise biometrics and a full search through a central reference database. In contrast, verification systems are generally faster and more accurate than identification systems.

According to the respective biometric characteristics, biometric systems can broadly be classified into two categories: physiological and behavioral. Physiological biometrics verify fingerprints or hand contours, i.e., characteristics that practically do not change over a person's lifetime. Behavioral biometric systems analyze the way a person is acting. Examples for monitored behaviors are signature movements or voice-prints. It should be noted that behavioral biometrics also have a physiological component, e.g., the articulatory system in the case of speech.

In general, biometrics is a pattern recognition task. A biometric system consists of the components signal acquisition, feature extraction and classification (pattern or model match), see Fig. 1. For new users reference templates or models have to be generated first (so-called enrollment). The classification is subsequently performed by comparing the actual recorded sample (image or signal) with the stored reference template and calculating the matching difference. Identification or verification is achieved by setting a threshold for the matching difference to discriminate between accepted and rejected trials.

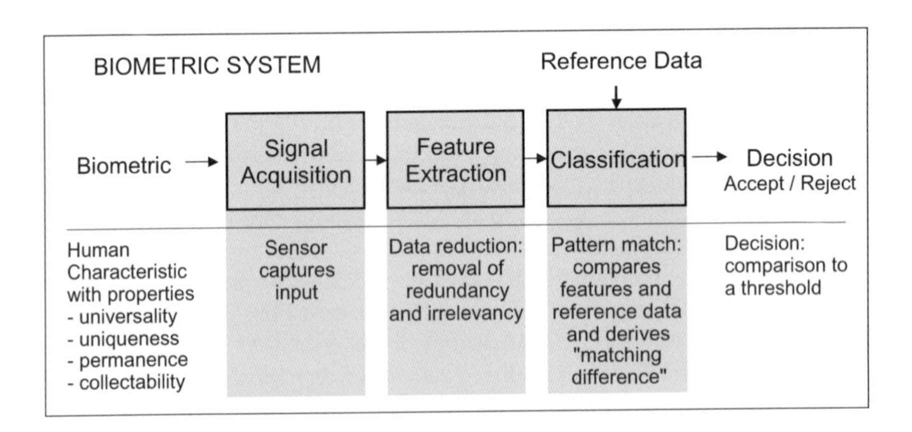

Fig. 1. Components of a biometric system. Individual features are derived from the sensor signal and compared to a reference data set. A biometric system performs pattern recognition

One possibility to measure the performance of biometric systems is the determination of the two error rates: the false acceptance rate (FAR) and the false rejection rate (FRR). The FAR is the probability of letting an unauthorized user pass for someone else. This error rate must be low enough to provide sufficient security for a given application. The FRR is the probability that the biometric does not recognize an authorized user and therefore denies him or her access. A low FRR is so important because this type of error can occur every time an authorized person uses the device. The FRR therefore is critical for user acceptance.

The working point of a biometric device, and thereby the effective value of the FAR and FRR, can be adjusted by increasing or decreasing the decision threshold of the device. How much each error rate is affected by altering the threshold is a characteristic of each manufacturer's biometric system. A system may exhibit an extremely low FAR, but the corresponding FRR may be totally unacceptable. The balance of the two error rates for a given application is critical for the success of a biometric installation.

The point at which the FAR equals the FRR is called the equal error rate (EER). The ERR is often used as a good indication of the overall performance, though the effective working point might be somewhere else.

1.1 Biometrics in SmartKom

1.1.1 Background

One intention of SMARTKOM is to create intelligent user interfaces that accept and support the natural communication style of humans and thus contribute to a user-friendly and user-centered design of information technology. In this context it is also necessary, of course, to provide not only safe and reliable but also natural access to knowledge, information and data. This implies a natural way of authentication: As we go about our daily lives, we are used to identifying other people by the way they look, speak or walk — we are performing a biometric authentication. Thus, in SMARTKOM biometrics is also the method of choice when talking about a convenient and natural way of authentication.

Which biometrics is most suitable depends, on the one hand, on the type of application and, on the other hand, on the personal attributes and preferences of a user:

- Applications make high demands on security, that are probably not satisfiable by every type of biometrics.
- In other cases, the application scenario advises a particular type of biometrics, e.g., speaker recognition in telephony applications.
- In addition, user acceptance is not the same for all kinds of biometrics but, for example, depends on the data acquisition device, hygienic aspects, intrusiveness and so on.
- Further on, a particular biometrics may be unsuitable for some users at some times, e.g., because of hoarseness in the case of a speaker recognition, unstable signatures in the case of a signature verification or an injury in the case of a hand contour recognition.

Therefore, because of the need for both an increased reliability and security and an enhanced user comfort and user acceptance, a multiple biometric approach was chosen. This approach consists of a hand contour recognition, a signature verification and a speaker verification. Such multiple biometrics are also easily integrated into a multimodal system like SMARTKOM.

Verification (one-to-one match) was chosen as the appropriate method of authentication, mainly because of performance reasons and privacy issues, especially in SMARTKOM-Public. Below, the particular biometric technologies either enhanced or developed in SMARTKOM are introduced.

1.1.2 Hand Contour Recognition

In SMARTKOM a gesture recognition is integrated, where the hand is used for pointing — so why not use it to recognize the user?

Biometric systems based on hand contour authentication are not new and have been available since the early 1970s. One of the most well known hand contour application is the INSPASS system in the Unites States, allowing frequent travelers to bypass waiting lines of immigration ("Your hand is your passport"). Hand contour verification was also used, for example, to secure the 1996 Olympic games in Atlanta.

Commercial hand recognition systems are small boxes that look like a trap. Up to five pegs ensure an accurate placement of a user's hand (Jain et al., 1999). This design limits usage by certain populations, e.g., people with very small or very big hands. Contrary to that, in SMARTKOM a hand contour recognition system was developed that does without such contraptions and allows free positioning of the user's hand within the work space. Here, also a life check can easily be performed to protect against simple fakes by asking the user to close his fingers and spread them again.

The silhouette of the hand as obtained from the SiViT video camera and processed from the device "realDocument" is used to extract invariant and characteristic features like palm size, finger lengths, finger thickness and curvatures. Although these metrics do not vary significantly across the population, they are distinctive enough for verification or usage in multiple biometric systems like SMARTKOM. Some of the advantages of hand contour verification are (Nanavati et al., 2002):

- It is nonintrusive.
- It is able to operate in challenging environments.
- It has a high user acceptance.
- It is the biometrics of choice in combination with gesture-based devices.

1.1.3 Signature Recognition

The signature is a widely spread method to authorize documents. Signing a document is a socially and legally accepted binding agreement.

Usually signatures are verified by comparison of images, i.e., by the form of the signature. This is called static verification. In dynamic signature verification the way of signing is also taken into account. Typical features are velocity, acceleration and pressure (Gupta and McCabe, 1997). Since the signature is a quite variable characteristic, five to six signatures are required during enrollment.

The main advantages of dynamic signature recognition are (Nanavati et al., 2002):

- It leverages existing processes.
- It is perceived as non intrusive.
- It has a high user acceptance.
- Users can change their signatures if needed.
- It has a "built-in" liveness test.

Dynamic signature verification is ideally suited for online applications. The signature may be performed on the paper document or pasted into an electronic document. Hardware expenses are comparably low: only an off-the-shelf PC tablet is required. In SMARTKOM, the flat surface of the tablet also serves as the projection and pointing area of the SiViT system.

The main field of application is the authentication of electronic documents or everywhere else where today simple signing is used. Financial transactions (from insurance to credit cards) and any form of contract may be safeguarded by this biometric technology.

1.1.4 Speaker Recognition

The voice is a good characteristic of every human. In communication we are used to identifying our partners from their voices, for example, on the telephone. In speaker verification the biometric system does the same.

Human speech depends on the anatomy of the vocal tract (mouth and nose cavities) of the person and on the linguistic behavior (pronunciation, speaking rhythm) the speaker has developed. Voice verification or speaker recognition systems model the acoustic features of speech that have been found to differ among individuals and that reflect both physical and learned behavioral patterns (Campbell, 1997).

In principle, speaker verification can operate in text-dependent as well as text-independent mode.

- Most speaker recognition systems use text-dependent input, which involves selection and enrollment of one or more voice pass-phrases. In this case the system knows what will be spoken and can perform a pattern match, but does not necessarily have explicit (phonetic) knowledge about the pass-phrases:
  - Language-independent input is possible if the user is free to choose the pass-phrase. In this case the system has no explicit (phonetic) knowledge about the input.
  - Text-prompted input is a version of text-dependent input which is used when there is concern about an impostor playing back a previously recorded voice pass-phrase. Here the system prescribes the pass-phrases to be spoken and can rely on explicit phonetic knowledge about them. Such a system is typically language dependent.

- Text-independent mode would allow free speech for verification where the variability of words and the natural personal variability have to be compensated. In this mode, the verification is performed independently of what is spoken. Explicit knowledge could only be gained after successful speech recognition and possibly language identification. This mode is still a research topic.

Generally, text-dependent recognition systems allow a more precise modeling and therefore yield higher security. Hence, also in SMARTKOM a text-dependent recognition was applied. The advantages of speaker recognition are:

- It is ideally suited in the context of communication applications and has the ability to leverage existing telephony infrastructure.
- It is perceived as nonintrusive.
- It allows for synergy effects in combination with other processes such as speech recognition.

To account for the natural variations in everyone's voice a minimum of three enrollments is required. Furthermore, to achieve a high accuracy of speaker recognition, also some quality measures were developed:

- An outlier detection ensures that a particular user does not change his pass-phrase between the separate enrollments.
- The length of the chosen pass-phrase is checked — utterances that are too short are rejected.
- Noise and channel compensation are integrated to minimize the effects of background noise or different channels.

2 Multiple Biometrics

2.1 Background

In the development of human computer interaction, the aspect of multimodality gains increasing importance. In the future small personalized mobile devices and public information and transaction terminals,such as those being developed in the SMARTKOM project, will allow the user to interact with different modalities such as voice, handwriting, gestures or mimic. This has also an effect on the usage of biometrics. By different input sensors like microphone, camera or touch screen, several biometric features such as voice, signature, hand contour or facial form can be recorded, interpreted and combined. This is called multiple biometrics and offers authentication with a higher level of security or comfort (Hong and Jain, 1999). Within SMARTKOM a new multiple and multimodal approach was developed to combine the three individual biometric technologies described above: speaker recognition, signature verification and hand contour recognition (Fig. 2).

Biometric systems are always two-class decision problems where the features of the authorized users, the genuines, must be separated from those of the impostors. The underlying similarity measurements are either costs or scores. Costs are the

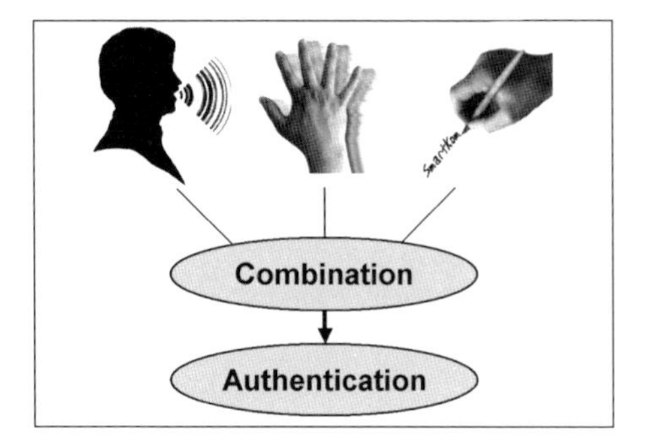

Fig. 2. Multimodal biometrics in SMARTKOM

calculated distances between reference and test patterns, whereas scores indicate the probability that the patterns stem from the same user. Multimodal biometrics are also binary classification problems, where the dimensions of the feature space depend on the number of single biometrics considered. The demanding task in multiple biometrics is to combine or fuse the expert outputs optimally in order to reach a higher performance than any single biometrics. An ideal multiple biometric system should meet the following requirements:

- higher degree of security than the best integrated individual biometric technology
- low and possibly selectable FAR
- low and possibly selectable FRR
- flexibility with respect to modification
- fast decision time

In addition, multiple biometric systems may also exhibit the following characteristics:

- Increase in robustness and reliability because of redundancy.
- Spoofing attacks become more difficult.
- Increase in user-acceptance (e.g., if a user can chose his or her preferred biometric).
- Enlargement of the population (e.g., if the universality of a particular biometric is insufficient).

In general, we distinguished between two main strategies to combine different biometrics: feature fusion and decision fusion (Ross and Jain, 2003).

Feature fusion (Fig. 3) or early fusion combines all features extracted from different modalities. As the new feature space has a higher dimensionality and all features can be weighted optimally, the strategy is generally believed to be more discriminative. However, this fusion method assumes feature access to all used biometrics and expert knowledge about the single biometrics and their features. As the type and number of features differs in different combinations and therefore the feature combination and weighting has to be readjusted, the feature fusion method is not well suited for flexible biometric combinations with varying types and numbers of biometrics.

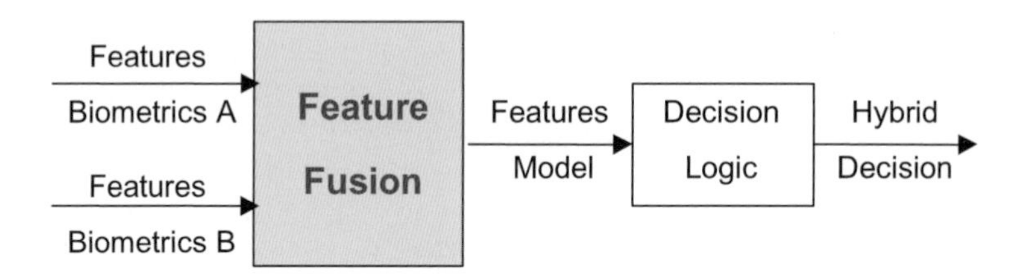

Fig. 3. Feature fusion

Most multimodal approaches are based on decision fusion or late fusion, which combines the output of all single biometrics (Fig. 4). This output of the so-called expert modules are either similarity measurements (costs, scores) or binary decisions. Usually the supervisor or decision module does not need any expert knowledge about the single biometrics and their features. Therefore this method is more flexible regarding the modification of the combination (changing number or type of biometrics). In some approaches the decision fusion unit can be trained and therefore adjusted automatically by training output data of the single biometrics.

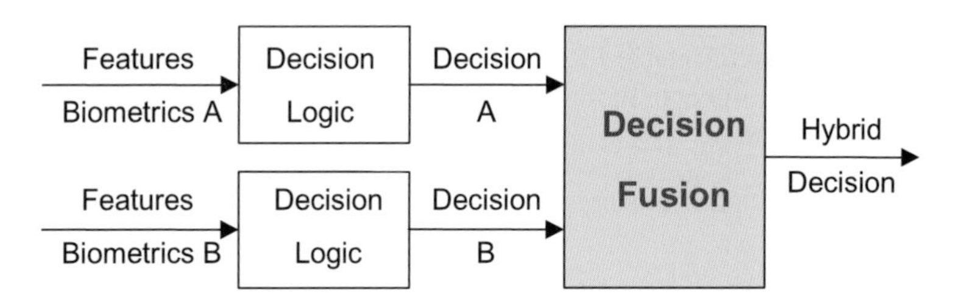

Fig. 4. Decision fusion

Typical decision fusion strategies are logic combination (Dieckmann et al., 1997), where the decisions of the single biometrics are combined by logic operators, or the calculation of a global weighted score from the costs or scores of the single biometrics (Brunelli and Falavigna, 1995). Further approaches are neural networks (Dieckmann et al., 1997), support vector machines (Ben-Yacoub, 1999), and Bayesian statistics (Bigün et al., 1997).

The existing methods have some disadvantages, however (Daugman, 1999). These may be inherent assumptions about the distribution of the costs or scores of the single biometrics, large training effort, insufficient error rates, or difficult modification for the usage of different types or numbers of biometrics as the combinations are optimized for a special configuration.

2.2 SmartKom Approach

The multimodal biometric approach developed in the SMARTKOM project is also a decision fusion method, but it does not use any of the above-described combination methods and does not make any assumption about the distribution of scores or costs like in Bayesian statistic approaches (Bigün et al., 1997). It is based on the multidimensional probability density of the similarity measurements for the genuines and imposters, which can be derived from the real measured cost or score distributions of the individual biometrics.

In the following the approach is described by an example, where biometrics 1, B_1, which is based on costs, and biometrics 2, B_2, which provides scores, are combined to an overall multiple biometric system. The cost or score probability densities of a single biometric are calculated by normalizing the integrals of the single distributions to unity. Figures 5 and 6 show the probability densities $p_g(x)$, $p_i(x)$ of B_1 and $p_g(y)$, $p_i(y)$ of B_2 for the genuines (g) and impostors (i).

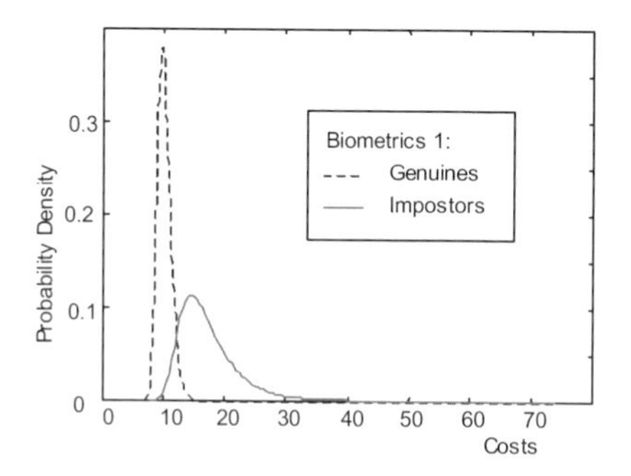

Fig. 5. Cost probability density B_1

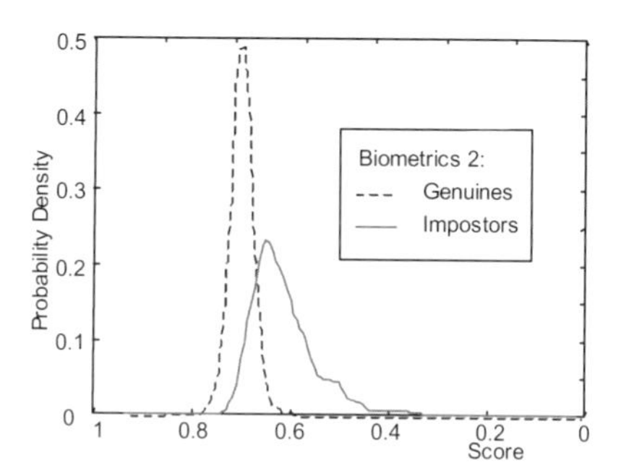

Fig. 6. Score probability density B_2

If the different biometrics can be considered to be statistically independent, the multidimensional probability densities of the costs or scores are given by the product of the probability densities of the single biometrics for both genuines and impostors. The assumption of statistical independence is fulfilled in most cases, especially with biometrics that use different modalities.

The multidimensional cost or score probability density of two statistically independent biometrics is calculated as follows:

$$p_g(x,y) = p_g(x) \cdot p_g(y). \qquad \text{Genuines} \tag{1}$$

$$p_i(x,y) = p_i(x) \cdot p_i(y). \qquad \text{Impostors} \tag{2}$$

Figure 7 shows the probability densities of the similarity measurements for genuines and impostors in the combination of B_1 and B_2.

The concept of our decision strategy is to find suitable borderlines in the multidimensional cost or score space that separate the class of genuines from the impostor

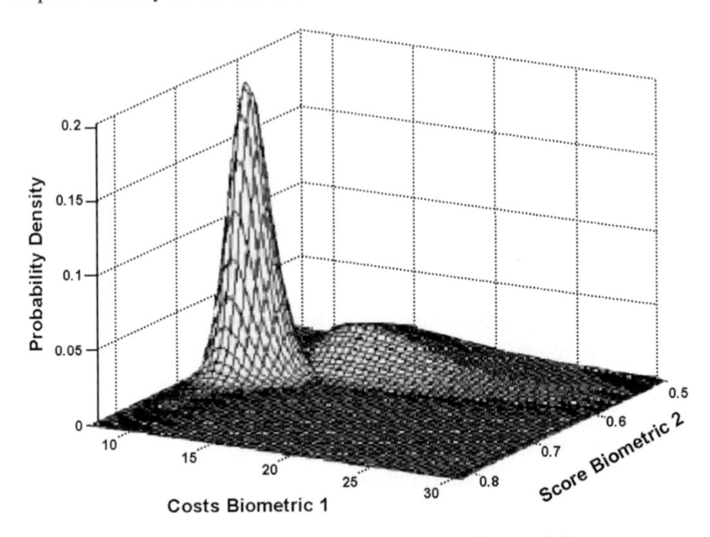

Fig. 7. Two-dimensional cost–score probability densities of B_1 and B_2

class in an optimal way and under special constraints, e.g., a given FAR or FRR. The multidimensional probability densities of the genuines and impostors are used to calculate these decision curves. There are several methods to find such borderlines based on the probability densities. The SMARTKOM approach is based on a strategy that determines the decision curves under the constraint of a given FAR. Within this method the decision borderlines are the curves that are described by the points where the ratio of the two probability densities has a constant value V. This value has to be adjusted properly by test data to yield the intended FAR value, FAR_{Target}. A score or cost point (x,y) is assigned to the class of genuines when:

$$\frac{p_g(x,y)}{p_i(x,y)} > V(FAR), \qquad FAR = FAR_{Target}. \tag{3}$$

In Fig. 8 the decision border and the area of acceptance is illustrated for an exemplary combination of B_1 and B_2 with this fusion method. In addition, lines of constant probability density of the similarity measurements for genuines and impostors are shown.

2.3 Experiments and Results

To examine the performance of the SMARTKOM multimodal biometric method, qualitative measurements were made for the combination example of B_1 and B_2. The corresponding error rates were calculated and compared with those of the single biometrics and logic combination methods, which are almost solely used in present multimodal biometric systems.

For that, the data of costs and scores from two of the three biometrics integrated in SMARTKOM, speaker verification and signature verification, were used for the

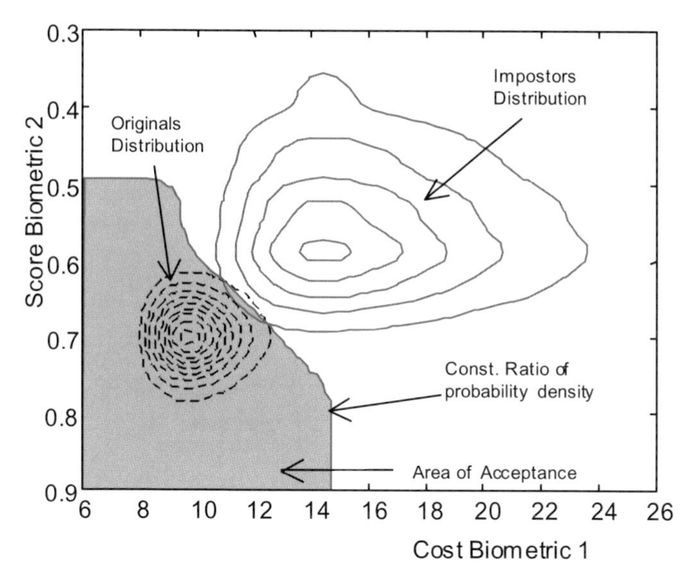

Fig. 8. Area of acceptance

determination of the single cost or score distributions, the borderlines in the decision space and for evaluating the combination method.

For speaker verification 1500 genuine and 70,000 imposter costs were available. The values result from a database of 47 persons and the SMARTKOM speaker verification system, which utilizes two-digit random numbers between 21 and 99. The SMARTKOM dynamic signature verification system and a signature database of 20 persons offered 2150 genuine and 2200 imposter scores of the second biometric.

All similarity measurements were divided into a training set for the calculation of the decision criterion and into a distinct evaluation set to evaluate the described combination method. For training and evaluation the cost and score values were combined randomly to more-dimensional score–cost vectors for genuines and impostors, because the signature and speaker data do not stem from the same persons. Thus, statistical independence of the data used is guaranteed.

Within the evaluation, the FRR values were calculated for several given FAR values. This was performed not only for the developed combination method, but also for the individual biometrics and logical AND and OR combination. The AND combination accepts the user solely when every single biometrics returns a positive result, whereas an OR combination requires only the acceptance from at least one biometric. Within the evaluation the best operating points and therefore the best FRR values for given FAR values were also calculated for these logic combinations.

In Fig. 9 the FRR values of the described combination method, of the two logic combinations, and of biometrics 1 and biometric 2 are illustrated as a function of given FAR values. This type of diagram is called receiver operating characteristics

(ROC), and is a standard method to compare different biometric systems (Biometric Working Group).

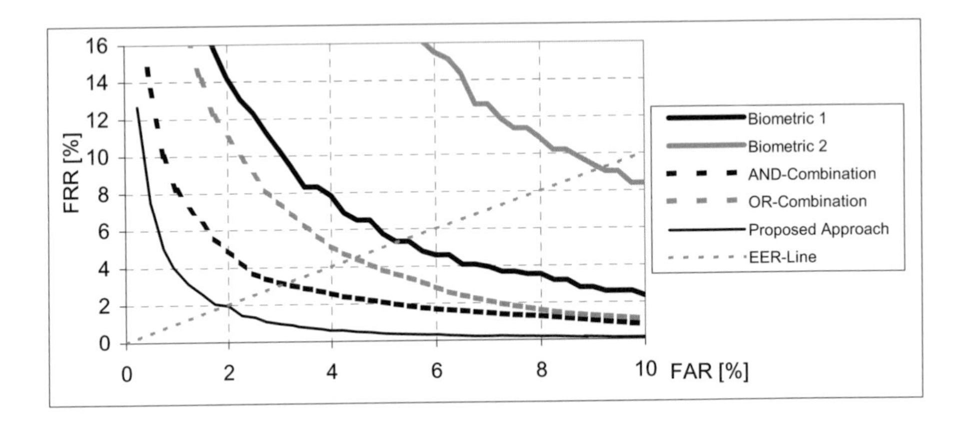

Fig. 9. ROC curves

The ROC diagram shows that biometric 1 has better performance than biometric 2. The logic combinations achieve better results than the single biometrics. As the best operating points of the logic combinations were calculated for given FAR values, the resulting threshold values differ from the thresholds of the single biometrics for the same FAR. Therefore a performance improvement compared to the single biometrics is also plausible for the OR combination. But the SMARTKOM combination yields the best results: The method based on the multidimensional probability densities outperforms the single biometrics and the logic combinations in all operating points.

3 Conclusion

Within SMARTKOM, biometrics ensure natural, comfortable and secure user authentication. Three biometric technologies are integrated: hand contour verification, signature verification and text-dependent speaker verification.

An approach for the combination of these individual biometrics to a multiple and multimodal biometric system was developed. It is based on the multidimensional probability densities of the costs or scores for genuines and impostors. There are several strategies to find decision borders in the space of the similarity measurements spanned by these probability densities. Here an example was described were two biometrics were combined with the constraint of a given FAR value. Possible decision borders can be curves where the ratio between the genuine and impostor probability density has a constant value. The resulting FRR values are compared to the corresponding error rates for each single biometric and also to the results from logical AND and OR combinations. In every operating point the SMARTKOM approach exhibits superior performance to the stand-alone biometrics and the logic combinations. Furthermore, the method flexibly allows modification and integration of different types and numbers of biometrics.

References

S. Ben-Yacoub. Multi-Modal Data Fusion For Person Authentication Using SVM. In: *Proc. 2nd Int. Conf. on Audio- and Video-based Biometric Person Authentication (AVBPA'99)*, pp. 25–30, 1999.

E.S. Bigün, J. Bigün, B. Duc, and S. Fischer. Expert Conciliation for Multi-Modal Person Authentication Systems by Bayesian Statistics. In: J. Bigün, G. Chollet, and G. Borgefors (eds.), *Audio and Video based Person Authentication – AVBPA97*, LNCS 1206, pp. 311–318, Berlin Heidelberg Germany, 1997. Springer.

Biometric Working Group. Best Practices in Testing and Reporting Performance of Biometric Devices. http://www.cesg.gov.uk/site/iacs/itsec/media/protection-profiles/BBP.pdf. Cited 5 April 2006.

R. Brunelli and D. Falavigna. Person Identification Using Multiple Cues. *IEEE Transactions on Pattern Analysis and Machine Intelligence (PAMI)*, 17(10):955–966, 1995.

J.P. Campbell. Speaker Recognition: A Tutorial. *Proc. IEEE-1997*, 85(9):1437–1462, 1997.

J. Daugman. Biometric Decision Landscapes. Technical Report TR482, University of Cambridge Computer Laboratory, Cambridge, UK, 1999.

U. Dieckmann, P. Plankensteiner, and T. Wagner. SESAM: A Biometric Person Identification System Using Sensor Fusion. *Pattern Recognition Letters*, 18:827–833, 1997.

J. Gupta and A. McCabe. A Review of Dynamic Handwritten Signature Verification. Technical report, James Cook University, Townsville, Australia, 1997.

L. Hong and A. Jain. Multimodal Biometrics. *Biometrics – Personal Identification in Networked Society*, pp. 327–344, 1999.

A.K. Jain, A. Ross, and S. Pankanti. A Prototype Hand Geometry-Based Verification System. In: *2nd Int. Conf. on Audio and Video-Based Biometric Person Authentication (AVBPA)*, pp. 166–171, Washington, DC, 1999.

S. Nanavati, M. Thieme, and R. Nanavati. *Biometrics — Identity Verification in a Networked World*. John Wiley, New York, 2002.

A. Ross and A.K. Jain. Information Fusion in Biometrics. *Pattern Recognition Letters*, 24(13):2115–2125, 2003.

Natural Language Understanding

Ralf Engel

DFKI GmbH, Saarbrücken, Germany
ralf.engel@dfki.de

Summary. This chapter presents SPIN, a newly developed template-based semantic parser used for the task of natural language understanding in SMARTKOM. The most outstanding feature of the approach is a powerful template language to provide easy creation and maintenance of the templates and flexible processing. Nevertheless, to achieve fast processing, the templates are applied in a sequential order that is determined offline.

1 Introduction

The task of the natural language processing module is to transform the word lattice sent by the speech recognizer into a list of hypotheses representing in an abstract way possible user intentions. For this task a new template-based semantic parsing approach called SPIN (Engel, 2002) was developed within the SMARTKOM project. This chapter motivates and describes this approach.

1.1 Module Input and Output

First, a closer look at the module input and output should be given. The input consists of spoken utterances, which are intended to interact with a computer system. Therefore, they are usually syntactically less complicated than in other surroundings, e.g., human–human interactions, and they are also limited in length. But the utterances are often realized syntactically incorrectly by the user (because the user has to think and speak at the same time). Additionally, they contain many speech recognition errors due to a word error rate that ranges between 7.9% and 21.1% (dependent on the test set).

The output hypotheses are expressed as instances of the SMARTKOM ontology, which means that the output representation is already on a high semantic–pragmatic level and encodes direct instructions for the action planner. This entails that during the analysis of a spoken utterance many details can be ignored, as long as the representation is detailed enough to trigger the desired action in the action planner.

1.2 Approach

For the described scenario a semantic parsing approach is best suited, and semantic parsers are also widely used within dialogue systems. Characteristic of semantic parsers is that they do no syntactical analysis or only perform a very limited one but build up more or less directly the high-level output structure from word level. Examples for semantic parsers based on variations of finite state transducers (FSTs, see Roche and Schabes (1997)) are Miller and Stallard (1996), Potamianos and Kuo (2000) and Ward (1991). Examples for semantic parsers based on context-free grammars (CFGs) are Allen (1994), Gavald (2000), Kaiser et al. (1999), Lavie (1996) and Seneff (1992).

Instead of using one of the existing approaches for SMARTKOM, a new approach was developed. The main reason to develop a new approach and not to use one of the existing approaches was that writing and maintaining of the rules should be simplified compared to existing approaches. This includes especially the writing of rules that are robust against speech recognition errors and rules that are robust against word reordering in free word order languages like German. Since the rules have to be manually created[1] and the number of rules is quite large (435 templates are needed to cover all SMARTKOM applications), easy creation and maintenance of the rules is one of the most important issues for parsers in dialogue systems.

This is achieved mainly by allowing more powerful rules. This results in a more direct and natural encoding of rules. Most rules look like partial or complete utterances, where some of the words are exchanged with conditions on instances of the SMARTKOM ontology. Hence the rules are called templates in the following. The increased rule power also reduces the need for tricky constructions caused by the lack of expressive power. Such constructions are often time consuming to write and test, and later on are hard to understand and maintain for other template writers. The increased expressive power includes set-based operation, more than one expression on the left side, optional parts, additional constraints and the possibility of subsequent modifications of embedded structures. A more detailed presentation of the template language is given in Sect. 2.

1.3 Processing

The increased rule power has the effect that conventional parsing strategies and optimization techniques are either not sufficient or not applicable. Therefore, several measurements were developed to nevertheless achieve fast processing. Fast processing is an important issue since a dialogue system is an interactive system. Furthermore, the speech recognizer delivers not only one hypothesis but many hypotheses encoded in a word lattice. The lower the parsing time of one hypothesis, the more hy-

[1] Data-oriented rule creation is still a topic of research, e.g., Minker et al. (1996); Schwarz et al. (1997); Maedche et al. (2002) and is not fully solved. Additionally, the collection and annotation of the data is also very time consuming and error-prone.

potheses can be processed. Together with a clever selection mechanism[2] to choose one of these hypotheses, this can enhance the overall performance of the dialogue system.

The reduction of parsing time is mainly achieved by trying to avoid wasting time with the generation of results that are probably suboptimal. Results regarded as suboptimal are, e.g., results that are nonmaximal or results that are generated using general templates in cases where more special templates are available. The generation of such results is avoided by a fixed application order of the rules and by grouping rules that lead to alternative results using the same input. The application of rules is discussed in Sect. 3, the preprocessing of the rule set, like the computation of the template order, is presented in Sect. 4.

2 Template design

The template design is inspired by production systems where rules operate on a working memory (WM), i.e., a rule selects some elements of the WM, may delete some of them and adds new elements. In the presented approach the idea is similar: the templates operate on a WM that contains as possible elements words and instances of the SmartKom ontology. Each template consists of a condition part, a constraint part and an action part. The actions are only applied if all conditions and constraints can be satisfied.

2.1 Conditions

The condition part contains conditions on words and on instances of the ontology. Conditions on words can include tests concerning orthography, stem, semantic category and part of speech. A test on the stem instead of the orthography helps to be more robust against wrongly recognized inflections, a frequent problem in the German speech recognizer. A test on the semantic category is useful to match all words of a specific semantic type, e.g., all actors. A test on part of speech allows writing of some kind of common syntactic rules and partially compensates the lack of a syntactic analysis. Conditions on instances include checking of the type and tests on slots and their assigned instances. For the type check the type hierarchy of the SmartKom ontology is regarded. Tests on the embedded instances can again contain tests on their embedded content.

Each condition can be marked as optional, i.e., the condition does not need to match. Robustness against speech recognition errors can be increased if nonessential conditions are marked as optional. Also, each condition can have an associated variable that the matched element is bound to. That way, the actions of the same template can refer to elements matched by conditions. Sets of alternative conditions

---

[2] In SmartKom a first selection is done in the parser itself; further selections are done by the multimodal fusion and the intention analyzer considering more context.

```
or([road map],[car map],[vehicle map])
roadMap()

$S=word(semCat:SIGHT)
sight(location(locationName:$S))

[or(the,this) $O=physicalObject($R=refProp())]
$O($R(det:det))

there
to() sight(refProp(det:udet))

there
sight(refProp(det:udet))

[$M=streetMap() %with $S=sight()]
$M(object:$S)

tell %me %more [%about $I=sight()]
informationSearch(pieceOfInformation:$I)

%how %do %i or(go,get) %$R=route() %[from $L1=sight()]
   %[to $L2=sight()] !atLeastOne($L1,$L2)
motionDirectedTransliterated(source:$L1,target:$L2,path:$R,
   meansOfTransportation:onFoot())
```

Fig. 1. Excerpt from the SMARTKOM template set. The templates are separated by empty lines; the first line of each template contains the conditions and the constraints (marked with *!*); the second line, the actions. Optional conditions are marked with %. List-based matching is activated using square brackets

can be expressed to reduce the number of required templates. Negation of conditions is also available, i.e., a condition must not match any element in the WM.

By default, conditions can match elements at any positions in the WM (further referred to as set-based matching). Writing templates for free word order languages is greatly simplified by set-based matching. This can be changed by grouping conditions in so called sequence conditions (further referred as sequence-based matching). In this case the matched elements in the WM have to be exactly in the same order as the corresponding conditions in the sequence condition and other elements must not be in between. Set-based matching and sequence-based matching can be mixed in one template.

2.2 Constraints

Constraints allow us to express conditions over the content of several variables after the conditions have matched. Currently, three different constraints exist. *atLeastOne* checks if at least one of the specified variables is bound. This is useful if a template has several optional conditions, but at least one has to be satisfied to produce a meaningful result. *checkOrder* checks the order of the elements bound by variables in the WM. This constraint can be used if the word order is relatively free but not completely negligible. In such a case the use of set-based matching together with *checkOrder* constraints is appropriate. *isOfType* can be used to check if the type of an instance was refined (see also Sect. 3.4).

2.3 Actions

Possible elements in the action part are constructors for words, constructors for instances of the SMARTKOM ontology and variables mentioned in the condition part. Constructors for words insert new words into the WM. This can be used to preprocess the WM so that a later applied template matches. This helps to reduce the complexity of templates. Constructors for instances insert new instances of the SMARTKOM ontology. Embedded content of the instances can be specified directly in the constructor. Within the embedded content it is possible to refer to variables of the condition part.

2.4 Robustness

Often the input utterance is not fully correct either due to speech recognition errors or errors produced by the user. The templates have to be robust against three different types of recognition errors: deletion of words, insertion of words and exchange of words.

Errors occurring as the deletion of words are addressed by marking less essential conditions as optional. Less essential conditions may include tests on modal verbs, pronouns, prepositions etc. It is only important that an utterance is not unintentionally matched by the template. Of course, if essential words are missing, this does not help.

The insertion of words does not harm in the case of set-based processing. Additional words are simply ignored. If list-based processing has to be used, erroneous processing cannot be avoided. But often the input can be analyzed at least partially since list-based processing is used often for modifying instances.

Errors occurring as the exchange of words can be divided into two categories. The exchanged words are either wrong inflections of the original words (often a problem with the German speech recognizer) or completely different words. The first case is addressed by using the stem of a word as default for matching instead the orthography. The second case can be reduced to a combination of the insertion and the deletion of words.

3 Processing

3.1 Major Internal Data Structures: Working Memory and Agenda

The templates are applied on a working memory (WM). There may exist several WMs at the same time, each representing an alternative result. A WM contains words and instances of the ontology. For each word and instance, the transitions of the word lattice that are used for creating and modifying the word or instance are stored. List-based matching uses this information to decide if two elements are directly in succession.

To support chart parsing the WM can also contain local alternatives. Local alternatives are structures that store words and instances that can be used only alternatively but not together. A sequence structure is available to store several words and instances within one alternative.

The second important internal data structure is the agenda. During processing at many points several possibilities exist how to proceed. In such cases one possibility is chosen, and the rest is stored in the agenda to be computed later.

3.2 Processing Overview

The WM is initially filled with the words of the best path through the word lattice. Before further processing, the words are enriched by syntactic and semantic information looked up in the lexical database. To also be able to process the lower scored paths in the word lattice, possible alternative branches in the word lattice are stored in the agenda.

The next step is the application of the templates in a predefined fixed order. The application order is not a sequential but a structured one. More details are given in Sect. 3.3.

In the case when a template can be applied in several ways, one way is chosen and all the necessary data to continue the other ways later on are stored in the agenda. More details on the application of a single template are given in subsection 3.4.

To build up a chart, the application of the next template is delayed in some cases, and first some of the agenda entries resulting from the last template application are continued. After the application the different results are merged, and local alternatives are created where necessary. The number of selected entries determines how fast the first result is found versus how effective the next results are generated.

After the application of all templates is finished, the first parsing solution is simply the current content of the WM, which is stored in a result list.

To produce the next solution one entry is selected from the agenda, [3] and the application continues where the processing was interrupted. This continues until the agenda is empty or a timeout (default 500 ms) is expired.

A scoring function (presented in Sect. 3.5) selects the *n* best results from the result list and sends them encoded in M3L to multimodal fusion (*n* defaults to 10).

Figure 2 gives an overview of the architecture; Fig. 3 shows the application of the templates in more detail.

[3] In SmartKom the agenda is only handled as stack (first in, last out). A more intelligent selection using a scoring function is the subject of current research.

3.3 Template Application Order

The application order is determined by grouping templates in so-called *blocks*, which can be nested. The blocks are somewhat similar to control structures in programming languages. Four different types of blocks are available:

- **Sequential block:** The embedded blocks and templates are applied in the specified order.
- **Repeat block:** The embedded blocks or templates are repeated until no block and no template can be applied anymore.
- **Alternative block:** The embedded blocks and templates are applied alternatively. If more than one block or template can be applied, the first one is applied and the alternatives are stored in the agenda.
- **Optional block:** The embedded block or template is applied optionally, i.e., in the case the block or template can be applied, the alternative of not applying this block or template is put on the agenda.

A simple sequential application of the templates is not sufficient since in such a case many relevant results could not be generated. Which block is used when and why is described in Sect. 4, in which the automatic offlinegrouping of the templates into blocks is described.

3.4 Application of One Template

A template is applied to the WM in four steps:

1. Each condition is tested to see if it is satisfiable with respect to the WM.
2. Alternative solutions omitting optional conditions are generated.
3. All constraints specified in the template are checked.
4. The tokens in the WM matched by the conditions are replaced by tokens generated by the corresponding actions in the action part of the template.

The steps cannot be presented in full detail here; only some features should be mentioned.

Refinement

Referring expressions are supported by allowing ontology instances to refine their types during matching. If a condition requires a more specialized type than the instance in the WM has, the instance can refine its type to match the condition. The instance retains the refined type during further processing.

This allows us to transform a referring expressing like *this one* first to an instance of the most common type for things (in SMARTKOM *physicalObject*). When a template later requires, e.g., an instance of type *avMedium*, the instance representing the referring expression can refine its type from *physicalObject* to *avMedium*. This refined type can be used later in the modality fusion for referential resolution, e.g., for querying the dialogue history.

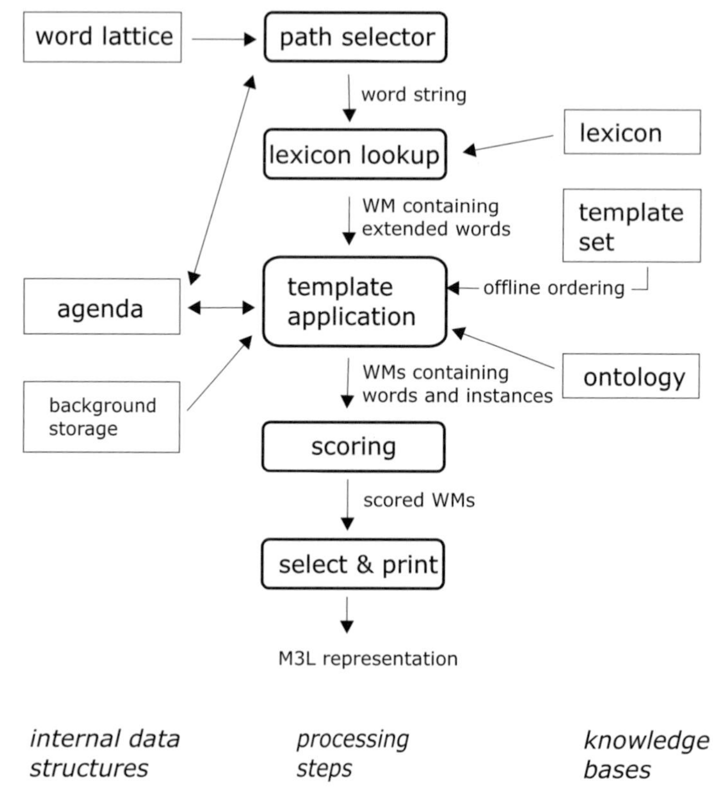

Fig. 2. The architecture of SPIN. The data flow is shown together with the knowledge bases and the most important internal data structures

Since it is possible that the user has meant an instance of a different subtype, alternative solutions with possible remaining types (original type without refined type) are also considered by adding a suitable entry to the agenda.

Optional Conditions

Templates may contain optional conditions. It is not obvious during the application of a template if any and which alternative solutions should be generated by omitting optional conditions. The generation of all possible solutions would lead to many that are suboptimal since nonmaximal solutions should be avoided to achieve faster processing. But it is also not always sufficient to generate only the maximum solution. For example, regard the case that several commands within one utterance are mentioned. Because of the sequential application order and the ability of set-based processing, an optional condition can unintentionally bind elements of the wrong command. To determine which solutions should be generated is checked using an offline-generated hash map if at least one succeeding template exists that can use this solution.

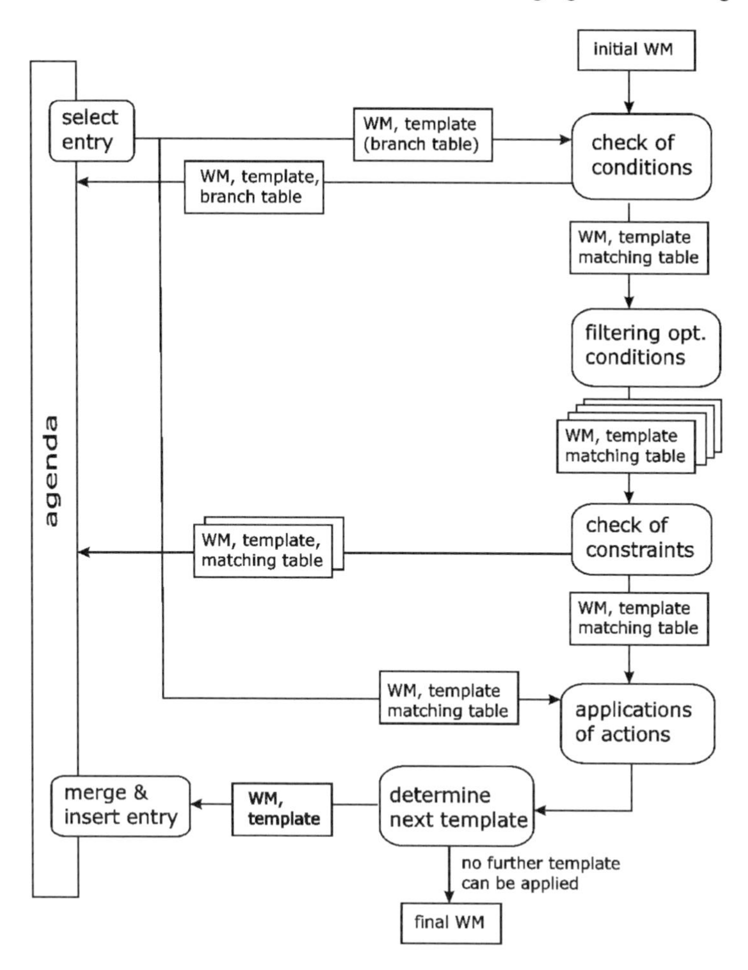

Fig. 3. The single steps within the application of the templates. additionally, the information passed between the single steps and the information that is stored in and retrieved from the agenda is shown

3.4.1 Considering Background Storage

The matching process uses instances in the background storage if the condition is not optional and no other element in the WM matches. The background storage is used for already recognized gestures to enhance robustness (see also Engel and Pfleger (2006)). If referring expressions are not recognized or are not uttered by the user, the problem can occur in which a template cannot be applied since an essential instance is missing. Putting the instances representing gestures in the background storage solves this problem.

3.5 Scoring Heuristics and Result Filtering

The scoring function has two tasks. First, the number of hypotheses sent to the modality fusion is limited. In the case that more results are generated than can be sent to modality fusion, the scoring function is used to determine which results are sent to the next module. Second, the value returned by the scoring function is used in the intention analyzer module together with scores from modality fusion, discourse modeling and situation modeling to select the best hypothesis to send to the action planner.

Two factors determine the score:

1. It is checked if either an event instance (an instance that triggers an action within the action planner) or an expected instance (e.g., a valid answer to a question) is part of the result. In this case, the score is in the interval $[0.5, 1]$. If not, the score is in the interval $[0, 0.5]$.
2. The portion of words that are used for the creation of event instances or expected instances in relation to all input words. If such instances are not present the remaining instances are regarded.

4 Offline Ordering and Grouping of the Templates

The order of the template application is determined offline in advance. Therefore, first a dependency graph is created describing which templates have to be applied before others. Subsequently, this dependency graph is transformed into the application blocks presented in Sect. 3.3.

Three different kinds of relations are possible between two templates. A relation between the templates T_1 and T_2 means that T_1 has to be applied after T_2 if all relevant results should be generated. It depends on the type of the relation if an additional application of T_1 before T_2 hinders the generation of all relevant results or leads only to additional suboptimal results. In the following, the relations are described informally minus some details; an exact definition is beyond the scope of this chapter.

- **require:** This relation expresses that template T_1 requires the application result of template T_2. For example, if T_1 modifies an instance of type X, it must be applied after a template T_2 that creates instances of type X. Another example is that T_1 deletes an instance of type X from the WM and T_2 modifies an instance of type X. T_1 must not be applied before T_2.
- **moreOrEqualGeneral:** The relation expresses that a template T_1 is more general than or equally general to a template T_2, i.e., for each condition in T_1 a corresponding condition in T_2 exists. The idea behind this relation is that it is assumed that the template writer wants that the more general template only be applied if the more specialized template cannot be applied. T_1 must not be applied before T_2.

- **modifyInstanceOfSameType:** The relation expresses that two templates modify the same instance and both templates use list-based processing. In this case, the order cannot be determined in advance, e.g., the input consists of a noun with two adjectives, and two templates are used to process the adjectives. The templates can be sorted, so that both can be applied, but as soon as the adjectives are swapped only one template matches. T_1 may be applied before T_2.

The dependency graph is built up by comparing the templates pairwise. Each node represents a template, and a transition between two nodes represents a relation. A direct linearization of the dependency graph is in most cases not possible due to cycles in the graph. The steps described in the following eliminate the cycles by combining nodes and assigning to the new node an application block containing the templates of the merged nodes. The steps are executed in the given order once.

1. All cycles according to *require* and *moreOrEqualGeneral* are detected. The nodes of each cycle are merged to one node, which is assigned a repeat block containing an alternative block. The alternative block in turn contains the templates of the merged nodes.
2. All cycles according to *require* and *moreOrEqualGeneral* and *modifyInstanceOfSameType* are detected. The nodes of each cycle are merged to one node, which is assigned a repeat block. The repeat block contains the blocks and templates in the order given by the relations *require* and *moreOrEqualGeneral.*
3. The final result is a sequential block containing the blocks and templates of all nodes. They are ordered according to the relations *require* and *moreOrEqualGeneral.* As a result of step 2 no cycles can exist.

The description of the steps lacks some introduced optimizations to avoid the generation of large alternative blocks that have a negative impact on performance. An example for an optimization is that some blocks or templates are removed from a cycle in step 1 by putting them in an optional block. An example of a dependency graph and the template application order derived from this dependency graph is shown in Fig. 4.

5 Conclusion

In this chapter SPIN, a template-based semantic parsing approach used for the task of natural language understanding in SMARTKOM, was presented. The design of the knowledge bases consisting of a lexicon, the SMARTKOM ontology and set of templates as well as the processing of the templates was motivated and described.

The approach was sufficient to produce the required output, while the templates remained clear and maintenance was relatively easy. Extensive system tests showed that the parser is fast (10 ms per path) and a timeout of 500 ms is used.

Further development will focus on the automatic optimization and extention of a given template set using machine learning methods. Optimization means here that conditions are detected automatically as nonessential and therefore can be marked as

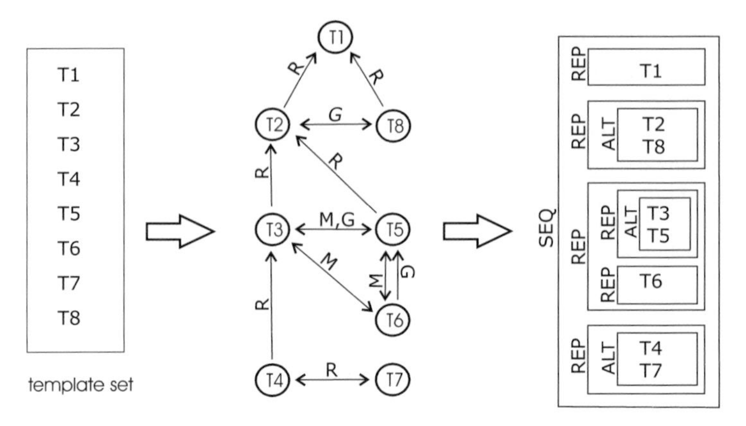

Fig. 4. Computation of the template application order starting from a set of templates and using a dependency graph as intermediate step. (*R* = Require, *G* = Moreorequalgeneral, *M* = Modifyinstanceofsametype, *Seq* = Sequential Block, *Rep* = Repeat Block, *Alt* = Alternative Block)

optional to increase robustness against speech recognition errors. Extension means here that the already existing templates are extended automatically when pairs of already working utterances and still nonworking utterances are provided.

References

J. Allen. *Natural Language Understanding*. Benjamin/Cummings, Menlo Park, CA, 2 edn., 1994.

R. Engel. SPIN: Language Understanding for Spoken Dialogue Systems Using a Production System Approach. In: *Proc. ICSLP-2002*, pp. 2717–2720, Denver, CO, 2002.

R. Engel and N. Pfleger. Modality Fusion, 2006. In this volume.

M. Gavald. SOUP: A Parser for Real-World Spontaneous Speech. In: *Proc. 6th Int. Workshop on Parsing Technologies (IWPT-2000)*, Trento, Italy, February 2000.

E.C. Kaiser, M. Johnston, and P.A. Heeman. PROFER: Predictive, Robust Finite-State Parsing for Spoken Language. In: *Proc. Int. Conf. on Acoustics, Speech, and Signal Processing (ICASSP-99)*, vol. 2, pp. 629–632, Phoenix, AZ, 1999.

A. Lavie. GLR*: A Robust Parser For Spontaneously Spoken Language. In: *Proc. ESSLLI-96 Workshop on Robust Parsing*, Prague, Czech Republic, 1996.

A. Maedche, G. Neumann, and S. Staab. Bootstrapping an Ontology-Based Information Extraction System. In: P.S. Szczepaniak, J. Segovia, J. Kacprzyk, and L.A. Zadeh (eds.), *Intelligent Exploration of the Web*, Heidelberg, Germany, 2002. Physica.

S. Miller and D. Stallard. A Fully Statistical Approach to Natural Language Interfaces. In: *Proc. 34th ACL*, Santa Cruz, CA, June 1996.

W. Minker, S. Bennacef, and J.L. Gauvin. A Stochastic Case Frame Approach for Natural Language Understanding. In: *Proc. ICSLP-96*, pp. 1013–1016, Philadelphia, PA, 1996.

A. Potamianos and H.K. Kuo. Statistical Recursive Finite State Machine Parsing For Speech Understanding. In: *Proc. ICSLP-2000*, Beijing, China, 2000.

E. Roche and Y. Schabes (eds.). *Finite State Language Processing*. MIT Press, Cambridge, MA, 1997.

R. Schwarz, S. Miller, D. Stallard, and J. Makhoul. Hidden Understanding Models for Statistical Sentence Understanding. In: *Proc. Int. Conf. on Acoustics, Speech, and Signal Processing (ICASSP-97)*, pp. 1479–1482, Munich, Germany, 1997.

S. Seneff. TINA: A Natural Language System for Spoken Language Application. *Computational Linguistics*, 18(1), 1992.

W. Ward. Understanding Spontaneous Speech: The Phoenix System. In: *Proc. Int. Conf. on Acoustics, Speech, and Signal Processing (ICASSP-91)*, Toronto, Canada, 1991.

The Gesture Interpretation Module

Rui Ping Shi, Johann Adelhardt, Anton Batliner, Carmen Frank, Elmar Nöth, Viktor Zeißler, Heinrich Niemann

Friedrich-Alexander Universität Erlangen-Nürnberg, Germany
{shi,adelhardt,batliner,frank,noeth,zeissler,
niemann}@informatik.uni-erlangen.de

Summary. Humans make often conscious and unconscious gestures, which reflect their mind, thoughts and the way these are formulated. These inherently complex processes can in general not be substituted by a corresponding verbal utterance that has the same semantics (McNeill, 1992). Gesture, which is a kind of body language, contains important information on the intention and the state of the gesture producer. Therefore, it is an important communication channels in *human computer interaction*.

In the following we describe first the state of the art in gesture recognition. The next section describes the *gesture interpretation module*. After that we present the experiments and results for recognition of user states. We summarize our results in the last section.

1 State of the Art

1.1 Applications of Gesture

Gesture can be used in a wide range of applications: gesture in conventional human computer interaction (HCI), interaction through linguistic gesture and manipulation through physical contact. We cover each of these in the following.

Gesture in Conventional HCI

Under the window, icon, menu and pointing device (WIMP) paradigm, the use of mouse and pen of a graphic tablet such as that of the Wacom [1] Company are typical example applications of gesture. This kind of gestures with the help of pointing devices is intensively employed in *computer aided design* (CAD) (Sachs, 1990) and online handwriting recognition (Buxton et al., 1985). In the literature this category of gestures is called pen-based gesture. Rubine introduced the GRANDMA system (Rubine, 1991), in which the user is allowed to define arbitrary gestures interactively. These user-defined gestures can be input either through a mouse or with the help of a pen. The system is able to learn the static and dynamic properties of the gestures on the basis of some training data and subsequently analyzes them in real time.

[1] http://www.wacom.com

Interaction Through Linguistic Gesture

American Sign Language (ASL) and audio–video–speech recognition represent applications of linguistic gestures. In the ASL there exist, as a general rule, strict syntax, semantics and their mapping in the gesture configuration similar to their "acoustic" counterpart — speech. The understanding of ASL takes place in the space in which gesture and its grammatical structure are expressed through the hand movement and posture. Humans also use facial expression as well as head and body posture to support their expressions. In Attina et al. (2003) a system is described in which speech recognition is supported by the conventionalized gestures, similar to ASL. This kind of application often deals with hearing-impaired patients. Lip-reading is the only reliable way for these patients to communicate with other people in daily life, assumed that they have no hearing device and have not yet learned some strict sign language like ASL. The speech accompanied by such gestures is referred to as Cued Speech (Cornett, 1967). Moreover, linguistic gestures can be used in HCI as artificial commands, which a computer can interpret and execute. The Morpha system in Lütticke (2000) can be controlled through dynamic gestures, which are learnde by the system offline. This kind of gesture is more intuitive and flexible in comparison with the gestures in ASL.

Manipulation Through Gesture

Through the commitment of data glove and touch screen, the user can physimechanically interact with (virtually) presented objects in 2D/3D space. Virtual reality is by all means the direct application of such gestures, in which the user gesticulates with the virtual environment with relatively free gestures, as if he or she were also a part of that world. The use of the data glove can even improve the impression of authenticity by providing the user with feedback in response to the gesture input such as through pressure and temperature.

1.2 Approaches in Gesture Recognition

There are different methods in the field of gesture analysis. These methods are shown in the following with respect to sensor and recipient.

From the View of the Sensor

The very first attempts were data gloves, which were equipped with light sensors made of fibreglass (Zimmerman and Lanier, 1987; Marcus and Churchill, 1988; Eglowstein, 1990). The light sensors installed on the fingers convert each finger movement, like bending, rotation and outstretching, into analog signals, which are in turn used to calculate the angles of the joints of the fingers, their respective positions and the orientation of the hand. Different configurations and postures of the hand can be interpreted as commands for the computer. However, the data glove has a big disadvantage due to its unwieldiness: The user has to carry a lot of hardware with him- or herself which consequently makes this kind of gesture interaction unnatural.

Table 1. An overview of gesture analysis systems

Author (see references)	Hardware	Interaction	System
Oviatt (1999)	Pen	Gesture, speech	Service transaction
Rubine (1991)	Mouse	Mouse-based	GRANDMA
Buxton et al. (1985)	Mouse	Mouse-based	Editor for music note
Waibel and Yang	Graphic tablet, pen	OCR	INTERACT
Raab et al. (1979)	Magnetic field	–	Person tracking
Bolt (1980)	ROPAMS	gesture, speech	"Put-that-there"
Azuma (1993)	Ultrasonic	–	Person tracking
Fels and Hinton (1993)	VPL DataGlove	Speech synthesis	Glove-Talk
Kurtenbach and Baudel (1992)	VPL DataGlove	Presentation	HyperCard
Wu and Huang	Camera	Hand posture	Paper–Rock–Scissors
Quek (1995)	Camera	Hand gesture	Finger mouse
Kettebekov and Sharma	Front camera	Gesture, speech	*i*Map
Akyol et al. (2001)	Infrared camera	Gesture	Car infotainment

Electromagnetic fields (Raab et al., 1979) for gesture localization and recognition are also popular. However, the high acquisition costs, the sensitivty to noise and the short working range are the negative factors, which must be accounted for. In contrast, the video-based gesture analysis with the help of a CCD camera and the corresponding image processing technique seems more promising on account of its uncumbersome hardware. By using a camera, a set of modalities in addition to hand gestures can be integrated into the HCI, e.g., lip, gaze direction, head movement and interpretation of facial expressions. Several typical systems of gesture analysis are listed in Table 1, some of which operate simultaneously with speech. There are also other methods, which use special hardware to achieve high throughput and efficiency, e.g., the SiVit (Siemens Virtual Touchscreen) unit introduced in Maggioni (1995). SiVit is also integrated in SmartKom.

From the View of the Recipient

Video-based gesture analysis is advantageous according to the comparison above, thanks to the ever-improving efficiency and capacity of the computer hardware nowadays. The current major problem lies in the increasing demand for algorithms, which should be fast, robust, traceable, efficient, reliable, modularized and fault tolerant. There are mainly three different methods in the video-based gesture analysis: marker-based, hand model-based and view-based. Due to the nonconvex volume of the hand, many researchers attach markers to the hands, which are placed at certain positions of the hand. Normally, they have a special color or geometric form, with which the detection of hand and fingers becomes easier such as in case of occlusion of some part of the hand without markers. This is an indirect method and thus makes the gesture interaction unnatural. The hand model provides a full-fledged modeling of the respective finger joints and postures. Therefore, this method is able, theoretically, to analyze any gesture, given enough training data. In practice, however, the computing complexity and the lack of efficiency hinder its spread, although it can shed

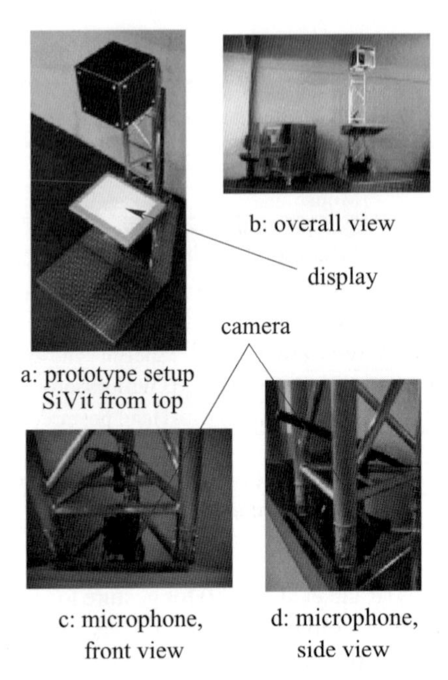

Fig. 1. SMARTKOM demonstration system. **a** prototype with integrated SiVit; **b** overall view, with camera for facial expression analysis and microphones for speech analysis (**c,d**)

light on the solution of many practical problems. The view-based method utilizes the pixel values as its starting point, which can be either directly used as features or be converted to a suitable form through some transformations. It has a relatively small computing intensity and is therefore preferred in practice.

2 Module description

2.1 Gesture in SmartKom

Figure 1 shows the set up of an intended SMARTKOM system with an integrated SiVit unit at the top of the machine. A similar version of this system was used to collect the gesture data in the Wizard-of-Oz experiments. The SiVit unit consists of a video projector, an infrared camera and a virtual touch screen, which is not sensitive to vandalism. The system works in the following manner: The video projector projects all the graphical user interface (GUI) information onto the display, where the user can use his hand to select or search for objects. The infrared camera captures the trajectory of the users hand for the gesture analysis. Gestures are captured together with the recording of the face via video camera, and speech through a microphone array. The positions of these components are pointed out in Fig. 1.

In SMARTKOM the hand gesture is used in two categories: object manipulation and contribution to user state recognition. An introduction to the latter subject is given separately in Streit et al. (2006), while experiments referring to user state recognition as well as object manipulation are shown in the following section.

2.2 Work Course of the Gesture Module

Apart from selecting virtual objects by gesture, the user state, which is expressed through gesture and describes the mood of the user, influences to some degree the way of gesturing: If the user gets annoyed, his gesture tends to be quick and iterating, while it becomes short and determined if the user is satisfied with the service and the information provided by the system. Both gesture and speech indicate the user state and both complement each other. Thus, we will base our experiment on a joint sample set of speech, gesture and facial expression. Since we deal with constantly changing user states, it is clear that the central point of this issue is concentrated on the dynamics of the gesture and its interpretation, instead of focusing on its segmentation from background. In SMARTKOM, all exchanged information packets are coded in an XML format.

2.3 Data flow in the gesture module

Figure 2 shows the data flow in the gesture module, which consists of two main input streams and two output streams, all coded in the XML format. Based on the two assignments of the gesture model, it reads from the data pool *generated.presentation* the geometric coordinates of the virtual objects on the GUI surface, and from *recognized.gesture* the trajectory of the gesture. After aligning the time stamps of these two packets and parsing these two XML packets, the module decides which object the user has chosen or manipulated with regard to the pointing gesture position and virtually depicted GUI objects. In this scenario, the hand gesture takes over the role of a mouse. Afterwards an object hypothesis will be generated in XML format and sent to the *gesture.analysis* pool, whose content can be evaluated further by other modules to respond to the user gesture input by calling the corresponding service such as cinema information or TV programs.

In the case of user state recognition a similar process happens, where the raw gesture data go through XML parsing, feature extraction and classification by Hidden Markov Models (HMMs). As a result, the recognized user state hypotheses will also be sent to the user state pool.

2.4 Hidden Markov Models and Gesture Analysis

HMMs are a suitable model to incorporate temporal continuity. Temporal continuity here means that a pixel of the gesture trajectory belongs to a certain category (state) for a period of time. If a pixel moves at a high speed at a given time, it is likely that this pixel will still keep moving fast at the next time step. HMMs are able to learn the observation distributions for different categories (hidden states) from the trajectory of the gesture. The training data are recorded in a system similar to the one depicted

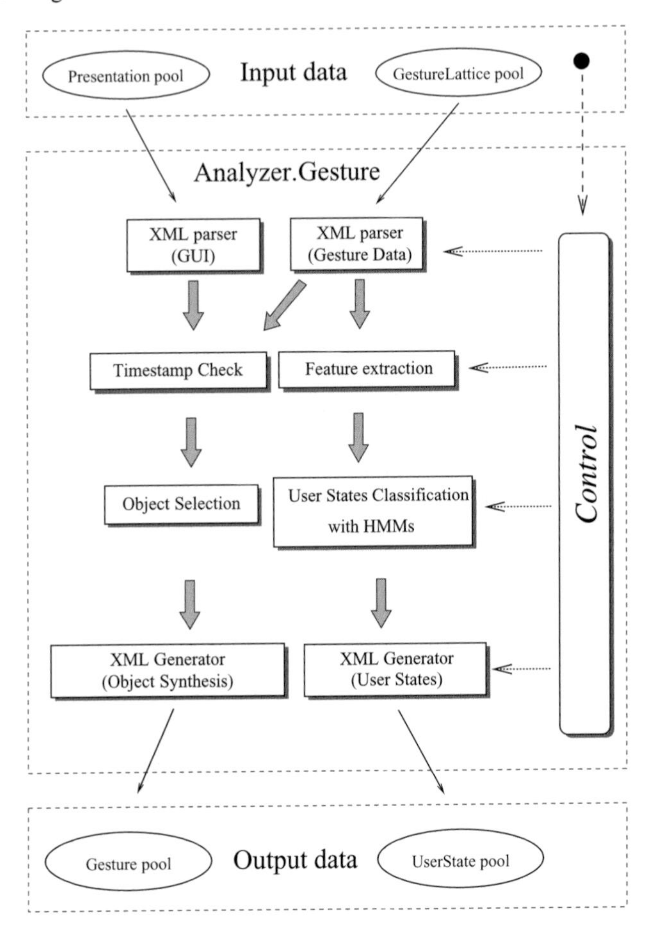

Fig. 2. Architecture of the gesture module

in Fig. 1. In this paper, each feature vector will be assigned to one of three or four hidden states of the HMM (see Sect. 2.7).

We use the standard *Baum–Welch* reestimation algorithm for the training, which is based on the expectation maximization (EM) algorithm (Rabiner and Juang, 1986), and the standard *Forward Algorithm* to solve the classification problem. A detailed description of these algorithms can be found in Rabiner and Juang (1986, 1993); an example of how to apply these algorithms can be found in Rabiner (1989). Here we use discrete HMMs because of their simplicity.

2.5 Feature Extraction

In order to incorporate the temporal continuity, we choose four features: trajectory variance, instantaneous speed, instantaneous acceleration and kinetic energy as the feature vector, which best represents the motion and the dynamics of the gesture. The

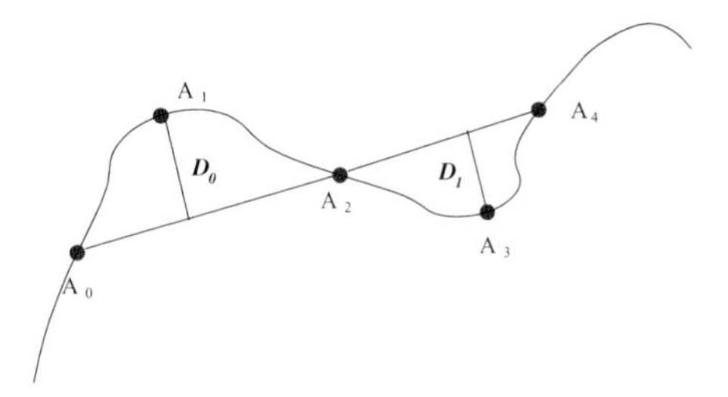

Fig. 3. Typical trajectory of a pointing gesture with sampling points $A_I, I = 0...N$ points and geometric variances $D_I, I = 0..N-1$, which show the intensity of fluctuation of the gesture

continuous two-dimensional coordinates (trajectories) and the time stamp, which are recorded by the SiVit unit, are the most important information on the dynamics of the gesture. The reason for computing the instantaneous velocity v over time is for the system to learn from the behavior of the user's gesture. That is, with simple data analysis, it would be possible to determine trends and anticipate future moves of the user. The next set of data points is the acceleration a of the gestures, which is easily computed by approximating the second derivative of the position coordinate. Kinetic energy K, which is just the square of the velocity while the mass is neglected, is also a significant factor.

In our feature set, the trajectory variance is also included. This is the geometric variation or oscillation of the gestures with respect to their moving direction. A large value of this variance can indicate that the user gesticulates hesitantly and moves his or her hand around on the display, while a determined gesture leads to a small variance. Figure 3 shows how the trajectory variance D is computed. So we have a feature vector

$$f = (v, a, K, D). \tag{1}$$

The vector D can be computed every N points along the gesture trajectory. Other possible features are, e.g., the number of pauses of a gesture, the transient time before and after a pause, the transient time of each pause relative to the beginning of the gesture, the average speed, and the average acceleration or change of moving direction. However, in this study we just consider the feature vector shown in Eq. (1).

2.6 Modification of User State Classes

As mentioned above, the goal of SMARTKOM is the combination of all three input modalities. Gesture, as one of the input channels, must define its own output to contribute to the fusion of the analysis of the three modalities. In contrast to facial and prosodic analysis, where four user states are defined, *neutral, angry, joyful* and *hesitant*, we define in gesture analysis only three user states: *determined, angry* and *hesitant*. The reason for making this mapping is the intuition that normally people cannot tell if the user is neutral or joyful by only observing his or her gesture. This

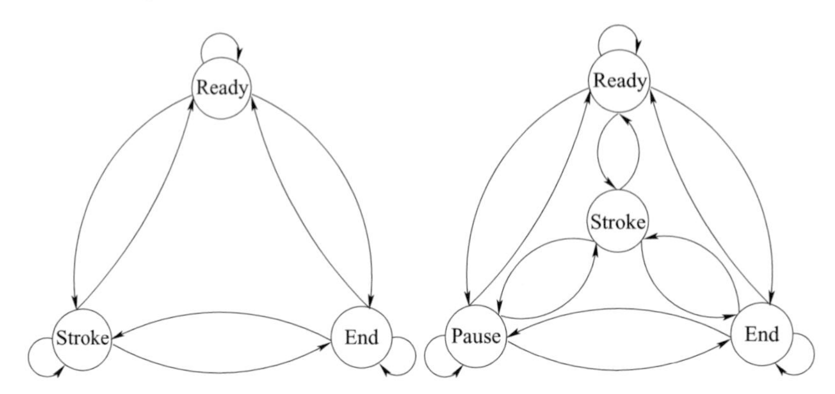

Fig. 4. Ergodic HMMs with different numbers of hidden states for gesture analysis

was confirmed by the preliminary experiments, where *neutral* and *joyful* had a high confusion. We decided thus in favor of the three states topology. The user state *determined* is given if the user knows what he wants from SMARTKOM, e.g., if he decides to zoom in a part of a city map on the GUI by pointing to it. If the user gets confused by SMARTKOM and does not know what to choose, his gesture will probably ponder around or zigzag among different objects presented on the SMARTKOM GUI. Finally, if he feels badly served by SMARTKOM or if the information presented is not correct, he can use gestures in such a way as to show a strong negative expression like a windshield wiper, which in our context corresponds to the user state *angry* in facial expression.

2.7 Choice of Different Topologies

For the HMMs, we evaluated different topologies; an HMM with three or four states gave the best results. We suppose that a gesture consists of some basic states such as *ready*, *stroke*, *end* and/or *pause*. This can be observed in the production of the gesture: The user moves her hand to a start position, then makes a gesture consisting of several strokes, probably with pauses in between, and finally ends her gesture. An alternative is to merge *pause* and *ready*. We also tried different connection schemata; the simplest one is an ergodic HMM, while a partially connected HMM better corresponds to the correct physical order of each state (Fig. 4). The conventional left–right HMM model is also an alternative that has been successfully used in speech recognition.

3 Experiments

Tables 2, 3 and 4 show the results of the gesture analysis. We can see that the user state *hesitant* is sometimes mismatched with *angry*. The reason is that some users, whose gestures are used in the training set, made similar gestures like those in *angry*

Table 2. Confusion matrix of user state recognition with gesture data using ergodic HMM (see Fig. 4, CL: classwise averaged recognition rate)

Reference user state	3 HMM states (%)			4 HMM states (%)		
	Determined	Hesitant	Angry	Determined	Hesitant	Angry
Determined	**61**	5	34	**80**	15	5
Hesitant	5	**72**	23	15	**77**	8
Angry	10	6	**84**	10	18	**72**
CL	72			76		

Table 3. Confusion matrix of user state recognition with gesture data using ergodic HMM (LOO, see Fig. 4, CL: classwise averaged recognition rate)

Reference user state	3 HMM states (%)			4 HMM states (%)		
	Determined	Hesitant	Angry	Determined	Hesitant	Angry
Determined	**62**	5	33	**75**	7	18
Hesitant	5	**74**	21	13	**74**	13
Angry	8	8	**84**	30	8	**62**
CL	73			70		

Table 4. Confusion matrix of user state recognition with gesture data using nonergodic HMM (CL: classwise averaged recognition rate)

Reference user state	3 HMM states (%)			4 HMM states (%)		
	Determined	Hesitant	Angry	Determined	Hesitant	Angry
Determined	**72**	16	12	**40**	49	11
Hesitant	32	**45**	23	2	**70**	28
Angry	60	12	**28**	2	24	**74**
CL	48			61		

states, in that the windshield wiper movement has the same zigzag only with different dynamics and speed. Probably, some persons gesticulate slowly while indicating anger, thus their corresponding gestures may have similar properties like those of a *hesitant* state.

Table 5. Confusion matrix of User state recognition using left–right HMM (CL: classwise averaged recognition rate)

Reference user state	3 HMM states (%)			4 HMM states (%)		
	Determined	Hesitant	Angry	Determined	Hesitant	Angry
Determined	**63**	4	33	**66**	6	28
Hesitant	6	**47**	47	13	**51**	36
Angry	20	4	**76**	30	4	**66**
CL	62			61		

Another reason for a wrong classification is that the training data for the user state *determined* consists of those from *joyful* and *neutral*; the latter makes the HMM for *determined* biased towards *hesitant* in Table 4 with four internal states. In general, the classification has a classwise (CL) averaged recognition rate of 72% for three internal states and 76.3% for four internal states, while the leave-one-out (LOO) test achieves 73% for three internal states and 67% for four internal states. Table 5 shows the recognition result when using a conventional left–right HMM model.

4 Conclusion

Gesture is an important communication channel in HCI, whose usage ranges from direct manipulation of object to indication of user states as shown above. These two models have been successfully integrated into the SMARTKOM demonstrator, which runs as an autonomous service agent between the user and different information sources through gesture, speech and facial expression. The user is, therefore, free to communicate with the system, similar to talking with another human. Furthermore, the gesture, speech and facial expression complement each other in a redundant way so that the demand of precise expression in each modality can be relaxed and thus extend the applicability with respect to prospective users.

References

S. Akyol, L. Libuda, and K.F. Kraiss. Multimodale Benutzung adaptiver Kfz-Bordsysteme. In: T. Jürgensohn and K.P. Timpe (eds.), *Kraftfahrzeugführung*, pp. 137–154, Berlin Heidelberg New York, 2001. Springer.

V. Attina, D. Beautemps, M.A. Cathiard, and M. Odisio. Toward an Audiovisual Synthesizer for Cued Speech: Rules for CV French Syllables. In: J.L. Schwartz, F. Berthommier, M.A. Cathiard, and D. Sodoyer (eds.), *Proc. AVSP 2003 Auditory-Visual Speech Processing*, pp. 227–232, St. Jorioz, France, September 2003. ISCA Tutorial and Research Workshop.

R. Azuma. Tracking Requirements for Augmented Reality. In: *ACM*, vol. 36, pp. 50–51, July 1993.

R. Bolt. "Put-That-There": Voice and Gesture. In: *Computer Graphics*, pp. 262–270, 1980.

W. Buxton, R. Sniderman, W. Reeves, S. Patel, and R. Baecker. An Introduction to the SSSP Digital Synthesizer. In: C. Roads and J. Strawn (eds.), *Foundations of Computer Music*, pp. 387–392, Cambridge, MA, 1985. MIT Press.

R.O. Cornett. Cued Speech. *American Annals of the Deaf*, 112:3–13, 1967.

H. Eglowstein. Reach Out and Touch Your Data. *Byte*, 7:283–290, 1990.

S. Fels and G.E. Hinton. Glove-Talk: A Neural Network Interface Between a Data-Glove and a Speech Synthesizer. In: *IEEE Transactions on Neural Networks*, vol. 4, pp. 2–8, 1993.

S. Kettebekov and R. Sharma. Multimodal Interfaces. http://www.cse.psu.edu/~rsharma/imap1.html. Cited 15 December 2003.

G. Kurtenbach and T. Baudel. Hypermarks: Issuing Commands by Drawing Marks in Hypercard. In: *ACM SIGCHI*, p. 64, Vancouver, Canada, 1992.

T. Lütticke. Gestenerkennung zur Anweisung eines mobilen Roboters. Master's thesis, Universität Karlsruhe (TH), 2000.

C. Maggioni. Gesture Computer — New Ways of Operating a Computer. In: *Proc. Int. Conf. on Automatic Face and Gesture Recognition*, pp. 166–171, 1995.

A. Marcus and J. Churchill. Sensing Human Hand Motions for Controlling Dexterous Robots. In: *The 2nd Annual Space Operations Automation and Robotics Workshop*, Dayton, OH, July 1988.

D. McNeill. *Hand and Mind: What Gestures Reveal About Thought*. University of Chicago Press, Chicago, IL, 1992.

S. Oviatt. Ten Myths of Multimodal Interaction. *Communications of the ACM*, 42 (11):74–81, 1999.

F. Quek. FingerMouse: A Freehand Pointing Interface. In: *Int. Workshop on Automatic Face- and Gesture-Recognition*, pp. 372–377, Zurich, Switzerland, June 1995.

F.H. Raab, E.B. Blood, T.O. Steiner, and H.R. Jones. Magnetic Position and Orientation Tracking System. In: *IEEE Transaction on Aerospace and Electronic Systems*, vol. 15, pp. 709–718, 1979.

L.R. Rabiner. A Tutorial on Hidden Markov Models and Selected Applications in Speech Recognition. In: *Proc. IEEE*, vol. 77, pp. 257–286, 1989.

L.R. Rabiner and B.H. Juang. An Introduction to Hidden Markov Models. *Acoustics, Speech and Signal Processin*, 3(1):4–16, 1986.

L.R. Rabiner and B.H. Juang. *Fundamentals of Speech Recognition*. Prentice Hall, Englewood Cliffs, NJ, 1993.

D. Rubine. Specifying Gestures by Example. In: *SIGGRAPH '91 Proceedings*, vol. 25, pp. 329–337, New York, 1991.

E. Sachs. Coming Soon to a CAD Lab Near You. *Byte*, 7:238–239, 1990.

M. Streit, A. Batliner, and T. Portele. Emotion Analysis and Emotion Handling Subdialogs, 2006. In this volume.

A. Waibel and J. Yang. INTERACT. http://www.is.cs.cmu.edu/js/gesture.html. Cited 15 December 2003.

Y. Wu and T.S. Huang. "Paper–Rock–Scissors". http://www.ece.northwestern.edu/~yingwu/research/HCI/hci_game_prs.html. Cited 15 December 2003.

T.G. Zimmerman and J. Lanier. A Hand Gesture Interface Device. In: *ACM SIGCHI/GI*, pp. 189–192, New York, 1987.

Part III

Multimodal Dialogue Processing

Modality Fusion

Ralf Engel and Norbert Pfleger

DFKI GmbH, Saarbrücken, Germany
{engel,pfleger}@dfki.de

Summary. In this chapter we give an general overview of the modality fusion component of SmartKom. Based on a selection of prominent multimodal interaction patterns, we present our solution for synchronizing the different modes. Finally, we give, on an abstract level, a summary of our approach to modality fusion.

1 Introduction

SmartKom provides access to a rich set of applications by giving the user the opportunity to employ various multimodal interaction patterns. This means that the user has the possibility not only to switch between the different modalities — depending on which modality is more suitable for a present situation — but also to use those modalities in an integrated and effective manner.

Starting with Bolt's "Put-that-there" system (Bolt, 1980), the history of multimodal dialogue systems has shown a wide variety of different approaches. However, nearly every system comprises a component that is capable of integrating the data streams of the different modalities into one combined representation (e.g., Cohen et al. (1997); Johnston et al. (2001); Wahlster (1991)). Basically, there are two major architectural approaches for the analysis part of multimodal dialogue systems: (i) early fusion — fusing modes already at signal or recognition level (Bregler et al., 1993; Pavlovic and Huang, 1998) — and (ii) late fusion – multimodal semantic processing (Johnston et al., 1997; Johnston and Bangalore, 2001). Even though the point within the processing pipeline and the type of processed data vary, it is generally accepted that fusing the separate data streams will improve system performance by reducing uncertainty and resolving ambiguous input hypotheses. Another interesting aspect of multimodal interaction is that users can adapt their strategies of how to use the available modalities according to their current needs. Users show, according to Oviatt et al. (1997), a strong preference for multimodal commands and requests during map tasks. For example, if a user wants to set the starting point for planning a route it, is easier and more precise to use a pointing gesture toward the desired starting point on a map and accompany it with a spoken utterance containing a deictic reference (e.g., "I want to start here") than giving "exact" coordinates by speech (e.g., "I want to start about two and a half centimeters left from the church").

In SMARTKOM the integration and disambiguation of the different modalities is handled by the modality fusion component, called FUSION. This component is also responsible for initiating the resolution of those referring expressions that cannot be resolved by gestural input.

1.1 System Context

The above-mentioned tasks of the modality fusion component require communication with several other SMARTKOM modules. The main data communication channels (Fig. 1) are, on the input side, two channels bringing in data from the modality-specific analyzers and, on the output side, the channel transmitting the integrated intention hypotheses to the intention recognition module (te Vrugt and Portele, 2006).

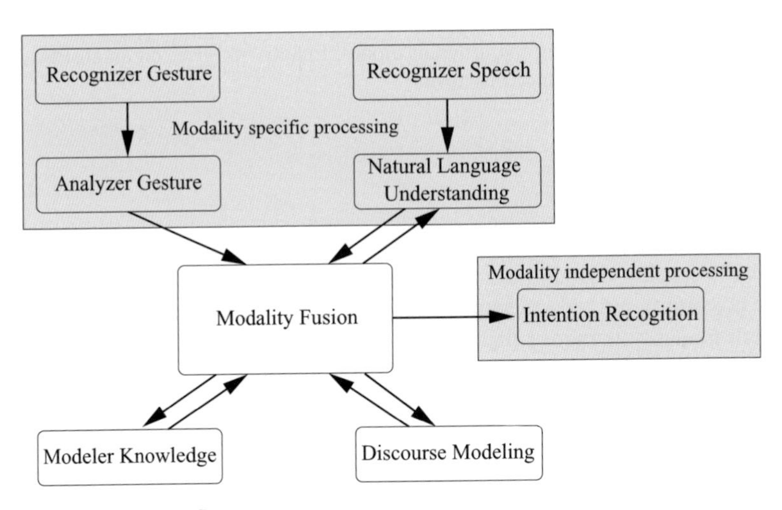

Fig. 1. Sketch of the dataflow

The gesture analysis component processes coordinates provided by a SiViT device (in case of SMARTKOM -Mobile by the pen interface) and a so-called *screen state* representing the currently displayed objects. The data FUSION receives from the gesture analysis component consist of a ranked list of probably referenced objects. The language understanding component provides a set of different hypotheses, each containing a semantic representation of a possible path within the word hypotheses graph provided by the speech recognizer (Berton et al., 2006). Important for FUSION is the fact that in SMARTKOM an "open microphone" approach is used and the gesture channel is always open, too. So, FUSION must be able to handle input at any time and provide meaningful ouput.

As we will see in the following section, FUSION needs access to information provided by two additional modules — modeler knowledge and modeler discourse.

The first one provides information about the current location of the system, which is needed for the resolution of expressions referring to the situational context, and the second component accomplishes the resolution of those referring expression that cannot be resolved without access to the discourse context.

As described in Gurevych et al. (2006), (nearly) all modules of SMARTKOM employ the same knowledge representation, and thus all meaningful data occur in the same format. For FUSION it is important that this data structure can be viewed in terms of *frames* (Minsky, 1975) or *typed feature structures* (Carpenter, 1992) since those data structures allow for recursive merging and matching of two data structures, thereby generating a fused or intergrated representation of them.

2 Multimodal Interaction Patterns: Phenomena and Processing Strategies

SMARTKOM enables the user to use speech and gestures in an integrated and efficient manner and provides wide coverage of multimodal utterances expressing the same intention. This freedom of interaction leads to a broad set of challenging interaction patterns that FUSION has to deal with. SMARTKOM incorporates three modalities (speech, gesture, and facial expressions). However, the task of FUSION is to combine only two of them, namely speech and gesture.

Besides the integration of the different modalities, FUSION has also the challenge to disambiguate the multiple hypotheses produced by the different recognizers. Therefore, FUSION computes for each integrated result a score based on a set of different parameters (Sect. 5). This score contributes to the final score for a hypothesis.

In the following section we will consider a smaller subset of prominent interaction patterns and show on an abstract level how FUSION handles them. For a detailed consideration of the underlying architecture and processing techniques, see Sect. 4.

2.1 Referring Expressions

An improtant feature of natural language is the possibility to refer to entities and events in the physical or mental environment verbally. These references occur often as deictic references, which means that the spoken referring expression (e.g., a demonstrative pronoun) is accompanied by a manual gesture selecting a particular object in the physical environment. Here, the task of FUSION is to combine the spoken and gestural expression into a single semantic representation of the communicative act (Johnston et al., 1997).

Consider, for example, the following sentence:

(1) User: *I want to go from here [↗] to there [↗]*

On an abstract level the processing strategy of FUSION for pointing gesture is dominated by the speech modality — it is assumed that each pointing gesture is accompanied by a spoken utterance (the only exception is called a direct gesture; see

Sect. 2.2). [1] The natural language understanding component (Engel, 2006) assigns to each referring expression a special marker (called *refProp*), thereby indicating that such an object is not yet fully specified and needs to be enriched with information of an appropriate referent. Then, FUSION tries to resolve those refProps by generating all possible combinations of refProps and recognized gestures that match the type of the expected referent (in this case a kind of LOCATION object).

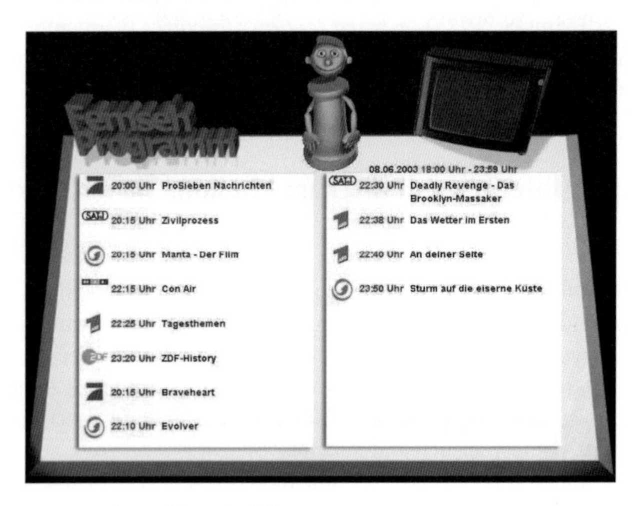

Fig. 2. System presentation of the television program

Resolving those multimodal references as shown in example (1) is quite straightforward. However, things become more complicated when the type of the referent inferred from the intrasentential context is a substructure of the referenced screen-object. Consider, for example, a dialogue situation where the user is browsing the television program (Fig. 2) and wants to tape the currently running broadcast on a particular channel. In SMARTKOM each object presented on the screen is internally annotated with a representation of itself in terms of the SMARTKOM domain model (for more details, please see Poller and Tschernomas (2006)). For broadcasts this representation includes a slot that represents the channel on which the broadcast is running. Utterance (2a) can be treated in the same way as utterance (1). However, in order to be able to handle utterance (2b), FUSION needs access to the substructures of the referenced screenobjects. Therefore, FUSION needs at least some domain knowledge for extracting the correct information from the referenced objects.

(2a) User: *Please tape this broadcast [↗]*
(2b) User: *Please tape this channel [↗]*

[1] However, it should be noted that treating the speech modality as the primary input mode is not acceptable for all kinds of multimodal dialogue systems. For example, such systems processing speech and pen input will fail to handle many types of spontaneous multimodal construction (Oviatt, 1999).

2.1.1 Selecting Sets of Objects: Encircling Gestures

Encircling gestures are a specialization of pointing gestures. The user selects a set of objects displayed on the screen with one continuous gesture. In contrast to pointing gestures, encircling gestures require a somehow more elaborate processing to be able to handle interdependencies with the previous dialogue context correctly. Additionally, the integration of multiple objects requires world knowledge about the involved objects and processes to be able to integrate the encircled objects in a sensible way. Consider, for example, a situation where the system displays a seat reservation plan and asks the user to select the desired seats.

(3) User: *I want to reserve these three seats [encircling gesture]*

The gesture analysis (Shi et al., 2006) sends weighted lists of all objects captured by the encircling gesture to FUSION. However, the ranking reflects only salience factors detached from a specific dialogue situation. So, FUSION has to deal with two possible problem sources:

(i) The number of selected objects does not necessarily correlate with the desired number of objects provided by speech.
(ii) The most probable objects are not necessarily those the user intended to select, nor do they have to make sense when viewed with respect to the current dialogue situation.

Usually, the first problem source is easily removed when the user has directly specified the desired number of objects in an utterance accompanying the encircling gesture (like in the previous example). However, another possible course of interaction is when the user has already provided the desired number of objects in a previous turn and now performs an encircling gesture without saying anything. In that case FUSION requests the discourse modeling component (Alexandersson and Pfleger, 2006) in order to retrieve the size of the last collection object.

The second problem source becomes obvious when considering Fig. 3. It is a difficult task to encircle exactly the desired seats out of a seating plan comprising those small seats. So, it is possible that FUSION receives a set of encircled seats where the most probable ones are located in different tiers (Fig. 3). To avoid presenting the user with a somewhat odd seat selection, FUSION applies a set of plausibility tests. These tests depend on the type of selected objects and comprise, for example, a constraint stating that as many seats as possible should belong to the same tier.

2.2 Unimodal Interaction: Using only a Single Modality

Besides the integrated usage of both modalities, the user has the possibility to select one of the modalities with respect to the current task or dialogue situation, for example, a gesture referencing an object on the screen that answers a system question. On an abstract level those interaction patterns seem to be quite easily manageable for FUSION. However, viewed on a detailed level it becomes clear that we need a sophisticated synchronization mechanism in order to be able to decide whether a gesture

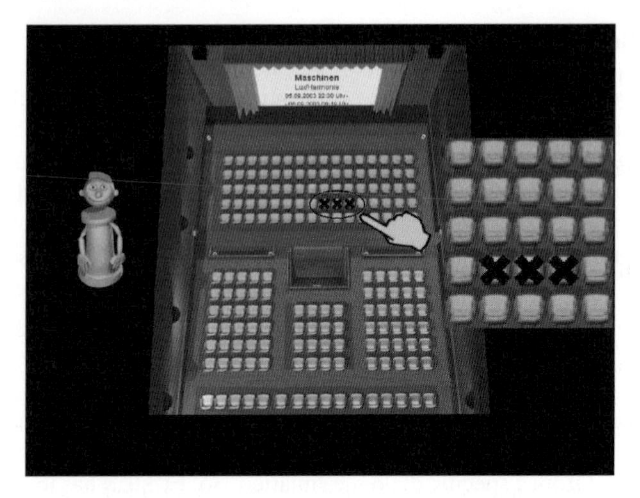

Fig. 3. Objects selected with an encircling gesture

event can be interpreted on its own or needs to be fused with a spoken utterance without adding any additional delays. Consider, for example, the following dialogue situation: a user is asked to select a point on a map to determine the starting point of a route plan and reacts with a pointing gesture selecting a particular object on the map. After a few seconds the user says something like: “Please zoom in”.

Since the two processing chains are not synchronized, FUSION first receives the result of the gesture analysis thread and after a few seconds (plus the processing time needed by the speech recognizer and the natural language understanding component) the result of the speech thread. Interpreting the gesture on its own in this dialogue situation would lead to a hypothesis stating that the user wants to start at this location, but as the user provides additional information on the speech channel this interpretation is not feasible. However, to obtain a fast system response it is only sensible to wait for a speech event if there are any hints suggesting that the user actually uttered something. Additionally, FUSION will only pass on gesture results that already match the expectations of the action planner (Löckelt, 2006). The synchronization mechanism coupling the input modules is descriped in Sect. 3.

2.3 Referring to the Situational Context

The referring expression “here” requires a specical treatment since it can be used either as referring to the situational context (User: “*I want to park here*”) or in combination with a pointing gesture like, for example, User: *“I want to start here,”* accompanied by a pointing gesture. If this referring expression is used in the first way FUSION needs to know the current location of the system in order to integrate it into the underspecified location object of the spoken request. This information is provided by the module called modeler knowledge (Porzel et al., 2006) and can be requested via a special data pool.

The challenge, however, is to identify when the adverb "here" occurs without any accompanying gesture whether it was used with respect to the situational context or with respect to the previous discourse context (e.g., when the user has requested during route selection dialogue information about a particular restaurant and utters in the next turn *"I want to park here."*

3 Synchronizing Different Modalities

In SMARTKOM the input channels are always open, i.e., the user can speak or perform gestures at any time. Additionally, the user can interact with the system using speech or gestures, alone and the user can also combine several gestures with one spoken utterance.[2] This implies the need for a mechanism deciding which gestures are combined with which speech input.

Since both speech and gesture messages contain time stamps indicating the start time, and the end time it is a simple task to decide which gestures and spoken utterances were performed at the same time. But often, the user does not speak and perform the gesture at the same time, instead, e.g., the user performs the gesture one second before starting speech or half a second after finishing speech. Therefore, gestures are also combined with spoken utterances if they occur at a certain time span before or after speech, further referred as t_{before} and t_{after}. Our own experiments have shown that 1.5 s for both time spans is a reasonable value.

To achieve the behavior described above, FUSION reacts depending on the type of input in a different way. The next two sections describe the respective procedure.

3.1 Messages Containing Speech Input

When a message containing speech input arrives, the module should let a time span $t_{after} - (t_{current} - t_{end\_of\_speech})$ pass waiting for new gestures until the combination of speech and gesture is started. But because of delay caused by the detection of the ending of speech and the processing time of the speech recognizer and natural language processing, $t_{current} - t_{end\_of\_speech}$ is in SMARTKOM always greater than t_{after}. Therefore, the computation of the combined speech and gesture hypotheses can start immediately. If no gestures are available for the time between $t_{start\_of\_speech} - t_{before}$ and $t_{end\_of\_speech} + t_{after}$ the speech hypotheses are passed to the intention analyzer without any changes.

3.2 Messages Containing Gesture Input

The treatment of a message containing a gesture is somewhat more complicated. As described above, it is not possible that speech input is already available for this

[2] Several spoken utterances need not be combined by FUSION since the silence detection within the audio module already decides the turn boundaries for speech.

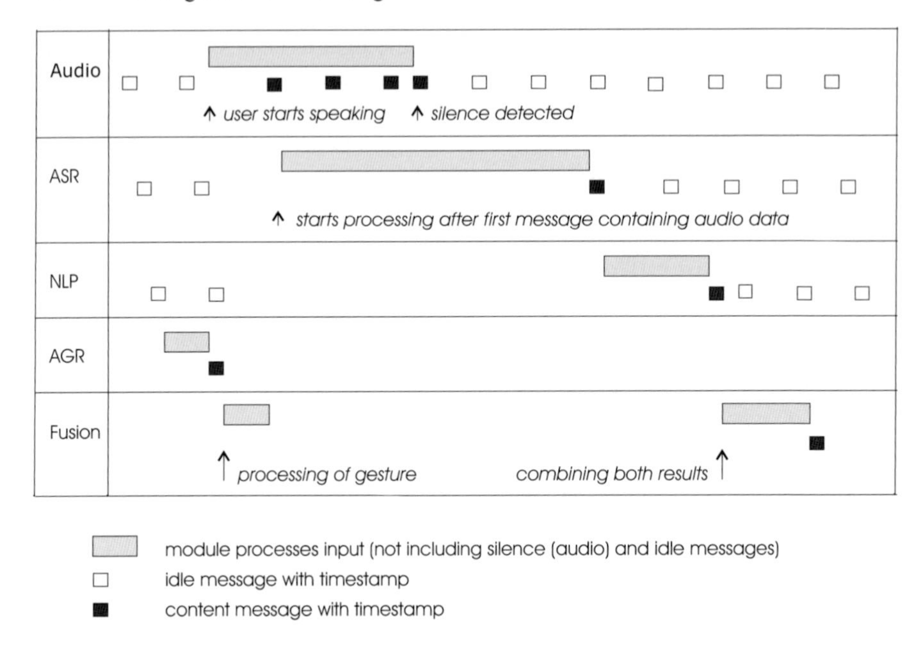

Fig. 4. Illustration of module activity and output messages

gesture due to the delays during the processing of speech.[3] Therefore, FUSION stores the gesture and waits for speech input.

Since it is possible for the user to perform gestures without any speech, it is not guaranteed that a corresponding speech message will arrive at all. A possible solution would be to wait for a timeout of $t_{before} - (t_{current} - t_{end\_of\_gesture})$ until the gesture is passed without any modifications to the intention analyzer. But as in most cases the processing time of the speech recognizer is not negligible, the timeout would be triggered falsely most of the time. At least two different modifications to the communication between the speech recognizer and FUSION are able to solve that problem:

1. The speech recognizer sends a message when it starts working. This message stops the timeout in FUSION. At the end the speech recognizer sends a result in any case, even if nothing meaningful was recognized. The drawback of this approach is that delays within the speech recognizer, e.g., the recognizer is currently updating its language model and does not process audio messages at the same time, as well as delays within the communication channels can nevertheless lead to erroneous timeouts.
2. FUSION does not use a timeout at all. Instead, when the speech recognizer does not process input, it sends idle messages to FUSION in regular intervals, e.g., ev-

[3] Otherwise it must be verified whether speech input is available for the time between $t_{begin\_of\_gesture} - t_{after}$ and $t_{end\_of\_gesture} + t_{before}$.

ery 500 ms. The idle messages contain time stamps so that they are independent from delays in the communication channels. As soon as FUSION receives an idle message it sends all gestures that satisfy $t_{end\_of\_gesture} + t_{before} < t_{idle}$ as gesture-only messages to the intention analyzer. The disadvantages of this approach are the increased communication overhead and the increased latency due to the time span between two idle messages.

In SMARTKOM the prevention of delay problems within the speech recognizer and the communication channels was considered to be more important than additional latencies and the increased communication overhead. Therefore, the second solution was chosen. An example for synchronization is shown in Fig. 4. Note that if the gesture recognition was more time consuming, such a synchronization mechanism would be required for gesture processing, too.

4 Integrating the Modality Hypotheses

A slot-filling approach was developed for the integration of speech and gesture. The term slot-filling approach means that the resulting hypotheses are produced by inserting the hypotheses of one modality into the hypotheses of the other modality. Such an approach is ideal in the context of SMARTKOM since it is powerful enough to handle all occurring tasks and requires no further knowledge bases besides the already existing SMARTKOM ontology.

4.1 Design Goals

Before the developed approach is presented in detail, the two most important issues that influenced the design should be mentioned: multimodal disambiguation and robust processing. Multimodal disambiguation means that ambiguities occur in one modality can be solved when the modalities are combined. Ambiguous input is represented as a sequence of several hypotheses in M3L. As a consequence the integration algorithm does not have to consider only the best hypothesis of each modality but also lower scored ones. Furthermore, something like a scoring function is required to detect within the generated results the ones in which disambiguation was successful.

Robust processing means that even in the case of input containing either recognition errors or errors produced by the user, FUSION still attempts to produce meaningful results. Robustness concerning speech input includes dealing with not-uttered or not-recognized referring expressions or additional referring expressions that were wrongly recognized. Robustness concerning gesture input includes imprecise pointing gestures that can be corrected using additional spoken information (e.g., the user points at a location close to a church and utters *information about this church*) and gestures that are performed unintentionally by the user. A good example for the latter case would be if the user rests too long with his finger on the same location when using SiViT gesture recognition.

4.2 Approach

The basic idea of our developed approach is first to generate all meaningful combinations considering all hypotheses and then to select the *n* best results, which are passed to the intention analyzer. In SMARTKOM *n* is set to 3. The advantage of this approach is, besides its relatively easy realization, that during processing no decisions based on local incomplete context have to be made as all results are generated. Potentially, local decisions can lead to solutions that are not optimal if seen from a global point of view. The disadvantage of this approach is that under adverse circumstances (especially in the case of recognition errors or unintended gestures performed by the user) the generation of all meaningful combinations takes too much time. To address this issue processing is stopped after a timeout. The probability to find the correct solution is increased in the case of a timeout by optimizing the algorithms to find more probable solutions first, e.g., processing last gestures first.

4.3 Implementation

A detailed presentation of the used algorithms is out of the scope of this chapter, but a rough sketch should give an impression of the realization of the approach. Keep in mind the initial task that one utterance represented as a set of speech hypotheses and one or more gestures consisting also of several hypotheses have to be combined.

To generate all possible solutions nested loops iterate over the gestures, the speech hypotheses and the different hypotheses the gestures contain.[4] Within the innermost loop one gesture hypothesis is tried to be inserted into one speech hypothesis, which may already contain gestures inserted in previous iterations. To insert the gesture hypothesis five different strategies are applied in sequence, beginning with the strategy getting later the highest score.

1. The gesture hypothesis is compatible with an embedded instance representing a spoken referring expression and replaces it. Compatible means that the type of the gesture hypothesis is of the same type or of a subtype of the referring expression. Furthermore, all slots specified in the gesture hypothesis must have corresponding slots with a corresponding content within the referring expression. Examples covered by this strategy include *Show me information about that church* and *I want to know more about this.*
2. Same as 1, except that the compatible embedded instance is not marked as referring expression, e.g., an uttered determiner was not recognized, like in *Show me information about movie.*
3. The gesture hypothesis is inserted into the speech hypothesis by filling a suitable empty slot. Empty slots are detected with the help of the ontology. This is useful in cases where the referring expression was not recognized or uttered at all, e.g., *Show me information.*

[4] To increase performance, loops may be aborted earlier if further iterations can only produced lower scored solutions.

4. Same as 3, except that the slot is not empty, but is already filled with an incompatible instance. This increases the robustness against severe speech recognition errors.
5. The gesture is not inserted at all. This alternative is necessary if the case occurs that more gestures are recognized than can be inserted.

To find all possible solutions even if one strategy is successful, the subsequent strategies are also applied and may lead to additional solutions. It is also possible that within one strategy several solutions are generated, e.g., there is more than one matching referring expression. The additional solutions are added to the set of speech hypotheses, so usually the number of speech hypotheses increases during processing.

5 Scoring the Generated Results

The scoring function is used for two tasks. First, the function decides which generated results are sent to the intention analyzer. Because of the large number of results, only a reduced number can be sent. In SMARTKOM a maximum of three hypotheses is sent to the intention analyzer. Second, the scoring function determines the so-called fusion score that is attached to each hypothesis. In the intention analyzer this score is used in combination with other scores, e.g., the acoustic score assigned by the speech recognizer, to select the one hypothesis which is passed to the action planner.

Within the scoring function several criteria are considered, all providing their own scores. The scores are listed in order of their importance:

1. how often temporal crossing occurs between modalities, e.g., an earlier gesture replaces a later referring expression and a later gesture replaces an earlier referring expression
2. the fraction of replacements of an instance marked as referring expression
3. the fraction of replacements of an instance not marked as referring expression
4. the score of natural language processing
5. the difference in time between the gestures and their corresponding instances in the spoken utterance, i.e., the instances that the gestures replace

To determine the results that are sent to the intention analyzer, the single scores are compared pairwise in the given order. Only if two single scores are equal is the subsequent score pair considered. The fusion score sent to the intention analyzer is computed as a weighted sum of the above-mentioned single scores.

6 Conclusion

In this chapter we presented the modality fusion component of SMARTKOM. Starting with a general survey of some prominent multimodal interaction patterns, we emphasized important aspects for the integration of the different modes. Then we showed on a detailed level how we solved the so-called synchronization problem and gave a general overview of the core integration algorithm.

The overall performance of the fusion component in SMARTKOM was verified in extensive system tests. Although the fusion component was not evaluated separately, the system tests showed that the fusion component performs very well.

Future research will aim for a more generalized application of the presented algorithms. Instead of representing the interaction patterns directly at the level of algorithms, we are amining for a knowledge-based approach where new interaction patterns can be added in a more flexible manner.

References

J. Alexandersson and N. Pfleger. Discourse Modeling, 2006. In this volume.

A. Berton, A. Kaltenmeier, U. Haiber, and O. Schreiner. Speech Recognition, 2006. In this volume.

R. Bolt. Put-That-There: Voice and Gesture at the Graphics Interface. *Computer Graphics*, 14(3):262–270, 1980.

C. Bregler, H. Hild, S. Manke, and A. Waibel. Improving Connected Letter Recognition by Lipreading. In: *Proc. Int. Conf. on Acoustics, Speech, and Signal Processing (ICASSP-93)*, Minneapolis, MN, 1993.

B. Carpenter. *The Logic of Typed Feature Structures*. Cambridge University Press, Cambridge, UK, 1992.

P.R. Cohen, M. Johnston, D. McGee, S.L. Oviatt, J.A. Pittman, I. Smith, L. Chen, and J. Clow. QuickSet: Multimodal Interaction for Distributed Applications. In: *Proc. 5th Int. Multimedia Conference (ACM Multimedia '97)*, pp. 31–40, Seattle, WA, 1997. ACM.

R. Engel. Natural Language Understanding, 2006. In this volume.

I. Gurevych, R. Porzel, and R. Malaka. Modeling Domain Knowledge: Know-How and Know-What, 2006. In this volume.

M. Johnston and S. Bangalore. Finite-State Methods for Multimodal Parsing and Integration. In: *Finite-State Methods Workshop, ESSLLI Summer School on Logic Language and Information*, Helsinki, Finland, August 2001.

M. Johnston, S. Bangalore, and G. Vasireddy. MATCH: Multimodal Access to City Help. In: *Proc. ASRU 2001 Workshop*, Madonna di Campiglio, Italy, 2001.

M. Johnston, P.R. Cohen, D. McGee, S.L. Oviatt, J.A. Pittman, and I. Smith. Unification Based Multimodal Integration. In: *Proc. 35th ACL*, pp. 281–288, Madrid, Spain, 1997.

M. Löckelt. Plan-Based Dialogue Management for Multiple Cooperating Applications, 2006. In this volume.

M. Minsky. A Framework for Representing Knowledge. In: P. Winston (ed.), *The Psychology of Computer Vision*, pp. 211–277, New York, 1975. McGraw-Hill.

S. Oviatt. Ten Myths of Multimodal Interaction. *Communications of the ACM*, 42 (11):74–81, 1999.

S.L. Oviatt, A. DeAngeli, and K. Kuhn. Integration and Synchronization of Input Modes During Multimodal Human-Computer Interaction. In: *Proc. CHI-97*, pp. 415–422, 1997.

V. Pavlovic and T.S. Huang. Multimodal Tracking and Classification of Audio-Visual Features. In: *AAAI Workshop on Representations for Multi-Modal Human-Computer Interaction*, July 1998.

P. Poller and V. Tschernomas. Multimodal Fission and Media Design, 2006. In this volume.

R. Porzel, I. Gurevych, and R. Malaka. In Context: Integrating Domain- and Situation-Specific Knowledge, 2006. In this volume.

R.P. Shi, J. Adelhardt, A. Batliner, C. Frank, E. Nöth, V. Zeißler, and H. Niemann. The Gesture Interpretation Module, 2006. In this volume.

J. te Vrugt and T. Portele. Intention Recognition, 2006. In this volume.

W. Wahlster. User and Discourse Models for Multimodal Communication. In: J.W. Sullivan and S.W. Tyler (eds.), *Intelligent User Interfaces*, pp. 45–67, New York, 1991. ACM.

Discourse Modeling

Jan Alexandersson and Norbert Pfleger

DFKI GmbH, Saarbrücken, Germany
{janal,pfleger}@dfki.de

Summary. We provide a discription of the robust and generic discourse module that is the central repository of contextual information in SMARTKOM. We tackle discourse modeling by using a three-tiered discourse structure enriched by partitions together with a local and global focus structure. For the manipulation of the discourse structures we use unification and a default unification operation enriched with a metric mirroring the similarity of competing structures called OVERLAY. We show how a wide variety of naturally occuring multimodal phenomena, in particular, short utterances including elliptical and referring expressions, can be processed in a generic and robust way. As all other modules of the SMARTKOM backbone, DIM relies on the a domain ontology for the representation of user intentions. Finally, we show that our approach is robust against phenomena caused by imperfect recognition and analysis of user actions.

1 Introduction

This chapter provides a comprehensive description of the central repository for contextual information in SMARTKOM: the *discourse modeling module* (DIM [1]). For man–machine dialogue systems, the functionality of DIM is often viewed and implemented as a part of the dialogue manager (Jönsson, 1993, 1997; Larsson, 2002). In SMARTKOM, DIM manages contextual information by recording or exchanging information with almost every backbone module in such a way that the analysis modules and the generation modules can interpret or generate with respect to the discourse state. Here, we will focus on analysis support but, as we will indicate, our approach has been designed in such a way that the generation of system reactions can be supported too.

In SMARTKOM, we have further developed the approach for discourse modeling taken in the speech-to-speech translation system Verbmobil (Alexandersson, 2003). In Verbmobil, there is no dialogue manager in the traditional sense; since

[1] Most of the material in this chapter is found in Alexandersson and Becker (2001, 2003); Pfleger (2002); Pfleger et al. (2003a,b, 2002), see also Alexandersson et al. (2006)

the Verbmobil system does not actively participate in the dialogue, dialogue management boils down to discourse processing whose task it is to passively track the dialogue. In SMARTKOM, we have added a three-tiered context representation based on the ideas developed by Luperfoy (Luperfoy, 1992) and some ideas from Salmon-Alt and Pfleger (Salmon-Alt, 2000; Pfleger, 2002). In SMARTKOM, DIM is closely connected to the action planner (Löckelt, 2006) and utilizes its predictions for the context-dependent interpretation for, in particular, short answers and partial utterances (Sect. 4).

1.1 The Challenge

DIM is the main repository for dynamic contextual information in SMARTKOM. During the course of a dialogue, each new user contribution has to be interpreted in the light of the previous discourse context. Compatible information needs to be incorporated into the representation of the new contribution through a process we call *enrichment*. DIM not only has to maintain a representation of the previous discourse context but also has to provide appropriate functionality for enrichment. Furthermore, since both the recognizers and the analysis components of the different modalities produce multiple hypotheses, DIM has to contribute to the selection of the most probable one on the basis of the previous discourse context — a process we call *validation*. The following example highlights the need for the enrichment and validation of intention hypotheses.[2]

U1: *What's on at the movies tonight?*

S1: (Displays a list of movies) *Here [↗]*[3] *you can see tonight's cinema program.*

U2: *And what's on TV?*

S2: (Displays a list of broadcasts) *Here [↗] are tonight's broadcasts.*

U3: *Is there a movie with Arnold Schwarzenegger?*

Turn U2 is a good example of where we need access to the previous discourse context to obtain the correct interpretation—the user wants to see tonight's television program and not the one that is currently running. Turn U3 exemplifies the need for a validation of intention hypotheses against the discourse context as the language understanding module will produce at least two hypotheses for this user contribution: one stating that the user wants to see a movie with Arnold Schwarzenegger at the movies and a second stating that the user wants to see a such a movie on TV. By considering the discourse context it becomes clear that the second hypothesis most probably reflects the intended meaning.

Besides enrichment and validation, DIM also has the task to integrate partial utterances into their previous discourse. Consider, for example, an elliptical expression, as in the following:

---

[2] The validation process is described in Alexandersson et al. (2006)

[3] [↗] stands for a pointing gesture.

U7: *What's on TV tonight?*

S7: *Here are tonight's broadcasts.*

U8: *And tomorrow?*

S9: *Here are tomorrow's broadcasts.*

Here, the *correct* interpretation is a request for showing the TV program for tomorrow night. In this case, DIM has to integrate the time expression of turn U8 into the representation of the previous turn, U7 thereby recovering the general request ("show me the TV program") and also the time interval 8:00 PM – 11:59 PM.

Another challenging task of DIM is the resolution of referring expressions like , e.g., anaphora or cross-modal references. Consider the continuation of the discourse above:

U10: *Tape the first one!*

To be able to resolve such a referring expression, DIM not only has to maintain a representation of the previous discourse context, but it also has to maintain a representation of the objects previously displayed on the screen together with some representation of their conceptual layout (e.g., their ordering).

To conclude, we pose the following requirements on discourse processing in SMARTKOM:

- We need a general and robust method for the enrichment and validation of user hypotheses.
- The context model must differentiate between modality-specific information and modality-unspecific information of a concept introduced into the discourse.
- The context model must reflect not only the temporal structure of the dialogue but also the topical structure.
- We need a robust and flexible method for resolving different kinds of referring expressions.

1.2 Our Approach

We use the following approach for these tasks:

Representation. Our approach for representing the discourse state is a compilation of three items:

- A unified modality-independent generic representation based on the three-tiered context representation presented in Luperfoy (1992). The three tiers are the *modality* layer, the *discourse object* layer and the *domain object* layer.
- The context structuring ideas of Salmon-Alt (2000). In particular, we deploy compositional information of collections.
- A simple and robust thematic-based generic focus mechanism, e.g., Carter (2000). Our focus mechanism contains a *local* and *global* structure organized around different *focus spaces* and accessible discourse objects of the focus space, respectively.

Ontology. The common ontology-based representation used in the SMARTKOM backbone (Gurevych et al., 2006). In DIM, the ontology is viewed and represented as typed feature structures (TFSs) which allows for the use of unification-based manipulation operations (see below).

Manipulation. Two fundamental operations for the manipulation of the ontology-based representation: unification and overlay (Alexandersson et al., 2006; Alexandersson and Becker, 2003).

The approach chosen enables resolution of different referential expression, e.g., anaphora, cross-modal referring expressions and a wide range of elliptical expression (see Sect. 1.1 for examples). Still, the deployed functionality is efficient and does not slow down the system. We have carefully designed the algorithms in such a way that we only use algorithms linear to the size of its arguments (unification and overlay). Additionally, the search space for resolving referential and elliptical expressions is drastically reduced by the topic structure.

The chapter is organized as follows: Sect. 1.1 provides a more thorough presentation of the challenges we are facing. Section 3 describes how we represent the discourse. In Sect. 4 the algorithms used are described. The description of overlay, however, is found in Alexandersson et al. (2006) Before we conclude the chapter in Sect. 6 we discuss our experiences gained in SMARTKOM in Sect. 5.

2 DIM in Context

As already indicated in the previous section, DIM needs to interact with many modules of the dialogue backbone. Figure 1 depicts the main communication from the viewpoint of DIM. In addition to the main data flow (solid lines), there are several auxiliary interfaces (dashed lines), which are used either to store contextual information being produced by action planner and output modules or to provide contextual information to the analysis modules.

The first module in the processing chain that interacts with DIM is the fusion component (FUSION, see Engel and Pfleger (2006)). In cases where FUSION encounters a referring expression that could not be resolved using gestural input, it requests DIM to resolve those referring expressions (the actual reference resolution is described in Sect. 4.2). In cases where DIM was able to determine the referent for those referring expressions, it sends the resolved object back to modality fusion, which in turn integrates it into the corresponding intention hypothesis. The next module that communicates with DIM is the intention recognition module sending the set of analyzed hypotheses to DIM. DIM in turn interprets them with respect to the current discourse context (the process of enrichment and validation; see Sect. 4.1). However, before DIM sends the enriched hypotheses back to the intention recognition modules, it requests the situation modeler to enhance the hypotheses with situational information (Porzel et al., 2006).

The analysis of a user contribution is finished when the intention recognition component eventually has selected an intention hypothesis. This intention hypothesis

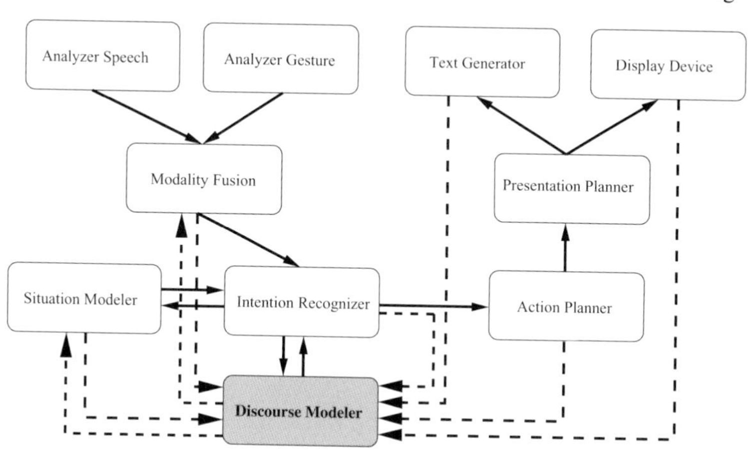

Fig. 1. The interfaces of DIM. *solid lines* depict the main data flow; *dashed lines* indicate auxiliary interfaces

is incorporated into the discourse history representing a user turn. The system turn is constructed by the expectations of the action planner (describing the anticipated user reactions to the next system turn), the screen state published by the display device (describing the objects presented on the screen) and concepts presented by the generation module (describing the system utterance).

2.1 Some Terminology

Before we continue with describing our approach to discourse processing, we have to introduce some terminology. The top level of the SMARTKOM ontology contains several *application objects*. The number of application objects is determined by the set of applications modeled by the action plans used in the system (Löckelt, 2006). Application objects describe events, e.g., a movie ticket reservation request, and can contain *subobjects* that are recursively composed of other subobjects. An application object describing a movie ticket reservation event includes, for example, subobjects describing the movie theater, the show time and movie information. Additionally, the type hierarchy is used to express relations between application objects. Subobjects that can appear in different application objects are introduced as early as possible in the type hierarchy. Additionally, some subobjects are called *slots*. Slots are characterized by their paths from the top level of an application object and represent objects in the real world that could be focused on during the conversation. Moreover, slots are predicted by the action planner via its expectations; a feature we rely on especially when it comes to the interpretation of short utterances (see below).

3 Discourse Representation

In this section we outline the concepts and realization of the structure of our discourse memory. Next, our extension to the three-tiered context representation of Luperfoy (1992) is described. We close this section by a brief description of our focus structure and the use of initiative response units.

3.1 A Three-Tiered Context Representation

Our approach to discourse modeling is based on a generalization of Luperfoy (1992) together with some ideas from Salmon-Alt (2000) and the Verbmobil dialogue module (Alexandersson, 2003; Wahlster, 2000). Following the ideas of Luperfoy (1992), we use a three-tiered context representation where we have extended her linguistic layer to a *modality* layer (Fig. 2). Additionally, we have adopted some ideas from Salmon-Alt (2000) by incorporating directly perceived objects and compositional information of collections. The basic discourse operations described in Alexandersson (2003) and Wahlster (2000) have been further developed (Alexandersson and Becker, 2003).

The advantage of our approach to discourse representation lies in the *unified* representation of discourse objects introduced by the different modalities. As we show in the next sections, this not only supports the resolution of elliptical expressions but allows for cross-modal reference resolution. The context representation of the discourse modeler consists of three levels: the discourse object layer, the modality layer and the domain object layer.

3.1.1 Discourse Object Layer

The discourse object layer is the central layer of the discourse representation. This layer comprises the concepts introduced into the discourse. An object on this layer is called a discourse object (DO, see Luperfoy (1992)). For each concept introduced during discourse there exists only one DO independent of how many modality objects (MOs, see below) mention this concept. Each DO is hence unique. A DO is created every time a concept is newly introduced into the discourse, independent of which modality was used. This holds not only for spoken concepts but also for visually perceivable concepts, e.g., graphically presented objects (Salmon-Alt, 2000).

DOs provide access to three classes of information:

1. modality-specific information via MOs
2. domain information by accessing substructures of the domain object layer
3. compositional information

Compositional information is available when a DO represents a collection or composition of objects. This information is provided by *partitions* (Salmon-Alt, 2000).

We use partitions to represent collections of objects. Collections are either based on perceptive information, e.g., the list of broadcasts visible on the screen, or discourse information stemming from grouping discourse objects. The elements of a

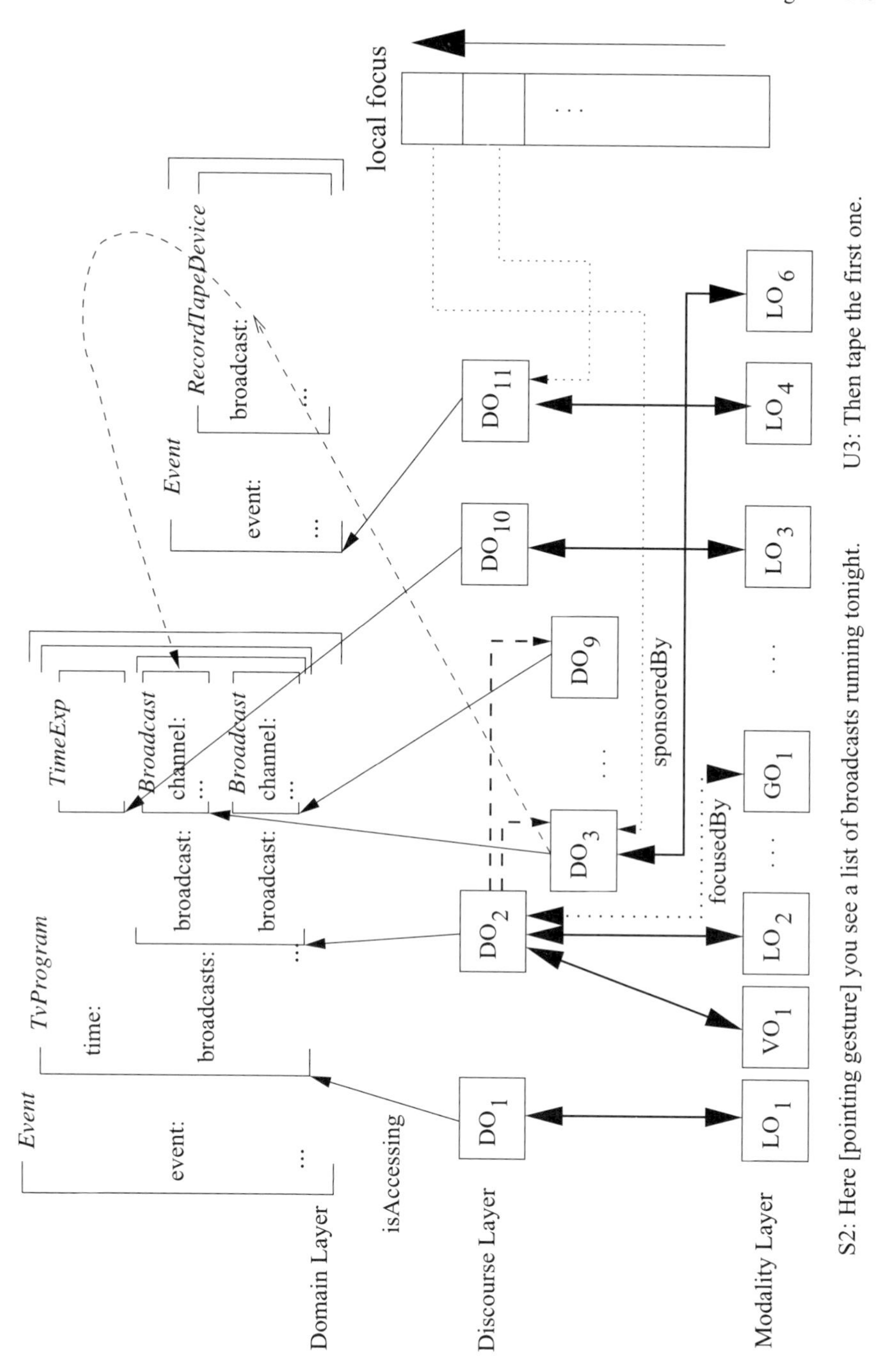

Fig. 2. Multimodal context representation. *Dashed arrow* indicates that the value of the broadcast in the (new) structure (to the *right*) is shared with that of the old one (to the *left*)

partition are distinguishable from one another by at least one *differentiation criterion*. Yet, one element alone of a partition may be in focus, according to gestural or linguistic salience. Figure 3 depicts a sample configuration of a discourse object (DO2) with a partition.

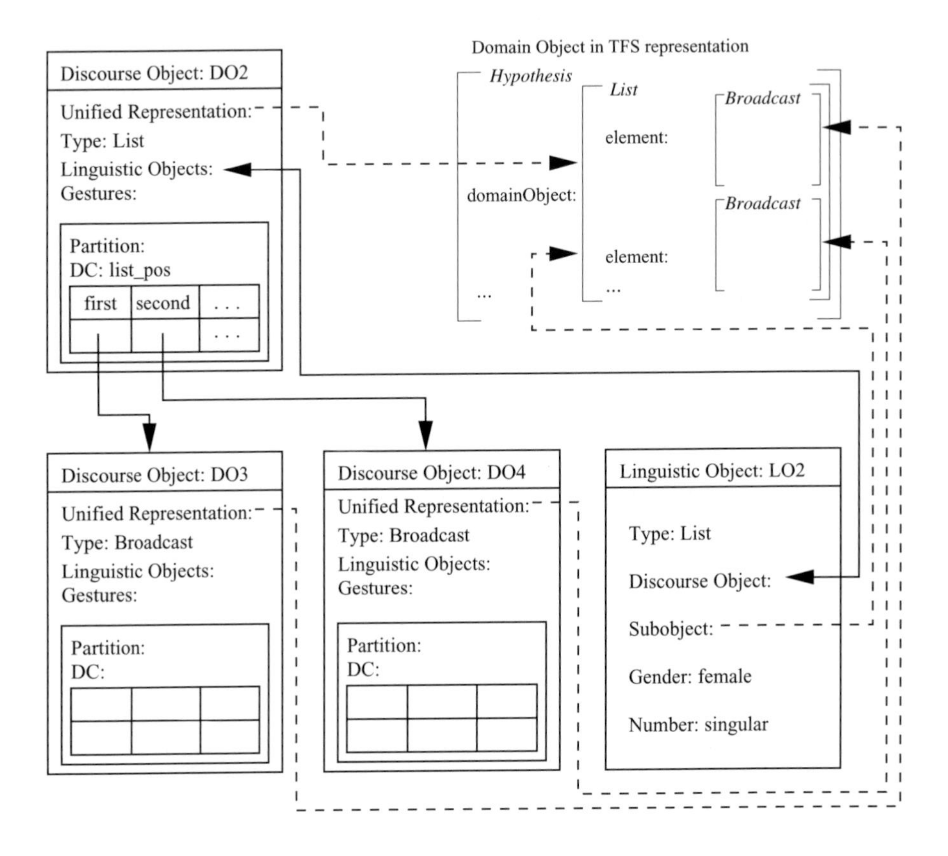

Fig. 3. Discourse objects

3.1.2 Modality Layer

Modality objects (MOs) are introduced to the modality layer every time a discourse object is mentioned in discourse (either by the user or by the system). Every MO is linked to its corresponding DO and encapsulates information about the surface realization of this particular mention of that DO. With respect to the three modalities that are stored by DIM, we differentiate three types of modality objects:

1. linguistic objects (LOs)
2. visual objects (VOs)

3. gesture objects (GOs)

Each type of modality object provides access to modality-specific information. A LO, for example, provides information about linguistic features (like number or gender) that were used during realization, whereas a VO provides information about the position the corresponding DO had on the screen.

Each MO is linked to a corresponding DO and shares its information about its concrete realization with the discourse object (Fig. 3). Note that a MO provides information about its original position within the event structure, i.e., its path in the corresponding application object in which it is embedded.

3.1.3 Domain Object Layer

The domain object layer encapsulates the instances of the domain model and provides access to the semantic information of objects, processes and actions. Initially, the semantic information of a DO is defined by a subobject (possibly embedded in an application object) representing the object, process or action it corresponds to. This information is only accessed by the DO. However, the semantic information of a DO might be extended as soon as the object is accessed again. Consider, for example, DO42 initially representing a movie with the title *The Matrix* (created by the user request *"When will the movie* The Matrix *be shown on television?"*). The system will respond by presenting a list of broadcasts of the movie *The Matrix* (accompanied with information specific to the different broadcasts, like time, database key, channel, etc). Now, if the user selects one of the movies — *"tape this [↗] one"* — the initial information of DO42 will be extended with this additional information.

3.2 Focus and Attentional State

We differentiate between a *global* focus structure and a *local* focus structure restricting access to objects stored in the discourse model. The former represents the topical structure of discourse and resembles a list of focused items—*focus spaces*. The items there are ordered by salience. In SMARTKOM, the global focus is imposed by the action planner providing a flat structure of discourse in terms of discourse topics. A focus space covers all turns belonging to the same topic and enables access to a corresponding local focus structure (see also Carter (2000)).

A local focus structure provides and restricts access to all discourse objects that are antecedent candidates for later reference. Also on this level, the content of the structure, i.e., discourse objects, are ordered by salience. For each user or system turn, the local focus structure for this topic is extended with all presented concepts (see also Fig. 2).

4 Discourse Processing

We now turn to the task of processing, e.g., referring expressions. In this section we describe how the structures presented in the last section are utilized for interpretation and validation based, on the one hand, on the discourse state and, on the other hand, on the state of the action planner (Löckelt, 2006).

4.1 Context-Dependent Interpretation

Our main operations for the manipulation on instances of our domain model is unification and a default unification operation, which together with a scoring functionality we call OVERLAY (Alexandersson, 2003; Alexandersson and Becker, 2003; Pfleger et al., 2003a; Alexandersson et al., 2006). The starting point for the development of the latter operation was twofold: First we viewed our domain model as typed feature structures and employed closed world reasoning on their instances. Second, we saw that adding information to the discourse state could be done using unification, at least as long as the new information was consistent with the context. This is, however, not always the case, as the user often changes his or her mind and specifies competing information, which causes standard unification to fail. There was thus a need for a nonmonotonic operation capable of overwriting parts of the old structure with the new information *and* at the same time keeping the information still consistent with the new information. The solution is default unification (Alexandersson and Becker, 2003; Carpenter, 1993; Grover et al., 1994), which has proven to be a powerful and elegant tool for doing exactly this: overwriting old, contextual information—*background*—with new information—*covering*—thereby keeping as much consistent information as possible.

U4: *What is on TV tonight?*

S4: [Displays a list of broadcasts.] *Here* [↗] *you see a list of the broadcasts running tonight.*

U5: *What is running on MTV?*

S5: (Displays a list of broadcasts for MTV tonight.) *Here* [↗] *you see a list of the broadcasts running tonight on MTV.*

U6: *And CNN?*

S6: ...

Fig. 4. Dialogue excerpt 2

We distinguish between full or partial utterances. An example of the former is a complete description of a user action, e.g., U4 in Fig. 4, whereas partial utterances often but not necessarily are elliptical responses to a system request. We handle these cases differently, as described below.

4.1.1 Full Utterances

For task-oriented dialogues, there are many situations where information can and should be inherited from the discourse history, as shown in the dialogue excerpt in

Fig. 4. Due to spatial restrictions on the screen it may be impossible for the system to display every broadcast for all channels, e.g., in S4. SMARTKOM therefore chooses some broadcasts playing on some channels. Clearly, the intention in U5 is to ask for the program on MTV *tonight*, thus requiring the system to inherit the time expression from U4. Full utterances are processed by traversing the global focus structure and picking the focused application object in each focus space (if any). In Fig. 4, default unifying U5 (covering) with U4 (background) results in *What is running on TV on channel MTV tonight.*

4.1.2 Partial Utterances

For the interpretation of partial utterances (henceforth partials) we gave a detailed description in Löckelt et al. (2002). For elliptical utterances (U6 in Fig. 4), the general idea is to extend the partial to an application object of the same type as the nearest focused application object—a process referred to as bridging—and then use this application object as covering and the focused application object as background.

There is one more challenge we have to face: resolving referring expressions. Given resolved referring expressions *and* the correct bridging, we can use the basic processing technique as described in Sect. 4.1.1. Next, we concentrate on the latter, whereas the former task is described in Sect. 4.3.

4.2 Resolving Referring Expressions

There have been many proposals in the literature for finding antecedents for referring expressions, e.g., Grosz et al. (1995). These approaches typically advocate a search over lists containing the potential antecedents, where information like number, gender agreement, etc., are utilized to narrow down the possible candidates. Our approach to reference resolution is a bit different. Additionally, in a multimodal scenario, the modality fusion first has to check for accompanying pointing gestures before accessing the discourse memory. In case of a missing gesture, DIM receives a request from modality fusion containing a subobject as specific as possible inferred by the intra-sentential context. This goes together, if possible, with the linguistic features and partition information. DIM searches the local focus structure and returns the first object that complies with the linguistic constraints in the respective LO and unifies with the object of the corresponding DO. We resolve three different referring expressions in the follwoing sections.

4.2.1 Total Referring Expressions

By a *total referring expression* (TRE) we mean that a referring expression corefers with the object denoting its referent (LuperFoy, 1991). Those referring expressions are resolved through the currently active local focus. The first discourse object that satisfies the type restriction and that was mentioned by a linguistic object with the same linguistic features is taken to be the intended referent. In this case, there is a linguistic sponsorship relation established between the referring expression and its referent.

If no linguistic sponsorship relation can be established, DIM tries to establish a discourse sponsorship relation. This condition is characterized by a mismatch between the linguistic features, but the objects themselves are compatible (i.e., in our case unifiable). However, if both conditions are not fulfilled, the focused discourse objects (but only the one most focused on) of the other global focus spaces are tested, thereby searching for a discourse object that allows for a linguistic sponsorship relation.

4.2.2 Partial Referring Expressions

By a *partial referring expression* (PRE), the user refers to a part of a discourse object, e.g., selecting a movie from a list of movies. In this case, the processing is the same as for total referring expressions with one exception: the local focus is scanned for a compositional discourse object that satisfies (i) the differentiation criterion specified in the partition feature of the request (i.e., *"the first"*), and (ii) has a discourse object in its value feature that is compatible (i.e., in our case unifiable) with the requested discourse object (see Fig. 3). The first such discourse object is returned.

4.2.3 Discourse Deictic Expressions

A *discourse deictic expressions* (DDE) occurs when the user refers to a complete system action, for example, *"that is good,"* where "that" refers to the entire last system action. The complete functionality for a sophisticated treatment of DDEs is not included in the SMARTKOM system. However, our approach to discourse processing allows for resolving these expressions, too (Pfleger, 2002). The DDE case is processed as follows: If the type of the referring expression cannot be identified, then the focused discourse object of the currently active local focus is tested as to whether it shares the linguistic features with the request. If it does, that discourse object is taken to be the intended referent, otherwise the referring expression is interpreted as being a discourse deictic one, in which case the discourse object representing the last system turn is returned.

4.3 Partial Utterances Revisited

We return to the interpretation of partials again. After being processed by FUSION, an intention is now guaranteed to be either an application object or, as we will focus on now, a subobject representing a partial. Processing partials consists of two steps (Löckelt et al., 2002):

1. Find an anchor for the partial. If present, the anchor is searched for and possibly found under (i) the list of expected slots, (ii) in the local focus stack or, finally, (iii) in the list of possible slots.

2. Compute the bridge by using the *path* in either expected or possible slots, or in the MOs in the local focus stack.

Conceptually, the process of the interpretation of partials is summarized in Table 1. Depending on who is having the initiative, conceptually different processing is performed. Although we *always* use overlay for the combination of new and old information, it suffices in some cases to use unification.

Table 1. Possible combinations of partial utterances and different expectations given user or system initiative

	Procesing	
	User initiative	System initiative
Expected slot	NA	Expected Unify Plan continues
Possible slot	Plausible Unify Replanning	Plausible Unify Replanning
Filled slot	Plausible Overlay Replanning	Plausible Overlay Replanning
Other	implausible recover strategies	

There are, however, some exceptions to this general scheme, of which we provide two examples in the following.

4.3.1 User Provides Too Much Information

If the system has the initiative, thus requesting for information, the user might provide more, still compatible information than asked for. A good example is the case where, during seat reservation for a performance, the system asks the user to specify where she would like to sit. The expected slot is in this case pointing to a seat. In our domain model the seat part of a set construction contains not just one seat but, e.g., a set of seats and cardinality. If the user contribution specifies something like "*Two seats here [↖]*"[4] then the user contribution will not fit the expectation.

A representation of the reservation is schematically depicted in Fig. 5. The system asks for a specification of a seat, i.e., a pice of information at the end of the path *f1:f2:. . . :seats:seat* that is a partial of type SEAT. The answer *contains* the expectation which, however, is embedded in an object consistent with the expectation. The correct processing of such an answer is to walk along the expectation until the answer is found. If this happens before the end of the expected path, the rest of the path (*seat*) has to be part of the subobject.

[4] . . . where [↖] stands for an encircling gesture.

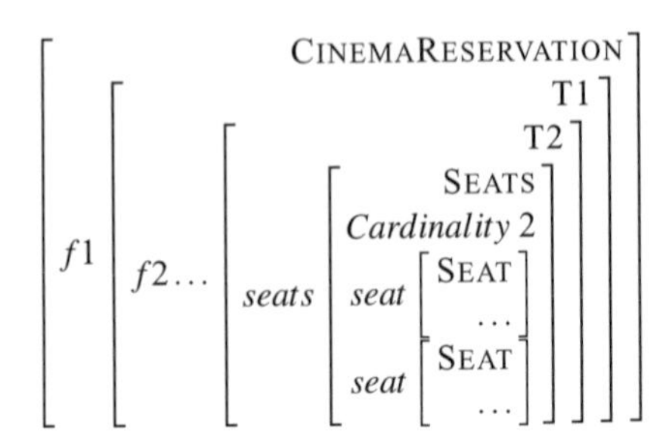

Fig. 5. An application object representing the reservation of two seats

4.3.2 Manipulations of Sets

For the correct processing of the example above, we had to extend OVERLAY with operations on sets, like union. Again consider a seat reservation scenario where we have the case where the user is not satisfied with the reservation and replies to the system request "*Is the reservation OK?*" with a modification containing manipulations of, in this case, a set of seats by uttering "*two additional seats here [↖]*". The processing consists of two steps: (i) SPIN marks the seats in the intention hypothesis with a set modification flag. (ii) FUSION requests the focused set of seats from DIM and computes the allowed set of, in this case, additional seats. (iii) The intention is now containing two seats (pointed at with the set modification flag), which are then processed by DIM in the same way as described above: the path from the root of the focused application object corresponding to the set of seats in the discourse memory. These are found in the local focus stack. After computing the covering, OVERLAY is performed where the sets of seats in the background and covering are unified with union.

5 Discussion

In the previous section we presented our approach to discourse modeling for a multimodal dialogue system. We showed how we achieve an enrichment and validation of user intention hypotheses with respect to their previous dialogue context using a single operation called *overlay*. It is important for the discussion here that we use overlay without any control mechanism to accumulate information stemming from previous dialogue states into a current user intention hypothesis. In many dialogue situations this accumulation of discourse objects is doing what we expect. However, by considering the example below, it becomes clear that there are situations where old information should be—or must be—ignored.

We showed that a least some of these overaccumulations can be avoided by inhibiting accumulation of previous information. In SMARTKOM, the speech interpretation component blocks parts of an event object if there are any explicit clues in the spoken utterance that suggest to do so (Engel, 2006).

The following dialogue excerpt is an example where the presented functionality of DIM will accumulate too much contextual information:

U1: What is running on TV tonight? (Context C1)

U2: ...and what is running on ARD? (Context C2)

U3: Show me information about this broadcast [↗]. (Context C3)

U4a: And is there something with Schwarzenegger?

U4b: Show me broadcasts with Schwarzenegger.

When considering U4a and U4b the question is which context to use. Possible contexts are C2 (channel: "ARD," evening) or context C1 (evening). An intuitive answer for U4a would be C2, as the syntactical structure of C2 and U4a both start with a conjunction. Utterance U4b, however, is a rather general request and should not be enriched with too much—if any—contextual information.

In general, enriching discourse contributions with respect to their previous discourse context is a complex venture. The approach used for SMARTKOM has—besides its robustness and generality—one important disadvantage caused by the fact that we deploy a straightforward use of overlay. For most dialogue situations this is feasable. As shown above, our current approach might lead to *overaccumulation*. Hence, there is a need for "forgetting" or "discarding" contextual information.

One can argue that this functionality should be placed in the action planner: in case the interpretation provided by DIM does not result in—in the example above—any broadcasts, the action planner could relax some of the contraints. This relaxation could, however, be provided by the discourse module by proving an ordered list of interpretations instead of a single contextual interpretation. In the example above, one could imagine the interpretations {OVERLAY(U4a,C2), OVERLAY(U4a,C1), OVERLAY(U4a,C3)...}. Now if the action planner fails to retrieve any broadcast using the first interpretation, the next—lower scored one—can be tested. This approach is only defeasable if the database is fast enough.

6 Conclusion

We provided a detailed description of the *discourse modeling module* of SMARTKOM. The resulting module is based on ideas and experience collected in several projects including the speech-to-speech translation system Verbmobil (Wahlster, 2000; Alexandersson, 2003), a typed front end to the ARIS database—NaRaTo—, and, finally, the multimodal dialogue system SMARTKOM. Hence our approach has proved flexible and powerful enough to meet the requirements of a wide range of dialogue systems. Current and future activities include the following topics:

Discourse segment recognition. As we have shown, there is a need for more detailed modeling of the topical focus structure. Several approaches based on a hierarchical structure have been suggested in the literature. The danger with such approaches is, however, that the intended referents might be shadowed and thus not found. Still, in many situations they provide a more precise access.

Granularity. In some cases OVERLAY should overwrite complete substructures of the background rather than include all information from the background. The keys for when this should be done are often found in the surface of the user actions.

Overlay. For setlike structures, i.e., seats in a movie-theater, it is unclear what operations should be used (see also Alexandersson et al. (2006)). The current version of DIM provides correct treatment for some examples. Ongoing work investigates how and when different set operations (i.e., adjoin, set difference, etc.) are used for the combination of new and contextual information.

Evaluation. The success of the overall system shows that our approach to discourse modeling is valid. Still, neither the overall system nor DIM has recieved a proper evaluation. Although there have been some attempts to create evaluation standards for dialogue systems, there exists—to our knowledge—no standard for how an intrinsic evaluation of DIM can be performed.

Forgetting. For longer dialogues (more than half an hour of discourse), the discourse memory runs out of memory. It is—even for most humans—necessary to forget information. How and when this can be done is still an open and challenging issue we will have to address in the future.

References

J. Alexandersson. *Hybrid Discourse Modelling and Summarization for a Speech-to-Speech Translation System*. PhD thesis, Universität des Saarlandes, 2003.

J. Alexandersson and T. Becker. Overlay as the Basic Operation for Discourse Processing in a Multimodal Dialogue System. In: *Workshop Notes of the IJCAI-01 Workshop on "Knowledge and Reasoning in Practical Dialogue Systems"*, Seattle, WA, August 2001.

J. Alexandersson and T. Becker. The Formal Foundations Underlying Overlay. In: *Proc. 5th Int. Workshop on Computational Semantics (IWCS-5)*, pp. 22–36, Tilburg, The Netherlands, February 2003.

J. Alexandersson, T. Becker, and N. Pfleger. Overlay: The Basic Operation for Discourse Processing, 2006. In this volume.

B. Carpenter. Skeptical and Credulous Default Unification With Applications to Templates and Inheritance. In: T. Briscoe, V. de Paiva, and A. Copestake (eds.), *Inheritance, Defaults, and the Lexicon*, pp. 13–37, Cambridge, CA, 1993. Cambridge University Press.

D. Carter. Discourse Focus Tracking. In: H. Bunt and W. Black (eds.), *Abduction, Belief and Context in Dialogue*, vol. 1 of *Studies in Computational Pragmatics*, pp. 241–289, Amsterdam, The Netherlands, 2000. John Benjamins.

R. Engel. Natural Language Understanding, 2006. In this volume.

R. Engel and N. Pfleger. Modality Fusion, 2006. In this volume.

B.J. Grosz, A.K. Joshi, and S. Weinstein. Centering: A Framework for Modelling the Local Coherence of Discourse. Technical Report IRCS-95-01, The Institute For Research In Cognitive Science (IRCS), Pennsylvania, PA, 1995.

C. Grover, C. Brew, S. Manandhar, and M. Moens. Priority Union and Generalization in Discourse Grammars. In: *Proc. 32nd Annual Meeting of the Association for Computational Linguistics*, pp. 17–24, Las Cruces, NM, 1994. Association for Computational Linguistics.

I. Gurevych, R. Porzel, and R. Malaka. Modeling Domain Knowledge: Know-How and Know-What, 2006. In this volume.

A. Jönsson. *Dialogue Management for Natural Language Interfaces – an Empirical Approach*. PhD thesis, Linköping Studies in Science and Technology, 1993.

A. Jönsson. A Model for Habitable and Efficient Dialogue Management for Natural Language Interaction. *Natural Language Engineering*, 3(2/3):103–122, 1997.

S. Larsson. *Issue-Based Dialogue Management*. PhD thesis, Göteborg University, Sweden, 2002.

M. Löckelt. Plan-Based Dialogue Management for Multiple Cooperating Applications, 2006. In this volume.

M. Löckelt, T. Becker, N. Pfleger, and J. Alexandersson. Making Sense of Partial. In: C.M. Johan Bos, Mary Ellen Foster (ed.), *Proc. 6th Workshop on the Semantics and Pragmatics of Dialogue (EDILOG 2002)*, pp. 101–107, Edinburgh, UK, September 2002.

S. LuperFoy. *Discourse Pegs: A Computational Analysis of Context-Dependent Referring Expressions*. PhD thesis, University of Texas at Austin, December 1991.

S. Luperfoy. The Representation of Multimodal User Interface Dialogues Using Discourse Pegs. In: *Proc. 30th ACL*, pp. 22–31, Newark, DE, 1992.

N. Pfleger. Discourse Processing for Multimodal Dialogues and Its Application in SmartKom. Master's thesis, Universität des Saarlandes, 2002.

N. Pfleger, J. Alexandersson, and T. Becker. Scoring Functions for Overlay and Their Application in Discourse Processing. In: *Proc. KONVENS 2002*, pp. 139–146, Saarbruecken, Germany, September–October 2002.

N. Pfleger, J. Alexandersson, and T. Becker. A Robust and Generic Discourse Model for Multimodal Dialogue. In: *Workshop Notes of the IJCAI-03 Workshop on "Knowledge and Reasoning in Practical Dialogue Systems"*, Acapulco, Mexico, August 2003a.

N. Pfleger, R. Engel, and J. Alexandersson. Robust Multimodal Discourse Processing. In: *Proc. Diabruck: 7th Workshop on the Semantics and Pragmatics of Dialogue*, Wallerfangen, Germany, 4–6 September 2003b.

R. Porzel, I. Gurevych, and R. Malaka. In Context: Integrating Domain- and Situation-Specific Knowledge, 2006. In this volume.

S. Salmon-Alt. Interpreting Referring Expressions by Restructuring Context. In: *Proc. ESSLLI 2000*, Birmingham, UK, 2000. Student Session.

W. Wahlster (ed.). *Verbmobil: Foundations of Speech-to-Speech Translation*. Springer, Berlin Heidelberg New York, 2000.

Overlay:
The Basic Operation for Discourse Processing

Jan Alexandersson, Tilman Becker, and Norbert Pfleger

DFKI GmbH, Saarbrücken, Germany
{janal,becker,pfleger}@dfki.de

Summary. We provide a formal description of the fundamental nonmonotonic operation used for discourse modeling. Our algorithm—overlay—consists of a default unification algorithm together with an elaborate scoring functionality. In addition to motivation and highlighting examples from the running system, we give some future directions.

1 Introduction

This chapter introduces the overlay operation—a default unificationlike operation enabling the assignment of consistent background information into a new structure. In order to assess how good the new structure fits to the older one, overlay generates a score reflecting the structural consistency of its arguments. Within the SmartKom project we used this operation for the task of discourse processing (Alexandersson and Pfleger, 2006), where intention hypotheses are enriched and validated with respect to their preceding discourse context.

1.1 Adding, Refining, Revising, and Changing

User utterances in dialogue systems usually continue a discourse by supplying new information that adds to the current context. In simple slot-filling applications, e.g., VoiceXML, the new information typically fills the next slot, often as an answer to a direct system question (case (i), see below). In more elaborate systems, a user utterance can also refine prior information, e.g., specify "tonight at 7 PM" as a refinement of a previous, less precise "today" (case (ii)). In both cases, the new information is compatible with the current context and—given an appropriate representation—*unification* can be used as the formal operation to combine a new utterance with the current context.

In systems like SmartKom, which allow for more natural, conversational dialogue that is not system-driven, users can revise their input freely and often do so

unexpectedly (cases (iii) and (iv)). In the following we will describe a formal operation to capture such revisions. Before presenting the details of our formal representation for utterances and current context, we give a set of examples with progressive complexity:
(i) A simple case of filling a previously unfilled slot:

U: *I want to see a movie.*

S: *When do you want to see it?*

U: *Tonight.*

(ii) Refinement of a previous, less precise slot-filler:

U: *What's today's TV program?*

S: *Here you see a list of movies running today.*

U: *What's on TV tonight at 7 PM?*

(iii) A case of changing the value of a slot:

U: *First, I'd like to make a reservation in this movie theater.*

S: *Sorry, this theater does not take reservations.*

U: *Ok, then I take [↗] this one.*

(iv) Finally, a more complex case of modifying the entire task:

U: *Which movies are showing tomorrow night on HBO?*

S: *Here's tomorrow night's program for HBO.*

U: *And how about the movie theaters?*

S: *Here's a list of movies in cinemas tomorrow night.*

Note how the system has kept the time information "tomorrow night," even though the task was changed from TV to movie theater information. This final case is picked up again in Sect. 3.

1.2 Representing the Current Context

In SMARTKOM we have chosen an ontology-based representation of the tasks that the user can solve through the system. Gurevych et al. (2006) give an in-depth description of the methodology. For the discussion in this chapter, we will assume that instances of the ontology can be represented as typed feature structures (TFS) with a type hierarchy that mirrors the ontology's class hierarchy.

1.3 Combining Current Context and User Utterance

The TFS-based representation allows us to discuss the four cases of *state update* introduced in the previous section in terms of *unification*, (Carpenter, 1992). In particular, (i) the user adding a value for an empty slot can be seen as unification in (untyped) feature structures. (ii) Refinement is a case of unification in typed feature structures, and (iii) revising and (iv) changing are cases of default unification or overlay (Carpenter, 1993). In Sects. 2 and 3, we formalize this approach. Sect. 3.2 describes an enrichment of overlay with a scoring operation that expresses how much information from the current context is retained.

2 Preliminaries: Typed Feature Structures and Unification

Almost 30 years frames or framelike formalisms have been used to represent knowledge (Minsky, 1975). Different requirements, such as processing speed, expressiveness, etc., have guided the development of the representation as such, as well as the algorithms for the manipulation of the representation into several directions. For representing linguistic knowledge and more abstract common-sense knowledge, typed frame-based formalisms are used, such as TFS (Carpenter, 1992) or description logics (Horrocks, 1998b). In parallel different inference operations meeting different needs have been developed. For linguistic analysis, different *unification* operations (Carpenter, 1992) and for description logics, the *classifier* (Horrocks, 1998a) are used.

In this work we are concerned with TFSs and, in particular, operations that ease the interpretation of discourse phenomena such as different elliptical expressions. The naïve standard unification is enough as long as consistent and not contradictory (conflicting) information is added to the discourse. However, as shown in Alexandersson and Pfleger (2006), there are a lot of cases where the user changes his/her mind and provides information that conflicts with the context. Still, it is desirable to obtain a complete description of the user intention that consists of the accumulated contextual information but where new overwrites the old competing information.

In what follows, we summarize our view on TFSs[1] in SMARTKOM (Gurevych et al., 2006) for a description of the actual ontology). We assume a type hierarchy where each pair of types has a least upper bound (LUB). A common characterization for type hierarchies is given in the following definition:

Definition 1. (Partial Order)
A relation $\sqsubseteq$ *on a set of types,* Type, *is a* partial order *where it is:*

- Reflexive, *i.e.,* $\forall x \in \mathsf{Type}$ *we have* $x \sqsubseteq x$.
- Antisymmetric, *i.e.,* $\forall x, y \in \mathsf{Type}$ *if* $x \sqsubseteq y \wedge y \sqsubseteq x$ *then* $x = y$.
- Transitive, *i.e.,* $\forall x, y, z \in \mathsf{Type}$ *if* $x \sqsubseteq y \wedge y \sqsubseteq z$ *then* $x \sqsubseteq z$.

[1] For the complete description, a curious reader is referred to Carpenter (1992) or Krieger (1995)

In our case, we restrict ourselves to bounded complete partial orders (BCPO), which are partial orders where each pair has exactly one LUB.

TFSs are defined based on a finite set of features Feat and an BCPO$\langle \mathsf{Type}, \sqsubseteq \rangle$:

Definition 2. (Typed Feature Structures)
A typed feature structure *(TFS) is a tuple* $F = \langle Q, \overline{q}, \theta, \delta \rangle$ *where:*

- Q *is a finite set of nodes with the root* $\overline{q}$.
- $\overline{q}$ *is the (unique) root node.*
- $\theta : Q \rightarrow \mathsf{Type}$ *is a total node typing function.*
- $\delta : \mathsf{Feat} \times Q \rightarrow Q$ *is a partial feature value function.*

We adopt the appropriate condition from Carpenter (1992), which states that a feature is appropriate for a type and all its subtypes.

Definition 3. (Appropriateness)
In an inheritance hierarchy $\langle \mathsf{Type}, \sqsubseteq \rangle$, *the partial function Approp:* $\mathsf{Feat} \times \mathsf{Type} \rightarrow \mathsf{Type}$ *meets the following conditions:*

- **Feature Introduction.** *For every feature* $f \in \mathsf{Feat}$, *there is a most general type—* $Intro(f) \in \mathsf{Type}$*—such that Approp(f,(Intro(f)) is defined.*
- **Downward Closure.** *If* $Approp(f, \sigma)$ *is defined and* $\sigma \sqsubseteq \tau$, *then* $Approp(f, \tau)$ *is also defined and* $Approp(f, \sigma) \sqsubseteq Approp(f, \tau)$.

In order to define unification and default unification, TFSs are ordered by the following subsumption relation which spans a lattice:

Definition 4. (Subsumption)
$F = \langle Q, \overline{q}, \theta, \delta \rangle$ *is said to subsume* $F' = \langle Q', \overline{q}', \theta', \delta' \rangle$, $F \sqsubseteq F'$, *iff there is a total function* $h : Q \rightarrow Q'$ *such that:*

- $h(\overline{q}) = \overline{q}'$.
- $\theta(q) \sqsubseteq \theta'(h(q))$ *for every* $q \in Q$.
- $h(\delta(f, q)) = \delta'(f, h(q))$ *for every* $q \in Q$ *and feature* f *such that* $\delta(f, q)$ *is defined.*

Definition 5. (Unification)
In an inheritance hierarchy as described above, the unification of two feature structures is their greatest lower bound (GLB) $\sqcup$.

It is worth noting that for treelike type hierarchies, two TFSs unify only in cases where they are type compatible. Finally, our default unification is characterized by the following elegant and powerful definition)taken from Carpenter (1993)):

Definition 6. (Credulous Default Unification)
$F \overset{<}{\sqcup}_c G = \{ F \sqcup G' | G' \sqsubseteq G$ *is maximal such that* $F \sqcup G'$ *is defined* $\}$.

3 Overlay = Default Unification + Score

In our main application, we are concerned with two challenges. We provided a motivation for the first one—enrichment—in Sect. 1. The second one—validation—is concerned with the task of assigning the most probable hypothesis among several a higher number than the competing ones. In SMARTKOM, i.e., systems processing user actions run though nonperfect recognizers and/or analyzers, the possible interpretations of a user action will almost never be unambiguous. The task of comparing the representation of the user action with context must involve some kind of scoring telling how well the representation fits the context. Our solution is to combine default unification with a scoring functionality so that the actual score is computed during the combination of new and old information. To this end we developed an efficient implementation of default unification and equipped it with a scoring mechanism. Below we provide a formal operational semantics of this default unification operation (Sect. 3.1), whereas the heuristic scoring function is described in Sect. 3.2 (see also Sect. 3.3 for a highlighting example).

3.1 Default Unification

Although the theoretical machinery presented above provides a precise characterization of unification and default unification, it says nothing about how to compute unification and default unification. Several researchers have provided efficient algorithms for unification for use in parsing, e.g., Tomabechi (1991). However, in the case of default unification there exists little about efficient implementations. The few exceptions are found in the area of robust parsing of NL texts, e.g., Imaichi and Matsumoto (1995); Fouvry (2003). This is due to the fact that for more complex inheritance hierarchies, the credulous default unification is, in the general case, regarded to be exponential, e.g., Copestake (1993); Grover et al. (1994). The main reason is that credulous default unification is ambiguous and the fact that in type systems providing multiple inheritance and reentrant structures (coreferences), the search for the least upper bound, or more precisely, the minimal upper bounds, is exponential. In our case, there are no reentrant structures, and in the type hierarchy we use unary inheritance only. Therefore even a naïve algorithm as presented below computes default unification with a complexity linear to the size of its arguments.

For this chapter we adapt the definition of overlay given in Alexandersson and Becker (2001). The basic schema for the computation of overlay is based on two steps. In the first—called assimilation—the general case is that the background is transformed to the type of the LUB of the covering and background. An important consequence of the transformation is that only the features appropriate for the LUB are kept. There are cases when this viewpoint does not hold, since it could be that the covering is more general than the background, in which case the two structures are kept. In the second step, an algorithm close to naïve unification is deployed with the small change, that in case of conflicts, the information of the covering overwrites the one in background (atomic values), or overlay is recursively called (complex values).

The type of the resulting structure is the type of the covering in case of conflicting types and the type of their GLB in case of compatible types.

Definition 7. (Assimilation)

Let co and bg be two TFSs (covering and background) such that $co = \langle Q_{co}, \overline{q}_{co}, \theta_{co}, \delta_{co}\rangle$ *and* $bg = \langle Q_{bg}, \overline{q}_{bg}, \theta_{bg}, \delta_{bg}\rangle$. *Then, the assimilation—*$assim(co, bg)$*—is defined as*

$$\begin{aligned}&if\ (co \sqsubseteq bg\ or\ bg \sqsubseteq co),\\ &then\ assim(co,bg) := \langle co, bg\rangle,\\ &else\ assim(co,bg) := \langle co, \langle Q_a, \overline{q}_a, \theta_a, \delta_a\rangle\rangle,\end{aligned}$$

where

- $t_a := LUB(co, bg)$,
- $\overline{q}_a = \overline{q}_{bg}$ *with* $\theta(\overline{q}_a) = t_a$,
- δ_a *is the same as* δ_{bg} *for* f *reachable from* $\overline{q}_a$ *(but for* $\delta_a(f, \overline{q}_a)$ *we require* $Approp(f, t_a)$*),*
- Q_a *is the set of nodes reachable from* $\overline{q}_a$,
- θ_a *is the same as* θ_{bg} *for the nodes reachable from* $\overline{q}_a$ *with* $\theta_a(\overline{q}_a) = t_a$.

In what follows, we assume a function α that returns the resulting cover and background, i.e., $\alpha_{co}(assim(co, bg)) = co$. Next, we define overlay:

Definition 8. (Overlay)

Let co and bg be two TFSs (covering and background) such that $assim(co,bg) = \langle co', bg'\rangle = \langle\langle Q_{co'}, \overline{q}_{co'}, \theta_{co'}, \delta_{co'}\rangle, \langle Q_{bg'}, \overline{q}_{bg'}, \theta_{bg'}, \delta_{bg'}\rangle\rangle$. *Then,* $Overlay(co, bg)$ *is defined as:*

$$Overlay(co,bg) := Overlay'(co', bg') := \langle Q_o, \overline{q}_o, \theta_o, \delta_o\rangle, \tag{1}$$

where

$$\overline{q}_o := \overline{q}_{co'}, \quad and \quad \theta_o(\overline{q}_o) := GLB(\theta_{co'}(\overline{q}_{co'}), \theta_{bg'}(\overline{q}_{bg'})), \quad and$$

$$\delta_o(f, \overline{q}_o) :=$$

$$Overlay(\delta_{co'}(f, \overline{q}_{co'}), \delta(f, \overline{q}_{bg'}))\ if\ f\ exists\ in\ co'\ and\ bg', \tag{2}$$

$$\delta_{co'}(f, \overline{q}_{co'})\ if\ f\ exists\ only\ in\ co', \tag{3}$$

$$\delta_{bg'}(f, \overline{q}_{bg'})\ if\ f\ exists\ only\ in\ bg'. \tag{4}$$

3.1.1 Related Work

Within AI there are different approaches to default reasoning, i.e., how to draw inferences despite inconsistencies in the knowledge base. Within these approaches, the distinction between credulous and skeptical inferences is made. The latter type of inferences means that inferences are not drawn in case of conflicts, whereas inferences of the former type mean that as many consistent conjectures as possible are drawn (Antoniou, 1999; Touretzky, 1986). A famous example related to this distinction is how in inheritance hierarchies one has to allow for overwriting in the case of describing a penguin. Since birds can fly, and penguins are birds but cannot fly, the information stating that birds can fly has to be overwritten. In our case we want to overwrite old contextual information with new conflicting information and at the same time be able to draw inferences—we are thus aiming for credulous default reasoning.

Within computation linguistics, the early work starts with Karttunen, e.g., Karttunen (1986), who used in his DPATR workbench a default unification algorithm called *clobber*. Kaplan (1987) introduces *priority union* as a technique for an elegant way of processing elliptical or morphological nonmonotonic constructions. These approaches are, however, based on nontyped feature structures. The processing of elliptical constructions is picked up by Grover et al. (1994), but because of the expense for computing priority union—it is exponential—the authors state that their implementation cannot be used in real applications.

In the beginning of the 1990s, techniques for the processing of typed feature structures were developed (Bouma, 1990; Calder, 1993; Carpenter, 1993). The application for default unification ranges from lexicon management (Villavicencio, 1999; Lascarides et al., 1996; Carpenter, 1993) over robust parsing (Imaichi and Matsumoto, 1995; Fouvry, 2003) to discourse modeling where, for instance, elliptical constructions have to be processed (Grover et al., 1994; Alexandersson and Becker, 2004).

3.2 Scoring

So far, we have seen that overlay can be used to transfer nonconflicting information from one structure to another—even in the case of type clashes or conflicting atomic values. What is missing is a way to assess the structural resemblance of the involved structures to be able to judge how well a background structure fits the covering. This scoring of the overlay result can be very useful, e.g., for tasks like disambiguation in discourse processing, where several ambiguous hypotheses have to be interpreted with respect to the discourse context. As a consequence, we extended the basic overlay operation with a scoring mechanism (Pfleger et al., 2002). This scoring mechanism is based on the collection of different parameters during the application of overlay. The actual scoring mechanism consists of two parts: (i) the collection of the scoring parameters, which is incorporated into the algorithm, and (ii) the actual scoring function, which compiles the collected parameters into one resulting score after the application of overlay is finished.

3.2.1 A Heuristic Scoring Function

For SMARTKOM we developed a heuristic scoring function that basically expresses the contrast between the amount of nonconflicting features and the amount of conflicting features (including type clashes). We identified four scoring parameters that need to be collected during the application of overlay:

co: a feature or a (atomic) value stemming from the covering is added to the result. **co** is incremented for each feature in the covering.
bg: a feature or a (atomic) value in the result occurs in the background; **bg** is incremented.
tc: type clash, i.e., the type of the covering and background was not identical. This is identified during the computation of the assimilation.
cv: conflicting values. This occurs when the value of a feature from the background is overwritten.

All these parameters are collected during the application of overlay, and in the end a function compiles that information into a single number reflecting the structural consistency of the two structures. The sum of **co** and **bg** minus the sum of **tc** and **cv** will be weighted against the sum of **co, bg, tc** and **cv**. This leads to a function (shown in the formula below) whose codomain is $[-1,1]$.

$$score(co,bg,tc,cv) = \frac{co+bg-(tc+cv)}{co+bg+(tc+cv)}.$$

The positive extremal ($score(co,gb,tc,cv) = 1$) indicates that the feature structures are unifiable. The negative extremal ($score(co,gb,tc,cv) = -1$) indicates that all information from the background was overwritten by information from the cover. A score within this interval indicates that the cover more or less fits the background: The higher the score, the better the cover fits the background. Negative values signal that conflicting and thus overlayed values outweigh unifiable values (and positive values vice versa).

3.2.2 The Distance Factor

The overall scoring of overlay should not only reflect the structural resemblance of the structures for particular tasks, but it is also necessary to incorporate the temporal distance of the overlayed structures into the scoring. In other words, it is necessary to weigh the overall scoring of the result against the amount of time that has passed since the occurrence of the background structure and the occurrence of the covering.

What kind of distance factor is best suited for weighting the score of overlay depends on the application in which overlay is employed. In an application like discourse processing, for example, the score described above has to be weighted by the distance between the two hypotheses in the discourse (Alexandersson and Pfleger, 2006).

3.3 An Example

The following example provides insight into the disambiguation as well as the enrichment functionality provided by overlay. As indicated above, one potential benefit of the scoring heuristic is the disambiguation of ambiguous user input or ambiguous recognition results. Consider, for example, a dialogue situation (see dialogue excerpt in Fig. 1) where the user has requested information about tonight's television program (see Fig. 2 for a TFS representation), resulting in a system presentation of tonight's television program. Now the user wants to reduce the number of displayed movies by uttering *Is there a movie with Nicole Kidman?*. This user request, if interpreted separately, leads to at least two intention hypotheses — one where the user requests information about the current TV program (see Fig. 3) and one where the user requests information about the current cinema program (see Fig. 4).

U: *What's on TV tonight?*

S: (Displays the TV program listings) *Here you see a list of movies running tonight.*

U: *Is there a movie with Nicole Kidman?*

Fig. 1. Dialogue excerpt

If we now overlay these two interpretation hypotheses with the latest user contribution we get for both a suitable result. However, one of the results is more likely than the other, and this is expressed by the scoring of overlay. Column 2 of table 1 shows the collected parameters and the overall scoring for overlaying the first hypothesis (TFS in Fig. 3) and the previous user contribution (TFS in Fig. 2), whereas column 3 shows the same for overlaying the second hypothesis (TFS in Fig. 4) and the last user contribution. As expected, the result of overlaying the second hypothesis over the last user contribution receives the higher score, namely 1.0, which is the highest possible score and indicates that the two structures are unifiable.

Table 1. Parameters collected during the application of overlay

	TFS 2 and **TFS 3**	**TFS 2** and **TFS 4**
co	10	10
bg	5	7
tc	2	0
cv	0	0
score	0.765	1.0

4 Conclusion

In this chapter we presented a formal description of overlay—the fundamental operation for discourse processing in SMARTKOM. In particular we described the two most important aspects of this operation—namely default unification and scoring—in detail. Furthermore, we outlined the application of overlay for the task of discourse processing and gave an example that highlights the value of the scoring mechansim.

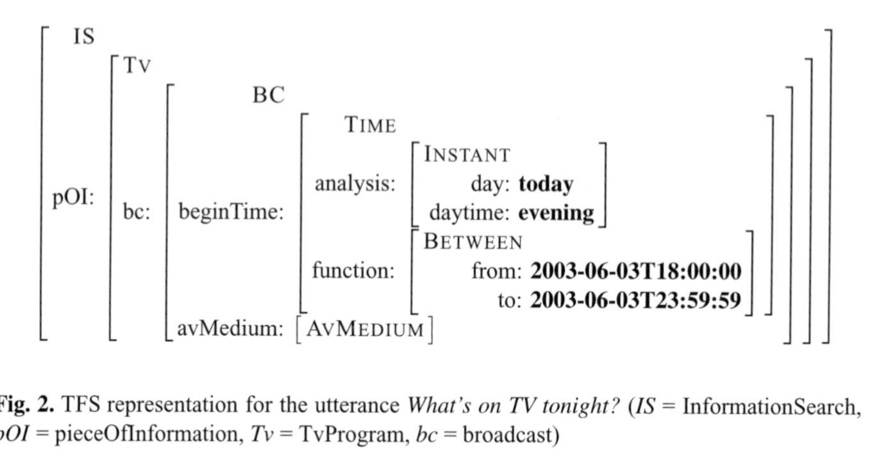

Fig. 2. TFS representation for the utterance *What's on TV tonight?* (*IS* = InformationSearch, *pOI* = pieceOfInformation, *Tv* = TvProgram, *bc* = broadcast)

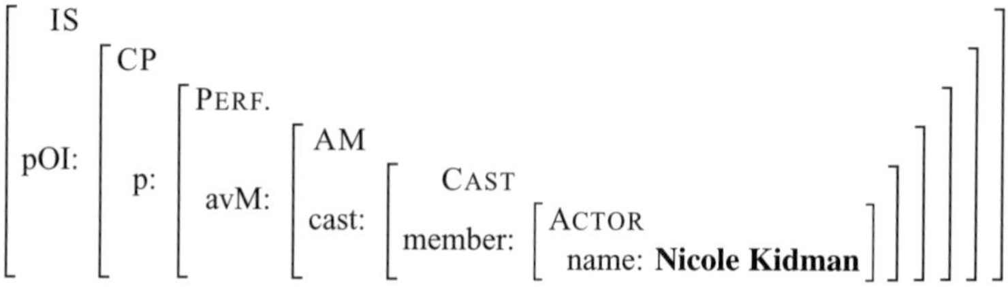

IS
pOI:
CP
p:
PERF.
avM:
AM
cast:
CAST
member:
ACTOR
name: **Nicole Kidman**

Fig. 3. One possible interpretation of the utterance *Is there a movie with Nicole Kidman?* (*IS* = InformationSearch, *pOI* = pieceOfInformation, *CP* = CinemaProgram, *p* = performance, *P* = Performance, *bc* = Broadcast, *Bc* = Broadcast)

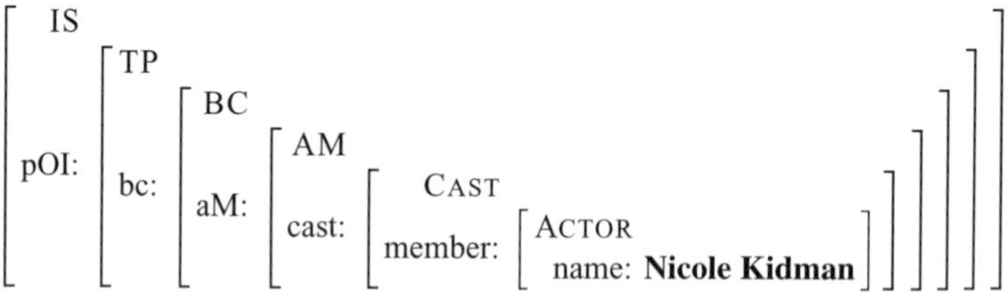

IS
pOI:
TP
bc:
BC
aM:
AM
cast:
CAST
member:
ACTOR
name: **Nicole Kidman**

Fig. 4. A second possible interpretation of the utterance *Is there a movie with Nicole Kidman?* (*IS* = InformationSearch, *pOI* = pieceOfInformation, *CP* = CinemaProgram, *p* = performance, *P* = Performance, *aM* = avMedium, *AM* = AvMedium)

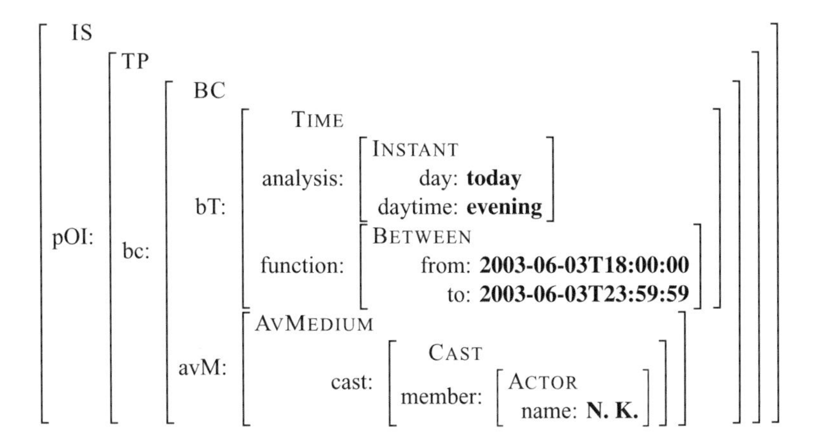

Fig. 5. Interpretation of the utterance *Is there a movie with Nicole Kidman?* with respect to the previous discourse context

For more information about the application of overlay in discourse processing, we refer to Alexandersson and Pfleger (2006).

Future work will cover the areas described below.

- This chapter provided a set-theoretical characterization of the behavior of overlay. The complexity of default unification for reentrant structures in complex inheritance hierarchies is regarded as exponential (Copestake, 1993). In Alexandersson and Becker (2003) we provide an efficient algorithm for the computation of default unification based on a multiple-inheritance type hierarchy.
- For SMARTKOM we decided to use typed feature structures (TFS) as the internal data representation for overlay. As we have shown above, TFS can be used to represent instances of an ontology, and thus we had the idea of an implementation of overlay which is directly based on ontological structures. To this end we are currently preparing an implementation of overlay for the Protégé ontology editor (`http://protege.stanford.edu/`).
- In some cases, the structures contain set-like objects, e.g., seats in a movie theater, which require a special treatment. For an ordered list it might be suitable to overlay the first element of the first list with the first element of the second one, and so on. However, in cases where the elements of a set are not ordered, it depends also on the objects themselves how overlay should be applied to the sets. Our current work outlines a framework for how and when what kind of operations should be used to combine such sets.
- Our present scoring function does not account for all available information, as it just counts the type clashes. There is, however, more information available that can contribute to the overall score of overlay. An example for this is the *informational distance*—the distance between the background type and the LUB

in case of a type clash (Alexandersson et al., 2004)—which assess how distinct conflicting covering and background are.

Acknowledgments

The authors would like to thank Massimo Romanelli for discussions and proofreading.

References

J. Alexandersson and T. Becker. Overlay as the Basic Operation for Discourse Processing in a Multimodal Dialogue System. In: *Workshop Notes of the IJCAI-01 Workshop on "Knowledge and Reasoning in Practical Dialogue Systems"*, Seattle, WA, August 2001.

J. Alexandersson and T. Becker. The Formal Foundations Underlying Overlay. In: *Proc. 5th Int. Workshop on Computational Semantics (IWCS-5)*, pp. 22–36, Tilburg, The Netherlands, February 2003.

J. Alexandersson and T. Becker. Default Unification for Discourse Modelling. In: H. Bunt and R. Muskens (eds.), *Computing Meaning*, vol. 3, Dordrecht, The Netherlands, 2004. Kluwer Academic.

J. Alexandersson, T. Becker, and N. Pfleger. Scoring for Overlay Based on Informational Distance. In: *7. Konferenz zur Verarbeitung natürlicher Sprache KONVENS-04*, Vienna, Austria, September – October 2004.

J. Alexandersson and N. Pfleger. Discourse Modeling, 2006. In this volume.

G. Antoniou. A Tutorial on Default Logics. *ACM Computing Surveys (CSUR) Archive*, 31(4):337–359, December 1999.

G. Bouma. Defaults in Unification Grammar. In: *Proc. 28nd Annual Meeting of the Association for Computational Linguistics*, pp. 165–172, University of Pittsburgh, PA, 6–9 June 1990.

J. Calder. Typed Unification for Natural Language Processing. In: T. Briscoe, V. de Paiva, and A. Copestake (eds.), *Inheritance, Defaults, and the Lexicon*, pp. 13–37, Cambridge, UK, 1993. Cambridge University Press.

B. Carpenter. *The Logic of Typed Feature Structures*. Cambridge University Press, Cambridge, UK, 1992.

B. Carpenter. Skeptical and Credulous Default Unification With Applications to Templates and Inheritance. In: T. Briscoe, V. de Paiva, and A. Copestake (eds.), *Inheritance, Defaults, and the Lexicon*, pp. 13–37, Cambridge, CA, 1993. Cambridge University Press.

A. Copestake. *The Representation of Lexical Semantic Information*. PhD thesis, University of Sussex, 1993.

F. Fouvry. Constraint Relaxation with Weighted Feature Structures. In: *Int. Workshop on Parsing Technologies - IWPT'03*, pp. 103–114, Nancy, France, April 2003.

C. Grover, C. Brew, S. Manandhar, and M. Moens. Priority Union and Generalization in Discourse Grammars. In: *Proc. 32nd Annual Meeting of the Association for Computational Linguistics*, pp. 17–24, Las Cruces, NM, 1994. Association for Computational Linguistics.

I. Gurevych, R. Porzel, and R. Malaka. Modeling Domain Knowledge: Know-How and Know-What, 2006. In this volume.

I. Horrocks. The FaCT System. In: *Automated Reasoning With Analytic Tableaux and Related Methods: Int. Conf. Tableaux'98*, pp. 307–312, Oisterwijk, The Netherlands, 1998a.

I. Horrocks. Using an Expressive Description Logic: FaCT Or Fiction? In: *Principles of Knowledge Representation and Reasoning: Proc. 6th Int. Conf. (KR'98)*, pp. 636–647, San Francisco, California, 1998b.

O. Imaichi and Y. Matsumoto. Integration of Syntactic, Semantic and Contextual Information in Processing Grammatically Ill-Formed Inputs. In: *Proc. 14th IJCAI*, pp. 1435–1440, Montréal, Canada, 1995.

R.M. Kaplan. Three Seductions of Computational Psycholinguistics. In: P. Whitelock, H. Somers, P. Bennett, R. Johnson, and M. McGee Wood (eds.), *Linguistic Theory and Computer Applications*, pp. 149–188, London, UK, 1987. Academic Press.

L. Karttunen. D-PATR: A Development Environment for Unification-Based Grammars. In: *Proc. 11th COLING*, pp. 25–29, Bonn, Germany, August 1986. Institut für angewandte Kommunikations- und Sprachforschung e.V. (IKS).

H.U. Krieger. *TDL — A Type Description Language for Constraint-Based Grammars. Foundations, Implementation, and Applications.* PhD thesis, Universität des Saarlandes, Department of Computer Science, september 1995.

A. Lascarides, T. Briscoe, N. Asher, and A. Copestake. Order Independent and Persistent Typed Default Unification. *Linguistics and Philosophy*, 19(1):1–89, 1996. Revised version of ACQUILEX II WP NO. 40.

M. Minsky. A Framework for Representing Knowledge. In: P. Winston (ed.), *The Psychology of Computer Vision*, pp. 211–277, New York, 1975. McGraw-Hill.

N. Pfleger, J. Alexandersson, and T. Becker. Scoring Functions for Overlay and Their Application in Discourse Processing. In: *Proc. KONVENS 2002*, pp. 139–146, Saarbruecken, Germany, September–October 2002.

H. Tomabechi. Quasi-Destructive Graph Unifications. In: *Proc. 29th ACL*, pp. 315–322, Berkeley, CA, 1991.

D. Touretzky. *The Mathematics of Inheritance Systems.* Morgan Kaufmann, San Francisco, CA, 1986.

A. Villavicencio. Representing a System of Lexical Types Using Default Unication. In: *Proc. 9th Conf. of the European Chapter of the Association for Computational Linguistics (EACL'95)*, Bergen, Norway, 1999.

In Context: Integrating Domain- and Situation-Specific Knowledge

Robert Porzel, Iryna Gurevych*, and Rainer Malaka

European Media Laboratory GmbH, Heidelberg, Germany
robert.porzel,rainer.malaka@eml-d.villa-bosch.de
iryna.gurevych@eml-r.villa-bosch.de
* Current affiliation EML Research gGmbH.

Summary. We describe the role of context models in natural language processing systems and their implementation and evaluation in the SMARTKOM system. We show that contextual knowledge is needed for an ensemble of tasks, such as lexical and pragmatic disambiguation, decontextualizion of domain and common-sense knowledge that was left implicit by the user and for estimating an overall coherence score that is used in intention recognition. As the successful evaluations show, the implemented context model enables a multicontext system, such as SMARTKOM, to respond felicitously to contextually underspecified questions. This ability constitutes an important step toward making dialogue systems more intuitively usable and conversational without losing their reliability and robustness.

1 Introduction

The human enterprise of answering or responding to conversational speech input in a suitable and felicitous manner is not imaginable without three essential features:

- the ability to recognize what was said by the questioner
- the ability to infer information that is left implicit by the questioner
- the ability to infer what constitutes a useful and felicitous answer

The realization of such abilities poses a formidable challenge in the development of conversational intuitive dialogue systems with more than one domain, modality or situational context. The SMARTKOM system has to deal with contextual dependencies as well as cross-modal references based on the system's symmetric multimodality (Wahlster, 2003). It can handle multiple requests in different domain contexts and features special scenario-specific situational contexts. Thus, *decontextualization* is needed to resolve the arising contextual ambiguities (McCarthy, 1986, 1990). In the case of restricted and controlled single-domain systems, the problem of contextually implicit information can be solved by generating full paraphrases out of the underspecified user utterances (Ebert et al., 2001). In systems with multiple contexts additional knowledge sources and dynamic context modeling are needed. Herein we describe the central contextual processing unit which combines ontological and situative knowledge. In the SMARTKOM system, discourse contextual influences are handled by unification-based operations, for example OVERLAY(Alexandersson and Becker, 2003; Löckelt et al., 2002), which operate on the schemas automatically created from the SMARTKOM ontology (Gurevych et al., 2003b; Porzel et al., 2003b). They also interact closely with the contextual processing module described herein, for the resolution of deictic expressions.

This integration of basic common-sense domain knowledge with situative context knowledge constitutes a necessary building block for scalable natural language understanding systems that facilitate felicitous cooperation and intuitive access to web-based, location-based and personal information. In single-context systems, such as train schedules or help desk systems (Aust et al., 1995; Gorin et al., 1997), this does not constitute a problem, since conversational phenomena such as pragmatic and semantic ambiguities or indirect speech acts do not occur (Porzel and Gurevych, 2002). A multidomain, multiscenario and multimodal system faces diverse usage contexts (e.g., at home or in mobile scenarios), conversational phenomena (e.g., indirect speech acts and pragmatic ambiguities) and multiple cross-modal references (e.g., gestural and linguistic discourse objects). A comprehensive understanding of naturally occurring discourse and of the often implicit questions embedded therein still has many unsolved issues in computational linguistics. In this work, we describe research on context and knowledge modeling components that enable dialogue systems with multiple contexts to realize the needed capabilities outlined above.

For example, in most conversational settings a passerby's response to a question such as:

(1) Excuse me, how do I get to the castle?

will most likely not be followed by asking where and when the spatial instructions should start. More likely, directions will be given. But, as the collected field data (see Sect. 5.1) shows, the felicity of spatial instructions is also dependent on contextual factors such as distance, mobility of the questioner or weather. Information concerning time or place, for example, is rarely explicated when given *default* settings, based on *common ground* (Krauss, 1987) hold. If not, however, such information is very likely to be expressed explicitely. In some cases, which are commonly labeled as *indirect speech acts* or *pragmatic ambiguities*, we are not only faced with implicit information, but also with implicit intentions.

It is, however, possible to resolve the ensuing ambiguities and determine appropriate default settings using additional context, dialogue and system knowledge. We will show how such knowledge can be based on collected data relevant to the domains and situations at hand. Next to the *Wizard-of-Oz* data collections and data collected in evaluations, based on the PROMISE framework (Beringer et al., 2002), we included existing lexicographic and ontological analyses, e.g., a model of frame semantics (Baker et al., 1998) as well as an ontological top-level (Russell and Norvig, 1995), and conducted additional experiments and data collections in *Hidden-Operator* experiments (Rapp and Strube, 2002) and the newly developed

Wizard-and-Operator paradigm (Gelbart et al., 2006; Porzel and Baudis, 2004). This collected, analyzed and modeled information, then became part of the ontological domain knowledge, i.e., the hierarchies, relations and cardinalities modeled therein.

Ontologies have traditionally been used to represent domain knowledge and are employed for various linguistic tasks, e.g., semantic interpretation, anaphora or metonymy resolution. In our case, the aggregate model of situative and domain knowledge contains the SMARTKOM ontology (Gurevych et al., 2003b; Porzel et al., 2003b; Gurevych et al., 2006). As follows from interfacing with automatic speech, gesture and emotional recognition systems, a significant amount of uncertainty is involved, which is probably best reflected in the ensuing intention lattices and their confidence scores. Whether one looks at intention, word lattices or n-best lists of hypotheses, the problem of facing several different representations of a single utterance arises. This remains, even though multimodal systems can use the individual modalities to disambiguate each other. Different hypotheses of what the user actually might have said, of course, lead to a different understanding and, in consequence, to potentially different requests to the background system. The role of the context model in this light is to assist in evaluating the competing intention hypotheses against each other to find out what was said. Then, such contextual domain and situational knowledge can be used for augmenting such intention hypotheses with implicit information, to spell out their underlying intentions and, finally, to define a common background representation for the processed content, i.e., intention lattices in the case of the SMARTKOM system. Summarizing, a context model, therefore, can be employed in the following tasks:

- The explication of situationally implicit information. This task can be further differentiated into two subtasks:
  - provision of information that is indexical — such as time and place — based on common ground and common sense defaults and their dynamic instances, e.g., the current position of the user
  - provision of information that is pragmatic, such as speech acts and intentions and their dynamic instances, e.g., the actual open or closed state (accessibility) of the goal object
- The scoring of individual interpretations in terms of their contextual coherence. Again, this task can be further differentiated into two subtasks:
  - using the ontological domain context to measure the semantic coherence of the individual interpretations, e.g., the ranking n-best lists or semantic interpretations thereof
  - using dynamic situational and discourse information, e.g., previous ontological contexts of prior turns

Additionally, we can use the ontological knowledge modeled therein as the basis for defining the semantics and content of the information exchanged between the modules of multimodal technology systems, as described by Gurevych et al. (2003b) and Gurevych et al. (2006)

After an overview of the state of the art of dialogue systems in light of their domain and context specificity, we discuss the nature of domain and situation models and their role in multidomain, multiscenario and multimodal dialogue systems. Finally, we describe the architecture and processing of the context-model in the SMARTKOM system (Wahlster, 2003).

2 Contextual Processing in Dialogue Systems

Earlier approaches to handle conversational natural language input produced only "toy" systems. Their respective aims were to cope with special linguistic problems and/or to model particular cognitive capacities of language users. Broad coverage of constructions, lexical information sources and semantic/pragmatic behaviors was not the primary concern and was also far outside the scope and capabilities of these systems. Today's linguistic development environments, representations and methodologies have shown that approximately complete coverage may be achieved in the areas of morphology and computational grammar. Furthermore, large lexical resources have been made available for linguistic applications in the area of parsing and also for natural language interfaces to application areas with restricted domains. Even though a broad coverage of frame semantic specification is still in the annotation progress (Baker et al., 1998), the handling of lexical semantics is still not set in stone (Allen, 1987; Gil and Ratnakar, 2002) and formal methods for dealing with pragmatic factors are in their beginning stages (Bunt, 2000; Porzel and Strube, 2002; Porzel and Gurevych, 2002). Systems are in development that can offer suitable natural language interfaces both on the reception and the production side. These systems can be (and are) employed in domain-specific applications or demonstrators where they are commonly linked to nonlinguistic applications (called the *background system*) such as databases (Gallwitz et al., 1998), geographic information systems (Malaka and Zipf, 2000; Johnston et al., 2002), task planning systems (Allen et al., 1995; Ferguson and Allen, 1998) or customer service systems (Gorin et al., 1997).

Some open issues in handling the multidomain problem are successfully beginning to be handled in the question-answering arena, by improved question parsing techniques coupled with more knowledge-based information understanding methods (Hovy et al., 2001; Prager et al., 2001; Moldovan et al., 2002). These information retrieval solutions assume more traditional desktop scenarios and more or less homogeneous content bases. While this is a reasonable thing to do for the type of information retrieval tasks with which the respective systems have to deal, conversational dialogue systems are faced with additional complications that are added to the general open domain problem. Next to the spontaneous speech recognition input, additional factors are the changing context/situation of a mobile user and system on the one hand, and the multitude of heterogeneous content bases that are needed to handle the topical informational need of a mobile user (e.g., a tourist) on the other. The content sources encompass, for example, rapidly changing online cinema information services, electronic program guides, or hotel reservation systems; slower changing remote geographic information systems; or relatively stable historical and architectural databases.

Natural language understanding in the area of parsing and pragmatically understanding questions as well as in terms of extracting their underlying *intentions* and finding suitable and felicitous answers is far from being solved. Still, a variety of robust parsers can deliver valuable contributions beyond part of speech and treebank tags (Pieraccini et al., 1992; Collins, 1996; Gavaldá and Waibel, 1998; Engel, 2002).

The fact that multidomain, multiscenario and multimodal conversational dialogue systems have so far been nonexistent in the real word is in part due to the fact that in all areas of natural language processing (NLP) we face a mixture of context-variant and context-invariant factors that come into play at every level of NLP pipeline, e.g., speech recognition, semantic disambiguation, anaphoric resolution, parsing or generation. Ensembles of techniques and experiments are therefore needed to identify whether a factor is context-variant or not, and to identify specific types of contextual settings on which a given context-variant factor depends. Based on these findings and their application, decontextualization can be performed based on the contextual knowledge extracted, learned and modeled from the collected data. The results of such decontextualizations, e.g., for semantic disambiguation, then in turn can be evaluated using existing evaluation frameworks (Walker et al., 2000; Beringer et al., 2002).

(2) Where is the cinema Europa?

In real situations seemingly "simple" questions such as (2) are difficult to understand in the absence of context. A felicitous answer often depends on the situational context in which the dialogue takes place. That is, as the data collected and analyzed shows (Porzel and Gurevych, 2002), *where is* questions are either answered by localizations — if the reference object happens to be closed or with an instruction — if the reference object, e.g., a cinema or store, is open. In such cases of pragmatic ambiguity, the model resulting from the analyses of the corresponding data has to embed the utterance at hand into a greater situational context, e.g., by computing a contextual coherence score for the competing interpretations. The situation model consequently has to monitor the corresponding situational factors relevant to resolving such pragmatic ambiguities. Additionally, these situational observations are also needed for the resolution of indexicals, e.g., in the case of spatial or temporal deixis (Porzel and Gurevych, 2003).

The contextual coherence computations that are needed for decontextualization have to be able to deal with a variety of cases:

- For example, if decisions hinge on a number of contextual features, e.g., the situational accessibility of referenced objects (Porzel and Gurevych, 2003), or domain-specific and pragmatic factors based on relations between referenced objects, as found in metonomyzation (Nunberg, 1987). Here both ontological factors as well as situational factors come into play, e.g., semantic roles, weather, and discourse factors, including referential status, as well as user-related factors, e.g., tourists or business travelers as questioners and their time constraints.

- Additionally, if decisions hinge the contextual coherence of sets of concepts and their relations by applying both dialogical as well as semantic coherence measurements (Gurevych et al., 2003a; Porzel et al., 2003a), e.g., for ranking speech recognition hypotheses or semantic ambiguities.

3 Context Modeling

Utterances in dialogues, whether in human–human interaction or human–computer interaction, occur in a specific situation that is composed of different types of contexts. A broad categorization of the types of context relevant to spoken dialogue systems, their content and respective knowledge stores is given in Table 1.

Table 1. Contexts, content and knowledge sources

Types of context	Content	Knowledge store
Dialogical context	What has been said by whom	Dialogue model
Ontological Context	World/conceptual knowledge	Domain model
Situational context	Time, place, etc.	Situation model
Interlocutionary context	Properties of the interlocutors	User model

Following the common distinction between linguistic and extralinguistic context[1] our first category, i.e., the dialogical context, constitutes the linguistic context, encompassing both cotext as well as intertext (Bunt, 2000). In linguistics the study of the relations between linguistic phenomena and aspects of the context of language use is called *pragmatics*. Any theoretical or computational model dealing with reference resolution, e.g., anaphora or bridging resolution, spatial or temporal deixis, or nonliteral meanings, requires taking the properties of the context into account.

As knowledge sources in dialogue systems, domain models are regarded to "hold knowledge of the world that is talked about" (Flycht-Erriksson, 1999). Following this general definition comes the observation that:

> Information from the domain model is primarily used to guide the semantic interpretation of user's utterances; to find the relevant items and relations that are discussed, to supply default values, etc. The knowledge represented in a domain model is often coupled to the background system, e.g., a database system ... the domain knowledge is used to map information in the user's utterance to concepts suitable for database search.

We propose a different definition of the role of domain models in NLP systems such as SMARTKOM. In our minds the knowledge contained in a domain model is to be modeled as an ontology proper, i.e., independent from the way an utterance or

[1] All extralinguistic contexts are also often referred to as the *situational context* (Connolly, 2001), however, we adopt a finer categorization thereof.

query is processed by the background system, that is, the knowledge about (going to) cinemas, (seeing) movies and (getting) tickets. The representation of this knowledge is the same whether the background system is a specific database, a set of Web-spidering agents or a combination thereof.

Statistical models based on specific corpora can serve to define *context groups* (Widdows, 2003) and allow us to differentiate between sets of distinct domain contexts that feature respective sense- and co-occurrence distributions. In our terminology, this formal context group function outputs a domain, i.e., the real-world utterance-based linguistic target of our definition. It is important to note that despite the multitude of domains that are to be encompassed by the SMARTKOM system, the central aim is to create a kernel NLP system capable of dealing with multiple and extensible domains, which ultimately can be added to the system during runtime (Rapp et al., 2000).

One of the central ideas embedded within the SMARTKOM research framework is to develop a kernel NLP system that can be used in a variety of situations, i.e., scenarios, domains and modalities, see Wahlster (2002), whereby:

- *Scenarios* refer to different manifestations of the system, i.e., a home, office and public (booth) manifestation as well as a mobile one.
- *Modality* refers to speech, gesture, mimics, affectives and biometry.
- *Domain* refers to the general backdrop against which dialogues can be pitted, i.e., areas such as train schedules, movie information or hotel reservations.

These additional scenario-specific contexts feature:

- dynamic mobility of the user - where traditional input modalities, such as keyboard- and mouse-based input, are highly unsuitable
- prolonged dialogues throughout sometimes hour-long spatial navigation tasks
- context-dependent intentions

Therefore dynamic, e.g., situational, context information has to be integrated together with the domain knowledge.[2]

Speakers may not always be aware of the potential ambiguities inherent in their utterances. They leave it to the context to disambiguate and specify the message. Furthermore, they trust in the addressee's ability to extract that meaning from the utterance that they wanted to convey. In order to interpret the utterance correctly, the addressee must employ several context-dependent resources. Speakers in turn anticipate the employment of these interpretative resources by the hearer and construct the utterance knowing that certain underspecifications are possible since the hearer can infer the missing information. In the same way, certain ambiguities become permissible due to shared common ground (Krauss, 1987). The role of the interlocutionary context is therefore also of importance in this process. Since it is assumed in the

[2] As noted in Porzel and Gurevych (2002), current natural language understanding systems need a systematic way of including situational factors, e.g., the actual accessibility of goal objects has been shown to be a deciding contextual factor determining whether a given `Where interrogative` at hand is *construed* as an instructional or a descriptive request.

SMARTKOM context, that general user model information is supplied via external sources, e.g., via a user's *SmartCard*, only the interaction preferences of the users are monitored actively by the system.

4 Decontextualization in SmartKom

In line with our proposal stated above to separate domain and application knowledge, the implementation within the SMARTKOM system exhibits a clear distinction between domain-specific knowledge and application-specific knowledge. This is consequently mirrored by respective modules: the domain and situation model (each can be addressed via separate blackboards/communication pools) implemented in a module called *modeler.knowledge*, and the function model, implemented in a module called *modeler.function*. This module can be described as the module that contains the knowledge of how specific plans are realized (given the actual software agents, databases and hardware devices). It therefore can be regarded as a translator between representations coming from the NLP and knowledge system and those of the background system.

4.1 The Modeler Knowledge Module

The running module for situational and ontological knowledge receives dynamic spatio-temporal information, e.g., global positioning system (GPS) coordinates and local times as well as (multiple) representations of user utterances in the form of *intention hypothesis* as input. It converts the incoming documents into document object models, on which it operates.[3] After processing, a decontextualized *intention hypothesis* document is returned as output.[4]

The context-dependent tasks performed by the context module implemented in SMARTKOM are:

- to know which information is ontologically required and provide the adequate situational and ontological information
- to detect situationally appropriate readings
- to compute contextual coherence scores for alternative intention hypotheses

4.2 Modeler Knowledge at Work

The first and foremost function is to add situation-specific discourse and dialogue knowledge. For example, no agent can check room vacancies without knowing the

[3] See http://www.w3c.org/DOM for the specification.

[4] The modeler.knowledge module features additional task- and domain-independent functionalities to probe and manipulate and compute on the ontology (http://www.w3c.org/OWL) as well as on the schema hierarchy (http://www.w3c.org/XMLS), the dynamic respective situational data and the static database information.

arrival date and duration of the intended stay, neither can a theater agent reserve tickets without knowing the seat numbers, etc. In human–human dialogues this knowledge is responsible for determining relevant answers to given questions. Consider the following exchange given in examples (3) and (4), where additional turns, asking the user to specify time and place, are avoided by decontextualizing the question and providing corresponding answers.

(3)	*Was*	*läuft*	*im*	*Kino*	*?*
	What	runs	in the	cinema	?

(4)	*Hier*	*sehen*	*Sie*	*was*	*heute*	*in*	*den*	*Heidelberger*	*Kinos*	*läuft*	*.*
	Here	see	you	what	today	in	the	Heidelberg	Cinema	runs	.

Table 2. Context-specific insertions into a sample intention hypothesis resulting from the intepretation of a speech recognition hypothesis

```
<informationSearchProcess>
  <entertainment>
    <performance>
      <cinema>
        <contact>
          <address>
            <town>
              here
            </town>
          </address>
        </contact>
      </cinema>
      <time>
        <beginTime>
          <at>
            now
          </at>
        </beginTime>
      </time>
    </performance>
  </entertainment>
</informationSearchProcess>
```

```
<contact>
  <x> 70.345 < /x>
  <y> 49.822 < /y>
  <town>
    Heidelberg
  </town>
</contact>
<time>
  <at> 19:00:00T26:08:03 </at>
</time>
```

```
<scores>
  <contextualCoherence>
    0.46
  </contextualCoherence>
</scores>
```

The SMARTKOM context model enables the system to act analogously, i.e., to provide hitherto implicit knowledge concerning what is talked about. The simplified structures given in Table 2 show insertions (in boldface) into an intention hypothesis made by the model in the case of a question such as that given in example (3). In this case, the insertions made via contextual knowledge are threefold:

- For the cardinally required time and place slot in the performance object respective defaults are inserted.

- These indexical defaults are contextually resolved[5] by means of accessing a GPS. This information can be used to resolve `here` with an appropriate level of granularity, e.g., town or spatial points, by means of a geographic information system in much the same way as `today` will also be replaced with granularity-specific temporal information, e.g., year, date or time.
- A contextual score for each hypothesis is computed indicating the contextually most adequate reading, as SMARTKOM processes *intention lattices* consisting of several intention hypotheses.

By means of explicating such information and providing topical and contextually adequate values, the system can retrieve appropriate information from Web sites or databases on what is currently playing in town, produce maps featuring cinema locations and then offer further assistance in navigation or reservation, for example.

We have therefore linked the context model to interfaces providing contextual information. For example, within both the SMARTKOM and the DEEP MAP framework (Malaka and Zipf, 2000), a database called the *Tourist-Heidelberg-Content Base* supplies information about individual objects including their opening and closing times. [6] By default, objects with no opening times, e.g., streets, can be considered always to be open. A GPS built into the mobile device supplies the current location of the user, which is handed to the geographic information system that computes, among other things, the respective distances and routes to the specific objects. It is important to note that this type of context monitoring is a necessary prerequisite for context-dependent analysis.

5 Application and Evaluation in SmartKom

5.1 Data and Annotations

We collected two types of data for demonstrating and realizing context-dependent effects in the SMARTKOM scenarios. First, we collected field data, by asking SMARTKOM-specific questions of pedestrians on the street and tracking the situational context factors and responses. The logged and classified field data was then used to train classifiers for recognizing specific intentions based on contextual factors. In a previous study another corpus of questions was collected and annotated in terms of their underlying intentions and was turned into a "gold standard" (Porzel and Gurevych, 2002). Second, we collected laboratory data, i.e., dialogues in *Hidden-Operator* and *Wizard-and-Operator* experiments (Rapp and Strube, 2002; Gelbart et al., 2006; Porzel and Baudis, 2004). All utterances were transcribed. Then specific

[5] For example, the topical resolution of *here* and *now* - enable the system to produce a suitable response, such as retrieving a map of the cinemas of Heidelberg and the specific performances. Therefore `here` and `today` constitute *placeholders* for defaults that are replaced almost immediately with actual values by the situation model or discourse model.

[6] Additional information extraction agents are able to gather data and information from the web, using ontological translators and updating the local database.

sets of the audio files were sent to the speech recognizer. We logged the speech recognition hypothesis (SRH), n-best lists of SRHs and the module's input and output for all utterances.

Using the laboratory data we created specific corpora for annotation experiments. In a first set of annotation experiments on a corpus of 1300 SRHs, the SRHs were annotated within the discourse context, i.e., the SRHs were presented in their original dialogue order. For each SRH, a decision had to be made whether it was semantically coherent or incoherent with respect to the best SRH representing the previous user utterance. In a second experiment the annotators saw the SRHs together with the transcribed user utterances. The task of annotators was to determine the best SRH from the n-best list of SRHs corresponding to a single user utterance. The decision had to be made on how well the SRH expressed the intentional content of the utterance (Porzel et al., 2003a). In the first experiment the interannotator agreement was 80%, and in the second 90%. Last, the annotators had to create corresponding gold standards by means of discussing the cases of disagreement until an optimal solution was found.

5.2 Evaluation in SmartKom

For evaluating the performance of the model described above we computed the task-specific accuracies as compared to the gold standards described above. The situational models trained on the field data of 366 subjects using a c4.5 machine learning algorithm (Winston, 1992) achieved an intention recognition accuracy of 88% as compared to baseline achieved by a context-insensitve model of 59% evaluated against the annotated gold standard of a corpus of dialogues with 50 subjects featuring various kinds of spatial interrogatives (Porzel and Gurevych, 2002).

For evaluating the contextual coherence scores of the model we logged the scores of all scoring modules (speech recognizer, parser and discourse model) that rank n-best lists of speech recognition hypotheses produced out of word graphs (Oerder and Ney, 1993) and those that rank the representations produced by the parser (Engel, 2002). As described above, these speech recognition hypotheses were annotated in terms of their coherence, correctness and *bestness* and turned into corresponding gold standards (Gurevych et al., 2002; Porzel et al., 2003a).

For computing contextual coherence in the evaluation, the module employed three knowledge sources, an ontology of about 730 concepts and 200 relations and a lexicon (3600 words) with word-to-concept mappings, covering the respective domains of the system and a conceptual dialogue history, including the concepts and relations of the previous best hypothesis. The final evaluation was carried out on a set of 95 dialogues. The resulting dataset contained 552 utterances resulting in 1.375 SRHs, corresponding to an average of 2.49 SRHs per user utterance.

The task of hypothesis verification, i.e., finding out what was said, in our multimodal dialogue system is to determine the best SRH from the n-best list of SRHs corresponding to a given user utterance. The baseline for this evaluation was the overall chance likelihood of guessing the best one, i.e., 63.91%.

The context- and knowledge-based system (Gurevych et al., 2003a; Porzel et al., 2003a) achieves an accurracy of 88%.[7] The knowledge-based system without the dialogical context features already exceeds that of the acoustic and language model scores produced by the automatic speech recognizer, reaching 84.06% on the same task.

The evaluation of the contextual coherence scoring in terms of its disambiguation performance meant calculating how often contextual coherence picks the appropriate reading, given an ambiguous lexicon entry such as *kommen* associated in the lexicon with both WatchPerceptualProcess and MotionDirectedTransliterated. For this evaluation we tagged 323 lemma with their contextually appropriate concept mappings, and achieved an accuracy of 85% given an aggregate majority class baseline averaged over the majority class baselines of each individual lemma of 42%.

6 Conclusion

The basic intuition behind explicating contextual dependencies originally proposed by McCarthy (1986) was that any given axiomatization of a state of affairs, meanings or relations presupposes an implicit context. Any explicit context model employed in processing information, therefore, needs to provide the information why specific meaning can be assigned to the underspecified information and, thus be, applied to its processing. This has often been called *fleshing out* and was considered impossible in its maximal form, e.g., Akman and Surav (1996) state that:

> It is seen that for natural languages a fleshing-out strategy — converting everything into decontextualized eternal sentences — cannot be employed since we do not always have full and precise information about the relevant circumstances.

Herein, we have presented a context model that performs a set of *fleshing out* tasks, which, as the successful evaluations show, suffice to enable a multicontext system, such as SMARTKOM, to respond felicitously to contextually underspecified questions. We have developed a corresponding system that integrates domain, dialogue and situative context in a multidomain, multiscenario and multimodal dialogue system. We have shown how:

- this knowledge can be used for improving the speech recognition reliability in the case of hypothesis verification, i.e., for finding out what was said
- this knowledge can be used to explicate contextually implicit information, i.e., for resolving indexical expressions
- this knowledge can be used to resolve context-dependent ambiguities, i.e., for lexical and pragmatic disambiguation

[7] This means that in 88% of all cases the best SRH defined by the human *gold standard* is among the best scored by the module.

We have therefore demonstrated that the inclusion of such contextual interpretation in natural language processing can enable natural language understanding systems to become more conversational without losing the reliability of restricted dialogue systems. Given the challenge to extract the underlying intentions from conversational utterances such as "is there a bakery close by" or "I don't see any bus stops", we presented the necessary knowledge stores and inferential capabilites necessary for their decontextualization, which is a prerequisite for understanding utterances and responding felicitously. This enables us to restate McCarthy's original claim to say that for natural languages a fleshing out strategy can be employed if we have sufficient and precise knowledge about the relevant contextual circumstances.

Acknowledgments

This work was partially funded by the German Federal Ministry of Education, Science, Research and Technology (BMBF) in the framework of the SMARTKOM project under Grant 01 IL 905 K7 and by the Klaus Tschira Foundation. The authors would like to thank Ralf Panse, Christof Müller and Hans-Peter Zorn for their valuable implementation and evaluation work.

References

V. Akman and M. Surav. Steps Toward Formalizing Context. *AI Magazine*, 17(3): 55–72, 1996.

J. Alexandersson and T. Becker. The Formal Foundations Underlying Overlay. In: *Proc. 5th Int. Workshop on Computational Semantics (IWCS-5)*, pp. 22–36, Tilburg, The Netherlands, February 2003.

J.F. Allen. *Natural Language Understanding*. Benjamin/Cummings, Menlo Park, CA, 1987.

J.F. Allen, L.K. Schubert, G. Ferguson, P. Heeman, C.H. Hwang, T. Kato, M. Light, N. Martin, B. Miller, M. Poesio, and D. Traum. The TRAINS Project: A Case Study in Building a Conversational Agent. *Journal of Experimental and Theoretical AI*, 7:7–48, 1995.

H. Aust, M. Oerder, F. Seide, and V. Steinbiss. The Philips Automatic Train Timetable Information System. *Speech Communication*, 17:249–262, 1995.

C.F. Baker, C.J. Fillmore, and J.B. Lowe. The Berkeley FrameNet Project. In: *Proc. COLING-ACL'98*, pp. 86–90, Montreal, Canada, 1998.

N. Beringer, U. Kartal, K. Louka, F. Schiel, and U. Türk. PROMISE: A Procedure for Multimodal Interactive System Evaluation. In: *Proc. Workshop on Multimodal Resources and Multimodal Systems Evaluation*, pp. 77–80, Las Palmas, Spain, 2002.

H. Bunt. Dialogue Pragmatics and Context Specification. In: H.C. Bunt and W.J. Black (eds.), *Computational Pragmatics, Abduction, Belief and Context; Studies in Computational Pragmatics*, pp. 81–150. John Benjamins, Amsterdam, 2000.

M.J. Collins. A New Statistical Parser Based on Bigram Lexical Dependencies. In: *Proc. 34th ACL*, pp. 184–191, Santa Cruz, CA, 1996.

J.H. Connolly. Context in the Study of Human Languages and Computer Programming Languages: A Comparison. *Modeling and Using Context*, 2116:116–128, 2001.

C. Ebert, S. Lappin, H. Gregory, and N. Nicolov. Generating Full Paraphrases of Fragments in a Dialogue Interpretation System. In: *Proc. 2nd SIGdial Workshop on Discourse and Dialogue*, pp. 58–67, Aalborg, Denmark, September, 1-2 2001.

R. Engel. SPIN: Language Understanding for Spoken Dialogue Systems Using a Production System Approach. In: *Proc. ICSLP-2002*, pp. 2717–2720, Denver, CO, 2002.

G. Ferguson and J.F. Allen. TRIPS: An Intelligent Integrated Problem-Solving Assistant. In: *Proc. 15th Nat. Conf. on Artificial Intelligence & 10th Conf. on Innovative Applications of Artificial Intelligence*, pp. 567–573, Madison, WI, 1998.

A. Flycht-Erriksson. A Survey of Knowledge Sources in Dialogue Systems. In: *Proc. IJCAI-99 Workshop on Knowledge and Reasoning in Practical Dialogue Systems*, pp. 41–48, 1999.

F. Gallwitz, M. Aretoulaki, M. Boros, J. Haas, S. Harbeck, R. Huber, H. Niemann, and E. Nöth. The Erlangen Spoken Dialogue System EVAR: A State–of–the–Art Information Retrieval System. In: *Proc. 1998 Int. Symposium on Spoken Dialogue (ISSD 98)*, pp. 19–26, Sydney, Australia, 30 November 1998.

M. Gavaldá and A. Waibel. Growing Semantic Grammars. In: *Proc. COLING-ACL'98*, pp. 451–456, Montreal, Canada, 1998.

D. Gelbart, J. Bryant, A. Stolcke, R. Porzel, M. Baudis, and N. Morgan. SmartKom English: From Robust Recognition to Felicitous Interaction, 2006. In this volume.

Y. Gil and V. Ratnakar. A Comparison of (Semantic) Markup Languages. In: *Proc. 15th Int. FLAIRS Conf.*, pp. 408–412, Pensacola, FL, 2002.

A.L. Gorin, G. Riccardi, and J.H. Wright. How May I Help You? *Speech Communication*, 23(1/2):113–127, 1997.

I. Gurevych, R. Malaka, R. Porzel, and H.P. Zorn. Semantic Coherence Scoring Using an Ontology. In: *Proc. Human Language Technology Conference / North American Chapter of the Association for Computational Linguistics Annual Meeting 2003*, pp. 88–95, Edmonton, Canada, 2003a.

I. Gurevych, R. Porzel, and R. Malaka. Modeling Domain Knowledge: Know-How and Know-What, 2006. In this volume.

I. Gurevych, R. Porzel, E. Slinko, N. Pfleger, J. Alexandersson, and S. Merten. Less Is More: Using a Single Knowledge Representation in Dialogue Systems. In: *Proc. HLT-NAACL'03 Workshop on Text Meaning*, pp. 14–21, Edmonton, Canada, May 2003b.

I. Gurevych, R. Porzel, and M. Strube. Annotating the Semantic Consistency of Speech Recognition Hypotheses. In: *Proc. 3rd SIGdial Workshop on Discourse and Dialogue*, pp. 46–49, Philadelphia, PA, July 2002.

E. Hovy, L. Gerber, U. Hermjakob, C. Lin, and D. Ravichandran. Toward Semantics-Based Answer Pinpointing. In: *Proc. 1st Int. Conf. on Human Language Technology*, 2001.

M. Johnston, S. Bangalore, G. Vasireddy, A. Stent, P. Ehlen, M. Walker, S. Whittaker, and P. Maloor. MATCH: An Architecture for Multimodal Dialogue Systems. In: *Proc. 10th ACM Int. Symposium on Advances in Geographic Information Systems*, pp. 376–383, Washington, DC, 2002.

R. Krauss. The Role of the Listener: Addressee Influences on Message Formulation. *Journal of Language and Social Psychology*, 6:91–98, 1987.

M. Löckelt, T. Becker, N. Pfleger, and J. Alexandersson. Making Sense of Partial. In: C.M. Johan Bos, Mary Ellen Foster (ed.), *Proc. 6th Workshop on the Semantics and Pragmatics of Dialogue (EDILOG 2002)*, pp. 101–107, Edinburgh, UK, September 2002.

R. Malaka and A. Zipf. Deep Map: Challenging IT Research in the Framework of a Tourist Information System. In: D.R. Fesenmaier, S. Klein, and D. Buhalis (eds.), *Proc. 7th. Int. Conf. on Information and Communication Technologies in Tourism (ENTER 2000)*, pp. 15–27, Berlin Heidelberg New York, 26–28 April 2000. Springer.

J. McCarthy. Notes on Formalizing Contexts. In: T. Kehler and S. Rosenschein (eds.), *Proc. 5th Nat. Conf. on Artificial Intelligence*, pp. 555–560, Los Altos, CA, 1986. Morgan Kaufmann.

J. McCarthy. Generality in Artificial Intelligence. In: V. Lifschitz (ed.), *Formalizing Common Sense: Papers by John McCarthy*, pp. 226–236, Norwood, NJ, 1990. Ablex Publishing Corporation.

D. Moldovan, M. Pasca, S. Harabagiu, and M. Surdeanu. Performance Issues and Error Analysis in an Open-Domain Question Answering System. In: *Proc. 40th ACL*, pp. 33–40, Philadelphia, PA, 2002.

G. Nunberg. *The Pragmatics of Reference*. Indiana University Linguistics Club, 1987.

M. Oerder and H. Ney. Word Graphs: An Efficient Interface Between Continuous-Speech Recognition and Language Understanding. In: *Proc. Int. Conf. on Acoustics, Speech, and Signal Processing (ICASSP-93)*, vol. 2, pp. 119–122, Minneapolis, MN, 1993.

R. Pieraccini, E. Tzoukermann, Z. Gorelov, J.G.E. Levin, C. Lee, and J. Wilpon. A Speech Understanding System Based on Statistical Representation of Semantics. In: *Proc. Int. Conf. on Acoustics Speech Signal Processing*, San Francisco, CA, 1992.

R. Porzel and M. Baudis. The Tao of CHI: Towards Felicitous Human-Computer Interaction. In: *Proc. Human Language Technology Conference / North American Chapter of the Association for Computational Linguistics Annual Meeting 2004*, Boston, MA, 2004.

R. Porzel and I. Gurevych. Towards Context-Adaptive Utterance Interpretation. In: *Proc. the 3rd SIGdial Workshop on Discourse and Dialogue*, pp. 90–95, Philadelphia, PA, 2002.

R. Porzel and I. Gurevych. Contextual Coherence in Natural Language Processing. In: P. Blackburn, C. Ghidini, R. Turner, and F. Giunchiglia (eds.), *Modeling and Using Context*, LNAI 2680, Berlin Heidelberg New York, 2003. Springer.

R. Porzel, I. Gurevych, and C. Müller. Ontology-Based Contextual Coherence Scoring. In: *Proc. 4th SIGdial Workshop on Discourse and Dialogue*, Sapporo, Japan, July 2003a.

R. Porzel, N. Pfleger, S. Merten, M. Loeckelt, I. Gurevych, R. Engel, and J. Alexandersson. More on Less: Further Applications of Ontologies in Multi-Modal Dialogue Systems. In: *Proc. 3rd IJCAI 2003 Workshop on Knowledge and Reasoning in Practical Dialogue Systems*, Acapulco, Mexico, 2003b.

R. Porzel and M. Strube. Towards Context Adaptive Natural Language Processing Systems. In: M. Klenner and H. Visser (eds.), *Proc. Int. Symposium: Computional Linguistics for the New Millenium*, pp. 141–156. Peter Lang, Berlin, 2002.

J. Prager, D. Radev, and K. Czuba. Answering What-Is Questions by Virtual Annotation. In: *Proc. 1st Int. Conf. on Human Language Technology*, 2001.

S. Rapp and M. Strube. An Iterative Data Collection Approach for Multimodal Dialogue Systems. In: *Proc. 3rd Int. Conf. on Language Resources and Evaluation (LREC 2002)*, pp. 661–665, Las Palmas, Spain, 2002.

S. Rapp, S. Torge, S. Goronzy, and R. Kompe. Dynamic Speech Interfaces. In: *Proc. Workshop Artificial Intelligence in Mobile Systems, ECAI 2000*, Berlin, Germany, 2000.

S.J. Russell and P. Norvig. *Artificial Intelligence. A Modern Approach.* Prentice Hall, Englewood Cliffs, NJ, 1995.

W. Wahlster. SmartKom: Fusion and Fission of Speech, Gestures, and Facial Expressions. In: *Proc. 1st Int. Workshop on Man-Machine Symbiotic Systems*, pp. 213–225, Kyoto, Japan, 2002.

W. Wahlster. SmartKom: Symmetric Multimodality in an Adaptive and Reusable Dialogue Shell. In: R. Krahl and D. Günther (eds.), *Proc. Human Computer Interaction Status Conference 2003*, pp. 47–62, Berlin, Germany, June 2003. DLR.

M.A. Walker, C.A. Kamm, and D.J. Litman. Towards Developing General Model of Usability with PARADISE. *Natural Language Engineering*, 6, 2000.

D. Widdows. A Mathematical Model of Context. In: P. Blackburn, C. Ghidini, R. Turner, and F. Giunchiglia (eds.), *Modeling and Using Context*, LNAI 2680, Berlin Heidelberg New York, 2003. Springer.

P.H. Winston. *Artificial Intelligence*. Addison-Wesley, Reading, MA, 1992.

Intention Recognition

Jürgen te Vrugt and Thomas Portele

Philips Research Laboratories GmbH, Aachen, Germany
{Juergen.te.Vrugt,Thomas.Portele}@philips.com

Summary. The intention recognition module identifies the analyzed representation of the user input to the SMARTKOM system that best represents this input in a collection of possible representations. The alternative representations are generated by recognition and analysis components, being enriched with knowledge, e.g., from the discourse and context. A probabilistic model combines various scores, based on features in the representations and computed by the SMARTKOM modules during processing to support the selection. The parameters inside the model are optimized on annotated data that has been collected using the SMARTKOM system using both a parameter study and a rank based estimation algorithm.

1 Introduction

The intention recognition in SMARTKOM (Wahlster et al., 2001) is part of the multimodal interaction subproject. In this working area the results of the analysis components for the various modalities are brought together, the user input is embedded into discourse and world context and the most likely interpreted reading of the user input has to be chosen. SMARTKOM has to react in an appropriate way. Reactions presented to the user are initiated by components located in this subproject, and the interaction of SMARTKOM with applications is also organized here. Due to the heterogeneous orientation of SMARTKOM into the scenarios “Home”, “Mobile” and “Public”, the components of the subproject must be easy to configure, independent of certain applications, and robust.

This chapter deals with the identification of the user’s intention. This is the computation of the most likely interpreted reading of the input provided by the user. The first part of this chapter describes how the intention recognition fits into SMARTKOM. The next part deals with the basic model for the combination of ratings from various sources. The following part is concerned with how data were collected and prepared in order to compute the parameters for the combination of the ratings, and finally the computation of the parameters is addressed.

2 Intention Recognition in the SmartKom Context

The decisions of the intention recognition are strongly depending on the preceding components in the information processing chain. On the other hand, the decision taken by the intention recognition has a strong influence on the actions executed and reactions toward the user.

2.1 Intention Recognizers' environment

Figure 1 shows the context of the intention recognition module in the SMARTKOM system as it is used in the three scenarios.

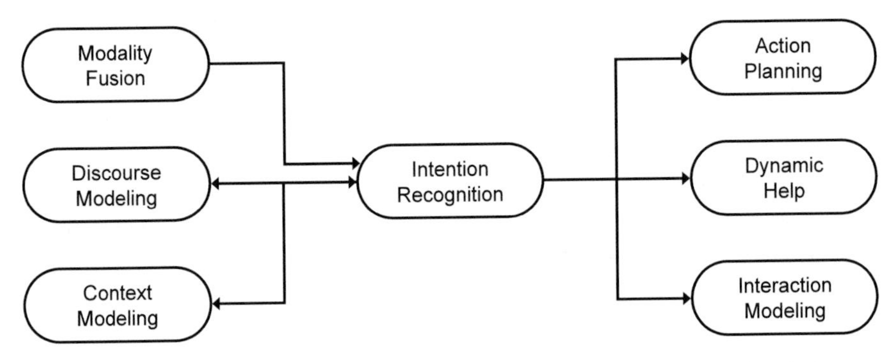

Fig. 1. Section of the multimodal interaction from the intention recognition point of view

2.1.1 Input to the Intention Recognition Module

The input of a user is initially processed by recognizers like a speech recognition or a gesture recognition component. Analyzers for the recognizers produce a semantic interpretation of the monomodal recognition results. For example, such a result is the interpretation of a grammar applied to the speech recognition result, or the representation of an element presented on the screen which is likely to be addressed by the user. Inside some of these components, stochastic methods are applied in order to compute readings of the respective input. Instead of computing one single "best" result, a number of hypotheses with a rating of the quality of each hypothesis might be computed.

The interpretations as produced by the monomodal recognition and analysis components are combined in the multimodal fusion. This results in a number of hypotheses for the multimodal input based on the monomodal hypotheses generated by the single recognition engines. The result of the multimodal fusion serves as the basis for the computation of the user intention. More details on the fusion can be found in Engel and Pfleger (2006).

The context model in SMARTKOM is divided into a model for the discourse (Pfleger et al., 2003) and a model for parts of the surrounding world (Gurevych et al., 2003). Before the intention recognition takes a decision on the set of hypotheses for the user input, the context models are requested to enrich the information in the hypotheses with context knowledge. These enrichments are evaluated by the discourse model and the context model to rate the quality of the augmentation of a hypothesis with knowledge from the discourse and the surrounding world, and to support the decisions to be taken by the intention recognition.

2.1.2 Processing of the Selected Hypothesis

After selecting the interpretation of the user input which is considered to be the best matching one, this interpretation is provided for further processing mainly to three components:

- The action planning takes decisions on the applications to contact and the next interaction with the user based on the hypothesis provided by the intention recognition.
- The dynamic help analyzes the hypothesis from the intention recognition to support the user in the interaction with the SMARTKOM system, e.g., the dynamic help identifies turns where the user asks for guidance by the system.
- The interaction model computes information on the user from the hypothesis reflecting the user input. This could for example be used to adapt the interaction style of the system to the interaction style of the user.

A more detailed description of these components can be found in the work of Löckelt (2006), Streit (2006) and Streit et al. (2006)

2.2 Communication with Surrounding Components

The communication of the modules in the "Modality Analysis And Fusion" part of SMARTKOM is based on intention lattices.[1] Initially, the intention lattice format coded the information transfered by the modules completely. This structure was separated into a hypothesis-related part and the content-related part. This separation was inspired by the introduction of an ontology in the context-modeling module. From this ontology, the event structure is derived. More details can be found in Porzel et al. (2006)

Basically, the intention lattice format contained the possibility to integrate a number of hypotheses (or — to be be more precise — hypothesis sequences) into the data exchanged by the modules. Each of these hypothesis sequences itself consists of one or more atomic hypotheses. The input of a user might be segmented into independent units, each unit represented by an atomic hypothesis. An example for such input with a possible reading using two atomic hypotheses in the "Home" scenario is a device control command followed by a database request like *"Turn on the TV and show*

[1] The specification of the interface is realized in M3L, see Herzog and Reithinger (2006).

me the movies of this evening!". While a collection of atomic hypotheses forms a hypothesis sequence, the hypothesis sequences stand for alternative readings of one and the same user input.
While the atomic hypotheses carry information on the content of the user input, the hypothesis themselves are equipped with technical information. Prominent examples are the ratings of the modules.

3 Internal Model

In the early stages of the SMARTKOM project, the development of the various models[2] had just started, or existing models had to be adapted for the use in SMARTKOM. The selection of the hypothesis considered to be the best matching one was mainly influenced by rule-based decisions. After more and more ratings showed up in the various modules, the rule-based decision was replaced by a rule-based preprocessing[3] and a stochastic model.

To illustrate the stochastic model motivated by Young (2000), the following nomenclature is used: The set of hypotheses generated by the modules preceding the intention recognition is denoted by $\mathcal{H} = \{h_1, \ldots, h_n\}$. The set $\mathcal{H}$ contains possible interpretations of one user input. Capital letters are used for random variables, where H is used for the hypothesis under consideration, and I_{ASR}, I_G, I_P and I_M are the input to the speech recognition, gesture recognition, prosody recognition and mimic recognition, respectively, provided by a user.

The primary role of the intention recognition is to identify the hypothesis $\hat{h}$ that is the best match of the given user input u:

$$\hat{h} = \text{argmax}_{h \in \mathcal{H}} P(H = h|U = u), \tag{1}$$

with $U = (I_{ASR}, I_G, I_P, I_M)$. If D, C and F denote the preliminary hypothesis created by the discourse model, context model and multimedia fusion, respectively, the distribution $P(H|U)$ can be rewritten to

$$\begin{aligned} P(H = h|U = u) = &\sum_c P(H = h|U = u, C = c) \\ &\sum_d P(C = c|U = u, D = d) \\ &\sum_f P(D = d|U = u, F = f) P(F = f|U = u). \end{aligned}$$

Assuming that the result of the intention recognition, discourse model and context model depends only on the preceding models and not directly on the user input U, this simplifies to

---

[2] Therefore, the realization of the models in SMARTKOM modules was also in an embryonic stage.

[3] Requirements for such a rule-based preprocessing have been formulated. The development of a proprietary language was discarded in favor of the upcoming XSL Transformations, which could easily be applied to the M3L documents (World Wide Web Consortium (W3C) / XSL Working Group).

$$P(H=h|U=u) = \sum_c P(H=h|C=c) \sum_d P(C=c|D=d) \sum_f P(D=d|F=f) P(F=f|U=u).$$

Since hypotheses in SMARTKOM are not recombined,[4] for a given hypothesis h for a given user input u there exists exactly one processing result of context model c, discourse model d and media fusion f with positive probability for all distributions. This leads to

$$P(H=h|U=u) = P(H=h|C=c)P(C=c|D=d) P(D=d|F=f)P(F=f|U=u)$$

for an outstanding sequence (c,d,f), depending on the hypothesis under consideration h and the user input u. If W, S, G, P and M denote the result of the speech recognition, speech understanding, gesture recognition, prosody recognition and mimic recognition component, respectively, it can be derived, that:

$$\begin{aligned} P(F=f|U=u) &= \sum_{s',g',p',m'} P(F=f|S=s',G=g',P=p',M=m',U=u) \\ &\qquad P(S=s',G=g',P=p',M=m'|U=u) \\ &= \sum_{s',g',p',m'} P(F=f|S=s',G=g',P=p',M=m') \\ &\qquad \sum_{w'} P(S=s'|W=w')P(W=w'|I_{ASR}=i_{ASR}) \\ &\qquad P(G=g'|I_G=i_G)P(P=p'|I_P=i_P) \\ &\qquad P(M=m'|I_M=i_M) \\ &= P(F=f|S=s,G=g,P=p,M=m) \\ &\quad P(S=s|W=w)P(W=w|I_{ASR}=i_{ASR}) \\ &\quad P(G=g|I_G=i_G)P(P=p|I_P=i_P)P(M=m|I_M=i_M) \end{aligned}$$

by using the assumptions:

- The result of the multimodal fusion depends only on the output of the recognition and analysis components (and not on the acoustic input, for example)
- The computations of the results of the recognition components rely only on the input for these components and are independent of the input for other components (e.g., the recognition of speech is independent of mimics)
- The result of the speech understanding unit depends only on the words recognized by the speech recognition engine and not on the acoustic input
- Each result of the fusion is created with positive probability by exactly one input constellation (s,g,p,m) and recognized word sequence w.

[4] That is, if one of these models decides to create more than one different hypothesis out of one hypotheses, these will be regarded as different hypothesis for the rest of the processing chain up to the intention recognition.

Putting it all together, this results in a model of the form:

$$P(H|U) = \prod_{\text{model } M} P_M(\cdot), \tag{2}$$

P_M being the distribution of model M that might depend on other models. The problem of finding the hypothesis that maximizes $P(H|U)$ can be transferred into the equivalent problem of maximizing scores, for example by applying the logarithm. That is,

$$S(H|U) = \sum_{\text{model } M} S_M(\cdot). \tag{3}$$

The measures used in the SMARTKOM modules differ a great deal, e.g., some modules use stochastic models while other modules create a rating based on counts of features. Some models might not fulfill the requirements of a mathematical measure at all, which is a prerequisite of a probability distribution (Billingsley, 1995). The ratings used in the intention recognition are finite, so these ratings could at least be normalized.

Taking this into consideration and denoting the rating of module M by R_M, the intention recognition combines the various models in SMARTKOM by a weighted sum of scaled ratings:

$$R_{\text{total}}(\cdot) = \sum_{\text{model } M} w_M f_M\left(R_M(\cdot)\right), \tag{4}$$

where w_M is the weight of model M, $f_M(\cdot)$ scales the ratings $R_M(\cdot)$ of model M. If necessary, $f_M(\cdot)$ renormalizes the rating of model M and transfers the ratings into the score space.[5] The introduction of weight parameters becomes necessary because of the approximations made in this model and since the ratings of the models might not be probability distributions at all.

While the model introduced in this section tries to identify the most likely hypothesis based on the user input and the preceding processing stages, the probabilistic model in Fleming and Cohen (2001) formulates benefits and costs to decide on actions to take in the interaction with the user. Another example of probabilistic dialogue management can be found in Roy et al. (2000), where the decisions are based on (partially observable) Markov decision processes.

4 Data Acquisition for Optimal Model Combinations

The computation of the hypothesis that describes the user input optimally is coupled to the model in Eq. (4). Initial setups and adjustments[6] of the parameters $(w_M)_{\text{model } M}$ have been done manually, but in order to reliable adjust the parameters, a data collection has been carried out.

Due to speed constraints, in the standard setup of SMARTKOM the intention recognition gets less than five hypotheses, usually two to three. The configurations

[5] In case of a probability distribution, f_M is the logarithm.

[6] For example, when new measures have been introduced.

of the speech recognition module and the speech understanding module have been modified for the data collection in order to obtain a large number of alternative hypotheses per user input.

The data collection was carried out in three steps that will be outlined in more detail below. The recordings focused on speech recognition as input modality. Speech was the first modality available for interaction with the SMARTKOM system.

4.1 Collection

In order to ensure sufficient coverage of the user input by the system (especially by the speech analysis grammar), a small data collection was carried out at Philips Research, Germany.Because of the modified settings for the speech recognition engine and the speech understanding component, it was not possible to collect run-time data from real interactions with the system. The resulting number of hypotheses led to a significant decrease of system performance in terms of processing speed. Therefore a number of exemplary interactions of a user with the system have been designed. These interactions are based on the examples provided by the SMARTKOM "Home" scenario managers and online interactions with the system (Portele et al., 2003). Two generic and seven "Home"-specific interactions that can be combined in different ways have been formulated. These interactions contain around 80 user-turns. Table 1 summarizes the data collection.

The decision to restrict the user utterances to a fixed set was triggered by the need of a basic grammatical coverage for these utterances, i.e., utterances the system cannot understand are worthless for the computation of optimal combination parameters. Utterances of five different persons have been recorded; four of them resulted in transcribed intentional structures. A close-talk microphone in two different positions was used. For the recordings, a simple SMARTKOM system running only the audio input module was used.

Table 1. Recording and transcription statistics

Observation in corpus	Occurrences
Recorded and orthographically transcribed user utterances	931
Recordings containing speech	919
Nonspeech recordings	12
Different formulations total	150
Formulations used once	66
Formulations used 2–10 times	39
Formulations used 11–20 times	42
Formulations used 21–26 times	3
Utterances with transcribed intentional structures	433

4.2 Preparation

After recording data from a user, each user input was fed into a SmartKom system containing the processing modules needed to create the input data for the intention recognition module in the context of the discourse so far, and the intention recognition module itself. This includes the recognition and analysis modules and the modules containing the domain and context models. The result of this operation is a set of intentional structures for each user input.

For each user input, the data preparation took place in two steps:

1. For each hypothesis produced, the semantic units created by the speech analysis component are extracted. These semantic units have been manually aligned to parts of the spoken input as annotated during the data collection.
2. For each hypothesis, the alignment of semantic units to text phrases can be used to create possible semantic interpretations from the annotated user input. These artificial interpretations are compared to the real hypothesis created by the system. The result of the comparison was used to resort the hypothesis.
   The sorted list of hypotheses was then presented to a human annotator (together with information on the "real" history in the discourse) who had to choose the best matching hypothesis created by the system given the transcription of the spoken user input. Due to the sorting, the hypothesis the annotator rated the best was, in most cases, contained in one of the top positions.

From this annotated data, equivalent hypotheses[7] are eliminated.The outcome of this procedure is an n-best list of hypotheses for each user input with one marked hypothesis being the best representation of the user input, see also Table 2.

Table 2. Statistics on the 433 transcribed intentional structures

Observation in corpus	Occurrences
Hypotheses per user input	8.9
Intentional structures with two or more hypotheses	221
Hypotheses per user input	16.4
Intentional structures with outstanding hypothesis marked by annotator and at least one alternative hypothesis	158
Hypotheses per user input (equivalent variants removed)	3.8

5 Computation of Combination Parameters

The 433 transcribed intentional structures have been split into a development and test set by taking every kth dataset out of the collected corpus and moving this dataset into the test set. To compute a setup for the parameter set $(w_M)_{\text{model } M}$ for the three

[7] Hypotheses that differ in parts not relevant for the computation of combination weights.

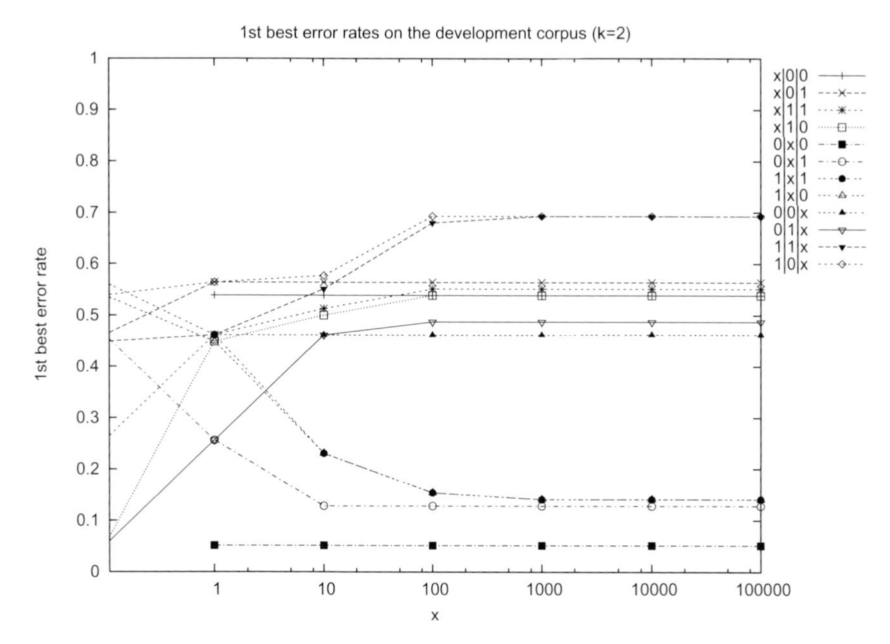

Fig. 2. First-best error rates on the development corpus ($K = 2$; $M_1|M_2|M_3$ denotes the shares of the models in the combination, i.e., model I gets the weight $M_I/(M_1 + M_2 + M_3)$)

models providing a score (speech recognition, speech understanding and discourse model), a parameter study was carried out. To allow an easier computation of optimal combination weights once data has been prepared, an algorithm to automatically estimate the parameters was also developed.

5.1 Parameter Study

Figure 2 allows for a glimpse on the parameter study. It gives a hint in which direction to search for a good parameter setup. From the studies, it turns out that the optimal configuration is a model based purely on the rating from the speech analyzer. [8] The speech analyzer is heavily dependent on the result of the speech recognition, therefore the score of the speech analysis already contains the relevant knowledge from the speech recognition engine. The measure provided by the discourse model was refined during SMARTKOM and might become more relevant. The intention recognition combines knowledge of modules based on different approaches. For example, models based on artificial intelligence compute a rating separately, while modules using a stochastic model have an inherent rating.

From the parameter study it turned out that the best weight combination was stable with respect to the different setups of development and test corpus. Table 3

[8] In the real SMARTKOM system, the number of hypotheses is strongly restricted (mostly 2, maximal 3 hypotheses per user input). Furthermore, these hypotheses are very similar.

Table 3. First-best error rates on development and test corpus for varying splits of the collected corpus

	Development		*Test*	
Corpus split parameter k	Entries in corpus	Error rate (%)	Entries in corpus	Error rate (%)
2	78	5.1	80	7.5
3	100	5.0	58	8.6
4	114	7.0	44	4.5
5	128	7.0	30	3.3
6	126	5.6	32	9.4
7	137	7.3	21	0.0

lists the first-best error rates on these corpora. For comparison, a small parameter study was also carried out on the test corpora. As far as this study can prove, the optimal setup computed on the development corpus was also optimal on the test corpus.

5.2 Automatic Estimation of Parameters

In order to automatically compute an optimal parameter setup $(w_M)_{\text{model } M}$ for the combination model formulated in Eq. (4), an algorithm motivated by rank-based statistics was developed.

5.2.1 Basic Idea

The initial situation for the algorithm is a processing chain that creates a list of n hypotheses describing the processing result for some input that has been fed into this chain (Fig. 3).

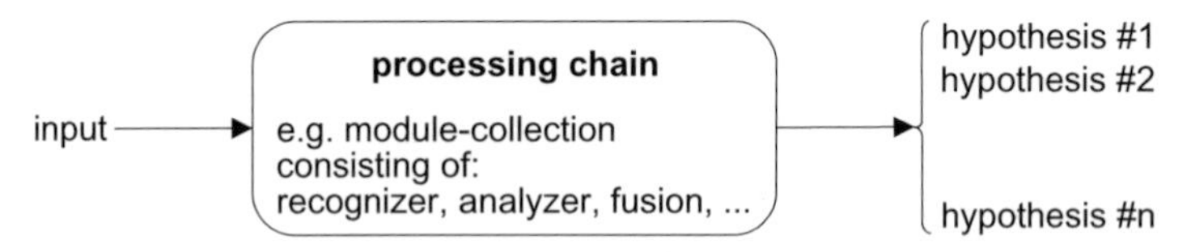

Fig. 3. Prerequisite situation for the automatic computation of the parameters $(W_M)_{\text{Model } M}$ in Eq. (4)

It is now assumed that one of these hypotheses can be regarded to be the "best" result of the processing for the input by the given chain. This hypothesis will be denoted as the true hypothesis in the remaining part of this section. For a fixed set of parameters $(w_M)_{\text{model } M}$ the true hypothesis is placed at a certain position in the n-best list of all hypotheses produced from the input by the processing chain according to the total rating as formulated in Eq. (4). For different sets of parameters $(w_M)_{\text{model } M}$, this rank of the true hypothesis relative to the competing hypotheses will differ. Given a collection of inputs to the processing chain, this results in a collection of n-best lists of hypotheses.[9] For each n-best list in this collection, one true hypothesis should be identifiable.

The goal of the algorithm is to compute parameter setups $(w_M)_{\text{model } M}$ such that the true hypothesis is positioned at the best possible rank simultaneously for all n-best lists in the collection. Instead of defining a setup $(w_M)_{\text{model } M}$ and computing the resulting quality measure for the simultaneous optimization, the algorithm places the true hypotheses on certain ranks in the n-best lists of the collection and tries to compute parameter setups that place the true hypothesis on exactly these predefined ranks. Such a parameter setup[10] for a given positioning of the true hypothesis may not exist. Considering all configurations of ranks for which a parameter setup exists, the ones with the best quality measure lead to the wanted parameter setups.

A canonical quality measure for the simultaneous optimization of all ranks of the true hypothesis in the collections is the mean rank of the true hypotheses:

$$\text{rank}_{\text{mean}} = \frac{\sum_{n-\text{best lists in collection } l} \text{rank}_{\text{pos of true hypo in list}}(l)}{\# \, n-\text{best lists in collection}}. \tag{5}$$

Of course, the $\text{rank}_{\text{mean}}$ quality measure could be replaced by another quality measure. For example, consider some application where it is necessary to place the true hypothesis within the first k hypotheses while a rank larger than k is neglected during further processing. The exact rank of the true hypothesis does not matter. Then the function $\text{rank}_{\text{pos of true hypo in list}}(\cdot)$ in Eq. (5) might be replaced by the function

$$f_{\text{rank of true hypo } \leq k}(l) = \begin{cases} 1, & \text{if } \text{rank}_{\text{pos of true hypo in list}}(l) \leq k, \\ penalty, & \text{else,} \end{cases}$$

with l being a n-best list containing a true hypothesis and *penalty* being some constant, usually larger than 0.

5.2.2 Outline of the Algorithm

Positioning all hypotheses of all n-best lists of a collection at certain ranks in these lists results in a certain arrangement of the true hypothesis. We will call this the configuration of ranks. Figure 4 sketches the basic flow of the search for the optimal parameter setup.

The computation of the quality measure for a given configuration $\mathtt{c}$ has been discussed before. It remains to compute a parameter setup for such a configuration $\mathtt{c}$. Let

$$\text{rank}(l,h) \stackrel{\text{def}}{=} \# \{ h' \in l; \text{rating}(\text{h}') > \text{rating}(\text{h}) \} + 1$$

[9] The value of n might depend on the input.

[10] Depending on the exact formulation of the algorithm, the trivial solution $(w_M)_{\text{model } M} = (0)_{\text{model } M}$ might be a possible solution for all configurations. Therefore only nontrivial solutions are searched for.

```
compute all possible configurations of ranks C

for all configurations c in C:
{
  if ( exists a parameter setup for configuration c )
  {
    compute quality measure qm for configuration c

    if ( qm is better than qm' of best configuration(s) so far )
    {
      empty the list of best configurations

      put c into list of best configurations
    }
    else if ( qm is equal to qm' of best configuration(s) so far )
    { append c to list of best configurations }
    else
    { ignore configuration c }
  }
  else
  { ignore configuration c }
}

return the list of best configurations
```

Fig. 4. Algorithmic framework of search for the optimal parameter setup

denote the rank of hypothesis h in the n-best list l. The rating $R_{\text{total}}(h)$ for hypothesis h is defined by Eq. (4), $R_{\text{total}}(h)$ is a linear polynomial in the variables $(w_M)_{\text{model } M}$. Therefore rank (l,h) depends on $(w_M)_{\text{model } M}$ if rating$(\cdot) = R_{\text{total}}(\cdot)$.

For a certain n-best list l, let h_k $(1 \leq k \leq n)$ be the true hypothesis. Consider the problem of computing a parameter setup such that the n-best list l has the ordering $(h_1,\ldots,h_k,\ldots,h_n)$ (without restrictions: rank $(l,h_k) = k$), i.e.,

$$rank(l,h_i) \begin{array}{c} < \\ \leq \end{array} rank(l,h_{i+1}), \text{ if } \begin{array}{l} i = k-1, \\ 1 \leq i < k-1, \text{ and } k \leq i \leq n. \end{array}$$

This can be reformulated by the set of inequalities

$$R_{\text{total}}(h_i) \begin{array}{c} > \\ \geq \end{array} R_{\text{total}}(h_{i+1}), \text{ if } \begin{array}{l} i = k-,1 \\ 1 \leq i < k-1, \text{ and } k \leq i \leq n, \end{array}$$

in the variables $(w_M)_{\text{model } M}$. A parameter setup for a configuration c can now be computed by the algorithm described in Fig. 5.

The main disadvantage of the algorithm presented is its complexity: n hypotheses of an n-best list can be ordered in $n!$ ways, therefore the complete collection of n-best lists contains

$$\prod_{n-\text{best lists } l} (\#\,\text{hypos in } l)!$$

```
let c be a configuration
let I be an empty set of inequalities in w

for each single n-best list l in c:
{
  create set of inequalites in w from l (see text)

  add inequalities to I
}

if ( exists nontrivial parameter setup w' that solves I )
{ return all nontrivial parameter setups that solve I }
else
{ return NULL }
```

Fig. 5. Computation of nontrivial parameter setups for a given configuration `c` ($W = (W_M)_{\text{Model } M}$)

possible configurations. For example, in a collection with p lists, each list containing n hypotheses, there are $(n!)^p$ collections.
To reduce the complexity of the algorithm, first improvements of the algorithm have been realized. Among these are

- a grouping of different collections
- the reduction to the problem of computing a solution, where the true solution has *at least* a certain rank (and not exactly such a rank), and
- a repeated application of the algorithm on the probabilistic preprocessed data taking into account the result of the previous application

5.2.3 Results on the In-House Corpus

The algorithm was applied on the collected data presented in Sect. 4. The results on various splits of the corpus conform completely with the results of the parameter study presented in Sect. 5.1. The improved version of the algorithm allowing up to 100 repetitions with preprocessed input data was applied. The algorithm converged to the solution after four repetions, at most. In addition to the parameter study, approaches based on ideas of the downhill simplex methods in multidimensions and a combination of simulated annealing together with the downhill simplex method (Press et al., 2002) verified the outcome of the presented algorithm.

6 Conclusion

This chapter described the role of intention recognition in the SMARTKOM project. It collects information from preceding sources and tries to identify the reading of the analyzed user input that fits best to the user input. The internal model for the combination of ratings was motivated and derived. To obtain reliable parameters, a data collection was carried out and tools have been provided to the annotator to ease the annotation of the data at the various annotation stages. A parameter study on development corpora led to optimal parameter setups for the combination model. The optimality of these setups was proven on the test corpus by a second parameter study on the test data. The setups were stable with respect to the segmentation of the collected corpus into development and test corpora. The application of the algorithm for the automatic computation of an optimal parameter setup confirmed the results of the parameter study. The complexity of the improved versions of the algorithm needs to be compared to the original algorithm. The application of the algorithm on other data, e.g., to combine acoustic and language model score in a speech recognizer, also looks promising.

Acknowledgments

The authors would like to thank Eva Lasarcyk for supporting us during the data collection and annotation.

References

P. Billingsley. *Probability and Measure*. John Wiley, New York, 3 edn., 1995.

R. Engel and N. Pfleger. Modality Fusion, 2006. In this volume.

M. Fleming and R. Cohen. Dialogue as Decision Making Under Uncertainty: The Case of Mixed-Initiative AI Systems. In: *Proc. NAACL-2001 Adaptation in Dialogue Systems Workshop*, Pittsburgh, PA, 2001.

I. Gurevych, R. Malaka, R. Porzel, and H.P. Zorn. Semantic Coherence Scoring Using an Ontology. In: *Proc. Human Language Technology Conference / North American Chapter of the Association for Computational Linguistics Annual Meeting 2003*, pp. 88–95, Edmonton, Canada, 2003.

G. Herzog and N. Reithinger. The SmartKom Architecture: A Framework for Multimodal Dialogue Systems, 2006. In this volume.

M. Löckelt. Plan-Based Dialogue Management for Multiple Cooperating Applications, 2006. In this volume.

N. Pfleger, R. Engel, and J. Alexandersson. Robust Multimodal Discourse Processing. In: *Proc. Diabruck: 7th Workshop on the Semantics and Pragmatics of Dialogue*, Wallerfangen, Germany, 4–6 September 2003.

T. Portele, S. Goronzy, M. Emele, A. Kellner, S. Torge, and J. te Vrugt. SmartKom-Home — An Advanced Multi-Modal Interface to Home Entertainment. In: *Proc. EUROSPEECH-03*, Geneva, Switzerland, 2003.

R. Porzel, I. Gurevych, and R. Malaka. In Context: Integrating Domain- and Situation-Specific Knowledge, 2006. In this volume.

W.H. Press, W.T. Vetterling, S.A. Teukolsky, and B.P. Flannery. *Numerical Recipes in C++ — The Art of Scientific Computing*. Cambridge University Press, Cambridge, UK, 2 edn., 2002.

N. Roy, J. Pineau, and S. Thrun. Spoken Dialogue Management Using Probabilistic Reasoning. In: *Proc. 38th ACL*, Hong Kong, 2000.

M. Streit. Problematic, Indirect, Affective, and Other Non-Standard Input Processing, 2006. In this volume.

M. Streit, A. Batliner, and T. Portele. Emotion Analysis and Emotion Handling Subdialogs, 2006. In this volume.

W. Wahlster, N. Reithinger, and A. Blocher. SmartKom: Multimodal Communication with a Life-like Character. In: *Proc. EUROSPEECH-01*, vol. 3, pp. 1547–1550, Aalborg, Denmark, September 2001.

World Wide Web Consortium (W3C) / XSL Working Group. The Extensible Stylesheet Language Family. http://www.w3.org/Style/XSL. Cited 6 April 2006.

S. Young. Probabilistic Methods in Spoken Dialogue Systems. *Philosophical Transactions of the Royal Society (Series A)*, 358(1769):1389–1402, 2000.

Plan-Based Dialogue Management for Multiple Cooperating Applications

Markus Löckelt

DFKI GmbH, Saarbrücken, Germany
loeckelt@dfki.de

Summary. The SmartKom dialogue manager implements the personality of the system and its behaviour. It plans and manages the task-oriented dialogue with the user and coordinates the operations of the applications to reach his goals. It also helps the analysis modules of the system by providing hints about the expected future dialogue.

1 Introduction

Users access the functionality of SmartKom by delegating their wishes to Smartakus, the system agent. Smartakus abstracts away the underlying structure of multiple interoperating applications by providing a unified and personalized interface. The dialogue manager module implements the behavior of Smartakus in two aspects: First, it plans and performs the dialogical interaction of Smartakus with the user. Second, it models the applications the system provides and triggers the interactions with the application and device modules needed to give access to the functionality offered by the system. The overall objective is to provide a flexible and convenient way for users to achieve their goals. The knowledge regarding the current state of the dialogue is also used to give the analysis modules feedback that aids in interpreting expected utterances.

Other important aspects for the SmartKom dialogue manager are scalability, incrementality, and reusability for other systems. The number of 14 integrated applications is quite large for a multimodal system, and many tasks of SmartKom can only be accomplished by concerted interoperation of several applications. The set of applications also was not integrated all at once, but was built up during the course of the project. Thus, to avoid impeding the parallel development of all modules of the system, it was essential to have a working prototype most of the time. As the system cannot be used sensibly without a dialogue manager, it was vital that the addition of new features could be done with minimal disruption. Two main design decisions helped in addressing these issues: taking advantage of the common ontological knowledge representation, and a flexible and extendable communication paradigm treating all communication partners uniformly.

Related Systems

The MATCH system (Johnston et al., 2002) implements one application similar to the route planner in SMARTKOM. It uses several modalities (speech, graphics, and pen input) that can be used alternatively or in combination. Its dialogue manager uses a toolkit similar to TrindiKit. MATCH features expectation rankings and a common representation across modalities.

Other systems derived from TrindiKit (e.g., IBIS (Larsson, 2002), TRIPS (Ferguson and Allen, 1998) and COLLAGEN (Rich and Sidner, 1998) use related approaches. However, most of the time a single application domain is covered at one time, which keeps the number of interacting modules low, and scalability requirements do not have that great an impact. TrindiKit uses a dialogue move engine and a rule-based approach to drive it and to update the *information state* of the engine. Information state distinguishes PRIVATE and SHARED information, but keeps both parts in one module. We chose to split the responsibility for these two parts in SMARTKOM, where the SHARED information is managed by the discourse modeler.

Variants of the dialogue manager were used in the MIAMM system (Reithinger et al., 2004), which allows multimodal access to information using an interface of speech, haptic interaction and graphics; and NaRATo (a system allowing natural-language database access).

2 Architecture

2.1 Communication with Other Modules

As a result of the complexity and concurrency of the system, it does not have a strict serial pipeline of input analysis, processing, and output generation. The dialogue manager is situated at the pivotal point between input analysis and output generation. It has to process input from three modules and four data pools, and it outputs messages going to six different modules (Fig. 1). The main data flow, along the thick arrows in the figure, consists of user input entering the module via the intention recognizer, and being answered by multimodal output for presentation planning. To use the functionality provided by the applications, additional communication is made via the function modeler module and various data pools. When an incoming message cannot be processed (e.g., an utterance was not understood), it is passed on to the dynamic help module, which will try to resolve the problem. Changes in the availability of modalities (e.g., inhibited graphical output in lean-back mode) are effected by communication with the interaction modeler module. Also, the dialogue manager publishes information about the state of the dialogue and its expectations about subsequent user inputs.

2.2 Internal Modular Structure

The dialogue manager module was developed for the SMARTKOM system, but was explicitly designed to be reusable for different dialogue systems. The module consists of two parts: the core dialogue move engine and a part that adapts it to the

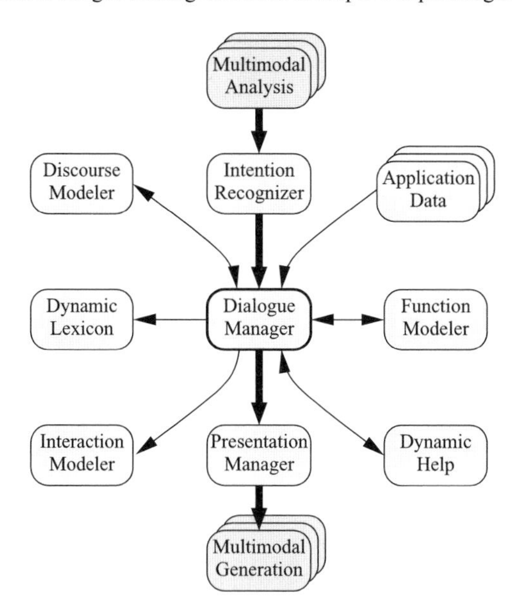

Fig. 1. Modules in direct communication with the dialogue manager

specific circumstances of the system, and also contains the plan knowledge bases for the applications. Their interrelation is shown in Fig. 2.

In the core engine, a planner module is responsible for devising the next actions, which are realized and monitored by the executor. The controller interprets incoming messages, maintains the goal stack, dispatches planning tasks to the planner, and computes the expectations (Sect. 5.3). It also initializes and shuts down the module.

The system-specific part contains data sources giving the plan operator specifications, which together with the process descriptions in the ontology define the behavior for the applications. To convert inbound and outbound messages from and to the data formats needed by the communication partners, modules implementing input and output channels are necessary. The dialogue manager is connected to the communication architecture of the system by a module interface. [1]

2.3 Channels

The various communication partners use different interfaces; however, the core engine of the dialogue manager represents all communications in terms of uniform intention structures, as will be described in Sect. 3.2. Therefore, the intention representation must be translated to and from the message format of the communication partner (except for module-internal communication). For example, when the user is

[1] In SMARTKOM, the Multiplatform blackboard architecture implements intermodule communication (Herzog et al., 2004); for MIAMM, a SOAP facilitator was used instead, while in NaRaTo, methods were called directly.

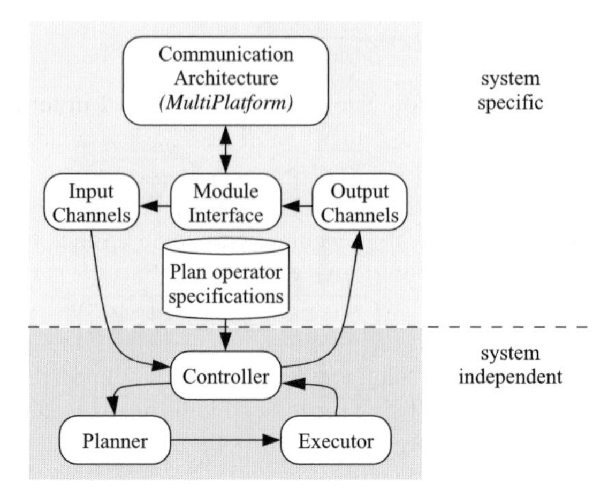

Fig. 2. The internal structure of the module

asked for a telephone number, the request will be represented by a presentation template requesting an ontological role, but the same request has to take the form of a database query if the address book application is addressed.

The communication channels are introduced to abstract away this mapping for the core engine. Each communication link is assigned a channel tailored to the specific message format that has to be dealt with. It then acts as a gateway for the messages, translating them to and from a particular outside representation as they cross the border of the module, inbound or outbound.

3 Action Model for Dialogue Planning

Following the dialogue metaphor underlying the system (Wahlster, 2002), the user delegates his wishes and goals to the animated agent Smartakus, which then collaborates with the user to accomplish the corresponding tasks. The dialogue planner, in determining its actions, realises the "personality" of the system agent. Smartakus was designed to be polite and helpful, but to never appear to be obtrusive. It guides the dialogue and cooperates with the user; however, if the user does not follow the leads strictly in its utterances, Smartakus attempts to accommodate the dialogue and integrate them. It will, however, not risk overinterpreting user utterances when they have not been clearly understood (a well-known cause of human irritation with software assistants). In these cases it will employ the help module instead to ask for clarification. Indicative for insufficient understanding are intentions that lack important information, or have been assigned low confidence scores by the analysis modules.

3.1 The Use of Ontological Processes

This section describes briefly how the dialogue is modeled in terms of processes provided by the SmartKom ontology. A more comprehensive description of the ontology representation and its use in our system is found in Porzel et al. (2003).

The domain the dialogue can be about is defined by the ontology underlying the backbone communication, which encompasses some entities to match general dialogue concepts, such as greetings, and others modeling the services offered by the applications of the system. The set of available applications varies across the different scenarios (public, home, and mobile). Some applications are not accessible directly by the user, but support other applications (e.g. the biometrics application). The utterances of the user, after being processed by the analysis modules, reach the dialogue manager in a data structure called *intention*, which contains an ontological description of the utterance content.

The development of the ontology has been closely tied to the development of the dialogue and application capabilities of SmartKom. The possible events that occur in the dialogue or the rest of the system are modeled as *processes* with associated *roles* in a frame-based approach. To meet the obligation of satisfying the user, the system has to determine which processes and roles will be involved, impose an ordering, conduct the dialogue with user and applications, and report the results back in presentations. For example, the wish to make a phone call will trigger a *PhoneTelecommunication* process involving (among others) the role of a *phoneNumber* to dial, and may possibly also incorporate subprocesses like an *AddressbookSearch* if needed, which might retrieve the *phoneNumber* based on a *callee* role, and report it back to the user for confirmation before making the call. It is also necessary that the system reacts gracefully to deviations from the planned course of action by the user or the system.

Each action, from the act of greeting the user to providing a complex service, is represented by a process instance. Requesting a service then translates to setting the goal to perform actions associated with that process. The actions are defined by plan operator specifications. They usually involve manipulating roles of the process, such as using them as parameters for some action, or finding instantiations for them. Also, frequently the results of other processes are required to achieve subgoals. To determine the appropriate processes and their execution ordering is the task of the planner component. An executor component then performs the subtasks and monitors the outcome.

3.2 Intentions

User utterances are represented in a data structure called an *intention lattice*, which may contain one or more *intention* segments that partition the utterance sequentially (in case of, for example, multisentence utterances). Each intention of the user can specify a set of dialogue acts, which in turn can be either *goal* or *slot manipulations*. These manipulations always occur in relation to a discourse object (or process in terms of the ontology). It also specifies whether an utterance was understood to be explicit positive or negative feedback. The constituents of each intention segment that are important for the dialogue manager are therefore:

- A discourse object derived from the ontological framework, containing information about processes and roles the dialogue contribution is about.
- Goal manipulations:
  - setting a goal to adopt a new task, usually putting one or more processes on the agenda
  - retracting a goal to abort a task currently pursued
  - retaining a goal (this confirms that it is still actively being talked about, and an utterance occurs in the context of the goal)
- Slot manipulations:
  - setting slots to establish new information
  - retracting slots to take back / invalidate current information,
  - retaining slots to confirm information

  Slot manipulations always occur in the context of a specific goal.
- Annotation of the intention as positive and negative feedback, which might either be an answer to a simple question, or an expression of the emotional state of the user.
- Confidence scores from the analysis modules.

These types of constituents, except the last two, also apply to messages from the applications and devices. Additionally, the intentions contain additional information not directly relevant for dialogue planning, such as lexical items occurring in the input, which is passed on when triggering system output. This enables, e.g., the text generator component to adapt to the user's choice of words.

3.3 Communicative Games

Communicative games are derived from the notion of *dialogue games* (see, e.g., Carletta et al. (1996); Carlson (1983)). In addition to the contributions of the user and the system in the actual surface dialogue, they also capture the communications between the system and the application modules, as well as intrasystem communication. As many games do, they consist of consecutive moves, attributed to one participant each, and rules apply as to whether a move is allowed, depending on the situation. Composite games are construed from atomic games (like request–response) by concatenating or nesting them.

From the point of view of the dialogue manager, the dialogue does not only take part between the user and the animated agent. Smartakus does not know the current TV programs and cannot do a biometric authentication all by itself. Instead, to accomplish the tasks at hand, the devices and applications also need to enter the conversation (but this remains transparent to the user). So, in a typical interaction like the one depicted in Fig. 3, the dialogue manager will have subdialogues with the user, but also with applications: Smartakus acts as a delegate on the user's behalf. System-internal exchanges like the lexicon update, which informs other modules of

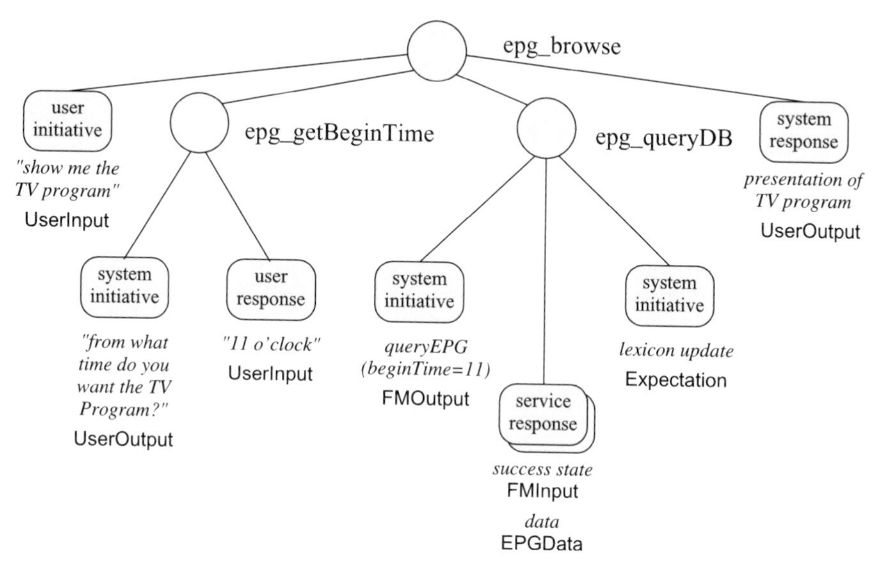

Fig. 3. Structure of a composite communicative game for television guide information search in SmartKom. The *circles* represent games; the *boxes*, game moves. The content of the messages and the channel used are shown below the boxes

new words likely to occur in future utterances (see Sect. 5.3.2), are also seen as moves of the communication game.

This model has the advantage that it captures the interactions with the user, the applications and devices, and other system modules in a uniform representation. Application protocols can be specified and treated as flexible as the user interaction. As an example, the database protocol prescribes that the results of voluminous database queries are returned to the dialogue manager in two parts, one indicating success or failure of the query, the other containing the actual data content. This interaction could be called request–$\{response_1, response_2\}$, wherer one request mandates two responses. The responses come from different sources and can arrive in arbitrary order. Also, the application requires that television program database requests should always entail a lexicon update (but not before the composite response has arrived). The *epg_queryDB* game in Fig. 3 bears this structure.[2]

4 Plan Operators and How Games are Played

Each application requires knowledge sources defining the set of plan operators to specify the services and the available dialogues of the application. The data for-

[2] As outlined in Sect. 5.3, an internal expectation message is sent (implicitly) every time the system releases the initiative. These moves were left out of the figure, except in the case where additional information — the lexicon update — is explicitly included by a plan operator.

mat of these specifications is given by XML schemata defining a plan operator language, which is based on the classic STRIPS language and extends it to effect actions through dialogical communication on several channels. Two types of definition files exist, one for each planning mechanism (see Sect. 5). A specification file consists of a header that relates a set of *slots* for the application to the ontological roles they are associated with, and a body defining a set of "states" similar to strips operators. The part corresponding to the body of the strips operator consists of a series of moves in communicative games. If an operator is applied, the moves will be placed on an execution stack. During the dialogue, the execution stack is reduced by the executor. This is done in two ways: If the topmost move is on an input channel and therefore an input is expected (e.g., an answer to a question, from the user or from an application), the system releases the initiative and waits for a contribution on the channel. If the move is on an output channel, a corresponding message is constructed and sent. For moves on output channels, the move specification also prescribes how the intention content of the corresponding communicative act is to be constructed.

An Example

Figure 4 shows a (slightly simplified) operator for the request–response action associated with the user browsing a TV program. The presence of a *goal* tag signifies that the result of this operator is a user-specifiable goal (i.e., "show me the TV program" or equivalent); the tag contains actions that will be done first upon adoption of the goal (the value of the slot *epg_broadcastSelection* is retracted). The *provides* tag specifies postconditions (the slot *epg_broadcastShown* will have a value when the goal is reached). This operator has no preconditions, which would otherwise be specified under a *needs* tag. The first *move* in the body says that this operator is to be triggered by an intention on the user input channel and should be answered on the user output channel. If the user sets the goal in an intention, both moves are put on the stack, and the first move is popped again immediately to "consume" the user utterance, leaving the "response" move on top of the stack. However, the "response" move has a precondition: It needs the slot *epg_broadcastSelection.*[3] Since this slot is retracted upon adopting the goal of *epg_browse*, it will not have a value after the first move, barring execution of the second. To proceed, this slot must be filled. The planner will be called to insert subgames (in this case, it will be another request-response game with the TV database) to satisfy the precondition. After the subgame has satisfied the precondition, the original "response" move can be made.

As can be seen from Fig. 4, the second move has two intention *templates* to cover different situations. If a move has intention templates, the first template whose preconditions are fulfilled is taken to realise the move intention. In this case, the templates define two alternative responses. The first template is selected if the slot *epg_broadcastSelection* has a value, but no textual content — which is the case if the TV database finds no results for the specific query. The answer of the system will then be an intention of type "noDataAvailable," generated as "I have not found

[3] Conditions test if a set of predicates holds. The condition *needs(slotname(X))* is really a convenience shorthand for *needs(slotIsSet(slotname(X)))*.

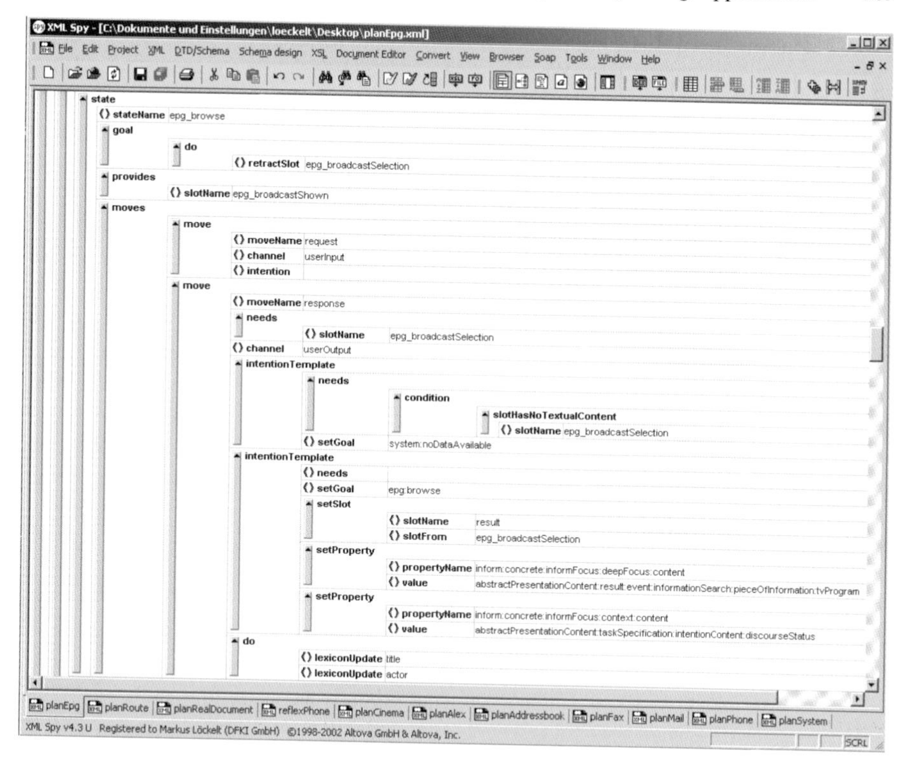

Fig. 4. Plan operator specification for EPG browsing

any data for your request." If, however, broadcasts matching the request are found, the second template (which has no preconditions) will be selected. The template results in an intention containing the results of the database query, taken from the *epg_broadcastSelection* slot, with some semantical markers for the graphical realization inserted (by the setProperty action). This intention is converted by the output channel to a TV program presentation, which is then passed on to the presentation manager. Finally, the *do* tag contains actions to be carried out during the move. In this case, entries are scheduled to be included in the next lexicon update after the system releases the initiative (Sect. 5.3.2).

Table 1 illustrates some of the keywords of the plan language, divided into structural keywords, predicates for conditions, and actions. In total, the language comprises 55 keywords; a complete descriptive listing is beyond the scope of this text.

5 Planning Mechanism

The dialogue manager uses two separate mechanisms to decide what to do next. For quick reactions to simple utterances or "stimuli," it employs *reflexes*, while inputs

Table 1. Descriptions of some of the keywords of the plan language

Structural keywords	
slotDefinition	Assigns a slot name to a role of the ontology for a particular application
needs	Defines preconditions in terms of required slots and Boolean conditions
provides	Defines postconditions
do	Specifies a series of actions to be done
doTemplate	Like *do*, but with preconditions. Given a set of templates, only the first template with fulfilled preconditions is executed
boolean slot conditions	
slotIsSet	True if a slot has some value
slotHasValue	True if a slot has a given value
predicate	True if a binary relation (e.g., "$\geq$") holds between two slot values
slotEqualsSlot	True if the value of one slot equals the value of another
actions	
setSlot	sets the value of a slot for an application
copySlot	Copies a slot value to another slot
genId	Generates a unique identifier
setProperty	Connects two parts of an intention by a link (as a semantical marker)
setGoal	Adopts a goal state. It is possible to specify whether a goal should be treated as a subgoal, or should replace the currently pursued active goal

that require a planned response are handled by a planning procedure. In this section, we first describe the method used for reflexes, before turning to planned actions.

5.1 Reflex Actions

Input events that do not need much planning, but instead require fast action from the system, are handled differently from normal input. We call this type of input and action *reflexive*, and they are treated somewhat analogously to animal reflexes in that the input is processed and reacted upon by a special, simple, and quick mechanism (the *spine*), and enters the planning process proper only afterwards. One example of reflexive input is entering a telephone number via a keypad, where the delay in displaying the typed numbers must be minimized to obtain adequate responsiveness, because the user will (reasonably) accept much less delay for this type of interaction than for more complex utterances. Reflex reactions are determined by a simple table look-up, and are therefore quite inflexible, but the advantage is that they can always be processed in small constant time. The operations that can be performed by the spine component are restricted: no planning is possible, and no parametrized intentions are generated. Of course it is also possible to pass on information, so that the planning component can deal with it (e.g., handing over the telephone number after dialing is complete).

Table 2. Top-level steps for sending a fax message

Step	Task
1	Clear screen
2	Obtain picture from realDocument application
3	Obtain fax number from phone application
4	Inform "fax being sent"
5	Convert picture to fax format
6	Send fax with picture attachment
7	Wait for confirmation of sending fax
8	Inform about completion of task

5.2 Planned Actions

In most cases, user utterances must be integrated into an overall plan of ongoing dialogue. The dialogue manager maintains a stack of goals that have been adopted during the dialogue. At any one time, one goal is focused for the planner, the *current goal*. A plan is created and then pursued that specifies a possible course of subsequent communicative acts to reach the goal. If the user fully cooperates with the system, the originally generated plan will be followed. However, frequently the user will make contributions that are not intended by the plan, or he will provide information in an order different from the planned sequence. In this case, the plan has to be revised and modified to account for changed circumstances, and to continue the dialogue.

An action plan specifys a series of plan operators that establish preconditions and ultimately lead to an operator representing the goal state. It is constructed by a backward search from the goal state. Where the plan needs to refer to another, subordinate application (e.g., the phone application needs a lookup in the address book application to fetch a telephone number), another hierarchical level is entered and the subplan is worked out in detail. The assumption is that the subordinate application guarantees that it can provide the service it offers (such as fetching a number), and therefore the planning can be deferred until the application is actually started. Thus, if the dialogue takes an alternative path (for example, the user might decide that she would like to enter the number via a keypad), the subplan will not need to be generated at all. After a plan has been generated, the series of its plan operators is followed, by playing the communicative games and executing the actions associated with the individual steps in order. During this process, the dialogue manager releases the dialogue initiative when it has to wait for an answer from either the user or one of the applications (i.e., the current game move is on an input channel).

Upon retaking the initiative, the planner checks if the assumed conditions still hold true at this point of execution. This might not be the case if something unexpected has happened, such as the user being uncooperative or changing his mind, or an error from an application. In this case, the plan is regenerated. In the SMARTKOM applications, the absolute plan lengths on each hierarchical level are not very long, so the overhead for full plan generation and regeneration instead of plan repair is small (the plan length is usually shorter than ten communication moves, and takes not more that one or two seconds to generate). Therefore it does not hamper the dialogue flow by causing an excessive wait for a system answer.

5.2.1 An Example Plan: Sending a Fax Message

In this section, we give an example of the plan that is generated for sending a fax when the user makes an utterance "I want to send a document". Upon this request, a *sendTelecommunication* process is initiated in the fax application. Planning backwards from a goal state, the top-level plan will be the one shown in Table 2.

However, to complete the task, several applications and services have to work together. If the initial plan is rolled out in full, it would contain the steps shown in Table 3. The current application changes as the different subdialogues are delved into. The table also shows the channels on which the communicative games are played. The games are *inform*, *instruct*, *request-response*, and the special type *graphicalAction* (clearing the screen, presenting the document scanning area). The channels include the user input/output channel and the function modeling channels for the devices realDocument and telephony. Note that the request to remove the document in 2-7 is answered not by the user, but by the realDocument device detecting the removal in 2-8. Not shown are the internal moves for the expectations, sent each time on giving up initiative to the user or some device (after steps 2-3, 2-5, 2-7, 2-9, 3-1, 5-1, and 6; the expectation mechanism is explained in the next section), and the reflexive moves during keypad interaction (collecting the number and updating the keypad display). The user presentations are in graphical-only (steps 1, 2-11), speech-only (2-2, 2-7, 4) and multimodal (2-1, 3-1, 8) form.

Before the execution of each plan step, its preconditions are rechecked to confirm that they still hold. It is possible that, during the course of the dialogue, the user would throw in "I want to send it to Anselm." It would then not be necessary to require the user to type in the fax number, since the number can be retrieved using the address book application. This would be noticed upon executing the plan step designated as "get the fax number," since that step has a branching condition that prefers using the address book instead of the keypad subdialogue if an addressee name is known. So, upon reaching this step, the dialogue is replanned and an address book search is used to expand step 3 instead.

5.3 Expectation Feedback

The fact that future moves of the dialogue are predicted by the dialogue plan can be used to help the analysis modules to resolve ambiguities or to prime the lexicon for words that may occur in the input. After each dialogue turn, the dialogue manager publishes an updated data structure, called an *expectation*, to inform other modules about likely subsequent utterances. The expectation mechanism is also used to inform the discourse memory when the user is being asked to confirm a past utterance. A more thorough description of the use of expectations is given in Löckelt et al. (2002).

Table 3. Fully expanded plan for sending a fax message

Step	Task	Channel	Application
1	Present clear screen	→ user (g)	fax
2-1	Present scanning area	→ user (g,s)	realDocument
2-2	Instruct to place the document	→ user (s)	
2-3	Initialize document scanner	→ realDocument	
2-4	Response initialization complete	← realDocument	
2-5	Request start scanning	→ realDocument	
2-6	Response scanning complete	← realDocument	
2-7	Request to remove document	→ user (s)	
2-8	Response document removed	← realDocument	
2-9	Request scanned image	→ realDocument	
2-10	Response scanned image	← realDocument	
2-11	Present scanned image	→ user (g)	
3-1	Present keypad and request number	→ user(g,s)	phone
3-2	Collect number	← user (reactive)	
3-3	Response number	← internal	
4	Inform "fax being sent"	→ user (s)	fax
5-1	Request transcribe picture	→ realDocument	realDocument
5-2	Response transcribed image	← realDocument	
6	Request sending of fax	→ telephony	fax
7	Response sending complete	← telephony	
8	Inform about completion of task	→ user (g,s)	

5.3.1 Slot and Goal Expectations

To give information about the state of the dialogue, the expectation structure specifies the currently active goal and the application to which it belongs. Additionally, if the system has released the initiative to the user, e.g., to answer a question, the analysis modules, especially the discourse modeler, receive indications about expected discourse contributions from the slot expectations, which come in three categories:

- *Expected slot*: If the user is expected to provide specific information, the associated slot is named. The discourse modeler can resolve ambiguities by assigning inputs referring to ontological roles that correspond to an expected slot a higher score.
- *Filled slots*: The slots related to the current task that have already been assigned a value. These will get preference, e.g., if the utterance indicates that the user wants to change something already said, such as in "and what's on TV tomorrow?". Such values can be present if either they were provided during the previous dialogue steps, or they were preset values (such as taking the fastest route by default instead of the shortest in a route planning application).
- *Possible slots* are either slots that would — according to the plan — be addressed in later steps of the dialogue, or that would not necessarily be mentioned at all. An example would be a genre restriction for a movie database query.

An additional flag of the expectation structure is used to inform the discourse modeler of the special case of asking for confirmation of the last input step. An elliptical "yes" in answer to, e.g., "Are you sure you want *X*?" can be put into context this way and be interpreted as a combination of "retainGoal" and/or "retainSlot" operations.

Since the expectations themselves are not limited to what is the single most likely utterance in the dialogue, but offer some fine-grainedness for different possibilities and utterance types, the user has freedom to deviate from the direct trail offered by the dialogue and can, for example, forestall a question not yet asked or change her mind about what he said earlier. The expectation structure will likely be helpful in putting the utterance into context. Dialogue contributions that are not explicitly focused on but are related to the current goal will likely still be of an expected category, although they will possibly get a score penalty.

5.3.2 Lexicon Updates

Applications can specify that they require dynamic lexicon updates during certain situations. For example, the lexicon does not contain composite movie titles or geographical landmarks at startup, since the necessary word list would be inappropriate in size and for actuality requirements, and even if a comprehensive list was available, the recognition rate would suffer if lots of words clogged up the lexicon. This type of lexical item is most likely to appear only in certain specific situations. So instead of statically holding all words that may occur at some point in the dialogue, the lexicon is adapted dynamically, inserting, e.g., movie titles in the context of the cinema application. During a cinema dialogue, the dialogue manager scans the results of movie database queries, and lexical items of several predefined categories (movie titles, actors, directors) are extracted. The next published expectation then contains an update for this word category. The mechanism also for the removal of words when the context changes. Analogous lexicon updates are performed for the television guide and route planning applications.

6 Conclusion

6.1 Summary

We have described the dialogue manager of SmartKom, which plans and executes user and application interactions in a large multimodal system comprising multiple applications in different scenarios. It features a flexible architecture suited for integration in different systems.

The actions of the system comprise dialogues with the user, the devices and applications, and other modules of the system, treated uniformly via communication channels. The ontological framework of SmartKom is used to model the problem domain in terms of processes and roles. We talked about how the plan operator specification language resting on this foundation is used to define the behaviour of SMARTKOM in a declarative XML-encoded knowledge source.

Two mechanisms for determining the actions of the system were described: the reflex mechanism used to generate fast responses to simple stimuli, and the planning mechanism to deal with inputs that require coordinated and planned actions. An added useful feature is the publication of expectations, which allow the analysis component of the system to take into account the current state of the dialogue in interpreting user utterances.

6.2 Further Work

The dialogue manager is currently being redesigned to extend its functionality. This will introduce the ability to conduct multiparty dialogues involving several virtual characters, and to use more sophisticated planning and plan accommodation mechanisms. A third focus of the ongoing work seeks to address the difficulties that were encountered in defining a nontrivial set of interconnected applications, where the dependencies can quickly become difficult to handle, and errors are easily overlooked. Separating dialogue management proper from the application functionalities as far as possible, we investigate possibilities to support the designer of the application in improving the overview of content and dependencies, as well as providing easier manipulating and testing facilities.

References

J. Carletta, A. Isard, S. Isard, J. Kowtko, and G. Doherty-Sneddon. HCRC Dialogue Structure Coding Manual. Technical Report HCRC/TR-82, HCRC, 1996.

L. Carlson. *Dialogue Games*. Synthese Language Library. Reidel, Dordrecht, The Netherlands, 1983.

G. Ferguson and J.F. Allen. TRIPS: An Intelligent Integrated Problem-Solving Assistant. In: *Proc. 15th Nat. Conf. on Artificial Intelligence & 10th Conf. on Innovative Applications of Artificial Intelligence*, pp. 567–573, Madison, WI, 1998.

G. Herzog, A. Ndiaye, S. Merten, H. Kirchmann, T. Becker, and P. Poller. Large-Scale Software Integration for Spoken Language and Multimodal Dialog Systems. *Natural Language Engineering*, 10, 2004. Special issue on Software Architecture for Language Engineering.

M. Johnston, S. Bangalore, G. Vasireddy, A. Stent, P. Ehlen, M. Walker, S. Whittaker, and P. Maloor. MATCH: An Architecture for Multimodal Dialogue Systems. In: *Proc. 10th ACM Int. Symposium on Advances in Geographic Information Systems*, pp. 376–383, Washington, DC, 2002.

S. Larsson. *Issue-Based Dialogue Management*. PhD thesis, Göteborg University, Sweden, 2002.

M. Löckelt, T. Becker, N. Pfleger, and J. Alexandersson. Making Sense of Partial. In: C.M. Johan Bos, Mary Ellen Foster (ed.), *Proc. 6th Workshop on the Semantics and Pragmatics of Dialogue (EDILOG 2002)*, pp. 101–107, Edinburgh, UK, September 2002.

R. Porzel, N. Pfleger, S. Merten, M. Loeckelt, I. Gurevych, R. Engel, and J. Alexandersson. More on Less: Further Applications of Ontologies in Multi-Modal Dialogue Systems. In: *Proc. 3rd IJCAI 2003 Workshop on Knowledge and Reasoning in Practical Dialogue Systems*, Acapulco, Mexico, 2003.

N. Reithinger, D. Fedeler, A. Kumar, C. Lauer, E. Pecourt, and L. Romary. MIAMM — A Multimodal Dialogue System Using Haptics. In: J. van Kuppevelt, L. Dybkjaer, and N.O. Bersen (eds.), *Natural, Intelligent and Effective Interaction in Multimodal Dialogue Systems*, Dordrecht, The Netherlands, 2004. Kluwer Academic.

C. Rich and C.L. Sidner. COLLAGEN: A Collaboration Manager for Software Interface Agents. *User Modeling and User-Adapted Interaction*, 8(3-4):315–350, 1998.

W. Wahlster. SmartKom: Fusion and Fission of Speech, Gestures, and Facial Expressions. In: *Proc. 1st Int. Workshop on Man-Machine Symbiotic Systems*, pp. 213–225, Kyoto, Japan, 2002.

Emotion Analysis and Emotion-Handling Subdialogues

Michael Streit[1], Anton Batliner[2], and Thomas Portele[3]

[1] DFKI GmbH, Saarbrücken, Germany
streit@dfki.de
[2] Friedrich-Alexander Universität Erlangen-Nürnberg, Germany
batliner@informatik.uni-erlangen.de
[3] Philips Research Laboratories GmbH, Aachen, Germany
Thomas.Portele@philips.com

Summary. The chapter presents the cognitive model–based approach of abductive interpretation of emotions that is used in the multimodal dialogue system SMARTKOM. The approach is based on Ortony, Clore and Collins' (OCC) model of emotions, which explains emotions by matches or mismatches of the attitudes of an agent with the state of affairs in the relevant situation. We explain how eliciting conditions, i.e., abstract schemata for the explanation of emotions, can be instantiated with general or abstract concepts for attitudes and actions, and further enhanced with conditions and operators for generating reactions, which allow for abductive inference of explanations of emotional states and determination of reactions. During this process concepts that are initially abstract are made concrete. Emotions may work as a self-contained dialogue move. They show a complex relation to explicit communication. Additionally, we present our approach of evaluating indicators of emotions and user states that come from different sources.

1 Introduction

For a long period, the concept of rational agents, which exchange rational arguments, was the predominant paradigm for research on dialogue systems. In the last decade the scientific community became aware of the fact that emotions, moods and other attitudes play an important role in natural communication. This insight in human behavior resulted in research on affective interfaces, development of believable agents, and in research that aims at supporting the recognition of problematic dialogue situations by analyzing the affective state of the user.

While there are considerable advancements in generating affective artificial agents that display believable emotions in appropriate situations (Picard, 1997), the recognition and interpretation of human emotions in dialogue systems is still in its infancy. In SMARTKOM we explore the handling of emotions by focusing on exemplary cases, which show how emotions work as dialogue moves or and how the emotional color of an utterance may change the literal meaning into the converse.

The term *emotion* normally aims at pronounced, clear forms of human states marked by strong feelings such as, e.g., anger, fear, sadness, joy, etc. — the so-called "full-blown, big" *n* (*n* typically ranging between 4 and some 20) — emotions. Under a close look, however, almost nothing is that clear-cut: the underlying (bodily and cognitive) processes are not yet fully understood, emotions do often occur in mixed, not in pure forms, their marking can be overtly suppressed due to social constraints and rules (Cornelius, 2000; Batliner et al., 2003a,c,b). Finally there is no full agreement as for a catalogue of emotions, and of pivotal characteristics, telling emotions apart from other states such as attitudes, mood, etc.

The OCC Model of Emotions

Research concerned with the generation of an affective and believable behavior of artificial agents is often based on the so-called OCC model of emotions (Ortony et al., 1988), which explains emotion by cognitive processes relating the user's goals, standards, likes and dislikes to the actions of other agents and the state of the world that results from these actions. This approach starts from clearly distinguished emotions, which result from a cognitively comprehensible *appraisal* of the situation by the agent. How intensive emotions are, or if they are displayed at all is determined by the intensity of the attitudes of the agent and by social display rules.

Though mixing or suppressing emotions is a problem for the recognition of emotions as well as for the fine-tuning of the artificial generation of emotional behavior, the OCC model provides a systematic account for relating a certain situation of an interaction to emotional states that fit to this situation. The logical structure of the situation that causes a certain emotion is not affected by the question of how intensive an emotion is or if it is displayed at all. This is not only attractive for the generation of affective behavior, but also for inferring appropriate reactions on emotions that are observed in a certain situation.

Emotions and Other Affective States

For research, which is concerned with the detection of problematic situations in communication by analyzing the user's behavior, not only emotions are relevant. This is the case, independently from the question of whether the catalogue of emotions is completely defined or not. For instance, if the user is hesitant, she may need help, or if she is tired this may affect the presentation of information, or the system may even request the user to stop some activity that needs high attention.

We use the term "(emotional) user states" to encompass all nonneutral, somehow marked behavior of the user within a human–machine communication. Thus, user states as bored, stressed, irritated, tired, etc., can and have to be addressed as well, irrespective of whether they belong to the one or the other psychological or physiological category. In contrast, the psychological or physiological category of a state is relevant for its interpretation. The SMARTKOM system will react to this match or mismatch, e.g., by trying to repair the mismatch or by promoting further matches.

The spirit of the approach, namely to consider what type of conditions elicit the affective state of the agent, extends to some nonemotional states, but not to all. For instance, the user may be hesitant, because she has a goal, but does not know how she can achieve it, or she does not know how to decide between certain alternatives, etc. Such considerations add up to a very similar kind of cognitively comprehensible appraisel of a situation, as elicit emotions in the OCC model. In contrast, the state of tiredness (in the literal meaning) cannot be accounted for by such an approach that analyses the *cause* of the user state. Rather, the system should consider the possible consequences of the user's state.

Preconditions for a Working System

Some important conditions have to be met, however, if one wants to deal with user states in an automatic system:

- It must be possible to classify the state correctly up to a satisfying extent.
- Thus there has to be a sufficiently large training sample.
- The respective user state can be processed within the whole system, not only at the classification stage.

The first condition means that we should start with user states that are clearly marked. This rules out such states as "slightly irritated," even if they might occur quite often and have a strong impact on the felicity of communication. We are thus left with those pure emotions like anger or joy, which do not, alas, occur often in real or in Wizard-of-Oz human–machine communications (Batliner et al., 2003c,a,b; Ang et al., 2002). Thus we decided to concentrate on some few user states and to collect acted data for our training sample.

Overview

The focus of the paper is to present the cognitive model–based approach of abductive interpretation of emotions as it is used in the SmartKom system. Additionally, we present our approach on the collection and evaluation of indicatons of emotions, user states, problematic situations and other states of the user or the interaction. Recognition of emotions from prosody and facial expressions is described in Zeißler et al. (2006) and in Frank et al. (2006).

We start in the second section with a brief description of the architecture of emotion analysis in SmartKom. In the third section we introduce the type of interaction we want to focus on. In the fourth section we describe the method for evaluating indicators from different sources. At the time, we use only a little part of the information provided there, but much more would be useful The remaining sections are dedicated to the interpretation of emotions and user states and the generation of reactions to these states. First, we introduce the OCC model of emotions, then we say a little bit about abduction and the problems that occur if we use the OCC model for an abductive interpretation. In the next step we show how we generate explanatory hypotheses by introducing abstract concepts as fillers of eliciting conditions. We introduce conditions for explanation selection and recovering operators for generating initiate reactions of the system. Conditions and generation of reactions are interwoven: usually conditions provide parameters that are used by the recovering operators.

2 Emotion Processing in the SmartKom System

SmartKom (http://www.smartkom.org) is a multimodal dialogue system that provides access to multiple applications (Wahlster, 2003). In addition to input-modalities that are used for intentional communication, the system accounts for the emotional state of the user as it is displayed by facial expression or by prosody. The processing of emotions and user states consists of three stages.

- At the first stage the emotional state of the user is recognized from facial expression and prosody.
- At the second stage indications of problematic situations and the emotional state of the user are collected from several sources and collectively evaluated. The component also analyzes the dialogue in respect to the style of interaction and the task and paradigm knowledge of the user (Portele, 2003).
- The interpretation of emotions and user states, and the generation of reactions to these states build the third stage. It is realized by so-called dynamic help. This component is dedicated to manage subdialogues, and to provide presentation specification and intention analysis in problematic situations that are not handled by the standard dialogue component of SmartKom.

3 The Use Cases

To demonstrate the added value of user state classification and its subsequent processing in the SmartKom system, we designed so-called **use cases**. The first use case is intended to show how a merely emotional reaction, without explicit communication, can work as a self-contained dialogue move. In this case, joy or anger are interpreted as positive or negative feedback, respectively. In the second use case emotion works as a semantic operation that turns a positive feedback into a negative one, which is considered as a form of sarcasm. In both use cases, the system suspects that the emotional reaction may be caused by a like or a dislike concerning the properties of the presented objects. If reasonable candidates of such likes or dislikes can be identified that are not already known by the system, it starts a preference update dialogue.

If the system knows positive or negative preferences, it first presents objects that contain a preferred feature. Objects that show a disliked feature will be shown last. [1]

user: *What's on TV tomorrow?*
system: Shows talk show at the top of the display, in the middle popular music, and crime at the bottom.
user: *And what's in the evening, in the first program?*
system: Shows a science fiction movie.

[1] It is possible that an object has both liked and disliked attributes, e.g., there may be a movie with a preferred genre, in which a disliked actor plays.

First constellation: emotion only
user: Displays joy via facial gestures.
system: *Do you like science fiction? Shall I account for that in future presentations?*

Second constellation: emotionally marked verbal communication
user: *That's really a beautiful program!* (She produces this sentence with an angry prosody. The positive feedback is analyzed as being sarcastic.)
system: *You don't like science fiction? Shall I account for that in future presentations?*

user: *Yes./No.*
system: *OK. I'll take care of that!*

(Suppose the user's answer was *yes*: In the first constellation **science fiction** will be presented at the beginning of a presentation, in the second constellation at the end.)

user: *Please, again tomorrow's program!*
system: Shows **science fiction** at the beginning (at the end) of a presentation.

Instead of answering *no* or *yes*, the user may also correct the supposed like or dislike, e.g., by saying *No, I like crime movies*, or she may just ignore the question, by moving to a different topic. In such cases, the system will simply not rearrange the order of presentation.

4 Indications of User States and Problematic States of Interactions

We introduced in SMARTKOM a component, the *interaction module*, that collects and evaluates indications of emotions, problematic situations and other aspects of the interaction. This approach is used, on the one hand, to estimate recognition results by taking other indicators into account that may support or devaluate the result. On the other hand, the component introduces its own results concerning states of the user or charcteristics of the interaction. It operates by analyzing a set of possible *indicators*.

Indicators can have values between 0 and 1 and these values may change in time. For example, the mimic analysis module delivers information about the user's current emotional state as a set of four probabilities (likelihood of anger, likelihood of joy, likelihood of dilatoriness, neutral). The set of indicators is fairly fixed and is defined by the capabilities of the modules in the system (Table 1).

The interaction module provides a set of *models* as output. Each model value is also in the range between 0 and 1. Several models support the recognition of emotions and try to detect problematic situations during a dialogue.

4.1 Indicators

The indicator values are mapped to the models by means of a matrix multiplication. One element of the matrix denotes the influence of one indicator value to one model. This design is motivated by the observation that most indicators can contribute to different models, and that the combination of simple indicators to complex models may be optimized by machine-learning algorithms. Furthermore, new models on demand from other modules can be constructed easily by combining the indicator set with a different weighting scheme.

Table 1. List of indicators

Source	Description
Mimic recognizer	Mimically conveyed anger
Prosody recognizer	Prosodically conveyed anger
Mimic recognizer	Mimically conveyed joy
Prosody recognizer	Prosodically conveyed joy
Mimic recognizer	Mimically conveyed dilatoriness
Prosody recognizer	Prosodically conveyed dilatoriness
Speech recognition	Linguistically conveyed anger
Speech understanding	Ratio of unanalyzable words
Intention analysis	Overall score of the best hypothesis
Intention analysis	Difference in score between first and second best hypotheses
Intention analysis	Number of possible hypotheses (depth of lattice)
Speech recognition	Score of the speech recognizer
Gesture recognition	Score of the gesture analyzer
Speech understanding	Score of the language analyzer
Media integration	Score of multimodal integration
Discourse history	Score of the discourse module
Domain model	Score of the domain module
Intention analysis	Final score of the intention module
Intention analysis	Number of elements in the user input
Discourse history	Number of new (not previously mentioned) elements
Speech understanding	Number of elements addressed by speech
Gesture analysis	Number of elements addressed by gesture
Media integration	Number of elements addressed by speech and gesture
Intention analysis	Importance of speech recognition score for overall score
Intention analysis	Importance of gesture analysis score for overall score
Intention analysis	Importance of domain model score for overall score
Intention analysis	Importance of language understanding score for overall score
Intention analysis	Importance of discourse model score for overall score
Speech understanding	Relative number of sentence-like units in one turn
Speech understanding	Relative number of words in one turn
Speech understanding	Relative frequency of pronouns
Speech understanding	Relative frequency of verbs
Speech understanding	Relative frequency of adverbs
Speech understanding	Relative frequency of nouns
Speech understanding	Relative frequency of content words
Speech understanding, language generation	Relative frequency of content words appearing in the system output
Speech understanding, language generation	Relative frequency of content words not appearing in the system output

4.2 Models

The module delivers four sets of models. The distribution of the models to the sets is somewhat arbitrary and is mainly governed by the intended use within the SmartKom system. Therefore the models for task and paradigm knowledge use many in-

Table 2. List of models

Set	Description
Problem	Likelihood of a problem
Problem	Likelihood of an analysis problem
Problem	Discourse progress rate
Problem	Likelihood of the user being angry
Problem	Likelihood of the user being happy
UserKnowledge	Estimation of user familiarity with task
UserKnowledge	Estimation of user familiarity with system
Modality	Ratio of spoken input content
Modality	Ratio of gestural input content
Modality	Ratio of multimodal input content
ModalityContrastive	Ratio of contrastive usage of multimodal input
ModalityRedundant	Ratio of redundant usage of multimodal input
Linguistic	Adaptivity of user's lexical choices to former system output
Linguistic	Likelihood of long turns
Linguistic	Likelihood of long sentences
Linguistic	Ratio of pronoun usage
Linguistic	Ratio of verb usage
Linguistic	Ratio of adverb usage
Linguistic	Ratio of noun and verb usage

dicators that also contribute to the problem detection models. This overlap reflects the design principle of the module.

- One set of model values reflects the assumed task and paradigm knowledge of the user. The task knowledge describes the user's knowledge of the current task (e.g., programming a VCR), while the paradigm knowledge indicates how well the user is accustomed to dealing with multimodal dialogue systems, and, especially, with SMARTKOM. These models can be employed by the dynamic help and the presentation/language generation to deliver an appropriate amount of feedback and assistance.
- A second set of models describes the linguistic behavior of the user regarding the number of, e.g., referential expressions, usage of complete sentences and average length of input. Some of these features can, in principle, be used by language models or parsers (although this is not the case in the current SMARTKOM system). All of them may help to adapt the language generation in order to reflect the user's style—based on the assumption that this is beneficial. Furthermore, the adaptivity of a user's lexical choices to former system output is estimated, which can helping adapting dynamic language models used in SMARTKOM and language generation in order to maximize this value as a measure of the common vocabulary.
- The third set compares the use of different modalities by the user (for instance, a preference for gestures or spoken input). The presentation and the behavior of the animated character can be adapted toward the user's preferred distribution of the different modalities — users who prefer pointing gestures could be supplied with an interaction display with more possibilities and details, while users interacting mainly through speech should get the most important system feedback also by spoken output.
- Problematic situations and user state information are expressed by three models in the fourth set of the module:

- One model describes the likelihood that the user is angry by combining scores from mimic analysis, emotion extraction from prosody and use of certain words. These indicators are either computed directly from other modules or are obtained by counting appearances of specific words (a related model assesses the user's satisfaction by looking for joyous emotional expressions; this and the anger model are used for setting content-based user preferences).
- A second model combines confidence values from recognizers and similar scores from speech analysis, domain model, discourse history and intention recognition as well as differences in the distribution of these values among concurring hypotheses; this model is supposed to indicate problems in the analysis part of the system.
- A third model estimates the dialogue progress. Here, the ratio of new information items to total information items (after completion by overlay in the discourse history) is employed as one indicator, another one being the ratio of overlayed (i.e., changed, interpretable as corrected) information to unaltered information. A further indicator is the overall number of information items in the user input. Because of the system design, all other indicators can contribute to the model values as well (and the indicators used here are also used for other models, e.g., about user experience with the task or the interaction paradigm), but the indicators named above are assumed to be the most important ones for the respective models.

These model values are part of the input of the dynamic help module.

5 Cognitive Model–Based Interpretation of Emotions

Our approach to the analysis of emotions is based on the OCC model of emotions developed by Ortony, Clore and Collins. Following the OCC model, emotions are characterized by their eliciting conditions. These conditions consist of a certain combination of

- the goals of the agent in this situation
- her attitudes to certain events (mainly likes and dislikes)
- the standards that she uses to (morally) judge an event
- the facts that hold in a certain situation
- the actions (of other agents) that caused these facts

For triggering an emotion, it is important to know how facts are related to the goals and the likes and dislikes of the user. In particular, it is interesting if they coincide or not. Standards are important for emotions such as anger or gratitude that contain criticism or praise of another agent based on her actions. Eliciting conditions can be viewed as expressing the cause of an emotion by providing a cognitively comprehensible explanation of an emotion. The following eliciting condition for anger is taken from Prendinger et al. (2002):

```
anger(Agent1,Agent2,State,Sit) if
   holds(did(Agent2,Action),Sit),
   causes(Action,State,Sit0),
   wants(Agent1, non_State,Sit),
   blameworthy(Agent1,Action)) ,
   (Sit0 < Sit)
```

This condition means that the agent is angry if she believes that another agent caused some state of affairs that contradicts her goals by performing an action that is not acceptable according to the user's standards (expressed by the *blameworthy* predicate). By the situation variables Sit, Sit0, one can express how the elements of the conditions are connected with respect to the sequence of situations that occur (subsequently we will omit situation variables).

Recognizing the intensity of emotions could provide additional valuable information, e.g., slight anger may occur at the beginning of a problem, while strong anger may indicate an enduring problem. But the recognition of the situation that caused the emotion and the generation of appropriate reaction is basically the same whether emotions are displayed slightly or strongly.

5.1 Abductive Interpretation of Eliciting Conditions

The OCC model is mainly used for the generation of the behavior of an animated agent. In this case, one can deliberately define the agent's likes, dislikes and standards in advance. If we want to interpret emotions that are displayed by an agent, we have to find out which combination of facts, attitudes and standards may have caused the emotion. Our approach is to achieve this by analyzing eliciting conditions in an abductive manner. Abduction as a form of practical inference is introduced by Peirce (1995). Abduction is often characterized as inference to the best explanation: Suppose we observe some fact A, which is surprising for us. If we know the rule

$$B,C \rightarrow A$$

(i.e., A is true if B and C are true), then we may suspect that also B and C are true, because this would plausibly explain A. If we know that there is another rule

$$D \rightarrow A$$

then D is another candidate for explaining A. Hence we need a criterion to decide which explanation is better. The quality of an explanation depends on two factors: Do we know all relevant rules (i.e., explanations)? Do we possess criterions to choose from explanations? With eliciting conditions we have the advantage of possessing schemata that claim to characterize all possible explanations of an emotion.

5.2 Problems with Abductive Interpretation

Eliciting conditions are abstract schemata that cannot be used directly to infer possible causes of emotions. To perform abductive reasoning on eliciting conditions, we have to identify concepts that could be filled into the schemata. Seemingly, we are in a problematic situation. The system has no information about the user's standards, likes and dislikes in advance. It can get information about her goals from the user's input. But, on the one hand, this information may be based on misunderstanding, and, on the other hand, the user may have goals which cannot be recognized from her utterances. Similar problems occur with the actions of the system. The concrete actions that are based on misunderstanding are not relevant for the analysis of the user's emotion.[2]

5.3 Abstract Goals and Actions for Emotion Interpretation

To overcome the problems mentioned in the last paragraph, we introduce metagoals concerning general principles of communication and abstract goals concerning user needs that (to some extent) depend on the application. For every metagoal or abstract goal we introduce an abstract action that satisfies the goal.

For instance, to account for misunderstandings, we introduce *understanding* as an action on the metalevel and *to be understood* as a goal on the metalevel. To account for user preferences, we introduce the concept that *a presentation accounts for the user's preferences* as an abstract action of the system — let it be called *presentByPreferences* — and accordingly the possible abstract fact or user goal *isPresentedByPreferences*.[3] This goal is abstract and underspecified because we do not know the concrete preferences of the user. Further, the relevant types of preferences depend on the type of the application.

Reasonable goals (facts, actions, likes, standards) have to be identified by careful analysis of general principles of communication and the needs of the user with respect to the type of applications with which she is working. This needs empirical validation, which could not be provided within the scope of the SMARTKOM project. Which set of concepts is chosen also depends on practical decisions: which goals will the system support at all, will the system possibly recognize goals that it is not able to handle, will the system react on any recognized emotion in some way (e.g., by regretting as a default in case of anger), or will it only react to emotions to which it can provide a repair or other meaningful cooperative reaction? We demonstrate the approach by the example of **anger**.

General Concepts

We first look for actions or facts that may contradict the user's wishes, likes, dislikes or standards on a general level. Important candidates for abstract actions that contradict the user's wishes are **misunderstanding**, **slow processing** and **requests with a negative or disliked outcome**. Accordingly, we stipulate abstract or general goals, e.g., the goal to be understood properly.

---

[2] Although the type of the action that the system wrongly performs may influence the intensity of the user's negative feelings.

[3] For convenience we often identify the name of the fact and the name of the goal to make this fact true.

Application-Dependent Concepts: Problematic Results of Database Queries

According to our use cases we concentrate on requests with liked or disliked outcome as a source of negative or positive emotions. We identified four types of disliked results:

- The result is empty.
- The majority of retrieved objects show features that are not liked by the user.
- The objects are presented in a way that is contrary to the preferences of the user, e.g., by presenting disliked objects first.
- The user query resulted in a recall, which is too large. The user may need help for further specification possibilities.

We assume for this list of topics that the disliked or problematic results are not due to misunderstanding. Misunderstanding is taken as evoking its own class of constellations. If misunderstanding is involved, the result is not relevant for the analysis.

User-Specified Goals and System-Initiated Actions

As far as no misunderstanding is involved, the SMARTKOM system will usually simply follow the user's specification. If this works, no anger should arise with respect to the fact that the system tries to achieve this goal (but perhaps instead *joy*). In specific situations the system may initiate actions that are necessary from the point of view of the system, but that may be disliked or even considered blameworthy by the user. For instance, the system may require a biometric verification, which the user dislikes. Such actions are relevant for explaining negative emotions but are not considerd in our implementation. Inappropriate or undesired reactions on emotions could also be a cause for anger (or for being bored). In fact, this is a subcase of disliked system-initiated actions.

6 Analyzing and Handling Pure Emotion

With the concepts introduced in the last section, we are able to build instantiations of eliciting conditions that allow us to infer combinations of goals, facts, actions, likes and dislikes that possibly explain the user's emotion. We call instantiations of eliciting condition schemata **eliciting constellations**. To get criteria for selecting the relevant constellation, we augment constellations with conditions and organize these conditions internally as a decision tree.[4]

Further, the system has to determine reactions that are appropriate for

- resolving the situation that caused the negative emotion
- avoiding negative emotions in future in similar situations
- promoting the occurrence of positive emotions in similar situations

[4] As mentioned in the conclusion, we could perform testing only in a limited way. Thus no training of the decision tree was possible.

It is also desirable to include methods that provide abstract underspecified goals and actions with presumable values. Such values are not only used for determining concrete system reactions, they serve as a part of the constellation conditions.

According to our use cases, we have to consider database queries that retrieve disliked objects. The system offers as repair that it will regard the likes and dislikes of the user in its presentations.

A constellation for handling anger according to our use cases is given below (leaving out some minor details) in a Prolog-style notation. It applies to browsing television programs or cinema programs. For these applications preferences are actually taken into account for the presentation. These rules are basically processed in the following manner: First the conditions are tested (internally the conditions are processed in a decision tree–like order). Then the cause of the emotion, which is represented by the clauses above the conditions, is considered as a reasonable explanation, whereby the variables are filled by the result of the condition processing. Then the system action is performed.

```
anger(thisConstellation,user,system) if
  holds(did(system, non_presentByPreference(dislike(user,X))),
  causes(non_presentByPreference(dislike(user,X)),
         non_isPresentedByPreference(dislike(user,X))),
  wants(user, isPresentedByPreferences(dislike(user,X))),
  blameworthy(user, non_presentByPreferences(dislike(user,X)))),
  conditions(thisConstellation,X),
  (proposed system action:) update(dislike(user,X)).
```

The constellation expresses that there is a concrete reading of the goal *presentByPreferences* that may be a goal of the user, that this goal is not satisfied and that ignoring the goal is against the standards of the user. The constellation contains facts and actions that are not concretely specified. For instance, we do not know whether the presentation contains some possibly disliked feature, and we do not know which feature it is.

We test the salience of the constellation by establishing the following conditions. The predicate *presentationEntriesContainCommonFeature(X)* also delivers a concrete presumable instance of the user's dislike.

```
conditions(thisConstellation,X)  if
     presentationEntriesContainCommonFeature(X)),     (1)
     non_specified(user,X),                            (2)
     non_knows(system,like(user,X)),                   (3)
     non_knows(system,dislike(user,X)).                (4)
```

Condition 1 verifies if the user perceives *too many* objects with the supposed disliked feature. (It also excludes the case that there is no result at all, which would support a different explanation for anger.) It is important for the other tests that the predicate delivers a hypothesis for the disliked feature. Condition 2 excludes that the user is angry about the occurrence of features that she has specified in her request (there is a possibility of misunderstanding). Condition 3 excludes, that the user is angry about a feature that she has already declared to like. Condition 4 excludes that the system in fact tried to present the disliked feature appropriately but just did not find other objects.

For emotions displayed by facial expressions, we prove if the emotion emerges in a certain time interval after the presentation was displayed. With prosodically displayed emotion we prove if the verbally expressed content was compatible with the explanation of the emotion. It turned out that it is not sufficient to test if there are already stored preferences. It should additionally be proved, if a user has not agreed with storing a preference. This has to be remembered, otherwise the system may propose the same preference repeatedly.

The action *update(dislike,user,X)*, which is attached to the constellation, initiates a subdialogue that verifies if the user has the supposed dislike. It is not only a repair action, but takes part in the explanation process.

The conditions mentioned so far are not sufficient to **discriminate competing explanations**. Such competing explanations have to be modeled, even if no reaction is foreseen for these cases. We distinguished three main sources of anger: misunderstanding, slow processing and requests with a negative or disliked outcome. Evidence for problems in the analysis part is detected by the interaction module (Sect. 4). Slow processing is a possible explanation for anger, if anger occurs during the analysis. Also, the absolute duration of processing is a criterion. These dates are accessible via a module (the so-called watchdog) that monitors the processing state of the system.

Extending the Method — Beyond the Use Cases

We present the handling of an example not covered by the use cases in order to show how the method extends. By using the relaxation functionalities, which are provided by the dynamic help component, one could add constellations that handle anger concerning the factual result of a query. In this case, it is debatable if the observed anger must be directed to the system. It may also directed to the television program. We stipulated as a possible source of anger (or, even more likely, as a source of disappointment) the fact that the recall on a query contains no objects that are liked by the user.

```
anger(user, system)
   (The relevant facts that hold in this constellation and
   the actions that caused the facts are omitted)
   if
     wants(user, getObjectsAccordingToLikes)
     blameworthy(user, NONgetObjectsAccordingToLikes)

     (conditions(X):)
     (
      knows(system,dislike(user,X)) ,
      NONspecified(user,X),
```

```
  ResultContainsProminentCommonFeature (X))
 )
  or
 (
  knows(system,like(user,X)) ,
  NONResultContains (X))
 )

(proposed system action:)
relaxRequest
```

Case: Anger on Empty Result

This is a simple case:

```
anger(user, system)
   if
   wants(user, getNonEmptyResult)
   blameworthy(user, getNonEmptyResult)
   (conditions:) emptyResult
   (proposed system action:)
   relaxRequest
```

7 Emotions and Communicative Acts

Emotions that are signaled by facial expressions do not need to be accompanied by additional communication at all. Emotions expressed by voice are naturally related to some acoustic output. In the extreme, this output is only a container for the expressed emotion, but usually it contains a certain semantic content. The analysis of the relation between semantic content and underlying emotions is in its infancy, compared, e.g., with the relation between verbally communicated semantic content and pointing gestures. The latter is sufficiently known to build practical application. We distinguish in the following between communicative acts with semantic content, which are provided by speech and gestures, on the one hand, and emotions, on the other hand. The interpretation of pointing gestures and verbal utterances can be conceived as a fusion process that unifies pieces of information. Semantic contradictions between pointing gestures and verbally provided information are indications for errors. The relation between emotions and communicative acts is much more complicated. We give a presumably nonexhaustive classification of types of interaction between displayed emotion and communicated semantic content.

Redundancy

Semantic content redundantly expresses a simultaneously displayed emotion such as *that makes me angry* or *I'm glad about that*, or semantic content expresses an attitude that corresponds to the direction of the emotion (whether it is positive or negative) such as *great*, *bad*.

Contribution to the Explanation of the Emotion

Semantic content expresses a concrete attitude (like or dislike) that is involved in triggering the emotion such as *I don't like thrillers* or *great movies*, or semantic content addresses the facts and actions that caused the emotion such as *you didn't understand me* or *that takes too much time*, or simply by uttering *thrillers* accompanied by a positive or negative emotion.

The *thriller* example contributes the concrete feature that may fill the abstract goal of being presented accordingly preferences. But this example does not necessarily express a like or dislike as *great movies*. With a negative emotion, the example may also belong to the topic *Semantic Content as Repair Action*.

Semantic Content as Repair Action

The semantic information is provided to repair the state of affairs that has caused the emotional state of the user. The example *thriller* works also here: *thriller* could be a correction of a misunderstanding of genre. There is no direct relation between the content of the utterance and the displayed emotion.

This is very common and important in human–machine dialogue as well as in human–human dialogue: The dialogue partner repeats or reformulates her request and concurrently displays a negative emotion. With overt anger, it could also be expected that the user cancels the interaction as a final form of repair.

Change of Semantic Content

The user displays a negative emotion and communicates verbally a positive attitude such as *marvelous, great movies*. The direction of the valenced attitude that is communicated verbally is changed by the direction of the displayed emotion. This is a simple form of sarcasm.

8 Conclusion

A complete implementation of the whole processing chain was available at the end of the project. There was no opportunity for systematic tests, which require high effort. For instance, the recognition of facial expression needs careful preparation of the environment in respect to lighting conditions in order to work. Our limited testing shows that, provided recognition is correct, the emotion interpretation generates the reactions that are requested by the use case specification.

We successfully implemented a cognitive model-based approach for analyzing emotions and other affective states of a user who participates in a multimodal human–machine dialogue. This is a success, but it will still take considerable effort to make it practically useful. The approach is based on an elaborated theory, which covers a broad range of phenomena. This is promising with respect to the extensibility of the approach. It is an important advantage of the approach that it generates conceivable explanations of emotions that allow for well-directed system reactions.

The approach is not restricted to handle classical emotions, but extends to other affective states. Also it is not restricted to states that are displayed nonverbally. Affective verbal feedback, such as *I like this*, can be explained along similar lines.

References

J. Ang, R. Dhillon, A. Krupski, E. Shriberg, and A. Stolcke. Prosody-Based Automatic Detection of Annoyance and Frustration in Human-Computer Dialog. In: *Proc. ICSLP-2002*, pp. 2037–2040, Denver, CO, 2002.

A. Batliner, K. Fischer, R. Huber, J. Spilker, and E. Nöth. How to Find Trouble in Communication. *Speech Communication*, 40:117–143, 2003a.

A. Batliner, C. Hacker, S. Steidl, E. Nöth, and J. Haas. User States, User Strategies, and System Performance: How to Match the One With the Other. In: *Proc. An ISCA Tutorial and Research Workshop on Error Handling in Spoken Dialogue Systems*, pp. 5–10, Chateau d'Oex, Switzerland, August 2003b.

A. Batliner, V. Zeißler, C. Frank, J. Adelhardt, R.P. Shi, E. Nöth, and H. Niemann. We Are Not Amused – But How Do You Know? User States in a Multi-Modal Dialogue System. In: *Proc. EUROSPEECH-03*, vol. 1, pp. 733–736, Geneva, Switzerland, 2003c.

R.R. Cornelius. Theoretical Approaches to Emotion. In: *Proc. ISCA Workshop on Speech and Emotion: A Conceptual Framework for Research*, pp. 3–10, Newcastle, Northern Ireland, 2000.

C. Frank, J. Adelhardt, A. Batliner, E. Nöth, R.P. Shi, V. Zeißler, and H. Niemann. The Facial Expression Module, 2006. In this volume.

A. Ortony, G.L. Clore, and A. Collins (eds.). *The Cognitive Structue of Emotions*. Cambridge University Press, Cambridge, UK, 1988.

C.S. Peirce. Abduction and Induction. In: J. Buchler (ed.), *Philosophical Writings of Peirce*, pp. 150–156, Berlin Heidelberg New York, 1995. Springer.

R.W. Picard (ed.). *Affective Computing*. MIT Press, Cambridge, MA, 1997.

T. Portele. Interaction Modeling in the SmartKom system. In: *Proc. ISCA Tutorial and Research Workshop on Error Handling in Spoken Dialogue Systems*, Chateau d'Oex, Switzerland, August 2003. ISCA.

H. Prendinger, S. Descamps, and M. Ishizuka. Scripting Affective Communication With Life-Like Characters in Web-Based Interaction Systems. *Speech Communication*, 16:519–553, 2002.

W. Wahlster. SmartKom: Symmetric Multimodality in an Adaptive and Reusable Dialogue Shell. In: R. Krahl and D. Günther (eds.), *Proc. Human Computer Interaction Status Conference 2003*, pp. 47–62, Berlin, Germany, June 2003. DLR.

V. Zeißler, J. Adelhardt, A. Batliner, C. Frank, E. Nöth, R.P. Shi, and H. Niemann. The Prosody Module, 2006. In this volume.

Problematic, Indirect, Affective, and Other Nonstandard Input Processing

Michael Streit

DFKI GmbH, Saarbrücken, Germany
streit@dfki.de

Summary. Natural communication is accompanied by errors, misunderstandings, and emotions. Although in the last decade considerable research has evolved on these topics, human–computer dialogue applications still focus on the exchange of rational specifications of the tasks that should be performed by the system. In this chapter we describe a component that is devoted to processing problematic input that is characterized by the lack of a clear specification of the user's request. The following topics are discussed: interpretation of emotions and verbally communicated likes and dislikes, interpretation of indirect specifications, clarification of underspecified input, help on demand, robust navigation in multimodal dialogue and generation of error messages.

1 Introduction

Natural communication is accompanied by errors, misunderstandings and emotions. Consequently, human agents spend much effort in avoiding and clarifying misunderstandings, in ensuring the formation of common ground, and in accounting for affective states. Although, in the last decade considerable research has evolved on these topics (Traum, 1994; Hirst et al., 1994; Traum et al., 1999; Streit et al., 2004), human–computer dialogue applications still focus on the exchange of rational specifications of the tasks that should be performed by the system. The plan-based intention recognition approach, which does not demand direct specification of actions, nevertheless relies on utterances that may be recognized as rational contributions to a plan. Though the approach has been extended, e.g., for handling clarification subdialogues, it is not suited to process affective communication or to explain communicative reactions in absence of shared intentions. On the other hand, because the approach is difficult to realize, practical dialogue systems usually take speech acts as expressing the user's intention more or less directly.[1] On the *main road* of dialogue processing, the user is expected to provide specifications of her request in terms of the actions she wants to be performed or the goals she wants to achieve.

[1] Dialogue systems usually do not expect a complete specification. They ask for necessary parameters.

The conglomeration of attributes in the title of this chapter points to types of interaction, that are — more or less — *off the main road*, and are characterized by lacking a clear specification of what the user exactly is expecting from the system. Some topics presented in this chapter are rather close to the *main road*, such as clarification of underspecified input and the provision of help, others are further away from standard processing, such as the handling of emotions and indirect specifications. By the latter we understand requests where the agent does not specify an action or goal by providing their inherent parameters (e.g., the begin time of an entertainment). Instead she will specify her goal indirectly by relating it to other goals or objects.

Which topics are considered as belonging to standard processing is partly a design decision. But it seems to be clear that topics such as indirect specification or emotion interpretation need specialized knowledge and inference capabilities to be processed adequately, independent from being assigned to some *official* module or not. In SMARTKOM the issues listed below are handled in a special *nonstandard* processing component. There a several alternatives to integrate the processing of these issues, which is discussed in the next section.

The following topics are discussed in subsequent sections:

(1) Interpretation of emotions and verbally communicated likes and dislikes
(2) Interpretation of indirect specifications
(3) Clarification of underspecified input
(4) Help on demand
(5) Shallow model for robust navigation in multimodal dialogue
(6) Generation of error messages

The processing of emotion is presented in more detail in Streit et al. (2006), but we say something about the handling of verbally expressed likes and dislikes.

The processing of underspecified and indirect specification may be considered, at the first glance, as a similar problem. But the problems exist on different levels of analysis. In the first case we are faced with input that is incomplete and cannot be completed by analyzing discourse history. For instance, the user utters time expressions, which fits to many goal specifications, but none of them occur in the salient context. In this case we are looking for reasonable completions, and usually ask the user which alternative she has in mind, after checking if a misrecognition has been occured. In contrast, indirect specifications are complete intentions but cannot be mapped to a specification of a task directly. For instance, if the user asks the system to *show more crime movies*, the user delegates to the system to find out how to adjust parameters to the effect that more crime movies are found.

In the analysis of most of these topics, the consideration of context is much more important than the analysis of the semantic structure of the communication act, which is another common feature of the phenomena that are discussed here. This is also true for the topic of *multimodal dialogue navigation*, where the simple request to leave a subdialogue requires a thorough analysis of the multimodal dialogue structure to determine the state to return to.

2 Architectural Issues

There is a fundamental architectural alternative for implementing the handling of the topics listed above. One alternative consists in devoting an assistant component to care about events (input events in the first place) that cannot be handled on the standard strand of processing. The assistance is consulted if a certain input cannot be processed on the main strand. This is the architecture chosen in SMARTKOM (with the exception of the handling of pure emotions, where the nonstandard component takes initiative).

The other approach installs a metadialogue component that inspects every input event and its processing — including the events processed on the main strand — whether there is some indication to modify the processing (Purang et al., 2000). Normal processing could be canceled, e.g., if a misunderstanding is suspected, if the result is such that it probably does not satisfy the user's wishes, e.g., if it is empty, or if the communication act shows a certain emotional color which indicates that it should not taken literally.

In graphical user interfaces, there is the possibility to handle (at least some of) the issues mentioned in the introduction as a parallel strand of interaction, e.g., by devoting a special window to them. This approach does not seamlessly fit into the paradigm of dialogue (though in a multimodal system the approach is possible in principle). One could speculate about transferring this approach from graphical user interfaces to multimodal dialogue by introducing several artificial agents that take care of different aspects of the interaction. But this far beyond the scope of SMARTKOM, which, e.g., lacks the turn-taking mechanism that would be needed if independent specialized agents interact. We only risk in the processing of pure emotion that the nonstandard component takes initiative independently.

In the metadialogue approach, it is basically the metadialogue component that decides whether there is a situation that needs special attention or not. In the other approach the standard processing dialogue component has to overtake the position to decide whether a problem occurs and when the subdialogue that handles the problem is finished.

For a self-contained understanding of this chapter, we sketch in one sentence the *main road* processing in SMARTKOM: The user provides a probably ambiguous or elliptic and even noisy multimodal specification of her goal and its parameters (i.e., the slots that belong to the goal). The user's specification is mapped to a formal specification of the task. This semantic specification is enriched by inheriting information from the discourse memory. The best scored hypothesis is taken as a specification of the task that has to be done. Input that does not represent an application goal or cannot be processed for technical reasons is delivered to the dynamic help module, which is the name of the module devoted for processing nonstandard input events.

The SMARTKOM approach brings about a formally strict control and clear division of initiative. It is decided by the dialogue management (i.e., the *action planning* module) if user input is in some sense nonstandard. The problem of this approach is that the knowledge about critical situations is situated in the nonstandard component.

Therefore, the responsibility for handling events is de facto based on a distinction of the structure and semantic content of the input. If it specifies an application task,[2] the action planner will handle it; otherwise it either addresses a nonstandard task, e.g., a help demand or an indirect specification, or it is incomplete (disregarding ill-formed representations).

In the SmartKom architecture, we must solve the problem of how the flow of processing resulting from subdialogues on the nonstandard strand can be controlled. We discuss this problem in Sect. 5.

3 Verbally Expressed Likes and Dislikes — Preference Subdialogues

Preferences for specific genres or actors are regarded by the presentation planning component.[3] Objects that contain features, which are explicitly preferred by the user, are presented at the start of a presentation, and features that are explicitly disliked are presented at the end. If the user verbally expresses a *like* or *dislike*, the semantic representation is passed to the nonstandard component. If the message explicitly addresses a feature that can be handled by the system's preference management (e.g., *I like crime movies*), the component will generate the offer to present objects in the future according to the preference. The system asks if it should store this positive or negative preference. If the dynamic help module detects that the preference is already stored, it does not newly generate a preference dialogue.

Verbally expressed likes and dislikes are closely related to positive or negative emotions, but they are not necessarily accompanied by an emotion (either the user has no emotion or displays no emotion). This is in agreement with the OCC model of emotions (Ortony et al., 1988), that we use for the interpretation of emotion. For instance, a dislike triggers anger if the user *blames the system* for showing her disliked objects, which is not necessarily the case. But even if the user does not blame the system for ignoring her (up to now) unknown preferences, it may promote further joy if the system proposes to regard her preferences in the future.

The problematic case is the handling of unspecific utterances such as "nice" or "I like this". On the surface, the unspecific case is elliptic or anaphoric. The identification of the objects of a like or dislike is not resolved by the SmartKom discourse analysis. The resolution is rather taken as a process of inferring an explanation of the positive or negative feedback, as we process in the case of pure emotions. [4] This is achieved by an abductive reasoning as is explained in more detail in the chapter on emotions. As with pure emotions, we use genres as the default feature for preferences.

[2] Including certain system-related tasks.

[3] Other possibilities to react on the user's likes and dislikes are discussed in Streit et al. (2006).

[4] In fact, the process of explanation may be interwoven with another anaphora resolution process, if one considers examples such as *I like this actor.*

4 Interpretation of Indirect Specifications (Relaxation)

Indirect specification, which is usually not an indirect speech act, is rather common in natural communication. For instance, it is quite normal to specify a goal not by its inherent parameters (which correspond in the SMARTKOM system to the slots of an object of the system's ontology), but by demanding that it obeys a certain relation to other objects or events. This leaves it to the addressee of the indirect (and usually vague) request to figure out appropriate actions for achieving the goal. We discuss indirect specifications that contain *vague* goals concerning database queries. These requests are basically handled by relaxation and occasionally by adding additional specifications which are provided by the user.

USER: Which crime movies are on TV this afternoon?
SYSTEM: *Shows the crime movies running this afternoon.*
USER: *(indirect request)* More of these!
SYSTEM: *System checks if there are any crime movies that afternoon on TV, which are not presented yet. Otherwise it relaxes the query specification*

In the example above, the user's indirect specification contains a reference to the set of results presented to the user beforehand. The specification adds up to extend this set. The user does not specify how this extension should be done, but there is a vague plea to find objects that are *similar* to the objects in the set.

We distinguish two basic reactions on the request. The first reaction only checks if there are any crime movies which are not presented yet. But there may be no more crime movies left. A more cooperative reaction is to extend the set by relaxing the specification. *Crime* is explicitly specified, therefore this genre is not a feature that should be relaxed. We use modification of time intervals as a default operation for achieving a modified result, which is similar to the objects that are already found.

There is another dimension, in respect to which *more* can be interpreted in two ways: as asking for a super set of the original set or as asking for additional items only. Taking into account the limited space for presentations, we take the second interpretation, by moving the time interval instead of extending it. [5]

The user may also combine an indirect request for relaxation with an additional direct specification:

USER: What's on TV this afternoon?
SYSTEM: *Shows a part of the afternoon program, containing one crime movie.*
USER: Show me more crime movies.
SYSTEM: *System checks if there are more crime movies that afternoon on TV, which are not presented yet. Otherwise it relaxes the query specification on the one hand and makes it more specific at the other hand.*

The dynamic help module constructs an new specification from the indirect (relational) request and the specification of the former request by using relaxation and

[5] This is a decision which is appropriate for SMARTKOM, but may be wrong for other systems in certain situations.

default unification [6] Query relaxation, as is described in the next section, is performed first. Default unification is used to add additional specifications provided by the user.

4.1 Query Relaxation

The concept of query relaxation is mainly explored in the context of relational databases and logic programming (Gaasterland et al., 1994). The basic operations are:

(i) rewriting a predicate with a more general predicate
(ii) rewriting a constant with a more general constant

We perform relaxation on constants, and also on sequences of constants and intervals, but not on predicates. Hence we distinguish three cases:

- atoms
- lists
- intervals

In SMARTKOM, predicates correspond to nonatomic types, constants correspond to atomic types. There is no hierarchy on constants (or atoms) in SMARTKOM. It is not possible to exchange atoms by more general atoms. Hence atoms may be deleted or exchanged in special cases. For lists, relaxation by deleting an element is an option. But experience with the list-valued features in the TV and cinema domain led us to choose the deletion or exchange of lists instead. The only intervals relevant for TV and cinema database queries are time intervals. As mentioned, intervals are moved rather than extended. This modification accounts for the *additional values* reading of more. This reading cannot achieved in general for atoms and lists.

5 Clarification of Underspecified Input

Semantic multimodal analysis in SMARTKOM delivers a set of alternative interpretations. The alternatives are either specified starting from the top of the ontology (*full* structures), or they are substructures that may be completed in many ways and may even be inconsistent. Discourse analysis extends these structures, but does not delete alternatives. In case of full structures, the intention recognition module selects one alternative as being the intention of the user. In case of substructures the intention module delivers a set of alternatives. Substructures may be completed in many ways, and sets of substructures may even be inconsistent. The main topic in this section is the handling of substructures.

Substructures are typically instantiation of types that are placed at a *low* level in the hierarchy. Substructures come either from incomplete parses, i.e., from partial

[6] Default unification as is used in the dynamic help module is different from the method used in the SMARTKOM discourse analysis and is described in Streit and Krieger (2004).

understanding, or from deliberately underspecified or highly elliptic user input. In the first case, misrecognition is a possible source of the partial understanding. The following utterances are examples of underspecified input:

(1) *Thrillers!*
(2) *Today evening!*
(3) *Heidelberg!*

In example 1 it is unclear if the user wants to get information about the TV program or if she is interested in cinema performances. In example 2 there is the same ambiguity.[7] In example 2 there is another ambiguity concerning the role of the substructure: It is unclear if the user refers to the begin time of an event or to the end time (or perhaps both). Similarly in example 3 it is unclear if, in the context of route planning, Heidelberg is the target or the start of the route. The role ambiguity in example 2 is solved by taking begin time as default. In example 3 the role ambiguity is considered as a topic of clarification dialogue.

In the first place, substructures are ambiguous with respect to the goal that the user wants to achieve by the help of the system. We understand *Goal* as it is used in the action-planning component. There the concept goal corresponds roughly to processes in the ontology (which is used for semantic representation). But a process may correspond to several finer-graded goals. Which goal comes into question depends on the specific substructures of the process.[8] For instance, an information search process concerning the TV program is assigned to a TV program browsing goal, while information search on cinema performances is assigned to a different goal. In most cases there are no role ambiguities or these are resolved by defaults. But in certain cases the role ambiguity is relevant (route planning).

5.1 Completion of Substructures

We want to generate the structures that are meaningful in terms of the goals that the user can pursue in the system. We want to achieve a minimal full structure that uniquely determines a goal. We do not add atomic values, except if this is necessary for determining the goal. The set of all full structures that contain the substructure has many members that represent no goal (in the sense of the action planning component). On the other hand, full structures that contain the substructure may be underspecified with respect to goals.

Technically, goals are related to ontological structures by paths of slots and types that must be contained in a structure to represent a goal. A path consists of slots and types; each slot is followed by the type of the fillers of the slot.[9] Though we

---

[7] In respect to the ontology there are many more ambiguities, but in respect to the possible applications these are the relevant ones.

[8] Alternatively, a process may be an abstract type, which has concrete subtypes that correspond to goals.

[9] This translates to feature structures if we take features instead of slots.

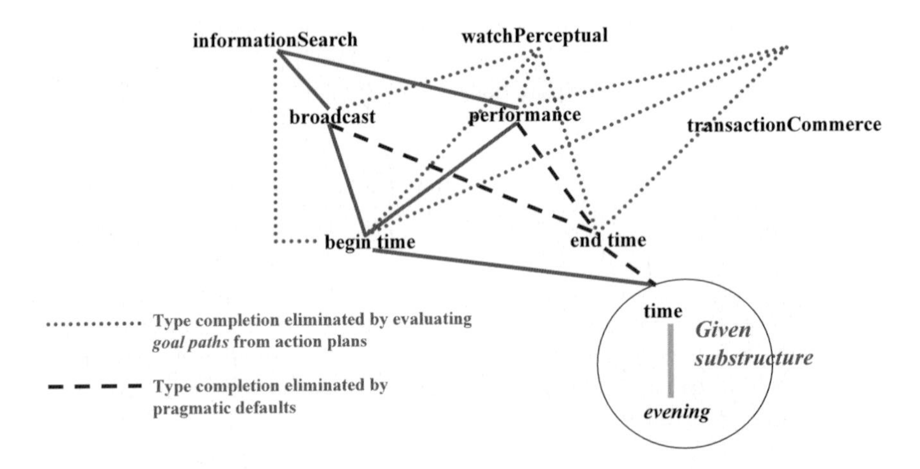

Relevant goal paths (simplified)

- **informationSearch/broadcast/**
- **informationSearch/performance/**

Alternative options, derived from the substructure (simplified)

- **informationSearch/broadcast/beginTime/evening**
- **informationSearch/performance/beginTime/evening**

Fig. 1. Ontology-based completion of substructures

distinguish here slots and types, they are not distinguished on the level of the internal representation of the ontology, due the internal definition of types by XML schemata.

Similarly, the parameters of a goal are characterized by paths. As a first step, we match parameter paths with substructures resulting in a parameter full structure by regarding defaults. Then we unify the minimal structure that is consistent with the goal path with the parameter structure.

In Fig. 1 we present a little cut-out of the ontology. The presented branches of the ontology are further simplified by omitting certain types and slots. Figure 1 shows how goal paths and defaults eliminate structures that are licensed by the ontology. If multiple substructures are present, we search for maximal consistent subsets of parameter structures and goal structures.

5.2 Clarification Subdialogue – Guided Interaction vs. Free User Initiative

Because underspecification is a possible indication for errors, we follow a cautious strategy[10] and first ask the user if the system understood correctly what she said, as is shown in Fig. 2.

[10] The dialogue strategy is determined in subdialogue plans of the dynamic help component and can easily be modified.

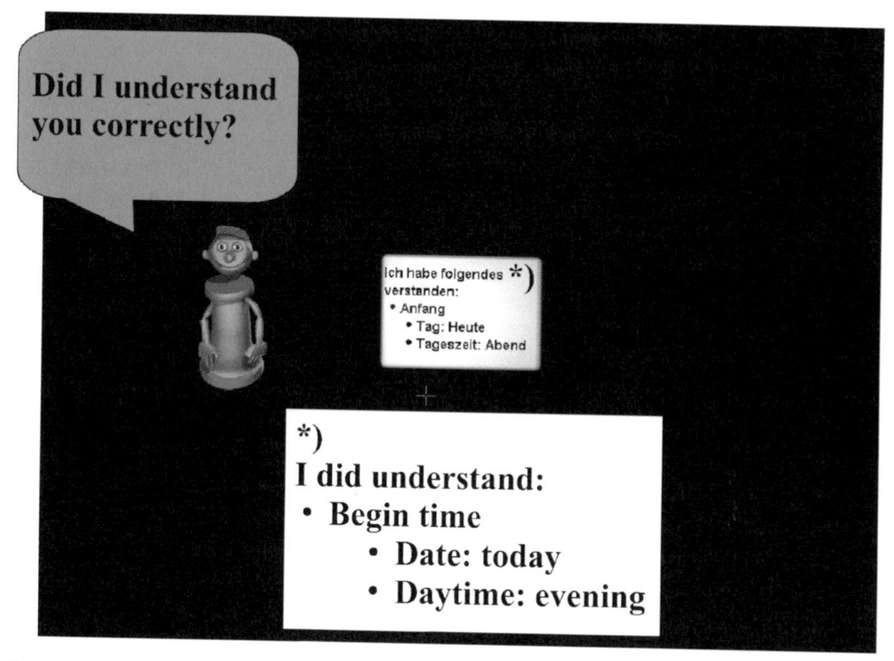

Fig. 2. First step of the substructure clarification dialogue

The user may react by stopping the subdialogue or by rephrasing her request. The user may correct the analyzed substructure by rephrasing. If there are more than one substructures she may select one by rephrasing it.

If the user has confirmed the analysis, the goals are presented that are compatible with the substructure (Fig. 3). In case of a role ambiguity we present alternative instantiations of the goal (Fig. 4).

We aim in performing the clarification subdialogue in a way that combines features of a guided dialogue paradigm with free user initiative. The user may:

(1) take the presented options as a menu to choose from
    (a) by pointing
    (b) by selecting the alternative by speech
(2) restate or modify her request
(3) express a new unrelated request
(4) stop the subdialogue

Stopping the subdialogue is processed by the dialogue navigation facility, as is presented later in Sect. 7. Alternative options are presented graphically by buttons. This is different from objects that are intended to be used by deictic pointing, which is normally accompanied by speech (Streit, 2001). The user may recognize from the graphic design that she may use them in a direct manipulative style.

Underspecification occurs if there is not sufficient context to solve the input fragment by discourse analysis. There is either no context at all, or discourse analysis has

Fig. 3. Second step — presenting goals that are compatible with the substructure

decided that the substructure does not fit to any salient context. In that, it is usually assumed that any goal that is compatible with the substructure is a reasonable option, provided that the user has confirmed that the system did understand her utterance correctly. There are some exceptions to this rule. If the system has requested an answer of the user, dynamic help reacts with a repetition of the request, rather than with an analysis of the under-specified substructure. Also, the coverage of the scenarios is considered. In the mobile system only substructures that fit to functionalities in this scenario are considered.

5.3 Processing of Clarification Subdialogue

The dynamic help component provides full semantic representations of the alternative, including the parameters that are due to the substructures. These representations are published as expectations, to allow the analysis of spoken input by matching it with these representations. If the user chooses an alternative, she will probably not specify the whole content of the alternative. She will instead provide elliptic information that is suited to select an alternative. The representations of the alternatives are further directly assigned to the graphical presentations. This is more convenient to use than expectations. It enables the multimodal fusion component to provide a semantic analysis directly, which unburdens discourse analysis.

5.4 Problems and Coverage

Help on underspecification is realized for the home scenario and the mobile scenario. Since the public and the home scenarios are often presented together, the provision of help then appears incomplete. The presentation of alternatives in the mobile scenario shows how the substructures are integrated in the goal parameters of that goal (Fig. 4). This is neglected in the home scenario, which makes the alternatives less understandable. If many different substructures are recognized, it would be valuable if the user could choose alternatives, not only on the goal level, but also on the level of substructure understanding by pointing.

6 Help on Demand

There are three basic types of information that should be provided by help on demand. These basic types are:

- goals to achieve
- slots or roles that further specify the goal
- values that serve as fillers for slots

We focused on the provision of goal information in the mobile scenario and on value information in the home scenario, neglecting the information about slots.

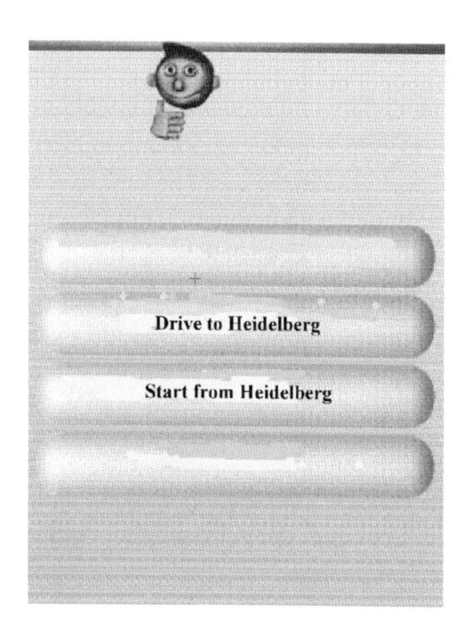

Fig. 4. Alternative readings of Heidelberg

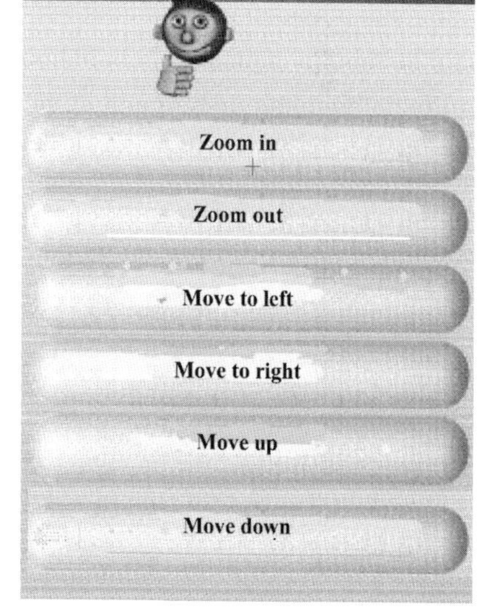

Fig. 5. Goals for navigating a map

6.1 Presentation of Goals

In SMARTKOM, the user can change her intention completely in almost any situation. In this respect, there is no context that strongly calls for presenting specific options, if the user asks nonspecifically for help. But there are certain goals that can always be performed as route planning, and others that can be performed only in a certain context, e.g., manipulating a map is only possible if there is a map to be manipulated. We call the first class of goals unconditioned goals, and the second class (pre)conditioned goals.

In the mobile scenario we realized the following strategy for the presentation of goal options: Conditioned goals are only presented if the preconditions are satisfied. In this situation conditioned goals are preferred. Unconditioned goals are not excluded, but are presented later, e.g., if the user repeats asking what she can do. Figure 5 shows the options for navigating in a map.

The conditioned vs. unconditioned classification is not a factual one, but is a classification in respect to the abilities of the main strand processing. For example, requesting a telephone call by specifying the name of the callee instead of her number needs access authorization to an address book. In this case the main strand cares about authorization and telephone number retrieval. If the user asks for reservation seats for a performance, standard processing does not care about identifying a unique performance, but it cares about identifying definite seats, by providing a seating plan.

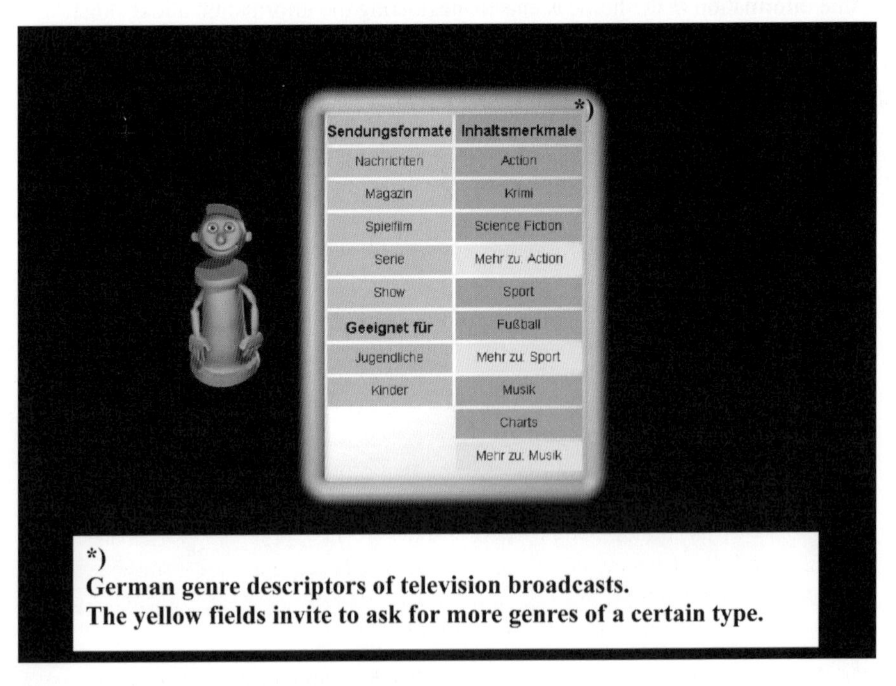

Fig. 6. Presentation of options for genres

6.2 Presentation of Values

Values are shown on explicit demand. Figure 6 shows the presentation of values for the role or type genre.[11] If the user asks nonspecifically for genres, the system provides genres for movies in a cinema program context and genres for TV broadcasts in a television context.

7 Dialogue Model for Navigation in Multimodal Dialogue

In a multimodal dialogue system, navigating is different in some respects from navigating in a pure speech dialogue system. If a subdialogue is finished in speech dialogue system, there is no artifact left. But in a multimodal system subdialogues leave artifacts on the screen. If the user wants to finish the subdialogue, these artifacts have to be removed. This is especially true if the user wants to finish the subdialogue not by proceeding with another topic, but, e.g., because she wants to see again some relevant information that was present before the subdialogue. It is obvious that restoring the presentations that belong to the state to which the user wants to return cannot be done by just restoring the state that precedes the current state. The subdialogue may consist of several successive presentations, the subdialogue may be preceded by another subdialogue, and subdialogues may be nested. Therefore a dialogue model is maintained that represents the subordination of dialogue parts. It was necessary to generate this model in a very short time. A shallow but robust model, which works very successfully, has been achieved by basically categorizing presentation acts into four classes: main contribution, information subdialogue contribution, help information, and clarification, and by imposing a subordination relation to these classes.

8 Generation of Error Messages

The generation of an error message is the default reaction to input that is not processed otherwise. In an earlier version of the system we distinguished error messages with respect to the stage of processing, where the error occured. For example, errors that occurred on the recognition level have been prompted by *I could not understand you acoustically*; errors on later levels of analysis have been prompted by *I could not understand you*. If the goal of the user had been analyzed successfully, but the actions to achieve this goal could not be identified or performed, this has also been commented upon. Also, the possibility that a result could not be presented has been considered. The stages where the error occurred are identified by the system's *watchdog*, which monitors the processing states of the module and infers an overall system state from the states of the single modules. Later, these error messages have been unified to the generic answer *I did not understand you*, except for presentation failures, which are commented by *an error has occurred*.

[11] The distinction between type (as filler of a slot) and slot (which is filled by items of a certain type) is not clearly marked in the internal representation of the ontology, but it also is not distinguished by the user as long as the relation between role and type is one-to-one.

9 Conclusion

Problematic, indirect, attentive nonstandard input has been explored by considering an extraordinarily broad range of phenomena. For certain problems, such as underspecified input, general solutions are developed. For some aspects, such as handling emotions, attentive user states, or indirect specifications, principled approaches are developed and implemented for a restricted range of phenomena. For other topics, such as navigation in multimodal dialogue, shallow but robust solutions are developed. One conclusion of the work is that architectural issues are relevant and may facilitate or impede a principled general approach for error recovery.

References

T. Gaasterland, P. Godfrey, and J. Minker. An Overview of Cooperative Answering. In: *Nonstandard Queries and Nonstandard Answers*, pp. 1–40, Oxford, UK, 1994. Clarendon Press.

G. Hirst, S. McRoy, P. Heeman, P. Edmons, and D. Horton. Repairing Conversational Misunderstandings and Non-Understandings. Technical report, University of Maryland, 1994.

A. Ortony, G.L. Clore, and A. Collins (eds.). *The Cognitive Structue of Emotions*. Cambridge University Press, Cambridge, UK, 1988.

K. Purang, D. Traum, D.V. Purushothaman, W. Chong, Y. Okamato, and D. Perlis. Meta-Reasoning for Intelligent Dialog Repair. Technical report, University of Maryland, 2000.

M. Streit. Why Are Multmodal Systems so Difficult To Build? About the Difference Between Ceictic Gestures and Direct Manipulation. In: *Cooperative Multimodal Communication*, Berlin Heidelberg New York, 2001. Springer.

M. Streit, A. Batliner, and T. Portele. Cognitive-Model-Based Interpretation of Emotions in a Multi-Modal Dialog System. In: *Proc. ESCA Tutorial and Research Workshop on Affective Dialog Systems*, Irsee, Germany, 2004.

M. Streit, A. Batliner, and T. Portele. Emotion Analysis and Emotion Handling Subdialogs, 2006. In this volume.

M. Streit and H.U. Krieger. Ellipsis Resolution by Controlled Default Unification for Multi-Modal and Speech Dialog Systems. In: *Proc. ACL Workshop on Reference Resolution and Its Applications*, Barcelona, Spain, 2004.

D. Traum. *A Computational Theory of Grounding in Natural Language Conversation*. PhD thesis, University Rochester, 1994.

D.R. Traum, C.F. Andersen, W. Chong, D. Josyula, Y. Okamoto, K. Purang, M. O'Donovan-Anderson, and D. Perlis. Representations of Dialogue State for Domain and Task Independent Meta-Dialogue. In: *Proc. IJCAI'99 Workshop on Knowledge And Reasoning In Practical Dialogue Systems*, 1999.

Part IV

Multimodal Output Generation

Realizing Complex User Wishes with a Function Planning Module

Sunna Torge* and Christian Hying**

Sony International (Europe) GmbH
Sony Corporate Laboratories Europe, Advanced Software Laboratory, Stuttgart, Germany

Summary. Recently, spoken dialogue systems became much more sophisticated and allow for rather complex dialogues. Moreover, the functionality of devices, applications, and services and the amount of digital content increased. Due to the network age (Internet, personal area networks, and so on) the applications one can control or wants to control change dynamically, as does the content. Therefore a dialogue system can no longer be manually designed for one application. Instead, a layer in between is required, which translates the functionalities of the devices such that they can be used via a generic dialogue system and which also resolves complex user wishes according to available applications. In this paper a module, called the function planning module, is described which builds a link between a dialogue system and an ensemble of applications. A planning component searches for the applications necessary to solve a user wish and the sequence of action which has to be performed. A formal description of the function planning module and its successful integration in the SmartKom prototype system are presented in this paper as well.

1 Introduction

Although human–machine interactions are ubiquitous, they are not always a very pleasing experience. An often-cited example is how inconvenient and error-prone it is to program a video cassette recorder (VCR). Another example is that currently users have a lot of trouble when they want to connect Bluetooth devices. The main problem with most of the user interfaces is that the user has to think in terms of devices and services (play, record). Instead, the user should be able to interact with a system in a natural way, as he would do with a human assistant. To give an example, the user might put a document on the table and simply say, "Send this to Mr. Green." In consequence, the system should deal with the devices and services for him. It should scan the paper with a document camera, find out how to reach Mr. Green by consulting an appropriate address book, and transmit the image either to an e-mail

* S. Torge is now with the Institute for Computer Science, Universität Freiburg, Germany; torge@informatik.uni-freiburg.de

** C. Hying is now with IMS, Institut für Maschinelle Sprachverarbeitung, Universität Stuttgart, Germany; christian.hying@ims.uni-stuttgart.de

client or a fax modem, possibly doing format conversion beforehand, etc. The user in this case has a complex wish involving several devices and several actions per device. He simply wants his wish to be solved; he does not want to care about the individual devices or the action sequences to be carried out. Similar problems arise in the case of copying videos or combining EPG access and personal calendars. Note that these tasks can only be solved by a system providing a central human–machine interface operating several devices.

Given a home network or personal area network, the problem we addressed within the SMARTKOM project is to enable a flexible and intuitive control of an ensemble of devices and services with a single intuitive human–machine interface. We focused on

- how complex user wishes can be served, and
- how plug and play can be realized on the side of the dialogue system.

In order to solve these problems different aspects need to be considered:

A traditional dialogue system used for the control of devices usually consists of an input understanding part, a dialogue manager, and the devices that are to be controlled. The simplest way to control the devices is to have a unique mapping of the user input to the appropriate control command. Given, e.g., a speech input "CD play", it can be uniquely mapped to the "play"*command* of a CD player (Rapp et al., 2000). However, this approach does not allow us to serve complex wishes like the above-mentioned examples, since the user input for complex wishes cannot be uniquely mapped to a single control command.

In order to gain more flexibility with respect to adding new applications, it is necessary to separate domain-dependent knowledge from domain-independent knowledge (Flycht-Eriksson, 2000; Thompson and Bliss, 2000; Han and Wang, 2000; Lemon et al., 2001).

Separation of domain-dependent and domain -independent knowledge can be obtained by introducing an additional module that serves as an interface between the dialogue manager and the devices to be controlled, including models of the devices (Pouteau and Arevalo, 1998). However, to serve complex user wishes it is not enough to consider models of single devices. Instead, some sort of reasoning on the provided services is necessary (Torge et al., 2002b,a; Heider and Kirste, 2002).

In Larsson et al. (2001), the use of precalculated plans for complex tasks is described. With this approach, however, the user is limited to those tasks where precalculated plans are available.

Concerning plug and play, there are existing physical solutions like Bluetooth and IEEE 1394 (e.g., iLINK, Firewire, see Hoffman and Moore (1995)), which offer plug and play on a hardware level. This has also to be reflected in the human–machine interface since new devices have to be controlled through the human–machine interface.

In order to overcome these sorts of problems within the SMARTKOM project, a multimodal dialogue system is enhanced with a planning component, called the function planning module. The function planning module consists mainly of a set of abstract models describing the functionalities of devices and services available in the network and a reasoning component. Using these abstract models the human–machine interface can infer how to control the available services and devices. Henceforth, we use the term "devices" instead of "devices and services" for reasons of brevity. This article is organized as follows: First, we give an overview about the function planning module, which is followed by a detailed description of the single parts of the function planning module and a formal description of the concept. The next part of the article focuses on the realization of the function planning module within the SMARTKOM prototype system. We conclude with a section about future work.

2 The Function Planning Module

2.1 Overview

As described above, our focus is on serving complex user requests which may involve several devices. In order to handle requests like "Please record the film XYZ on Saturday," the system needs to find out

- *which* devices are necessary to serve the request
- *how* to control the devices

For this purpose a planning component and appropriate data structures for planning are needed.

We introduce a new module, the so-called function planning module, in order to solve the above described problems. The purpose of the function planning module is to allow the integration of the functionalities of several different devices and to support plug and play. In a traditional dialogue system (von Kuppevelt et al., 2000) the devices to be controlled by the system are controlled by the dialogue manager (Fig. 1). In contrast, the enhanced system (Fig. 2) provides a sort of intelligent interface between the dialogue manager and the devices to be controlled, namely the function planning module. The function planning module consists of

- an abstract model of the functionalities of a device (the so-called functional model) for each device in the network
- a reasoning component
- a plan processor

Furthermore, the function planning module is always aware of the current state of each device in the network.

Instead of controlling the devices directly from the dialogue manager, this allows us to formulate the requests given by the dialogue manager on an abstract level. Based on this abstract request, the function planning module first calculates a plan to serve the request and then performs the plan.

The dialogue manager is therefore independent of the real devices and is robust against changes of devices. The overall functionality of the given devices does not

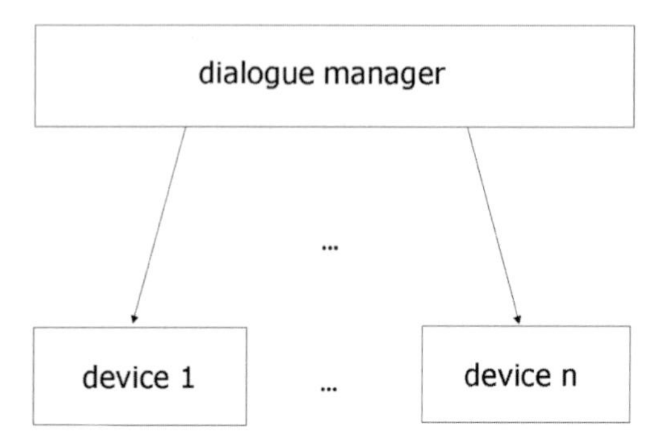

Fig. 1. Traditional dialogue system

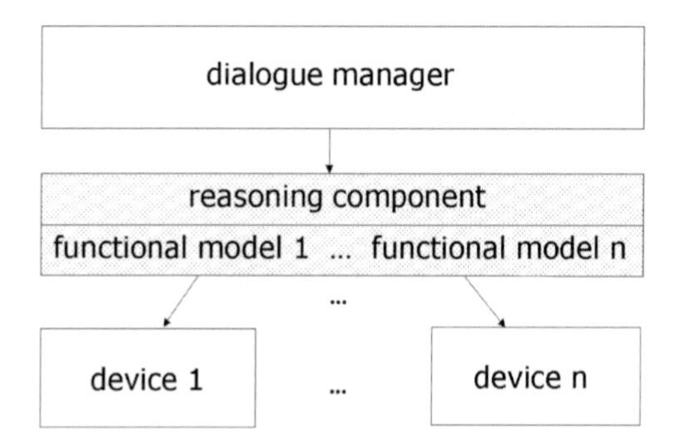

Fig. 2. Dialogue system with function planning module

need to be known in the dialogue manager, but it is deduced from the functional models of the given devices. This approach enables the system to serve complex user requests, and to be flexible and robust against changes in the network of controlled devices, i.e., it supports plug and play. In the following, we give a detailed description of the different parts of the function planning module, i.e., the functional model, the reasoning component, and the plan processor.

2.2 The Functional Model

The functional model is a two-layered model. It consists of an external model and an internal model, where the external model describes the knowledge necessary to find appropriate devices and the internal model describes the single device in more detail. The external and internal models are connected; however, they are used separately. This approach was chosen in order to reduce the search space.

2.2.1 External Model

The external model of a device models the knowledge that is necessary but also sufficient to decide whether a device is appropriate to serve a given request or not. Furthermore, the external models of the devices allow the system to draw conclusions on which devices are to be combined in order to fulfill a request. The external model of a device describes the input data and the output data of the device. It is modeled as a binary relation. Each external model of a device corresponds to a functionality of the device. Note that a single functional model may consist of several external models, since one device may provide several functionalities. As an example consider, a VCR, which provides the functionalities "recording" and "playback." The external models corresponding to "recording" and "playback" are depicted in Figs. 3 and 4, respectively. However, a VCR as a single device is not capable of recording a film. A second device, e.g., a tuner, is necessary in order to provide the data to be recorded. That is, the user wish "record the film" is a complex wish, such that the system needs to combine two devices, namely a tuner and a VCR, in order to fulfill the wish. This information is encoded in the external models of the tuner and the VCR.

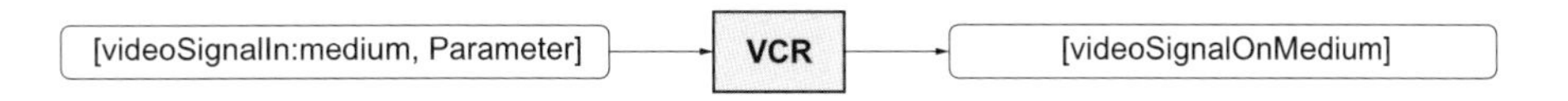

Fig. 3. External model of the VCR describing "recording"

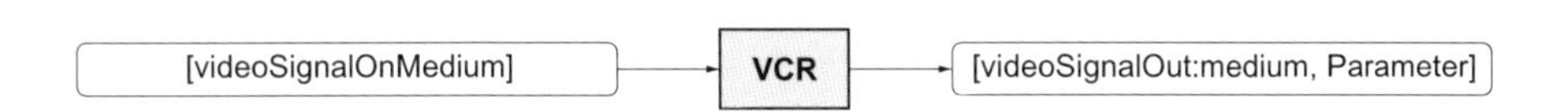

Fig. 4. External model of the VCR describing "playing"

2.2.2 Internal Model

The internal model of a device describes the possible states of the device and the actions that are necessary to bring the device from one state into another. Regarding the knowledge to be modeled, i.e., states and actions, finite state machines (FSM) are an appropriate means to build up an internal model.

The states in the internal model are partly annotated with incoming or outgoing data, respectively. This is an important point since these annotations ensure the connection between external and internal models of a device. In addition, the states also might be annotated with pre- or postconditions of each state. The internal model of a VCR is depicted in Fig. 5.

2.2.3 Formal Description of the Functional Model

The functional model can be formally described as follows:

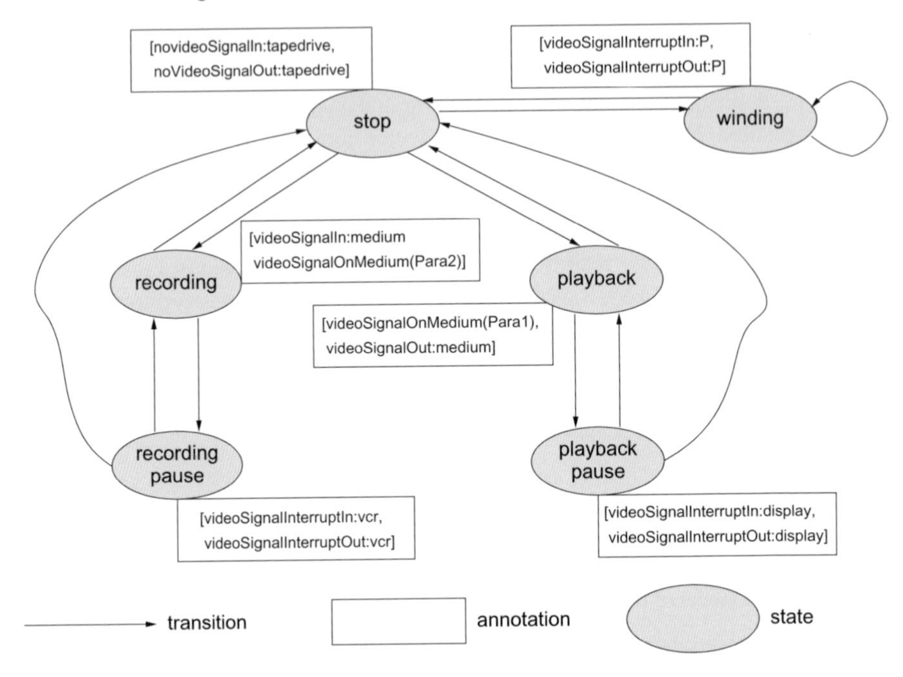

Fig. 5. Internal model of the VCR

Definition 1. (Functional Model)

A functional model $\mathcal{F}$ is a tuple $\mathcal{F} = (\mathcal{D}, \mathcal{S}, \mathcal{T}, i, o, pre, post)$ where $\mathcal{D}$ defines the external models and $\mathcal{S}, \mathcal{T}, i, o, pre, post$ define the internal model.

- *$\mathcal{D}$ is a finite set of* external models *with*

$$\mathcal{D} = \{(d^i_j, d^o_j) \mid d^i_j \in \mathcal{D}^i,\ d^o_j \in \mathcal{D}^o\}, \mathcal{D} \subseteq \mathcal{D}^i \times \mathcal{D}^o$$

.

- *$\mathcal{S}$ is a finite set of* states.
- *$\mathcal{T} : \mathcal{S} \times \mathcal{A} \rightarrow \mathcal{S}$ is a* transition function *and $\mathcal{A}$ a set of* actions *(including the empty action).*
- *$i, o, pre, post$ are* functions, *which informally spoken map additional information to a state $s \in \mathcal{S}$. They are defined as follows, where ε is the empty element:*

$$i : \mathcal{S} \rightarrow \mathcal{D}^i \cup \{\varepsilon\}$$
$$i : s \mapsto d^i$$
$$o : \mathcal{S} \rightarrow \mathcal{D}^o \cup \{\varepsilon\}$$
$$o : s \mapsto d^o$$

Let $\mathbf{F} = \{\mathcal{F}_1, ..., \mathcal{F}_n\}$ be a finite set of functional models with $\mathcal{F}_i = (\mathcal{D}_i, \mathcal{S}_i, \mathcal{T}_i, i_i, o_i, pre_i, post_i)$, $i = 1, ..., n$ and $\mathcal{F} \in \mathbf{F}$, $\mathcal{F} = (\mathcal{D}, \mathcal{S}, \mathcal{T}, i, o, pre, post)$.

Let $\mathbf{D^i} = \bigcup_{\mathbf{j=1}}^{\mathbf{n}} \mathcal{D}_j^i$ *and* $\mathbf{D^o} = \bigcup_{\mathbf{j=1}}^{\mathbf{n}} \mathcal{D}_j^o$*. Then pre and post are defined as*

$$pre : \mathcal{S} \rightarrow \mathcal{P}(\mathbf{D^i} \times \mathbf{D^o}) \cup \{\varepsilon\} \times \{\varepsilon\}$$
$$pre : s \mapsto \{(d_j^i, d_j^o) \mid j = 1, ..., l, d_j^i \in \mathbf{D}^i, d_j^o \in \mathbf{D}^o\}$$

$$post : \mathcal{S} \rightarrow \mathcal{P}(\mathbf{D^i} \times \mathbf{D^o}) \cup \{\varepsilon\} \times \{\varepsilon\}$$
$$post : s \mapsto \{(d_j^i, d_j^o) \mid j = 1, ..., k, d_j^i \in \mathbf{D}^i, d_j^o \in \mathbf{D}^o\}$$

Note that the definitions of i and o only refer to the external models which are part of the functional model to be defined, whereas *pre* and *post* refer to the external models of all other functional models under consideration. Note furthermore that the definition of pre- and postconditions, namely *pre* and *post*, corresponds to the definition of the external models.

2.3 The Reasoning Component

Each of the devices of the given network is modeled by its functional model. In order to use this knowledge, the function planning module includes a reasoning component. This is necessary since, given a complex user request and given a network of devices, the following steps need to be executed in order to serve the user request:

1. Find out the device/devices which are necessary to perform the request.
2. Find out how to combine and to control the devices, i.e., generate an appropriate plan of actions to be executed.

Using the functional model of each device in the given network, the reasoning component first searches for the necessary devices and then generates a plan of actions for every device that is involved.

The functional models allows us to describe how devices can to be combined. This is done via the input and the output data of each device. Furthermore, it is possible to describe a temporal order in which the devices are to be controlled. This is done by formulating pre- and postconditions of single states of a device. Let us have a closer look to the algorithms.

In a first step, a complex task T is mapped to a tuple $(d_T^i, d_T^o) \in \mathbf{D^i} \times \mathbf{D^o}$ (see Definition 1). Let $\bigcup_{j=1}^{n} \mathcal{D}_j$ be the set of all external models which are currently available. Then, using the representation (d_T^i, d_T^o) of the task, the device search algorithm returns a set of external models $\mathcal{D}^T$ of the necessary devices:

$$\mathcal{D}^T \subset \bigcup_{j=1}^{n} \mathcal{D}_j, \mathcal{D}^T = \{(d_T^i, d_1^o), ..., (d_j^i, d_T^o)\}.$$

There are two cases to distinguish:

Case 1: $|\mathcal{D}^T| = 1$, i.e., one single device is returned and planning starts and ends with this single device.

Given $\mathcal{D}^T$ and the internal model of the device, during the next step the initial and the final states of the device are searched for. This search is based on the annotation functions i and o, which establish the connection between external models and the appropriate states of the internal model. As mentioned before, the functional model also includes the knowledge of the current state of each device. Therefore at that point the system knows about the current, the initial, and the final state to be reached by the device.

Thus, the next step is to search for a path from the current to the initial to the final state in the internal model of the device. Since the transitions of the internal model of a device are annotated with actions, a sequence of actions, i.e., a plan, is returned.

While searching for a path (i.e., a plan), pre- and postconditions of the states are checked. Informally speaking, the pre- and postconditions model the necessity of using further devices and how to use them. Since pre- and postconditions are formulated like the external models (see Definition 1), the same planning algorithm as described here is applied to fulfill them, which returns additional plans. These additional plans, corresponding to other devices that are necessary to fulfill the given complex task, finally are integrated in the overall plan.

Case 2: $|\mathcal{D}^T| > 1$, i.e., an ordered set of external models is returned. Each model uniquely corresponds to one device. And for each device the planning algorithm is performed as will be described for case 1.

It should be pointed out that the concept described above allows two possibilities to model the necessity of using several devices. First, it can be modeled by mapping the complex task to $(d^i_T, d^o_T) \in \mathbf{D^i} \times \mathbf{D^o}$, such that the device search algorithm returns an ordered set of external models (and also devices) as described above. Second, the concept of pre- and postconditions allows us to model a more flexible temporal order.

2.4 The Plan Processor

As described above, the reasoning component generates a plan, which consists of a sequence of actions. In order to fulfill a requested user wish, the generated plan needs to be executed. This is done by the plan processor, as we discuss next.

3 Realization in the Prototype

The function planning module was realized in the SmartKom prototype system (Wahlster et al., 2001). It provides an intelligent interface between the dialogue manager (called the action planning module, see Löckelt (2006)) and the application interface of the devices (Michelitsch et al., 2000).[1] For the current realiza-

[1] The application interfaces for the devices used in the SmartKom-Home scenario (Portele et al., 2006) were realized in the Embassi project, [2], and adapted to the SmartKom prototype.

tion these devices are loudspeakers, microphone, VCR, television, document camera, telephone, telefax, biometric authentication, address book, calendar, e-mail, car navigation, pedestrian navigation, and internal scheduler. According to the concept described above, the implementation of the planning module consists of: (i) the functional model (i.e., the external and the internal model) for each device, which contain information about the basic functionalities of the devices and about the commands which are required to invoke them; (ii) a representation of the current state of each device; (iii) a reasoning component, which involves a device search algorithm and a planning algorithm; and (iv) a plan processor, which executes the generated plans.

In the following sections the processing flow is first described. Second, the plan language, which is used in the functional models, is explained in detail. Finally, the main features of the planning algorithm and the plan execution are presented.

3.1 Processing Flow

As already mentioned, the function planning module provides an intelligent interface between the action planning module and the devices. In the following, the processing flow of requests issued by the action planning module will be described.

The action planning module requests an abstract and complex user wish to be fulfilled by the function planning module. The planner of the function planning module generates a plan to serve the request. It starts off from the current state of the devices and finds a path through the functional models that yields a configuration which provides the requested functionality. The path determines a sequence of commands which define the actual plan. Its execution is controlled by the plan processor.

The interaction between the function planning module and the devices is based on a command and response mechanism, where the initiative always lies by the function planning module. As a consequence, devices can only change their state according to the functional model, if they can relate the change to a command issued by the function planning module. Note that this does not mean that there is a one-to-one correlation between commands and state changes. A single command can also trigger a sequence of changes. Changes of states are reported to the action planning module only if there are according definitions in the functional models.

Because of the abstraction from specific devices, it is hidden from the action planning module which devices are involved in serving a particular request and how they are controlled. Nevertheless, we can distinguish between three complexity classes of plans within the function planning module. Many requests can be served by just sending one command to one device, e.g., looking up an entry in the address book. Some requests can be served by one device, but more than one command is needed. This is the case for biometric authentication: If an authentication method for a particular user is requested, the function planning module first checks whether that method is available for that user. If the test has been positive, it sends the command to initialize that authentication method. Otherwise the whole request is rejected.

The most complex treatment is required by those requests that are served by a couple of devices that need to interact in a well-defined way. Such a complex request is, for example, to program a VCR. It involves the internal scheduler, the calendar application, a tuner, and the VCR. The programming request contains a time interval specifying when the recording procedure should be activated and information about the channel to be recorded. The function planning module checks by requesting the calendar application whether the VCR is programmed for the given time interval. If not, the request can be served. The scheduler is instructed to activate the recording procedure at the requested time interval with the given parameters. In order to carry out the recording procedure, the function planning module looks for an appropriate TV signal tuner, switches it to the specified channel, and then instructs the VCR to record the signal from the tuner.

3.2 Plan Language

The plan language is used to annotate the transitions of the internal model of a device. These annotations model the actions which have to be carried out successfully to pass the transition. The plan language consists of the following statements:

- action
- wait
- repeat_while_pending
- expect
- goal
- reached

The statement**action** models the actions to be taken in order to fulfill a complex wish. **wait** models the necessity of feedback of a device. **expect** is comparable with **wait**, but depending on the feedback of a device, the performance of the calculated plan will be performed, aborted, or alternative plans will be generated and performed. **repeat_while_pending** simulates a self-transition within a FSM, i.e., **repeat_while_-pending** is a loop of the very same sequence of actions. It causes a repetition of the specified actions as many times as the application involved informs the function planning module that it is still working on the current request, i.e., the status of the request is pending. While processing a multiple response the current state of the device is the one before the transition annotated with a **repeat_while_pending** statement. The **repeat_while_pending** statement was introduced since the SMARTKOM system needs to handle requests with an *unknown* number of responses (e.g., navigation service). Multiple responses with a *fixed* number of responses can be modeled with a fixed number of **wait** statements. **goal** and **reached** are used for comparing the current state of the device while performing a plan and the state based on which the plan was calculated. This is necessary since it might happen that some devices change their states independently during the performance of a plan. If so, the calculated plan must be updated (plan repair).

A generated plan consists of a sequence of statements in the plan language. The length of a plan is defined as the number of statements in the plan. The length of a plan is always a finite natural number.

3.3 Planning and Plan Processing

The planning algorithm and the plan processor implemented in the SMARTKOM system allow conditional planning, parallel processing, scheduling a task, and interrupting the performance of a task. These features are described in this section.

The planning algorithm comprises a depth-first search. It allows conditional/heuristic planning by using the **expect** statement of the plan language. A generated plan will be performed as long as the feedback of a device is the one as expected. Otherwise the calculated plan will not be performed any longer. Instead an alternative plan will be performed. This alternative plan either already exists, i.e., is precalculated, or will be generated based on the current states of the system and devices that are involved (plan repair). As a simple example, consider a biometrics application. The generated plan is based on the expectation that the authentication is successful. This is modeled via an expect statement. If it is not successful, the generated plan is aborted, and a precalculated plan is performed in order to inform the user about the rejection.

The SMARTKOM system provides some services that last over a certain time period like, e.g., the navigation service. Using such a service, the user should still be able to request other services from the SMARTKOM system. In order to do that, the function planning module has the capability to serve different tasks in parallel. Given a task, the reasoning component calculates a plan to serve the task. If at some point the performance of the plan depends on some feedback of a device, the calculated plan is stored as long as the system is waiting for feedback. While waiting for feedback the system is able to serve other user wishes, i.e., the reasoning component calculates another plan and the plan processor is executing the calculated plan. In this case the plan processor is performing two plans in parallel. As an example, consider a tour guide application, which provides information to the user during some time. While still using the tour guide, the user wants to call a friend, which is another task performed by the system. Without interrupting the tour guide, the user is able to call and continue using the tour guide after finishing the telephone call.

Another feature of the function planning module that is realized in the SmartKom system is the possibility to serve a task at a certain time, i.e., scheduling a task. For this purpose the function planning module contains a scheduler, which is regarded as an additional virtual device. This implies that the scheduler is modeled with a functional model, consisting of an external and an internal model, exactly like all the other devices and applications controlled by the system. Based on this concept, scheduling a task is modeled as a functionality that arises by using one or several devices together with the scheduler. As an example, consider the task "programming a VCR." The functionality "recording" is combined with the functionality "scheduling a task X." The process is as follows:The device scheduler causes sending the request "record(Parameter)" to the function planning module at time when the recording should start. As usual, the function planning module then generates a plan, which is performed in order to fulfill this task.

Plans can be interrupted before they have been executed completely. The possibility of interruption has to be defined explicitly in the functional models by introducing extra states and transitions. Those extra transitions bring the according device into the original state. (By *orginal state* we mean the state which the device was in before the execution of the plan started.) Furthermore, they define actions that take the rest of the plan from the agenda, and if necessary, also bring other devices into their original state.

4 Future Work

Future work includes several topics. Planning and performing a plan within the function planning module could be refined and enhanced with additional features. The existing concept of the function planning module supports plug and play. However, there is still some work to do in order to realize plug and play in an existing system. A more theoretical research topic is to consider complexity. In the following, future work will be described in more detail.

Planning and performing a plan are implemented as two subsequent procedures. Making them interactive with each other would enhance the functionality of the function planning module. If a device reported a change of state, the plan processor should check the consistency of all plans with respect to that updated state. If necessary, an inconsistent plan should be substituted by a new plan calculated by the reasoning component. As an effect, the devices would gain the possibility to change their state autonomously. Furthermore, this change would enable self-transitions in the functional model.

The interruption of plans is realized by extra states and transitions in the functional model. A more general concept is desirable, such that a plan can be interrupted and discarded at any point and that its effects are taken back automatically, so that the original states of the devices are restored.

In a complex network of devices and applications, a given user request might be accomplishable in several ways. First of all, the function planning module could generate alternative plans representing alternative solutions. These solutions either could be presented to the user, such that the user could choose the solution he likes the best. Or, alternatively, the actions that are the components of the generated plans could be annotated with costs. Depending on the calculated costs of a generated plan, the system itself decides on the most appropriate plan to be executed.

When designing the function planning module, one of the goals was to support plug and play. This caused a strict separation between the functional models of devices and any reasoning component, such that any device can be enhanced with its functional model and plugged into the network. However, in order to support plug and play within a network of devices controlled with a multimodal dialogue system, the systems not only needs to know about functionalities of a given device but also, e.g., about words and pronunciations the user can use in order to ask for additional functionalities. Therefore additional information needs to be added to the functional models of the devices.

Another issue is to consider complexity. In Domshlak and Dinitz (2001) complexity results are shown for a multiagent system. The models used in that work are comparable to the functional model. Thus the complexity results may also apply to the work described in this paper.

Acknowledgments

This research was conducted within the SMARTKOM project and was partly funded by the German Federal Ministry of Education and Research under Grant 01 IL 905 I7. The authors would like to thank their colleagues within Sony, Stefan Rapp, Martin Emele, and Ralf Kompe, for fruitful discussions about the ongoing work, helpful hints concerning the implementation, and helpful comments on earlier drafts of this paper. They would also like to thank their colleagues within the SMARTKOM project, Horst Rapp, Kerstin Reichel, and Victor Tabere (MID Dresden), and Axel Horndasch (Sympalog Erlangen) for the pleasant and successful collaboration.

References

C. Domshlak and Y. Dinitz. Multi-Agent Off-Line Coordination: Structure and Complexity. In: *Proc. 6th European Conference on Planning*, 2001.

A. Flycht-Eriksson. A Domain Knowledge Manager for Dialogue Systems. In: *Proc. 14th ECAI*, Berlin, Germany, 2000.

J. Han and Y. Wang. Dialogue Management Based on a Hierarchical Task Structure. In: *Proc. ICSLP-2000*, Beijing, China, 2000.

T. Heider and T. Kirste. Supporting Goal-Based Interaction With Dynamic Intelligent Environments. In: *Proc. 15th ECAI*, Lyon, France, 2002.

G. Hoffman and D. Moore. IEEE 1394: A Ubiquitous Bus. In: *Proc. COMPCON*, San Francisco, CA, 1995.

S. Larsson, R. Cooper, and S. Ericsson. menu2dialog. In: *Proc. Workshop Knowledge and Reasoning in Practical Dialogue Systems at IJCAI*, Seattle, WA, 2001.

O. Lemon et al. The WITAS Multi-Modal Dialogue System 1. In: *Proc. EUROSPEECH-01*, Aalborg, Denmark, 2001.

M. Löckelt. Plan-Based Dialogue Management for Multiple Cooperating Applications, 2006. In this volume.

G. Michelitsch, C. Settele, P. Hilt, and S. Torge, S. und Rapp. EMBASSI: Overview of the Architecture for a Multimodal User Interface for Future Home Entertainment Systems. In: *Proc. 10th Sony Research Forum*, pp. 156–161, Tokyo, Japan, 2000.

T. Portele, S. Goronzy, M. Emele, A. Kellner, S. Torge, and J. te Vrugt. SmartKom-Home: The Interface to Home Entertainment, 2006. In this volume.

X. Pouteau and L. Arevalo. Robust Spoken Dialogue System for Consumer Products: A Concrete Application. In: *Proc. Int. Conf. on Speech and Language Processing*, Sydney, Australia, 1998.

S. Rapp, S. Torge, S. Goronzy, and R. Kompe. Dynamic Speech Interfaces. In: *Proc. Workshop Artificial Intelligence in Mobile Systems at the 14th ECAI (AIMS 2000)*, Berlin, Germany, 2000.

W. Thompson and H. Bliss. A Declarative Framework for Building Compositional Dialog Modules. In: *Proc. ICSLP-2000*, Beijing, China, 2000.

S. Torge, S. Rapp, and R. Kompe. The Planning Component of an Intelligent Human Machine Interface in Changing Environments. In: *Proc. Workshop on Multi-Modal Dialogue in Mobile Environments*, Bad Irsee, Germany, 2002a.

S. Torge, S. Rapp, and R. Kompe. Serving Complex User Wishes With an Enhanced Spoken Dialogue System. In: *Proc. Int. Conf. on Speech and Language Processing*, Denver, Colorado, 2002b.

J. von Kuppevelt, U. Heid, and H. Kamp (eds.). *Best Practice in Spoken Language Dialogue Systems Engineering*, vol. 6(3,4) of *Journal of Natural Language Engineering*. Cambridge University Press, Cambridge, UK, 2000.

W. Wahlster, N. Reithinger, and A. Blocher. SmartKom: Multimodal Communication with a Life-like Character. In: *Proc. EUROSPEECH-01*, vol. 3, pp. 1547–1550, Aalborg, Denmark, September 2001.

Intelligent Integration of External Data and Services into SmartKom

Hidir Aras[1], Vasu Chandrasekhara[2], Sven Krüger[2], Rainer Malaka[1], and Robert Porzel[1]

[1] European Media Laboratory GmbH, Heidelberg, Germany
{hidir.aras, rainer.malaka, robert.porzel}@eml.villa-bosch.de
[2] Quadox AG, Walldorf, Germany
{vasu.chandrasekhara,sven.krueger}@quadox.de

Summary. The SMARTKOM multimodal dialogue system offers access to a wide range of information and planning services. A significant subset of these are constituted by external data and service providers. The work presented herein describes the challenging task of integrating such external data and service sources to make them semantically accessible to other systems and users. We present the implemented multiagent system the corresponding knowledge-based extraction and integration approach. As a whole these agents cooperate to provide users with topical high-quality information via unified and intuitively usable interfaces such as the SMARTKOM system.

1 Introduction

SMARTKOM is a multimodal dialogue system that offers access to a wide range of information and planning services. In contrast to many previous systems, it integrates services from multiple domains and in its current state covers over 50 different specific tasks (Wahlster, 2003, 2002; Wahlster et al., 2001). Some of these tasks are simple control applications such as that of sending a fax. Other applications are more complex and introduce open discourse areas. For instance, in the tourism domain, sights, streets, and locations can be subject to novel and unanticipated user requests. In the TV application, all sorts of topical and constantly changing broadcast information, actors, and genres are part of the SMARTKOM world. This information cannot be held locally by the system but rather must be gathered ad hoc from various Internet resources.

If Internet information resources are to be made accessible for a dialogue system such as SMARTKOM, which allows for natural language and gesture interaction, it is not sufficient to represent information as flat HTML or text pages. Rather, semantic information is needed, because users can refer to presented information and may ask additional questions such as "tell me more about this." In fact, we need to access the Web resources as a knowledge base in the sense of the semantic Web (Berners-Lee et al., 2001). This means that the unstructured and semistructured data in the Web must be semantically annotated while it is integrated into the SMARTKOM system. The integration of semantic Web information typically involves multiple levels:

- representation of content (usually XML-based) and locations (URLs)
- knowledge modeling/ontologies
- integration/translation of content using different ontologies
- brokering of actual content with respect to user queries

There are a number of XML-based content languages. For SMARTKOM, M3L as a multimodal markup language has been used as a system-wide format (Gurevych et al., 2003a; Herzog et al., 2003). The URLs for particular Internet resources are predefined and contain a variety of third-party information on the SMARTKOM domains such as electronic program guides (EPG) for TVs, cinema programs, and movie information. Such resources on the Web have become increasingly popular, leading to incredibly large amounts of information published online.

However, finding the information on a particular subject, and evaluating the quality and correctness of the data is still a difficult process. Several services ranging from search engines to manually preprocessed catalogues address the problem of mapping the wealth of online information. But because of the distributed and rapidly changing nature of the Internet, different ontological information models, and unclear separation of data and format in the underlying Web-publishing technology, there is no trivial solution for an intelligent knowledge broker.

While brokers typically select and compile data from different sources, translators are needed to encode the meaning of the varying information originating from different repositories in a commonly used scheme, such as SMARTKOM's Multimodal Markup Language (M3L). Knowledge repositories and ontologies must be fused for this purpose. While the SMARTKOM ontology (Gurevych et al., 2002; Porzel et al., 2003) defines the target knowledge representation, each data source may come up with its own respective legacy data model.

In this paper, we describe the integration of external data into the SMARTKOM system. For this purpose, we developed a multilayer architecture that realizes a semantic Web-like access to selected third-party information providers that do not yet provide such metadata. The architecture is implemented as an agent system that also allows for caching mechanisms to enable SMARTKOM to respond to user queries in an offline mode.

2 Semantic Data Integration

Information sources are already available online in global information systems and have been developed independently. Consequently, semantic heterogeneity in terminology, structure, and the context of information can arise. This has to be properly managed to effectively exploit the available information independently of the source from which it has been extracted. The goal of information extraction and integration techniques is to construct synthesized, uniform descriptions of the information extracted from multiple, heterogeneous sources and to provide the user or interface system with a uniform query interface. Moreover, to meet the requirements of global Internet-based information systems, it is important to develop tool-based techniques to make information extraction and integration semiautomatic and as scalable as possible. With today's tools, a tourist planning a trip to Heidelberg cannot simply download a map from the Web showing all the Italian restaurants close to the Castle, even though the information may well be available on the Web as maps of Heidelberg, lists of restaurants, and of tourist sites. The problem is not information distribution but information integration. Building tools to simplify access to the wealth of available information constitutes a significant challenge to computer science.

2.1 Logical Layers of Data Integration

In our prototypical implementation within the SmartKom demonstrator, we realized such a knowledge and service brokerage system. Our approach for an intelligent knowledge and service broker system within SmartKom is organized on three layers (Fig. 1):

- the interface layer for translating queries and responses
- a mediation layer that links to brokers and translators to the repositories
- a service layer that accesses the individual Web repositories and databases

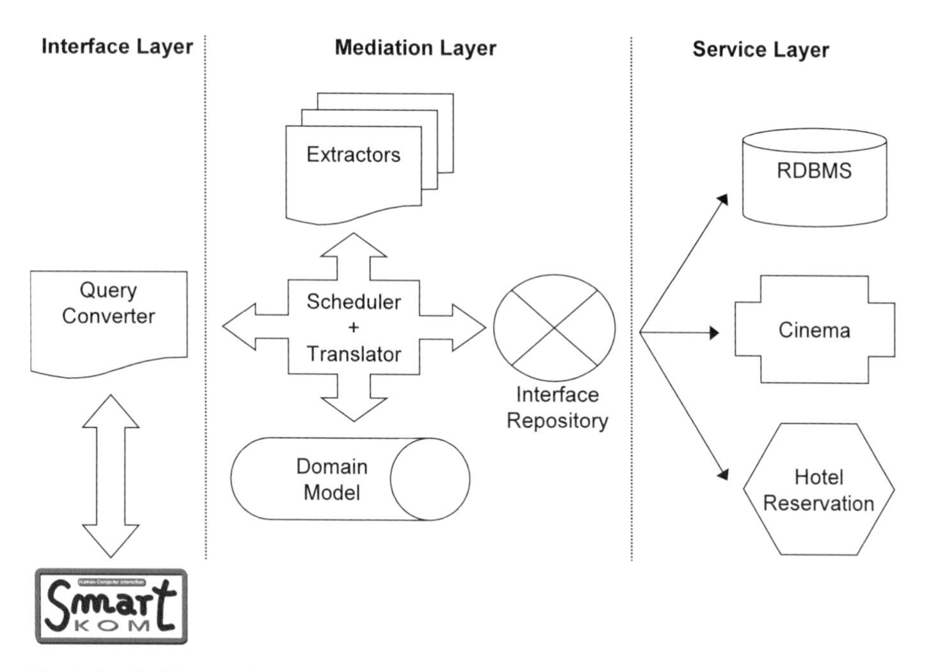

Fig. 1. Logical layers of data integration

On the interface layer, the information integration system is linked to the SMARTKOM semantic interfaces. Therefore, we prepared an in-depth domain model from which we derived a database schema. Furthermore, the model was also integrated into SMARTKOM's M3L standardization effort. This allows for efficient translation of user requests into XML-based schema instances that represent user queries and can be further analyzed by the integration layer.

Since different sources on the same information type may apply quite different taxonomic models on their information structure, the mediation layer must include semantic translators to integrate those external models. After the extraction and translation process, the question of the quality of the information is raised. By using sampled manual feedback and a computed confidence score, the more trustable information is passed back to the SMARTKOM system. The broker system exposes a uniform query interface; the queries are specified in M3L. We have constructed a generic query engine, adoptable to any ANSI SQL 92 compatible database. Queries must include selection and projection criteria. Metadata information for the construction of schema conform markups can be externally supplied or can be automatically extracted from the database schema.

The service layer includes a multiple-protocol handler and an interface repository to technically access external resources. For our system, we allowed for more than just Web resources that connect HTML pages via HTTP. The framework is more flexible and also integrates other protocol and content formats such as a CORBA-enabled hotel reservation system or standardized relational databases that come along with their own metadata model. However, standard Web pages without metadata are also included. It should be noted that in HTML, the term *markup* only describes annotations within the document that instruct a compositor how a particular passage should be printed or laid out. As the formatting and printing of texts was automated, the markup language was extended to cover all sorts of special markup codes inserted into electronic texts to govern formatting, printing, or other processing. In order to segregate the important information data from the printing or processing data we included source-specific extractor scripts, based on a high-level regular expression language. As more Web resources offer content data in XML, the requirement for data segregation will diminish.

2.2 Agent-Based Realization of the Layers

The three layers describe the information flow in the system. For the practical realization, however, it has to be considered whether these layers should be built into a static or more flexible component-based software architecture (Ding et al., 2001; Wooldridge, 1994). It has been shown in many scenarios that an agent-based software design is advantageous in many ways (Poslad et al., 2001). Introducing an agent-oriented paradigm to the system, we can model the tasks as cloneable agents as user, broker, and resource agents. The user agents locate brokers on the basis of costs and evaluate user preferences. In the middle layer, broker agents find the best resources for a request, evaluate match-rates and costs. The resource agents extract and translate the appropriate source, evaluate hit-rates, and remember costs.

The advantages of the agent approach are as follows (Wooldridge, 1994):

- Autonomy and adaptivity: Tasks delegated to the agent are solved independently, without immediate demand for user attention. By incorporating specialized learning algorithms, an agent can intelligently adapt to requests. Complex inquires can be solved dynamically using different partial tasks depending on external availabilities, thus leading to goal-oriented behavior.
- Cooperation: An intelligent agent is able to communicate with the surrounding environment, with other agents, with infrastructure facilitators, and, if equipped with a graphical interface, also with the user. Agent cooperation can be achieved by direct or broadcasted communicative events— speechacts— which are formulated in a high-level Agent Communication Language (ACL), but more importantly also by indirect dependencies leading to self-organization models based on stigmergy.
- Proactivity: An agent is able to start an action without having been initiated by direct requests. Proactive functions could be based on environment changes or internal state changes.
- Mobility: In a distributed environment, entities communicate mainly by request/response schemes. With a mobility-enabled distributed agent platform, agents can move dynamically from system to system, transferring code and interior states. This gives rise to a new application paradigm, for example, roaming agents with delegated responsibilities.

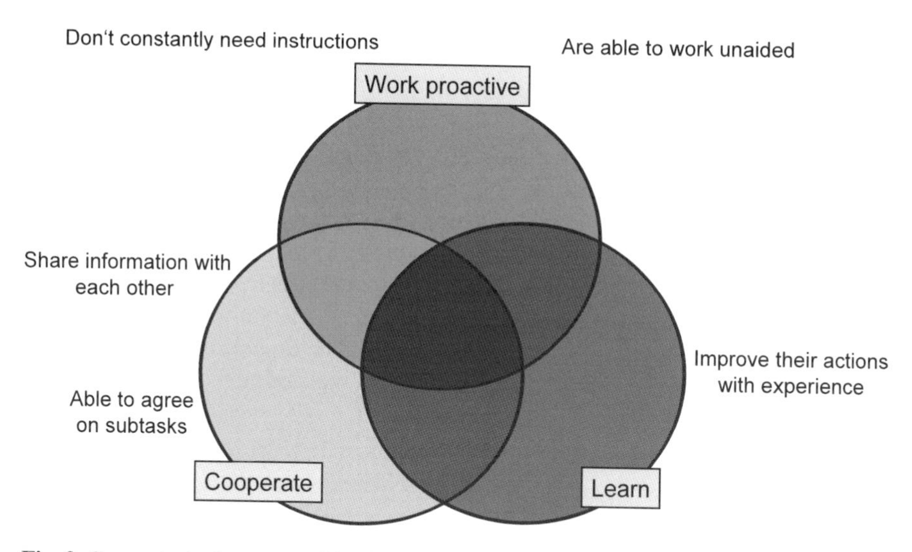

Fig. 2. Conceptual advantages of intelligent agents

The conceptual advantages of intelligent agents make them particularly interesting for the complex task of integration of data from heterogeneous sources that consist of several subtasks as described here. The implemented architecture of such a system consists of a set of software agents to allow cooperative, adaptive, and proactive behavior of the overall system (Fig. 2). With this approach, we aim at developing intelligent tools for the integration of information extracted from multiple heterogeneous sources. In our system, we work with a fixed set of external information sources. However, in order to effectively exploit the numerous sources available online in global Internet-based information systems, some challenges remain to be solved. The main problems involved are:

- the identification of semantically related information (that is, information related to the same real-world entity in different sources)
- the heterogeneity of semantically related information

For the SmartKom demonstrator, we rolled out a working knowledge and service broker system completely implemented in a Java environment. Redundant resources for cinema, film, actors, EPG, yellow pages, ticket reservation, hotel information, and reservation domain were connected. Furthermore, parts of the Deep Map historic content and a geo-map agent for Heidelberg were integrated (Malaka et al., 2000; Malaka and Zipf, 2000).

The actual integration of external data in SmartKom realizes a transparent bridge from SmartKom to Web services and databases. As described in the previous sections, the system is organized in three layers, each of which, in turn, is implemented in multiple software agents.

2.2.1 Interface Layer Agents

Although the agents on the interface layer are implemented very differently, the service provided by them is the same, i.e., offering interfaces to external clients enabling them to benefit from different agent system services. Users have various options available for using the agent system. They can choose between using JSP sites, mail requests, or a graphical user interface. The SmartKom system communicates via a blackboard pool architecture and is based on a PVM architecture (Herzog et al., 2003). A PVM bridge and a connector agent were constructed to enable both systems to communicate. This agent required special adaptation to the SmartKom pool architecture. In future systems, all agents should be capable of dynamically locating the optimal broker agent on a cost basis and registering the user's habits for the adaptation.

2.2.2 Mediation Layer Agents

The agents assigned to this middle layer (so-called broker agents) have to address the following tasks:

- solve complex inquiries dynamically by decomposing them into separate partial tasks depending on external availabilities
- locate agents that provide the requested service

- eliminate redundancies in the answers of different service agents by using sampled manual feedback and a computed confidence score
- provide a facility for translating various metadata models of external sources into a global domain schema

The current broker agents implement three main components. First, the XML–Java data-binding component dynamically translates XML into Java objects and back by using marshaling and unmarshaling mechanisms. The second component realizes a resource location protocol. The purpose of this is to determine the best resource agents by using a catalogue service, the so-called directory facilitator, where all resource agents have to register with the respective costs involved, e.g., response time, quality of content, etc. The third part of the broker agent is a merging and integration facility developed to eliminate redundant data from the answers of service agents.

2.2.3 Service Layer Agents

The service layer encompasses two different kinds of agents: database agents and agents responsible for providing access to different Web sources. Such wrapping agents implement source-specific translators to generate HTTP requests from the incoming XML requests that are understandable for the Web source. It is on the basis of these requests that the URL requester retrieves the corresponding HTML-based data. In order to segregate important information from printing or processing data, we included source-specific extractor scripts based on a high-level, regular, and fault-tolerant expression language. In the future, as an increasing number of Web resources offer content data coded in XML the requirement for data segregation will lose its importance. The database integrated into our system at present provides content periodically extracted from various Web sources about films, actors, cinema performances, and broadcasts. We have integrated this database into the system to enhance request performance, to avoid redundant operations, and to ensure the complete functionality of various services for offline operations as well. To provide an interface consistent with the SMARTKOM domain representations, a generic SQL–XML query engine was developed. This engine makes it possible to answer an XML query containing selection criteria and projection statements defining the resulting XML syntax structure. By using the metadata structure of the database, we were successful in developing an absolutely generic solution adaptable to any ANSI SQL 92 compatible database. This means that we can reuse this generic XML–SQL engine to integrate other SQL-conformable databases, e.g., sights or architectural databases.

3 Implementation of the Data Extraction and Integration Process

We will now give some more implementation details and examples of the methods implemented in the various agents on each layer. The prototypical scenarios of the SMARTKOM system have a special emphasis on the cinema and EPG domain, where most examples originate.

3.1 Development of the Domain Models for Messages und Databases

The first step for semantically integrating SMARTKOM external data is the provision of internal domain models. For the fields of cinema and EPG, these models were modeled such that they share many concepts such as actors or genres. The initial version of the model was built for the XML Document Type Definition (DTD) language.[1] An analogous version in XML schema[2] was developed for the current version of XSD, taking into account the W3C specifications from the working draft of the XML Schema Definition Language. The necessary translation between DTD and XMLS can be done using propriety parsers and tools. The initial schema-based models for the respective domains were also transformed into the SMARTKOM ontology (Gurevych et al., 2003b). From the ontology, in turn, tools allow for automated generation of corresponding XML schema (Gurevych et al., 2003a) from the ontology.

Based upon this model, a relational database has been set up configured using Cloudscape, a Java RDBMS. This database seemed appropriate because it supports portability, has a small footprint, and allows the persistent storage of Java objects.

For the domain of geographical data, necessary for many tourist applications, the data integration is done by specific geospatial agents who access geoservers that follow OpenGIS standards, which are also semantic models and also conform with the ontology.In this way, objects from one domain like a "cinema" can be linked to objects in the other domain in space (and time).

3.2 Building the Infrastructure and Middleware

As discussed above, we chose a multiagent system (MAS) for building the necessary infrastructure and middleware for the information integration platform. We evaluated a number of Java-based MAS and decided to use JADE,[3] which supports the important OMG[4] standard (known from CORBA) and the FIPA[5] standards that aim for promotion of a commercial standardization of the Foundation for Intelligent Physical Agents (FIPA). Distributed objects using, e.g., CORBA, RMI, DCOM, or RPC, communicate with each other based on the principle of local transparency, while the agent communication is done asynchronous by exchanging messages that are semantically divided into three layers:

- **communication layer:** describes the communication parameters, i.e., sender, receiver, unique communication identification, etc.
- **message layer:** defines the *performatives* (borrowed from *Speech Act Theory* (Searle, 1969), for example, *ask-if, tell, reply,* etc.
- **content layer:** the "real" message that is exchanged and encapsulated by the communication protocol

[1] http://www.w3.org/XML/1998/06/xmlspec-report-19980910.htm
[2] http://www.w3.org/TR/xmlschema-0
[3] http://sharon.cselt.it/projects/jade
[4] http://www.omg.org
[5] http://www.fipa.org

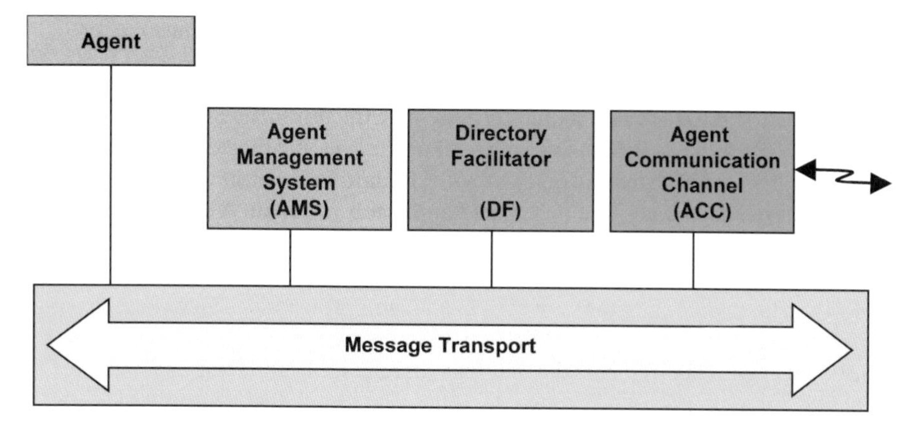

Fig. 3. Message transport

Using semantic messaging one can map multiagent contract negotiations, as in "Call-For-Proposal," without having specific API knowledge about the counterpart (Fig. 3). The communication layer is mostly defined by the MAS itself; the content layer reflects the domain models used here. The message layer links content to communication by describing for the agents what action they should perform or telling other agents what action has been performed.

3.3 Programming of Specific Data Extractors

The World Wide Web has evolued into a public information space. It no longer consists only of simple, hyperlinked documents, but more and more of queryable sources that deliver their responses in semistructured form. Product catalogs, financial services, weather news, commented publications, software collections, conference calls, etc., represent typical examples. The structured information provided by these resources are "pressed" into a pattern that is based upon the navigate-and-show paradigm (hyperlinking) of the Web. Such sources contain mostly overlapping or complementary information that is presented with different syntax and semantics. In general, one can identify the following types of access to data (Web content):

- **static HTML pages**, where information is provided by prepared pages in the HTML layout.
- **dynamic HTML pages** that are similar to static ones, but data and layout are mixed up, and target content can be queried in a directed way (using <form GET/POST/>).
- **static or dynamic XML pages** present content in XML-structured data format, either statically or dynamically (analogously to static and dynamic HTML). HTTP is used technically, and target content mostly refers to the underlying DTD.
- **distributed (object-oriented) interfaces** (RPC, IIOP/CORBA, DCOM, Java-RMI). The access is performed through well-defined, standardized interfaces.

- **direct database connectivity** over DB-API (Oracle Call Interface, ODBC, or JDBC). The specific query statement can be formulated in SQL.

The unstructured HTML sources require an extensive manual preprocessing to allow reasonable data extraction. As mentioned above, the most interesting data is today available in HTML sources. Different groups work on heuristics to analyze semistructured documents, but this is only applicable to certain scenarios (for example, XWRAP[6] or DERG[7]). A typical example of such HTML text from a cinema Web site is:

```
  [...]
  <TR><TD VALIGN="top" WIDTH="10"><IMG SRC="/img/dot.gif"
WIDTH="10" HEIGHT="10"></TD><TD ALIGN="left"
WIDTH="100%"><FONT FACE="Arial, Helvetica" SIZE="2"><A
HREF="/filminfo/a/abbuzze.htm" TARGET="inhalt">Abbuzze - Der
Badesalzfilm</A></FONT></TD></TR>
  <TR><TD VALIGN="top" WIDTH="10"><IMG SRC="/img/dot.gif"
WIDTH="10" HEIGHT="10"></TD><TD ALIGN="left"
WIDTH="100%"><FONT FACE="Arial, Helvetica" SIZE="2"><A
HREF="/filminfo/a/abgefahren.html"
TARGET="inhalt">Abgefahren</A></FONT></TD></TR>
  [...]
```

The information is provided by prepared pages in HTML layout. The example shows how layout and data are mixed. The specific extractor has to use a directory of URLs to intelligently extract the right target content. In a typical XML document the information is well-structured and content is semantically annotated:

```
<channel>
  <title>
    moreover ... Indian subcontinent news
  </title>
[...]
  <description>
    News headlines harvested every 15 minutes
  </description>
  <item>
    <title>
      China Denies Charges of Supplying Missile Tech to Pakistan
    </title>
    <link>
      http://c.moreover.com/click/here.pl?r8036262
    </link>
    <description>
      New York Times Jul 4 2000 11:25AM
    </description>
```

[6] http://www.cc.gatech.edu/projects/disl/XWRAPElite

[7] http://www.deg.byu.edu

```
    </item>
    <item>
  [...]
```

The XML.ORG[8] site tries to support standardization efforts for different vendors and application domains. The XML Tree[9] indexes content providers who provide their contents in pure data formats, unfortunately approximately 50 vendors for German content, mostly in WML form for WAP devices (WML is structured, but also mixes up layout and data).

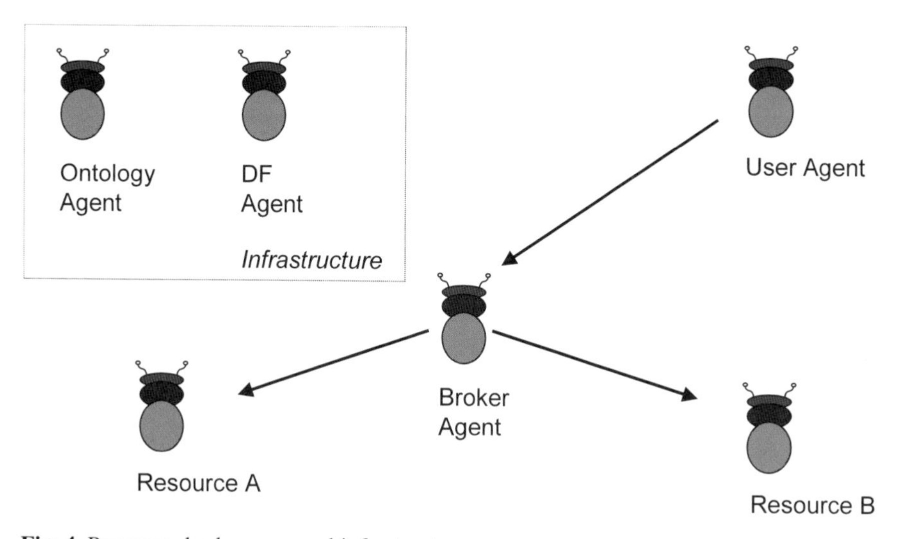

Fig. 4. Resource, broker, user and infrastructure agents

After a detailed feasibility study we decided to use JEDI 1.2.1,[10] a Web extraction toolkit (Huck et al., 1998). In order to use mediators and wrappers, JEDI only requires standard Web browser technology. All system components have been implemented fully in Java. The wrapper consists of a rich, error-tolerant parser. The source structure of documents (nesting included) can be described using attributed rules. The execution of these rules results in a network of objects that represent the documents that are the subject of the data extraction process. The parser is able to process incomplete or ambiguous source descriptions by using a simple approach, i.e., efficiently choosing the most specific rule combination out of the entire applicable set of rules. If a part of the document does not allow the parser to apply a rule, only the necessary part is skipped, and not the whole document. Beside the rules for structuring, the parser also provides methods for navigation and form-based queries

[8] http://xml.org
[9] http://www.xmltree.com
[10] http://www.darmstadt.gmd.de/oasys/projects/jedi

to information sources. It can also be extended by Java predicates and methods. Several tools are available for mediation in XML. Exolab's Castor tool was chosen to bind XML schema to Java objects.[11]

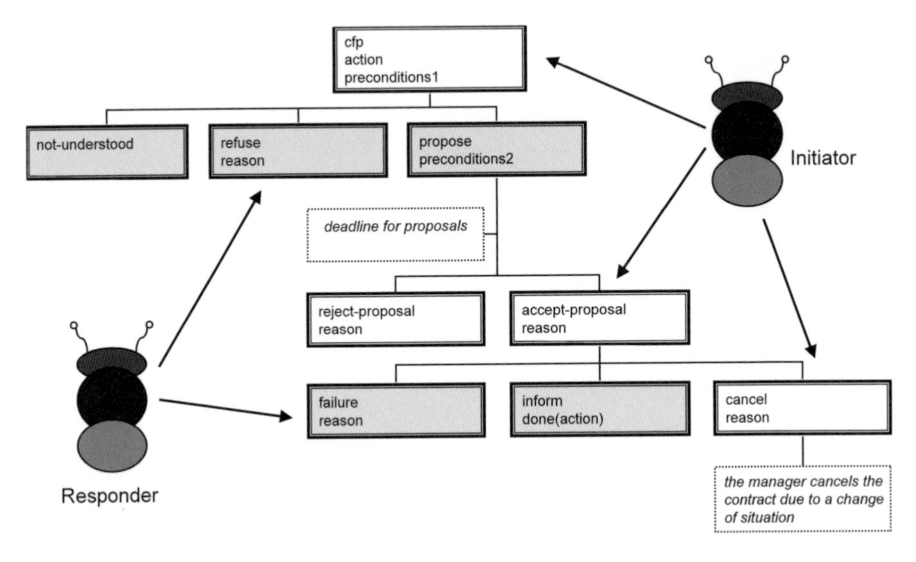

Fig. 5. The call for proposal protocol

3.4 Formulation of the Processing Logic of Queries

Information and external services from different sources from the Web are represented by resource agents (Fig. 4). A broker agent then initiates the negotiation process to access the service by using the call for proposal (CFP) protocol (Fig. 5). Available service agents can be found using the directory facilitator (DF) service. This procedure allows for individual handling of each resource. The scheduler, respectively, the query engine is distributed among the different agents. The broker is also the bridge to the interface/user agents. An ontology agent provides the system with ontological information.

The CFP protocol allows two agents to negotiate service interaction. A broker can start to search for service agents by broadcasting a CFP for a requested action and can parameterize the CFP with a set of preconditions. Responding agents have the choice between a proposal to react or to refuse the call in case they understand the request. If they propose a response, they also give a set of preconditions. After a deadline for the CFP, the initiating agent collects the responses for the CFP and decides to accept or reject certain offers from responding agents. The agents who get

[11] Xerces, http://xml.apache.org; XQL, http://xml.darmstadt.gmd.de/xql; Castor, http://castor.exolab.org.

an accept-response for their offer will either deliver a result or fail, and will inform the initiator accordingly. If the response takes too long or due to other circumstances is not needed any more, the initiator can also asynchronously cancel the execution of the request.

3.5 Definition of the Query and Response Syntax and Semantics

We have connected the resources into our prototypical demonstrator that feature content about cinematography, hotel reservations and geographical information modeled in the corresponding actor, film EPG, hotel, and map query DTDs. We will illustrate the function of these models with examples from the cinema domain. Consider, for example, a user asking whether the film *Gladiator* is running in the city of Viernheim tomorrow. The corresponding query syntax is defined by a generic DTD shown below:

```
<?xml version="1.0"?>
<!ELEMENT cr          (day,time?,town,title?,actor?)>
<!ELEMENT day         (#PCDATA)>
<!ELEMENT time        (#PCDATA)>
<!ELEMENT town        (#PCDATA)>
<!ELEMENT title       (#PCDATA)>
<!ELEMENT actor       (#PCDATA)>
```

Based on this syntax, the M3L-corresponding representation of the user's intent (see also Wahlster et al. (2001); Wahlster (2003)) is transformed into the specific query given below:

```
<?xml version='1.0'?>
<!DOCTYPE cr SYSTEM "CRquery.dtd">
<cr>
     <day>morgen</day>
     <time>14</time>
     <town>viernheim</town>
     <title>Gladiator</title>
     <actor/>
</cr>
```

Based upon such a query, the processing logic described above yields answers that are annotated with their source and contain the corresponding information, as shown in the example below:

```
<entry>
  <url>
    http://www.film.de/asp/content/film.asp?film_id=4361
  </url>
  <titel>
    Gladiator
  </titel>
  <ort>
```

```
    Viernheim
  </ort>
  <kinos>
    <kino zeit="8. Juli 2000 - 16.45, 20.00, 23.15 Uhr">
      Kinopolis Rhein-Neckar
    </kino>
  </kinos>
</entry>
```

Employing the ontology-based semantic translators, this answer is translated back into M3L and enabled the action and presentation planning component of SMARTKOM to respond to the user's query with topical and trustworthy information.

4 Conclusion

We have shown herein how the SMARTKOM dialogue system can offer its users access to a wide range of information and planning services that stem from external sources, and how an intelligent external data and service integration platform can provide users with topical high-quality information via an unified and intuitively usable multimodal dialogue system. It has been demonstrated that this can be achieved in a prototypical implementation based on a mutliagent system which can cooperate on extracting, integrating, and ensuring the quality of the services and information requested by the user.

While there is still room for improvement, e.g., better adaptivity toward the quality of the resources, more intelligent fetching strategies, and the employment of dynamic and learning semantic schema translators for handling HTML resources, we anticipate to connect more semistructured, XML-based Web resources in the future. Still, HTML resources will continue to pose a difficult challenge, as their content is quickly outdated and their presentations often undergo redesign. To further stabilize and increase the quality and quantity of the information content at our disposal, agreements with networks of content providers, or federations of agent-based service providers such as in the Agentcities[12] framework, constitute promising future directions.

Projects such as Agentcities are aiming to build worldwide, publicly accessible test beds for the deployment of agent-based services. We, therefore, joined such an open initiative, where the core effort will be devoted to interoperability at the service level, in particular the development of ontologies and service descriptions defining a new open standard. Finally, the SMARTKOM approach to external data and service integration described herein shows that important features such as quality negotiations and redundancies brought about by the employment of a layered multiagent system as well as the semantic extraction and integration capabilities thereof yield the desired added value of homogeneous and intuitive access to heterogeneous legacy systems and sources.

[12] http://www.agentcities.org

References

T. Berners-Lee, J. Hendler, and O. Lassila. The Semantic Web. *Scientific American*, 284(5):34–43, 2001.

Y. Ding, C. Kray, R. Malaka, and M. Schillo. RAJA — A Resource-Adaptive Java Agent Infrastructure. In: *Proc. 5th Int. Conf. on Autonomous Agents*, pp. 332–339, Montreal, Canada, 2001.

I. Gurevych, S. Merten, and R. Porzel. Automatic Creation of Interface Specifications from Ontologies. In: H. Cunningham and J. Patrick (eds.), *Proc. HLT-NAACL 2003 Workshop on Software Engineering and Architecture of Language Technology Systems (SEALTS)*, pp. 59–66, Edmonton, Canada, 2003a. Association for Computational Linguistics.

I. Gurevych, R. Porzel, E. Slinko, N. Pfleger, J. Alexandersson, and S. Merten. Less Is More: Using a Single Knowledge Representation in Dialogue Systems. In: *Proc. HLT-NAACL'03 Workshop on Text Meaning*, pp. 14–21, Edmonton, Canada, May 2003b.

I. Gurevych, R. Porzel, and M. Strube. Annotating the Semantic Consistency of Speech Recognition Hypotheses. In: *Proc. 3rd SIGdial Workshop on Discourse and Dialogue*, pp. 46–49, Philadelphia, PA, July 2002.

G. Herzog, H. Kirchmann, S. Merten, A. Ndiaye, P. Poller, and T. Becker. MULTIPLATFORM Testbed: An Integration Platform for Multimodal Dialog Systems. In: H. Cunningham and J. Patrick (eds.), *Proc. HLT-NAACL 2003 Workshop on Software Engineering and Architecture of Language Technology Systems (SEALTS)*, pp. 75–82, Edmonton, Canada, 2003. Association for Computational Linguistics.

G. Huck, P. Fankhauser, K. Aberer, and E.J. Neuhold. JEDI: Extracting and Synthesizing Information From the Web. In: I.C. Society (ed.), *Proc. 3rd IFCIS Int. Conf. on Cooperative Information Systems*, pp. 32–43, New York, 1998.

R. Malaka, R. Porzel, A. Zipf, and V. Chandrasekhara. Integration of Smart Components for Building Your Personal Mobile Guide. In: *Proc. Workshop Artificial Intelligence in Mobile Systems, ECAI 2000*, Berlin, Germany, 2000.

R. Malaka and A. Zipf. Deep Map: Challenging IT Research in the Framework of a Tourist Information System. In: D.R. Fesenmaier, S. Klein, and D. Buhalis (eds.), *Proc. 7th. Int. Conf. on Information and Communication Technologies in Tourism (ENTER 2000)*, pp. 15–27, Berlin Heidelberg New York, 26–28 April 2000. Springer.

R. Porzel, N. Pfleger, S. Merten, M. Loeckelt, I. Gurevych, R. Engel, and J. Alexandersson. More on Less: Further Applications of Ontologies in Multi-Modal Dialogue Systems. In: *Proc. 3rd IJCAI 2003 Workshop on Knowledge and Reasoning in Practical Dialogue Systems*, Acapulco, Mexico, 2003.

S. Poslad, H. Laamanen, R. Malaka, A. Nick, P. Buckle, and A. Zipf. CRUMPET: Creation of User-Friendly Mobile Services Personalised for Tourism. In: *Proc. 2nd Int. Conf. on 3G Mobile Communication Technologies*, pp. 28–32, London,UK, 2001.

J. Searle. *Speech Acts: An Essay in the Philosophy of Language*. Cambridge University Press, Cambridge, UK, 1969.

W. Wahlster. SmartKom: Fusion and Fission of Speech, Gestures, and Facial Expressions. In: *Proc. 1st Int. Workshop on Man-Machine Symbiotic Systems*, pp. 213–225, Kyoto, Japan, 2002.

W. Wahlster. SmartKom: Symmetric Multimodality in an Adaptive and Reusable Dialogue Shell. In: R. Krahl and D. Günther (eds.), *Proc. Human Computer Interaction Status Conference 2003*, pp. 47–62, Berlin, Germany, June 2003. DLR.

W. Wahlster, N. Reithinger, and A. Blocher. SmartKom: Multimodal Communication with a Life-like Character. In: *Proc. EUROSPEECH-01*, vol. 3, pp. 1547–1550, Aalborg, Denmark, September 2001.

M. Wooldridge. Intelligent Agents. In: G. Weiss (ed.), *Multi-Agent Systems — A Modern Aproach to Distributed Artificial Intelligence*, pp. 17–20, Cambridge, MA, 1994. MIT Press.

Multimodal Fission and Media Design

Peter Poller and Valentin Tschernomas

DFKI GmbH, Saarbrücken, Germany
{poller,tscherno}@dfki.de

Summary. This chapter describes the output generation subsystem of SMARTKOM with special focus on the realization of the outstanding features of the new human–computer interaction paradigm. First, we start with a description and motivation of the design of the multimodal output modalities. Then we give a detailed characterization of the individual output modules, and finally we show how their collaboration is organized functionally in order to achieve a coherent overall system output behaviour. ,

1 Design of Interaction and System Output in SmartKom

One of the main tasks of the SMARTKOM project was the investigation of the coordinated use of different modes and modalities in human interaction aiming at symmetrically exploring them to support the intuitive multimodal access to knowledge-rich services in a mixed-initiative system. SMARTKOM merges three user interface paradigms: natural language dialogue, graphical interface, and gestures (incl. facial expressions) in a symmetric use (Wahlster, 2003), i.e., seamless fusion and mutual disambiguation of multimodal input, as well as multimodal fission and adaptive presentation layout on the output side. According to these overall system requirements, the project first concentrated its activities on the design of a superordinated completely new human–computer interaction metaphor that allows for natural multimodal access to all the different applications in the three scenarios.

Interactions with the SMARTKOM system are based on the *situated delegation-oriented dialogue paradigm (SDDP)*, which means that the user delegates a task to a virtual communication assistant that is visualized as a life-like character. This dominating completely new SDDP-style of natural collaborative interaction with a virtual communication assistant in SMARTKOM, in conjunction with the three different scenarios, implies a series of essential requirements on the design (and their implementation in the system) of various prominent characteristic features of the SMARTKOM system:

- There is no direct interaction of the user with any output element independent of its respective medium and modality.

- The system persistently conforms to the design principle "no presentation without representation" (Wahlster, 2003), i.e., all output elements are (independent of their modality) internally represented using corresponding Multimodal Markup Language (M3L) expressions potentially referrable by the user.
- Consequently, all output graphics are visual widgets for output presentation elements, also potentially referrable by the user.
- The planning and realization of output presentations natually depends on the respective scenario and the available output media and modalities.
- The system constantly presents adaptive perceptual feedback that represents its internal processing state.
- All system output is based on context-dependent modality fission.
- SMARTKOM realizes symmetric multimodality, i.e., all input modalities are also available for output, and vice versa.
- In conformity to the SDDP paradigm, the communication assistant Smartakus is a de facto personification of the system.
- All multimodal capabilities of the system are integrated into the same reusable and efficient dialogue shell: the core dialogue processing engine.

In the following sections we show the various explicit and implicit impacts of the above-mentioned features on the design of interaction metaphor of SMARTKOM as well as various aspects of that metaphor. We give special focus to the respective consequences on the design of the multimodal system output, i.e., its modalities and media and their coordinated use.

1.1 SDDP

The before-mentioned conditions imply that system interactions in SMARTKOM must move beyond the traditonal windows, icons, menus, pointing device (WIMP) interaction with keyboard and mouse on a desktop. Rather, user input is based on natural human conversation means: speech, gestures, and mimics. In addition to that, in line with the guidelines elaborated in Oviatt (2003), SMARTKOM goes yet another step further as it breaks radically with traditional mouse- and keyboard-based interaction via desktop. Instead, interactions with the system are based on the SDDP, which means that the user delegates a task to a virtual communication assistant that is visualized as a life-like character. This, in turn, is done by natural human communication means, i.e., the user input is exclusively performed in a purely natural way, namely by the coordinated use of speech, gesture, and mimics, just as humans do in real life.

Only this way is it possible that the user gets a homogeneous and pleasing interaction experience through communication with the anthropomorphic personalized interaction agent Smartakus to whom the user virtually delegates the task to be solved. In SMARTKOM, both communication partners collaborate during the problem-solving process in which the personalized agent accesses the background services. The agent may ask back for more information and finally presents the results on the output channels.

This kind of human–machine interaction cannot be done in a simple command-and-control style for more complex tasks. Instead, a collaborative dialogue between the user and the agent elaborates the specification of the delegated task and possible plans of the agent to achieve the user's intentional goal.

Consequently, human–computer interaction in SMARTKOM is put on a completely new footing. The focus of interaction rests on the cooperative communication with the animated agent about the fulfillment of the respective service the user is interested in. Unlike the traditional desktop interaction metaphor, which is based on WIMP interaction, the content of the graphical user interface of SMARTKOM is radically reduced to only those elements (e.g., graphics) that are relevant to the user to the extent that the visible presentation contents on the screen exclusively consist of the graphical presentation elements such that the screen background becomes transparent. The presentation agent Smartakus itself guides the user through a presentation and comments on and points to objects as needed.

The user delegates a task to Smartakus and helps the agent, where necessary, in the execution of the task. Smartakus accesses various digital services and appliances on behalf of the user, collates the results, and presents them to the user. Therefore, all further design aspects must naturally be subordinated to this fundamental interaction style.

1.2 No Presentation without Representation

Interactions in SMARTKOM are managed based on representing, reasoning, and exploiting models of the user, the domain, the task, the context, the media, and the modalities themselves. Consequently, the system follows the underlying principle not to process and present anything without representing it. So, for all multimodal inputs and outputs a common representation approach based on the XML language M3L is used. Whatever the communication with the system is about (including the output widgets themselves that are used for graphical presentations), it is always internally represented.

This way, the SMARTKOM system properly follows the principle that — independent of the respective modality — user references to all system output elements must be generally possible. With respect to the design of visible output, this rules out graphical elements or fonts that are too small to be unambiguously referenced by a user gesture on the respective display.

1.3 Smartakus as the Personification of the System

The key design feature of SMARTKOM is the realization of the SDDP paradigm, which results in using the life-like character Smartakus as a kind of virtual "personification" of the system the user is communicating with. In this sense, the communication metapher opens up a large range of performances that constitute full system output presentations that may even go beyond human capabilities. Consequently, a SMARTKOM system presentation is usually designed as a kind of presentation show that is organized as follows: Depending on the current position of Smartakus on the screen, the graphical elements on it, and the space and positions of the graphical elements to be presented, Smartakus first "jumps" to a new position (leaving enough space). Then the graphical elements are brought up, and finally Smartakus comments on the graphic content, and performs appropriate pointing gestures and speech animations that are synchronized with the audio output signal.

1.4 The Three SmartKom Scenarios

There are three different instances/scenarios of SMARTKOM that differ in the respective applications and hardware. In this section we shortly recapitulate them in order to show their respective key influence factors on design aspects of different output elements.

SMARTKOM "Public" is a system that can be used as a public multimodal information kiosk. Contactless gestures of the user are rendered possible by an infrared camera (the Siemens Virtual Touch screen, SiViT), which tracks the user's hand moving across a video projection of the graphical output on a tablet screen. Naturally, this constellation may contain the complete spectrum of application types, modalities, and hardware functionalities. All input devices and input modalities are available. Because of the large display, the dominating output element in system presentations is graphics. Furthermore, there is enough space to show Smartakus with its full body and let it move around on screen to optimize its positioning during output presentations.

In the SMARTKOM Home scenario, a comparatively small touch-sensitive Tablet PC (Fujitsu Stylistic) is used to show the visual output of SMARTKOM. Here, the system is used as an information system at home, e.g., as an EPG and for device control (TV set, video cassette recorder). Input gestures are performed with a pen on the touch screen. There is no camera intergrated to interpret facial expressions or for scanning. SMARTKOM Home is a reduced system instance whose output device permits mobile interaction at home.

SMARTKOM Mobile is an instance of the SMARTKOM system with its display on a touch-sensitive PDA (Compaq iPAQ) that is used as a mobile travel companion to access tourist information and navigation (car and pedestrian). The very small display requires us to reduce Smartakus to only those body parts of it that are animated, i.e., one hand (for pointing gestures) and the head (for lip-synchronized speech output). Naturally, there is a correspondingly adapted gesture repertoire such that Smartakus' multimodal output capabilities are fully preserved, i.e., moving around, synchronized lip movements, and pointing gestures.

1.5 Adaptive Visual Perceptive Feedback

The dominating role of the communication assistant Smartakus in the interaction with the SMARTKOM system requires the realization of a high degree of adequacy, naturalness, and liveliness (not particularly during output presentations but rather at times of output inactivity) that is constantly perceivable over time. This does not that much of an effect on the agent behavior during system output presentation. Much more time passes by due to communication pauses or while the system is processing the last user input.

Simple observation of human–human communication shows that as long as temporary communication pauses are accompanied with feedback signals, it is even natural and polite to give the communication partner enough time to think about his next utterance. Transferring this observation to human–computer interaction, particularly to multimodal dialogue systems, means that especially those systems that have slower reaction times of several seconds have to focus on giving the user permanently adequate and adaptive perceptual feedback during the processing of the user input about the respective state or progress of this processing. Fortunately, concerning the realization of such a feedback, the SMARTKOM system is equipped with various multimodal output capabilities that can be used for that purpose in a natural way.

Accordingly, apart from the entire interaction facilities, the SMARTKOM system additionally provides constantly updated adaptive perceptual visual feedback. In line with the SDDP paradigm, the feedback is given indirectly by the animated agent, which simultaneously performs conformable gestures (Poller and Reithinger, 2004). In parallel to that, adequate graphic elements, e.g., a flashing light during input processing, are additionally presented on the display to reinforce the desired effect.

All in all, the realization of perceptual feedback is mainly achieved by the animated agent Smartakus by its capability to dynamically adapt its behavior depending on the respective system state. Figure 1 shows the eight most important behaviour patterns of Smartakus in form of the respective main characteristic posture.

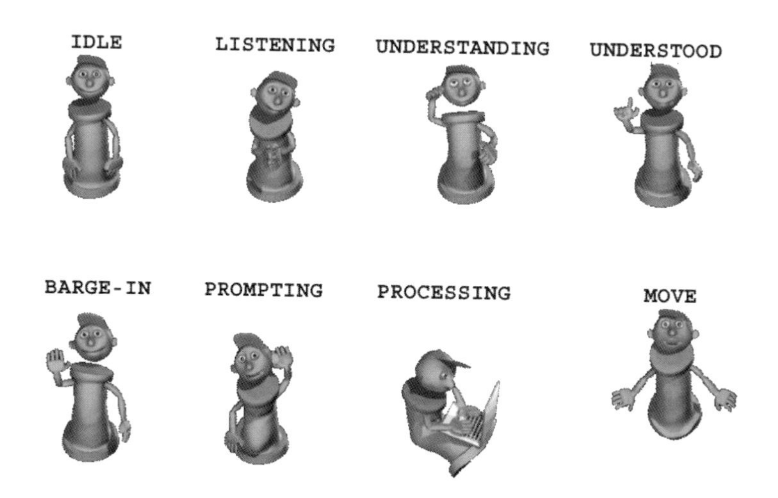

Fig. 1. Some state-reflective behavior patterns of the animated agent Smartakus

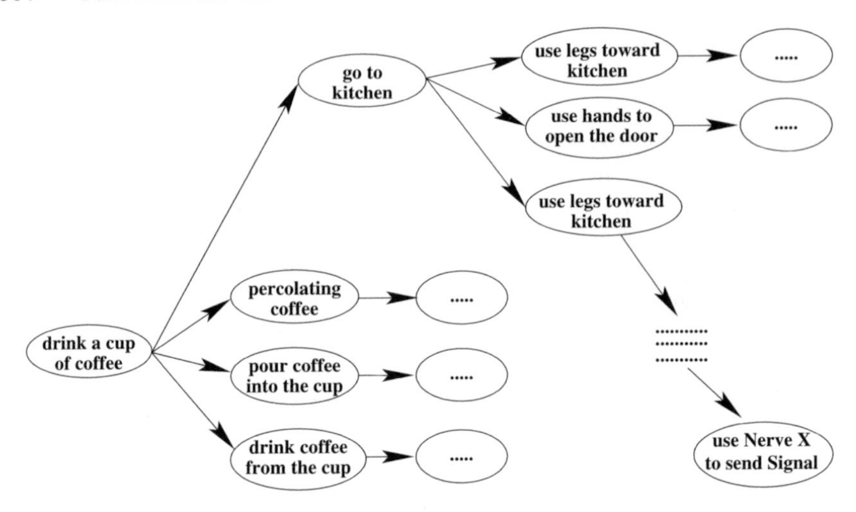

Fig. 2. Creating a plan by dividing the task by human mind

1.6 Dynamic Context-Dependent Modality Fission

Multimodal system output of SMARTKOM consists of synchronized presentations of appropriate graphics and speech output accompanied and moderated by corresponding Smartakus gestures and postures, including cross-modal references.

A crucial computational step is the distribution of the output elements over the available modalities (fission) that is based on a dynamic, context-dependent procedure taking care of several influencing factors, e.g., the dialogue context, the available media and modalities, or user preferences.

For example, in the lean-backward mode of the Home scenario, the user does not look at the display, which means that neither graphics nor gestures can be used. Consequently, all output elements must be adapted to be presented monomodally by speech only without any cross-modal references. The display itself is cleared in such a case, and a special "deactivation" screen is shown.

1.7 Symmetric Multimodality

An important and novel feature of the user interaction in SMARTKOM is symmetric multimodality, which means that all input modalities that may be part of a user input (speech, gesture, facial expression) are, in turn, available as modalities of system output presentations as well. The system uses speech, graphics, gestures, and facial expressions on the output side as well as on the input side.

In accordance with that, for users interacting with SMARTKOM, the system must be "personalizable" in the sense that there is the humanlike communication partner of the user. This is realized as the animated agent Smartakus to which — according to the SDDP paradigm — all tasks are delegated to by the user.

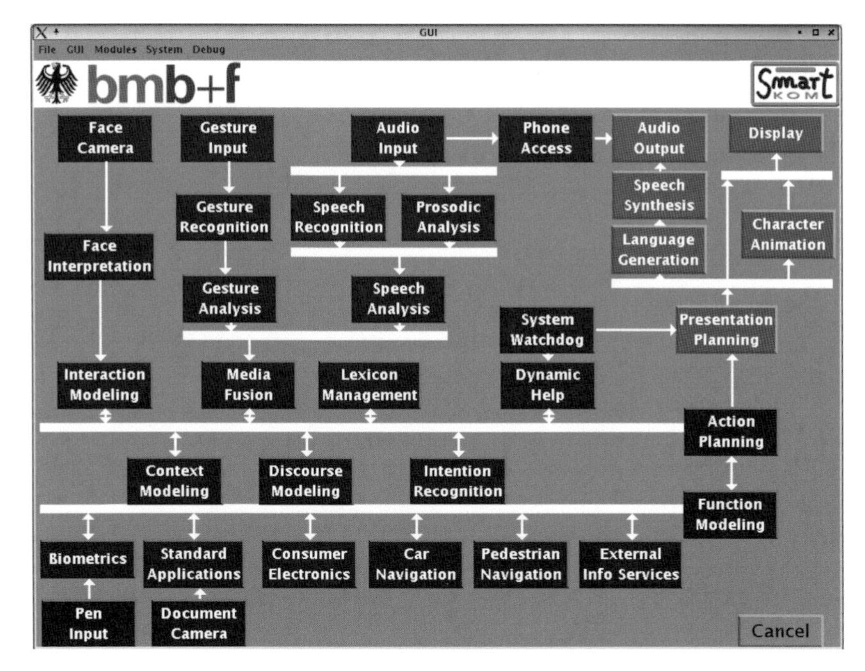

Fig. 3. Fission modules highlighted on the SMARTKOM GUI

2 Media Fission and Presentation Planning

The best way to communicate with a human is to use different information media simultaneously. For example, the use of graphics and speech in a presentation gives a better understanding level than a graphics-only presentation. The partitioning of a presentation into tasks for different media is called "multimedia fission."

2.1 Multimodal Media Fission

For better understanding of multimedia fission, take a look at the human being itself. The example task is to drink a cup of coffee, but there is no atomic human action like this. The human mind first divides this task into appropriate subtasks like going into the kitchen or percolating coffee. Each of the subtasks is divided further into sub-subtasks, and so on until the tasks become atomic, like sending signals over a nerve X to muscle Y. So a complete plan for elementary actions is computed (see Fig. 2 for an example) to be used for completing the given task.

In the same way, the human mind would act if the task were to inform someone verbally about something. In this case, the main subtasks are the modality-specific parts for gesture, mimics, and speech. So the mind would perform a "multimodal presentation" of the information that the human has to tell.

On different levels, some of the subtasks can, of course, overlap with each other, so they have to be synchronized. For example, while speaking the mouth should be used simultaneously with the tongue. Otherwise nobody can be understood.

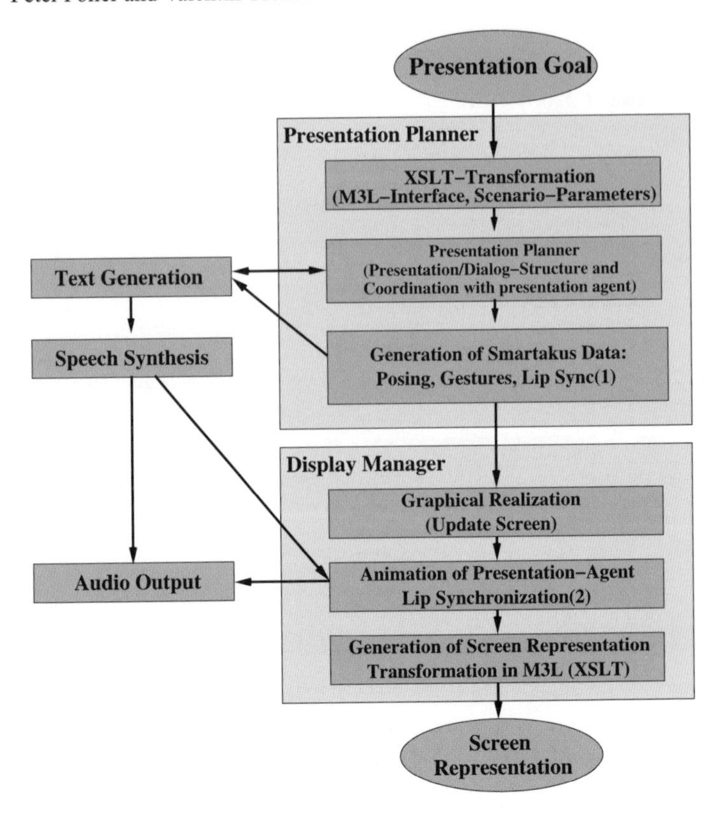

Fig. 4. Architecture of the fission modules

Accordingly, the task of multimodal fission consists of two major steps:

Partitioning: This is the division of the tasks into appropriate subtasks, including the generation of the corresponding evaluation plan. In SMARTKOM this task is performed by the presentation manager (Sect. 2.2).

Evaluation and synchronization of the subtasks: After that, the atomic subtasks are evaluated according to the generated plan. While evaluating the subtasks are synchronized to avoid presentation failures. This task is job of the display manager (Sect. 3)

SMARTKOM has more ways to present information than the human. Besides gestures, mimics, and speech, SMARTKOM also uses graphics, which can be presented on different graphic devices. This makes multimodal fission more complex such that the tasks are distributed between several modules as follows (see also Sect. 3): The presentation manager generates complete plans for graphics, gesture, and mimics output, while the plan for speech output is generated on an abstract high level only. The text generator creates a middle-level plan, i.e., a plan containing content-to-speech representations of the sentences to speak. Finally, the speech synthesis gener-

```
<presentationTask>
 <subTask>
  <presentationGoal>
   <inform> ... </inform>
    <abstractPresentationContent>
     <taskSpecification>
      <intentionContent>...</intentionContent>
     </taskSpecification>
     ...
     <result>
      <broadcast id="bc1">
       <channel> <name>EuroSport</name> </channel>
       <beginTime>
        <time> <at>2000-12-05T14:00:00</at> </time>
       </beginTime>
       <endTime>
        <time> <at>2000-12-05T15:00:00</at> </time>
       </endTime>
       <avMedium>
        <title>Sport News</title>
        <avType>sport</avType>
        <rating>0</rating>
       </avMedium>
      </broadcast>
     </result>
    </abstractPresentationContent>

   <interactionMode>leanForward</interactionMode>
   <goalID>APGOAL3000</goalID>
   <source>generatorAction</source>
   <realizationType>graphicAndSpeech</realizationType>

  </presentationGoal>
 </subTask>
</presentationTask>
```

Fig. 5. M3l document containing input for presentation

ates the low-level actions, i.e., the audio data (plus phoneme information that is used for lip synchronization, see Sect. 4.1).

Also, several modules are used for the evaluation of a presentation plan. The display manager is used to evaluate graphics, mimics, and gestures. For presenting graphics, dynamic and static data like pictures and output texts are combined on the graphics output device. For gestures and mimics, the character animation computes animation sequences for the animated agent Smartakus in such a way that the human mind can perceive them as natural, humanlike gestures and mimics. Speech and audio tasks are evaluated by the audio module. The synchronization of the particular output modality streams is done by the display manager (Fig. 4).

Fig. 6. Widgets for presentations in SmartKom-Public and Home

2.2 Presentation Manager

The main task of the presentation manager is the planning of a multimodal presentation that consists of two parts: static gesture-sensitive graphical elements and a corresponding multimodal animation of Smartakus, including gestures referring to objects with aligned audiovisual speech output.

All information that flows between modules is represented as an XML document conforming to the schema-based language M3L, which was developed in SmartKom (Herzog et al., 2003). The input of the presentation planner consists of a M3L document that is sent by the action planner or the dynamic help component. It contains an abstract representation of the system intention to be presented multimodally to the user.

2.2.1 Input for Presentations

The first computational step being performed on the M3L input document (see an example in Fig. 5, which specifies a "Sport News" broadcast as presentation goal) is a transformation of the document into the special input format of the core planning component PrePlan (André, 1995; Tschernomas, 1999) by application of an appropriate XSLT stylesheet. The use of stylesheets ensures flexibility with respect to the input structure of the presentation planner. Similarly, different situation-dependent XSLT stylesheets that reflect different dialogue situations are used (one stylesheet per situation). A dialogue situation is defined by the set of parameter for the presentation planner.

Fig. 7. Presentation of touristic sights in SmartKom-Mobile

Some examples for the presentation parameters are the current scenario (SmartKom-Mobile, SmartKom-Home, SmartKom-Public), the display size (SiViT and Tablet PC have a resolution of 1024×768 pixel while the PDA has only 240×320 pixels), output language (German, English), available user interface elements (e.g., lists for TV or cinema movies, seat maps, virtual phone or fax devices), user preferences (preference for spoken output while using SmartKom-Mobile in a car), available output modalities (graphics, gestures, speech) and design styles (e.g., different background colors for Mobile and Public, see Figs. 6 and 7).

The stylesheet transformations add scenario-specific or language-specific data to the knowledge base of the presentation planner. For example, translations for labels and a logical description of the available screen layout elements are inserted.

2.2.2 Presentation Tasks

The presentation planner starts the planning process by applying a set of so-called presentation strategies (see Fig. 8 for an example), which define how the facts are presented in the given scenario. Based on constraints, the strategies decompose the complex presentation goal into primitive tasks, and at the same time they do the media fission step depending on available modalities, which means they decide which part of the presentation should be instantiated as spoken output, graphics, or gestures of our presentation agent Smartakus. Also, they choose appropriate graphical elements like lists or slide shows.

```
(define-plan-operator
 :HEADER (A0 (Show-Sights ?sightlist))
 :CONSTRAINTS
   (BELP (layout ?layout canshow picture))
 :INFERIORS (
    (A1 (SetLayout ?layout))
    (A2 (seq-forall (?url) with
      (BELP (objectpicture ?sightlist ?url))
         (AI (Add-Picture-URL ?layout ?url))))
    (A3 (Add-Description ?layout ?url))
    (A4 (Persona-Speak))
  ))
```

Fig. 8. A presentation strategy in SMARTKOM

For simple presentation tasks, like showing the phone, the presentation planner creates commands for a graphical element (the phone itself), a speech comment, and a gesture script for Smartakus for pointing at the phone while commenting. In contrast to simple presentations, there are complex presentation tasks that depend on more input and modality parameters. For example, while planning of a TV program, the presentation planner chooses the appropriate graphical element depending on the amount and type of input data. If there is only one broadcast to show, a lot of background information about it can be presented (e.g., directors, actors, and description). The presentation planner must also take into account what the user likes (e.g., if the user likes action films but not the channel ARD, action films will be shown first while broadcasts on channel ARD will be shown at the end or not shown at all). If some information parts are not available, the place reserved for this information can be used to extend other information parts (e.g., if there is a film description, only some actors are shown, but if there is no description available, all actor names are presented. In Mobile, the whole space is used for text output on absence of pictures for a sight description).

After choosing the graphical presentation, appropriate speech and gesture presentations are generated. The gesture and speech form is chosen depending on the graphically shown information. That is, if the graphically presented information is in the focus of a presentation, only a comment is generated for speech output. The goal of the gesture presentation is then to draw the users's attention to the appropriate graphical element. If there is no graphically presentable information or it is insufficient to completely fulfill the presentation goal, more speech is generated.

In SMARTKOM-Home, there is a special mode (called the lean-backward mode) where the user may turn away from the screen, e.g., for watching TV. In that case the visual output modality is not available anymore, so the presentation planner switches to exclusively use speech output to inform the user. However, the presentation agent remains visible on screen just to indicate that the system is still active and that the display can be reactivated.

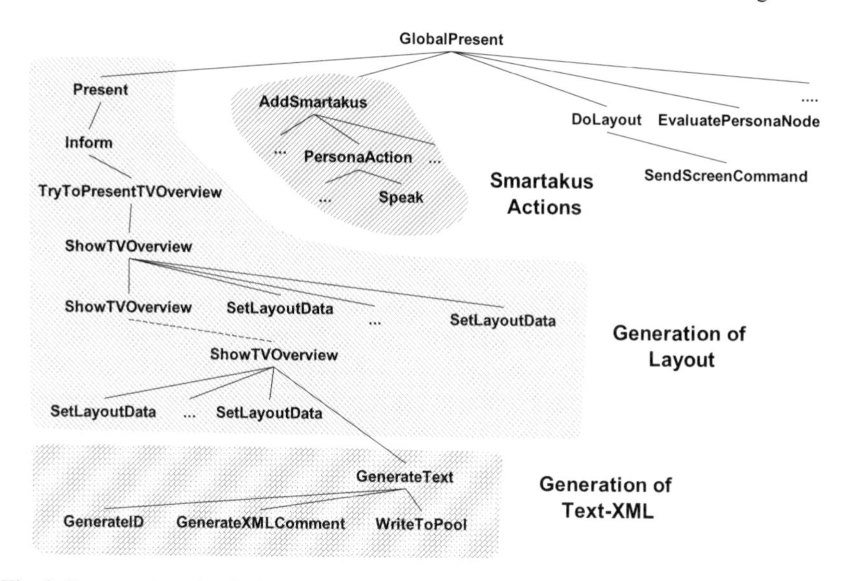

Fig. 9. Presentation plan in SMARTKOM

The presentation planner takes also the size of the graphical display into account. While in SMARTKOM-Home and in SMARTKOM-Public multiple graphical elements can be shown on the screen at the same time, in SMARTKOM-Mobile there is only space for one central graphical object to show on the screen due to the small screen.

In order to solve a presentation task, the Presentation Planner uses a top-down approach with declarative knowledge bases: one static and one dynamic knowledge base. The static knowledge base describes, for example, existing presentation alternatives and dictionaries for text translations. The dynamic knowledge base is derived from the presentation planner input on every presentation task. The presentation planner automatically decomposes the presentation task into subtasks, sub-subtasks, and so on. An important subtask of this planning process is the generation of animation scripts for the presentation agent Smartakus, which have to be related and coordinated with the graphical presentations. The main task of the presentation planner here is the selection of gestures that appropriately supplement the graphical output. The presentation planner composes a special script for Smartakus that is based on the script-language of the PERSONA system (Müller, 2000) on an abstract level. In order to generate speech output, a corresponding speech generation request is sent to the text generator (Becker, 2002) to produce the utterance. The speech synthesis module (Schweitzer et al.) reads this representation and generates the audio data plus the corresponding phoneme sequence for lip synchronization.

The final result of the planning process is a hierarchical plan that reflects the structure of the presentation (Fig. 9). This resulting planning tree is evaluated in order to create a presentation script that is finally passed on to the display manager, which renders the resulting presentation on the available output media and modalities.

3 Graphics and Gesture Generation

To show the generated presentation on a graphic-based medium, the display manager must first generate the layout of all graphical elements as well as configure the presentation agent Smartakus.

3.1 Generation and Layout of Graphical elements

After the presentation script is generated, it is sent to the display manager. The display manager evaluates this script and performs the computational steps as follows.

The display manager first chooses the appropriate graphical layout elements from the knowledge base. According to the planned presentation, these layout elements are dynamically filled with actual data (e.g., movie titles, channel logos, maps) with respect to space restrictions of the chosen graphical elements. This also includes the update of status information that is graphically on-screen, like the display of the virtual cellular phone or the virtual console for TV and video devices.

Since in SMARTKOM multiple graphical objects can be shown at the same time, a layout manager must be used to compute and optimize the arrangement of the objects on the screen and to remove unnecessary objects from the preceding presentation. For that purpose, all layout elements are classified with constraints as follows:

MUST-objects: objects that must be shown in the current context
SHOULD-objects: objects that are not essential but are useful in the current context
CAN-Objects: objects that are currently placed on screen, but are already out of the current context

The classification of the graphical objects is done by the presentation planner while generating the presentation tree (see Sect. 2). The presentation planner defines MUST- and SHOULD-objects for the current presentation task. All other objects that are already on the screen are classified as CAN-objects.

Additionally, constraints for every graphical object can contain boundary information for one and up to four sides. So, some objects can only be placed on a fixed position while others may float within a region. The former is, e.g., essential for the document scanning application, in which the scanning camera has only a limited rotation angle such that all documents must be placed inside the same predefined area. In the document scanning application, a video camera is used to scan a document. After this, the document data is sent via e-mail or a fax machine.

For some objects like status bar for TV and video recorder, it is useful to exclude them from the layout process, because they must always be visible at the same position independent from other objects. This feature can also be set through layout constraints.

Finally, after finishing the graphical presentations, the display manager computes a screen representation in M3L format (see an example in Fig. 10). This representation is published to the SMARTKOM system in order to permit the identification of all objects visible on screen, especially for the gesture analysis to relate gestures to on-screen objects to the real-world objects they represent.

```
<?xml- version="1.0"?>
<presentationContent>
  [...]
  <abstractPresentationContent>
    <performance id="PP325">
      <beginTime> 2003-06-04T20:15:00</beginTime>
      <avMedium>
        <avType>action</avType>
        <avType>scienceFiction</avType>
        <title> Matrix: Reloaded </title>
      </avMedium>
      <cinema>
        <name>Saal 1</name>
        <partOf>
          <movieTheater>
            <name>Lux/Harmonie</name>
          </movieTheater>
          [....]
        </partOf>
      </cinema>
    </performance>
  </abstractPresentationContent>
  [...]
  <panelElement>
    <label id="PM14">
      <boundingShape>
        <leftTop>
          <x>0.0546875</x>
          <y>0.6484375</y>
        </leftTop>
        <rightBottom>
          <x>0.40625</x>
          <y>6888020834</y>
        </rightBottom>
      </boundingShape>
      <contentReference>PP325</contentReference>
      <text>Matrix: Reloaded;  Action, Science Fiction;
          (Lux/Harmonie,    20:15 Uhr)</text>
    </label>
  </panelElement>
  [....]

  <abstractPresentationContent>
    <mapLocation id="PP717">
      <locationName> Lux/Harmonie </locationName>
      <objectTypes>
        <objectType> cinema </objectType>
      </objectTypes>
      <geometries>
        <point>
          <x> 3478601.0 </x>
          <y> 5474947.0 </y>
        </point>
      </geometries>
    </mapLocation>

    [...]

  </abstractPresentationContent>

    [...]

    <label>
      <boundingShape>
        <leftTop>
          <x>0.732421875</x>
          <y>0.39322916666666663</y>
        </leftTop>
        <rightBottom>
          <x> 0.7421875 </x>
          <y> 0.40625 </y>
        </rightBottom>
      </boundingShape>
      <contentReference>PP717</contentReference>
      <text> Lux/Harmonie </text>
    </label>

    [...]

</presentationContent>
```

Fig. 10. Simplified example of the screen representation

Fig. 11. Some life-like characters used in different systems

3.2 Presentation Agent Smartakus

Life-like characters offer great promise for a wide range of applications (Rist et al., 1997; Cassell et al., 2000; Perlin and Goldberg, 1996; Noma and Badler, 1997; Ball et al., 1997) since they make presentations more lively and entertaining and allow for the emulation of conversation styles common in human–human communication. In virtual environments, animated agents may help users learn to perform procedural tasks by demonstrating them (see different agent characters in Fig. 11). Furthermore, they can serve as a guide through a presentation and can release the user from orientation and navigation problems.

In SMARTKOM, according to the SDDP interaction metaphor (Sect. 1.1 above), the presentation agent Smartakus acts as a dialogue partner and represents the whole SMARTKOM system for the user. Smartakus has the role of a real presenter. He points to objects, speaks to the user, and shows mimics, i.e., realizes the multimodal communication means of the systems.

The presentation agent Smartakus is based on the PERSONA system (Müller, 2000). He is instructed by the presentation planner at an abstract level. Abstract complex commands are dynamically converted to simple agent scripts, which are used to control the animated character.

While converting, constraints are used to generate special so-called BETWEEN-gestures if needed. BETWEEN-gestures are automatically generated gestures for a smooth transition between two agent postures. These gestures are used to smoothly combine different gestures within one fluent animation.

The knowledge source for the behaviours and gestures of Smartakus is a large catalog of predefined agent behaviour patterns. These categorized agent behaviours

Fig. 12. Animation of Smartakus in Home/Public scenarios

form the knowledge base of the presentation subsystem for Smartakus. Smartakus itself is statically modeled with 3D Studio Max (Autodesk, US) as a 3D life-like character. But for efficiency reasons, the deeply 3D-modeled gestures, postures, and behaviours are technically rendered as animated GIF files. During run time, the animation of Smartakus is dynamically generated using the elements of this catalog by visually combining elementary gestures to larger animation sequences (see Figs. 12 and 13) in conjunction with inserting appropriate BETWEEN-gestures wherever needed.

Because of the different display sizes in the SMARTKOM scenarios, two Smartakus characters were developed:

Full Smartakus is a complete character. It has a head, body, and two arms with hands. There are more gestures for it in the catalog, and it can do gestures more precisely which makes the animation computationally more difficult.

Reduced Smartakus consists only of a smaller head and one hand. This minimizes the data size to be transferred and makes more places available on the screen for showing information data without cutting any modalities. Lip-synchronized speech output, including gestures and mimics, are equally possible.

Both characters use the lip and gesture synchronization for presenting contents. And both characters can move on the screen to avoid overlapping with information.

4 Output Synchronization and Media Coordination

All deictic gestures have to be synchronized with the display of the corresponding graphical output to which they are related. Also, the acoustic speech output has to be synchronized by appropriate lip movements of the communication assistant Smartakus.

The display manager is responsible for the realization of visible output, while the audio output module realizes speech output. But audio output and visual output are

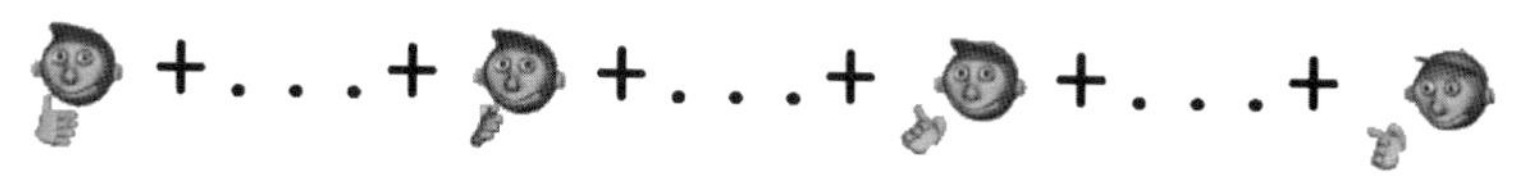

Fig. 13. Animation of Smartakus in Mobile scenario

performed in principle independently from each other and can even be processed and realized on different machines. Therefore to present both modalities simultaneously and synchronously, we have to coordinate them over time to get lip-synchronized output.

4.1 Lip Synchronization

The lip synchronization in SMARTKOM (Poller and Müller, 2002) is based on an underspecified mapping between acoustic and visual units (so-called phonemes and visemes). In our system a viseme is defined as a specific mouth position picture. Due to the cartoonlike character of Smartakus (neither tongue nor teeth are visible) only a limited variety of mouth/jaw positions or movements are possible at all (Fig. 14). Consequently, the only parameters that are relevant in our case to describe mouth positions are the lip rounding and the jaw opening.

Fig. 14. Possible visemes for lip synchronization

We found that almost every phoneme has one single corresponding viseme, while only a few of them (plosives and diphtongs) have to be mapped to at least two visemes to visualize their articulation appropriately (Schweitzer et al.). Thus, in such cases the mapping partly becomes a one-to-many mapping in the sense that one phoneme can be mapped to more than one viseme. Furthermore, the mapping has to be partly underspecified in lip rounding and jaw opening as well to be able to take articulation effects into account. Since the audio output module and display manager are two separate modules in SMARTKOM that in principle work independently from each other, the idea to synchronize lip movements with speech output is to synchronize the individual time points at which corresponding acoustic and visual events occur as exactly as possible.

The speech synthesis module in SMARTKOM not only produces audio data but also produces a detailed representation of the phonemes and their exact time points (in milliseconds) inside the audio signal. Based on this representation and the phoneme–viseme mapping mentioned above, the character animation module generates a lip animation script for Smartakus that is then executed during speech output.

The animation script is generated by a procedure iterating over the phonemes to consecutively specify the concrete viseme(s) and the exact point in time at which they have to be shown in the output animation stream. In terms of Fig. 14 the procedure always tries to select a viseme that has a common borderline with the previous viseme whenever possible (also by inserting intermediate visemes, if necessary).

4.2 Coordinated Lip and Body Animation of Smartakus

During presentation, Smartakus often points to graphical objects on the screen to draw the user's attention to them. To make such presentations more lively and natural, Smartakus is pointing while speaking. For this purpose the body animation (which is performing the pointing gestures) is separated from head (and visemes) animation (which is performing the lip animations). Additionally, heads with visemes for left, front, and right side exist. During pointing the body often moves into a posture for which the head must also be moved horizontally or vertically to realize a natural posture switch of the entire agent. Therefore, for every body posture and every head posture an image with one ground point is generated to be able to calculate body and head coordinates. Such ground point images can be easily generated in advance from the 3D model of Smartakus, such that the synchronization of heads and bodies can be done fully automatically at run time (Fig. 15).

Since the display manager, character animation module, and the audio device of the SmartKom system may run on different computers, the start of the output of the audio data and the start of the lip animation must be synchronized in a special manner. A fixed point in time at which the audio signal has to start is defined dynamically, and at exactly the same time the animation is started by the PERSONA player.

4.3 Visual Perceptive Feedback

In order to increase the liveliness of the character, the character shows idle-time actions like breathing or waving. Another job of the presentation agent is to show the actual system state (also called global system working state) to the user (see Fig. 1 above again for details).

Here is a short explanation of the agent behaviour patterns that are assigned to some of the global system working states (Poller and Reithinger, 2004).

IDLE: In idle state, the agent shows its liveliness by breathing regularly and blinking its eyes.

LISTENING: In the listening state, the animated agent purposefully turns toward the user in order to give him its best attention while he is, e.g., speaking to the system.

UNDERSTANDING: In the understanding state, the agent signals the user that the system is currently analyzing his input. The agent moves its right hand to its temple to scratch with its index finger.

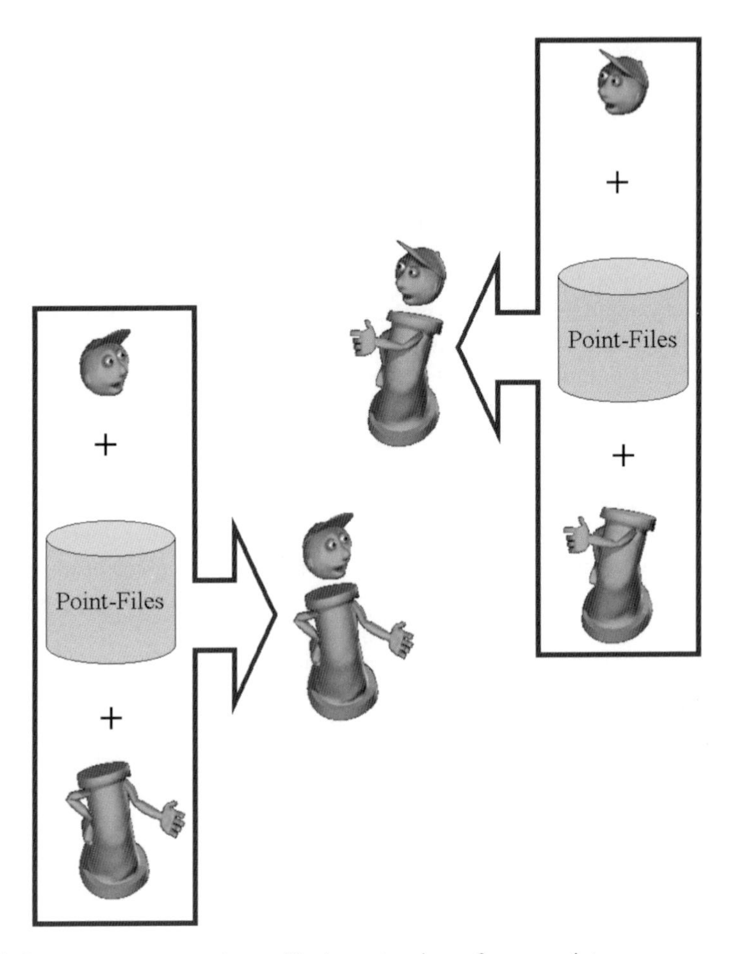

Fig. 15. Gestures are generated out of body parts using reference points

UNDERSTOOD: The upraised index finger is used to signal that the user input was comprehended (global system state "understood").

BARGE-IN: In the "bargeIn" interaction state, an upraised hand is used to show the user that the system is still processing the last input and will not accept barge-in data. Currently, our system cannot interact in a barge-in mode.

PROMPTING: If there were communication pauses, this gesture is used in the idle state to prompt the user implicitly to continue his interaction with the system.

PROCESSING: This gesture is used in the "processing" state to indicate that the system is currently accessing its knowledge sources to fulfill the user request.

MOVE: This gesture is used as part of the output presentations to enable the agent to move around on the screen.

These agent behaviors are shown as soon as the the respective state of the system/conversation is reached.

5 Conclusion

In this chapter we described the design of the novel human–computer interaction in SMARTKOM in line with the *situated delegation-oriented dialogue paradigm (SDDP)* and presented in a detailed compilation the multimodal system output planning and generation process. As shown, this is a very complex procedure containing the modality fission, i.e., the distribution of the output elements over the available output modalities, the generation of the respective modality streams, and finally the coordination and synchronization of all output streams within one multimodal presentation. Consequently, this procedure was divided into a bundle of actively communicating and cooperating modules.

The SMARTKOM solution of coordinating and synchronizing multiple output modality streams in a pleasing and natural user interaction situation was significantly determined by the principle of "interacting though delegation." Having a conversational communication with a virtual communication assistant on screen to which all tasks are delegated brings completely new human-computer interaction methods to light for further investigation.

References

E. André. *Ein planbasierter Ansatz zur Generierung multimedialer Präsentationen.* PhD thesis, Universität des Saarlandes, 1995.

G. Ball, D. Ling, D. Kurlander, J. Miller, and D. Pugh. Lifelike Computer Characters: The Persona Project at Microsoft. In: J.M. Bradshaw (ed.), *Software Agents*, Menlo Park, CA, 1997. AAAI/MIT Press.

T. Becker. Practical, Template–Based Natural Language Generation With TAG. In: *Proc. TAG+6*, Venice, Italy, May 2002.

J. Cassell, J. Sullivan, S. Prevost, and E. Churchill. *Embodied Conversational Agents.* MIT Press, Cambridge, MA, 2000.

G. Herzog, H. Kirchmann, S. Merten, A. Ndiaye, P. Poller, and T. Becker. MULTIPLATFORM Testbed: An Integration Platform for Multimodal Dialog Systems. In: H. Cunningham and J. Patrick (eds.), *Proc. HLT-NAACL 2003 Workshop on Software Engineering and Architecture of Language Technology Systems (SEALTS)*, pp. 75–82, Edmonton, Canada, 2003. Association for Computational Linguistics.

J. Müller. *Persona: Ein anthropomorpher Präsentationsagent für Internet-Anwendungen.* PhD thesis, Universität des Saarlandes, Saarbrücken, Germany, 2000.

T. Noma and N.I. Badler. A Virtual Human Presenter. In: *Proc. IJCAI-97 Workshop on Animated Interface Agents: Making Them Intelligent*, pp. 45–51, Nagoya, Japan, 1997.

S. Oviatt. Multimodal Interfaces. In: J.A. Jacko and A. Sears (eds.), *The Human-Computer Interaction Handbook: Fundamentals, Evolving Technologies and Emerging Applications*, pp. 286–304, Mahwah, NJ, 2003. Lawrence Erlbaum.

K. Perlin and A. Goldberg. Improv: A System for Scripting Interactive Actors in Virtual Worlds. *Computer Graphics*, 28(3), 1996.

P. Poller and J. Müller. Distributed Audio-Visual Speech Synchronization. In: *Proc. ICSLP-2002*, Denver, CO, 2002.

P. Poller and N. Reithinger. A State Model for the Realization of Visual Perceptive Feedback in SmartKom. In: *Proc. Interspeech 2004, (ICSLP, 8th Int. Conf. on Spoken Language Processing)*, Jeju Island, Korea, 2004.

T. Rist, E. André, and J. Müller. Adding Animated Presentation Agents to the Interface. In: *Proc. 1997 Int. Conf. on Intelligent User Interfaces*, pp. 79–86, Orlando, FL, 1997.

A. Schweitzer, G. Dogil, and P. Poller. Gesture-Speech Interaction in the SmartKom Project. Poster presented at the 142nd meeting of the Acoustical Society of America (ASA). http://www.ims.uni-stuttgart.de/∼schweitz/documents.shtml. Cited 11 April 2006.

V. Tschernomas. *PrePlan Dokumentation (Java-Version)*. Deutsches Forschungszentrum für Künstliche Intelligenz, Saarbrücken, Germany, 1999.

W. Wahlster. SmartKom: Symmetric Multimodality in an Adaptive and Reusable Dialogue Shell. In: R. Krahl and D. Günther (eds.), *Proc. Human Computer Interaction Status Conference 2003*, pp. 47–62, Berlin, Germany, June 2003. DLR.

Natural Language Generation with Fully Specified Templates

Tilman Becker

DFKI GmbH, Saarbrücken, Germany
becker@dfki.de

Summary. Based on the constraints of the project, the approach chosen for natural language generation (NLG) combines the advantages of a template-based system with a theory-based full representation. We also discuss the adaption of generation to different multimodal interaction modes and the special requirements of generation for concept-to-speech synthesis.

1 Introduction

The natural language generation (NLG) module of SmartKom is part of the kernel dialogue system and as such is used in all scenarios and applications. Its design is guided by the need to (i) adapt only knowledge sources when adding a new application and (ii) generalizing the knowledge sources from the applications. The NLG module generates not just plain text but rather generates complete syntactic structure and discourse information to supply highly annotated information structures for the concept-to-speech approach of SmartKom see also Schweitzer et al. (2006) on Speech Synthesis.

A further project constraint was the need to adapt the kernel dialogue system quickly to new applications. Initially, this could be achieved by generating fixed system responses for simple dialogues. However, these then needed to be extended to a flexible, "full" generation system.

The obvious choice for the initial system is a so-called template-based NLG system. Usually such a system is perceived as the opposite of full generation. However, as we will argue in the next section, a template-based NLG system can indeed be seamlessly extended into a full system. In SmartKom, we have taken this approach. The perceived differences between so-called full and template-based NLG systems are also discussed, e.g., in van Deemter et al. (1999).

The NLG system of SmartKom is based on lexicalized tree-adjoining grammar (LTAG) with full feature structures as its syntactic representation formalism. Development in the NLG module was mainly for German, but for some applications the templates and rules were also ported to English.[1]

[1] This work was carried out by Johno Bryant and David Gelbart at ICSI, Berkeley.

2 Fully Specified Templates

The design decision to use template-based generation in SMARTKOM is driven by practical considerations: We wanted generator output as early as possible in the project, and there were various applications with a considerable amount of nonoverlapping genres and relatively constrained linguistic coverage, down to a fixed list of utterances in the first systems. On the other hand, simple string concatenation is not sufficient. For example, the integration of concept-to-speech information, especially in the way the synthesis component of SMARTKOM is designed, calls for an elaborate syntactic representation, i.e., phrase structure trees with features, to guide the decisions on prosodic boundaries. At least since Reiter (1995) (also see Becker and Busemann (1999)), the use of templates and "deep representations" is not seen as a contradiction. Picking up on this idea, the generation component in SMARTKOM is based on fully lexicalized generation (Becker, 1998), using whole parts of a sentence together as one *fully specified template* that is then represented not as a string but rather as a partial TAG derivation tree (Fig. 1).

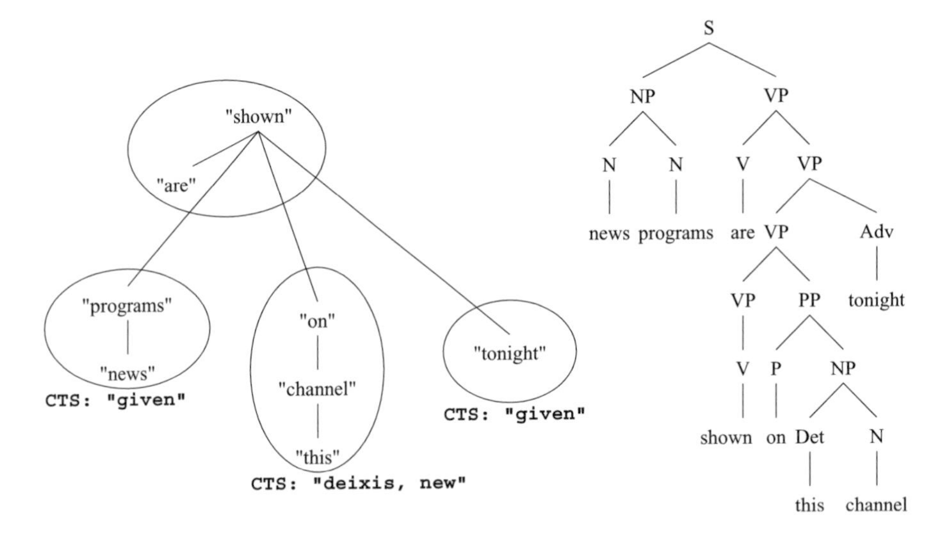

Fig. 1. Derivation tree with partial CTS (concept-to-speech) markup. Each ellipse is a fully specified template. The sentence-planning process combines such templates to a complete derivation tree, which when executed results in the derived tree shown to the *right* (shown without the feature structures attached to each node)

2.1 Specifying Only Relevant Representations

With this approach to specify intermediate levels of representation (which commonly only exist inside the NLG module), the question remains whether *all* levels of representation have to be specified. Clearly, this is desirable but not necessary in SMARTKOM. Thus only the levels of dependency and phrase structure are represented fully. Dependency is necessary to guide the top-down generation approach, and phrase structure is necessary for (concept-to-)speech synthesis. However, there is nothing preventing the later inclusion of other levels of representation, e.g., adding a level of semantic representation (Scheffler, 2003; Joshi and Vijay-Shanker, 1999) to allow for the inclusion of a separate realization module and which might also be used, e.g., by a sentence aggregation component.

3 From Templates to a Full Grammar

The TAG formalism (Joshi et al., 1975; Joshi and Schabes, 1997) offers an elegant way to reconcile the two notions of templates and deep representations. A TAG grammar consists of a set of partial phrase structure trees, so-called elementary trees. These trees are then combined to form the final, so-called derived tree by two elementary operations, adjunction and substitution. The applied operations are recorded in a graph called the derivation tree.

In a lexicalized TAG (LTAG), each elementary tree contains only one leaf node with a lexical element. There, the derivation tree has one node for each lexical element of the sentence and represents a dependency structure between the words. Note also that in an LTAG the grammar is, in fact, a kind of extended lexicon.

However, the pure TAG formalism has no constraints on the number of lexical elements in one elementary tree. In fact, even in LTAG, idiomatic expressions are usually represented as elementary trees with multiple lexical elements. So, for the initial version of the NLG system, we used large elementary trees that contained the complete expressions of the templates. Later versions of the grammar and generation rules then used ever smaller trees until eventually all elementary trees were, in fact, lexicalized in the sense of an LTAG.

3.1 Development Steps

In this section, we describe the intermediate steps that led the development over the time of the project from the initial version with large templates (typically full sentences) to the final version using a fully lexicalized TAG grammar. The work in Scheffler (2003) describes an approach to further extend the lexicalized TAG grammar with a flat semantics to allow for an extraction of a separate syntactic realizer from the NLG module of SMARTKOM.

The development from the initial version went from (i) simple templates to (ii) phrasal templates to (iii) feature-structure-enriched trees to a (iv) system based on a fully lexicalized TAG grammar:

(i) **Simple templates:** The initial SMARTKOM generation component was an instance of a simple template-based NLG system. Our "grammar" consisted of fully specified phrase structure trees at the sentence level that were augmented

with attributes specifying, e.g., sentence mode and part of speech for CTS purposes. Note that no derivation beyond tree selection is necessary when the selected tree is already a complete utterance.[2]
Of course, even the first version went beyond a "canned text" system, and we already needed a planning operator that could instantiate the simple templates (actually, overwrite leaf nodes in the derived tree with strings taken from the generator input XML document).

(ii) **Phrasal templates:** As soon as we encountered scenarios that required more variability, we began to disjoin some sentence-level templates into smaller templates, keeping most features untouched.
Accordingly, the corresponding planning rules grew more complex. In order to capture linguistic generalizations, we began to delegate the planning of noun phrases and (possibly optional) modifiers to subrules and began to extract the corresponding NP elementary trees.

(iii) **Feature-based TAG:** As a next step, all trees were annotated with top and bottom feature structures (Vijay-Shanker, 1987), which include, e.g., case, argument status (subject/object/adjunct/ppobject) and discourse information for the synthesis component. The simple realizer was enhanced with a unifier, enabling us to make use of feature checking, percolation and structure sharing (co-indexing, reentrancy). A morphology module was prepared that could easily be integrated.

(iv) **A full grammar:** The final step then was moving to a fully lexicalized grammar, further disjoining the TAG templates. The resulting set of lexicalized TAG trees can be considered a partial grammar for the current SMARTKOM scenarios. For better maintainability, we decided to name and group them in the vein of the XTAG grammar (XTAG Research Group, 2001). Since we have implemented the standard definition of (L)TAG, the free-word-order phenomena of German are captured only by including all ordering variants as separate elementary trees in our grammar.

Our architecture left room for all of these extensions that were introduced gradually while the generation component was maintained in a working condition. There was no need to fork an experimental or "next generation" development branch or even to start a new architecture from scratch. Linguistic knowledge was added monotonically, i.e., once it was encoded, it did not become obsolete.

The description of our approach so far has concentrated on the Realizer part of the NLG system for which we have executed the development from fully specified templates to a "full" grammar rigorously. Input to the NLG component, however, is on a language- independent conceptual level. Mapping directly from this nonlinguistic input to fully specified German dependency (derivation) trees proved to be quite a challenge for the planning rule authors. As our templates developed into lexicalized TAG trees, planning rules became fine-grained, and a principled way of writing and organizing them had to be found.

[2] An utterance in the SMARTKOM generator is either a single sentence or a dialogue turn that consists of sentences connected by discourse functions.

3.2 Dependency and Speech Markup

Templates in template-based NLG systems are commonly surface strings with gaps that are filled from the input information. In the SMARTKOM NLG system, templates are specified on the level of derivation trees rather than on derived trees (which contain the surface string on their leaves) or even surface strings directly. This has several advantages. In the context of concept-to-speech synthesis, it is necessary to add markup to entire phrases which need not even be contiguous in the surface string. This can be done easily by adding the information to the corresponding node in the derivation tree from where it is percolated and automatically distributed to the corresponding parts of the utterance when constructing the derived tree and the string from the derivation tree. Such markup relates to parts of the output that have to be (de)emphasized, parts that refer to objects in the graphical presentation and must be coordinated with a pointing gesture.

Figure 1 shows a derivation tree with speech-relevant markup on some nodes. Besides mere convenience in the markup,[3] the additional power of TAG allows us to distribute semantically connected markup to discontinuous parts in the final string. Since formal complexity is a very different issue in generation than in parsing, we are open to use extensions from the standard TAG formalism as in Rambow (1994) or Gerdes and Kahane (2001), which might be necessary for German.

4 Adding Levels of Representation

As we have shown in the previous sections, the templates for template-based generation need to have a representation that is as detailed as possible, in order to make generation very flexible, while at the same time preserving its advantages. For relatively restricted input, on the other hand, it is not necessary to fully specify all the levels of representation that are relevant to natural language generation. In this section we will discuss how a modularized generator architecture can be extended by adding new levels of representation.

As the number of possible outputs increases, so does the complexity of the set of planning rules that map from the conceptual input to the level of TAG derivation trees. At some point, this rule set needs to be organized by adding additional representational levels. An extended new architecture is shown in Fig. 2. The new architecture contains two new, linguistically well-defined levels of representation, semantics and argument structure. This helps the modularization of rules, because they can be strictly linguistically motivated (i.e., levels of linguistic representation are not processed all in one step, as previously).

The individual steps and the representational levels of the new generator will be developed as follows:

[3] For example, XML-style opening and closing parentheses can be integrated into the trees and thus are realized by a single marked node vs. the situation in a classical context-free–based string-expanding template generator, where opening and closing elements have to be given independently—a typical source for errors.

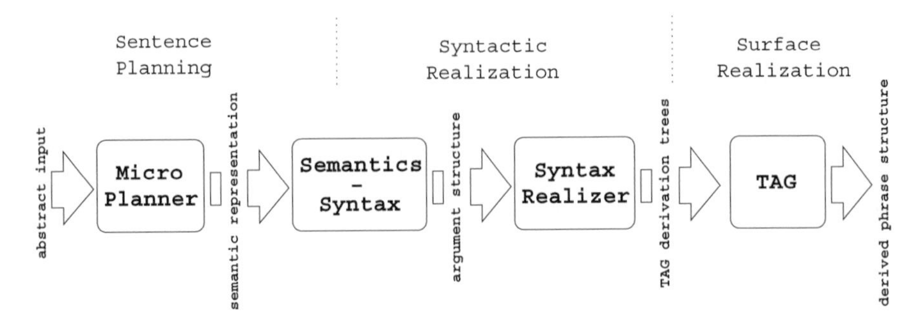

Fig. 2. The improved architecture of the system, with additional levels

Microplanner

Output of the Microplanner will be a flat semantic representation, such as MRS (Copestake et al., 1999). The planner maps the domain-dependent relations of the abstract generator input onto more general semantic relations. The semantic representation must also include additional information of pragmatic and/or situational content, such that this kind of information is available in further steps.

Semantics–Syntax Interface

The representation at the semantics–syntax interface level is argument structure, i.e., a dependency tree enriched with feature annotations at the nodes. Edges are labeled with the kind of dependency relation that holds between the two nodes, e.g., "temporal adjunct," "argument1," "argument2." The arguments are understood more as underlying deep syntactic roles (agens, patiens, etc.) than as concrete forms of realization of those (subject, object, etc.).

During this phase, lexicalization must be carried out, at least as far as part of speech and dependency structure are affected. The heuristics that guides us in defining what specifications are made during the lexicalization and in the argument structure, and which are left open as syntactic or morphological variants to be decided later in the syntactic realizer, is to prefer later specification to earlier specification, as long as this does not inhibit the generation process in any significant way. Choices that will be left for later steps in the generator architecture include synonyms, auxiliaries, active/passive, prepositions, etc.

Syntactic Realization

In this phase, the concrete syntactic representation for an utterance is generated. Argument structure is traversed and appropriate TAG elementary trees are found. This traversal can be done top-down (as, for example, in McCoy et al. (1992)). Edges of the argument structure usually translate into edges in the derivation tree (a tree is substituted or adjoined into its mother of the argument structure), but not always, as is clear for adjectives (multiple adjunctions), etc.

Surface Realization

The derivation trees and derived trees are formally equivalent, because all operations and their order are fully specified in the derivation tree. So in this step, the tree selections, adjunctions and substitutions planned previously and encoded in the derivation tree are actually executed, and feature unification and thus feature percolation (needed for morphology) takes place.

5 Adaptive Multimodal Generation

An important decision in the presentation of system output in a multimodal dialogue system like SMARTKOM is its distribution over the available modalities. Some of the relevant factors are:

- *Appropriateness of media:* Larger sets of systems results, e.g., from a database query, are most conveniently displayed as tables or lists in the graphics modality, leaving a commenting function for the speech modality. Short feedback on the execution of user commands, e.g., switching the TV set to another channel, can best be presented in speech.
- *User preferences:* Given an appropriate user model, personal preferences, e.g., about the level of verbosity in speech output, can be used.
- *Redundancy:* Whenever appropriate, information should be repeated in other modalities, which can resolve ambiguities for the user.

An additional interesting property of SMARTKOM in this context is the definition of various *interaction modes*:

- *Lean-forward mode:* in the home scenario, when the user is sitting on the couch, leaning forward, i.e., looking at and touching the screen, listening and talking
- *Lean-back mode:* in the home scenario, when the user is not looking at the screen, i.e., the interaction is restricted to audio only
- *Theater mode:* in the mobile scenario, when the user uses only the screen
- *Privacy mode:* also in the mobile scenario, when the user uses gestures (and potentially also speech) for input but wants only screen but not audio output from the system, which could be overheard by others

Our implementation of lean-forward and lean-backward in SMARTKOM puts the presentation design into the NLG module: The current mode is given as a parameter, and the initial planning stage selects the presentation templates and the corresponding data. Consider the following example from the home scenario:

User: *What's on TV tonight?*

In lean-forward mode, the display will show a selected list of films, using the screen as the preferred modality for mass data, filling it with the maximum of programs. This maximum depends on the graphical design decisions and available screen area. In this case, speech is used only as a redundant metainformation, explaining the status of this list of films and implicitly giving feedback about the system's understanding of the user request: *Here you can see a list of the films that are on TV tonight.*

In lean-back mode, the display cannot be used, and the communicative constraints of speech have to be considered. In this case, speech is again used to give feedback about the system's understanding of the user request, but only the first three[4] programs are presented: *I have, for example, the following three programs: "Das Boot" is shown on channel PRO7 at 8pm, ...*".

The selection between these presentations is made within the NLG module, which contains separate sets of planning rules for the different presentation modes. Note that the output differs in the amount of information conveyed, thus making a request for more information more likely in the lean-back case. This could be made explicit in changing the expectations published by the action planner, however, SMARTKOM's ability to deal with mixed initiative dialogue allows the system to handle all possible user reactions in an equal manner.

6 Tools

Currently we have editors available[5] for the planning rules and for the TAG-tree templates. Both build on XML representations of the knowledge bases and present them in an easily accessible format: a directory structure as known from the Windows Explorer for the set of trees and a graphical tree editor for the TAG trees.

In an ideal development environment, with a (partial) TAG grammar in existence, the rule editor can be enhanced with a TAG-parser: To add a new template to the generator, the user can type in an example sentence, have it parsed, select the correct parse, mark (delete) the variable parts, keeping the fixed part and add the remainder of the rule. Thus rules can be created without ever writing trees by hand, thereby avoiding inconsistencies in the set of templates.

7 Conclusion

Modular design and a reasonable amount of additional work (for fully specifying the templates) make template-based generation systems scalable, even up to a "full" system, while still keeping the advantage of rapid development of initial systems. We showed that extension of the generator is possible along multiple axes: making the templates smaller to get a grammar, extending the syntactic representation and adding new representational levels to the generator architecture.

---

[4] The exact number is a parameter to be determined by experiments and also a possible parameter for a user model.

[5] Thanks are due for the implementation and also further work on the NLG system by Jan Schehl, Quan Nguyen, Michael Kaißer, Christian Pietsch, and Tatjana Scheffler.

Thus, the SMARTKOM NLG system supports rapid prototyping without closing the path to a step-by-step development of full system. The TAG grammar formalism is particularly well suited to this approach since all relevant information is kept in the elementary trees, which can contain either just one lexical item (LTAG) or multiple words up to entire sentences. This setup also supports the needs of the concept-to-speech synthesis approach in SMARTKOM. Finally, the multimodal setup raises new issues in adapting the NLG system in its multimodal setting to the various interaction modes defined for SMARTKOM.

Acknowledgments

The author is indebted to Tatjana Scheffler and Christian Pietsch for their work on the NLG module and their contributions to early versions of parts of this chapter.

References

T. Becker. Fully Lexicalized Head-Driven Syntactic Generation. In: *Proc. 9th Int. Workshop on Natural Language Generation*, Niagara-on-the-Lake, Canada, August 1998.

T. Becker and S. Busemann (eds.). *May I Speak Freely? Between Templates and Free Choice in Natural Language Generation*, Bonn, Germany, September 1999. Workshop at the 23rd German Annual Conference for Artificial Intelligence (KI '99).

A. Copestake, D. Flickinger, I.A. Sag, and C. Pollard. Minimal Recursion Semantics. An Introduction. Draft, 1999.

K. Gerdes and S. Kahane. Word Order in German: A Formal Dependency Grammar Using a Topological Hierarchy. In: *Proc. ACL 2001*, Toulouse, France, 2001.

A. Joshi and Y. Schabes. Tree-Adjoining Grammars. In: G. Rozenberg and A. Salomaa (eds.), *Handbook of Formal Languages*, vol. 3, pp. 69–124, Berlin Heidelberg New York, 1997. Springer.

A.K. Joshi, L.S. Levy, and M. Takahashi. Tree Adjunct Grammars. *Journal of Computer and System Science*, 10(1), 1975.

A.K. Joshi and K. Vijay-Shanker. Compositional Semantics for Lexicalized Tree-Adjoining Grammars. In: *Proc. 3rd Int. Workshop on Computational Semantics*, Utrecht, The Netherlands, January 1999.

K. McCoy, K. Vijay-Shanker, and G. Yang. A Functional Approach to Generation With TAG. In: *Proc. 30th ACL*, pp. 48–55, Newark, DE, 1992. Association for Computational Linguistics.

O. Rambow. *Formal and Computational Models for Natural Language Syntax*. PhD thesis, University of Pennsylvania, Philadelphia, PA, 1994. Also vailable as Technical Report ICRS-94-08 from the Institute for Research in Cognitive Science.

E. Reiter. NLG vs. Templates. In: *Proc. 5th European Workshop in Natural Language Generation*, pp. 95–105, Leiden, The Netherlands, May 1995.

T. Scheffler. A Semantics Interface and Syntactic Realizer for TAG. Master's thesis, University of the Saarland, Germany, 2003.

A. Schweitzer, N. Braunschweiler, G. Dogil, T. Klankert, B. Möbius, G. Möhler, E. Morais, B. Säuberlich, and M. Thomae. Multimodal Speech Synthesis, 2006. In this volume.

K. van Deemter, E. Krahmer, and M. Theune. Real vs. Template-Based Natural Language Generation: A False Opposition? In: T. Becker and S. Busemann (eds.), *May I Speak Freely? Between Templates and Free Choice in Natural Language Generation*, KI-99 Workshop, Bonn, Germany, 1999.

K. Vijay-Shanker. *A Study of Tree Adjoining Grammars*. PhD thesis, University of Pennsylvania, Philadelphia, PA, 1987.

XTAG Research Group. A Lexicalized Tree Adjoining Grammar for English. Technical Report IRCS-01-03, The Institute For Research In Cognitive Science (IRCS), Philadelphia, PA, 2001.

Multimodal Speech Synthesis

Antje Schweitzer[1], Norbert Braunschweiler[1], Grzegorz Dogil[1], Tanja Klankert[1], Bernd Möbius[1], Gregor Möhler[1], Edmilson Morais[1], Bettina Säuberlich[1], and Matthias Thomae[2]

[1] Institute of Natural Language Processing, University of Stuttgart, Germany
{antje.schweitzer,braunnt,dogil,klankert,moebius,moehler,emorais,bettina.saeuberlich}@ims.uni-stuttgart.de
[2] DaimlerChrysler AG, Research and Technology, Ulm, Germany
thomae@ei.tum.de

Summary. Speech output generation in the SmartKom system is realized by a corpus-based unit selection strategy that preserves many properties of the human voice. When the system's avatar "Smartakus" is present on the screen, the synthetic speech signal is temporally synchronized with Smartakus visible speech gestures and prosodically adjusted to his pointing gestures to enhance multimodal communication. The unit selection voice was formally evaluated and found to be very well accepted and reasonably intelligible in SmartKom- specific scenarios.

1 Introduction

The main goal of the speech synthesis group in SmartKom was to develop a natural sounding synthetic voice for the avatar "Smartakus" that is judged to be agreeable, intelligible, and friendly by the users of the SmartKom system. Two aspects of the SmartKom scenario facilitate the achievement of this goal. First, since speech output is mainly intended for the interaction of Smartakus with the user, most of the output corresponds to dialogue turns generated by the language generation module (Becker, 2006). Therefore, most speech output can be generated from linguistic concepts produced by the language generation module ("concept-to-speech synthesis", CTS) instead of from raw text ("text-to-speech synthesis", TTS). The advantage of CTS over TTS is that it avoids errors that may be introduced by linguistic analysis in TTS mode. Second, the CTS approach narrows down the SmartKom synthesis domain from a theoretically open domain to a restricted domain, which makes unit selection synthesis a promising alternative to diphone synthesis for the SmartKom application.

Multimodality introduces additional requirements for the synthesis module. The visual presence of Smartakus on the screen during speech output requires lip synchronization. Furthermore, Smartakus executes pointing gestures that are related to objects which are also referred to linguistically. These pointing gestures influence the prosodic structure of the utterance and necessitate temporal alignment of the gestural and linguistic modes. Another momentous requirement was that the graphical design of Smartakus was given before the voice database was recorded. This entailed that the appropriateness of the speaker's voice for Smartakus could be included as an important factor in the speaker selection process.

In developing the synthesis voice for Smartakus, we pursued the following strategy: After the speaker selection process, a diphone voice was developed first. This voice served both as a starting point for implementing a unit selection voice by the same speaker tailored to the typical SMARTKOM domains, and as the default voice for external open domain applications that require TTS instead of CTS. The diphone voice and the unit selection voice were both evaluated in the progress of the project.

This chapter is organized as follows. We focus on the prosody generation in CTS mode in the subsequent section. The speaker selection process is described in Sect. 3. Sect. 4 concentrates on the unit selection voice. Lip synchronization and gesture-speech alignment are discussed in Sect. 5. Finally, the two evaluation procedures are described in Sect. 6.

2 Concept-to-Speech Synthesis

The motivation for CTS synthesis is the view that the linguistic content of an utterance determines its phonological structure and prosodic properties. It has been shown that prosodic structure can reflect aspects of syntactic structure, information structure, and discourse structure (Culicover and Rochemont, 1983; Selkirk, 1984; Cinque, 1993; Hirst, 1993; Abney, 1995; Büring, 1997; Mayer, 1999). The challenge in TTS is that text represents only a very reduced version of the full linguistic content of an utterance. It not only lacks marking of higher-level linguistic structure, but may also be ambiguous with respect to syllabic and segmental structure due to abbreviations and homographs. All these properties have to be inferred from the text in TTS. The idea of CTS is to use the full linguistic structure of an utterance, i.e., the original "concept," instead of its raw textual representation. This structure is available in dialogue systems that generate utterances dynamically. In SMARTKOM, it is available with some exceptions: many utterances contain material retrieved from external databases, such as movie titles or geographical names. Although the overall structure of such utterances is known, the internal structure of the retrieved items is unknown. They may contain abbreviations, material in unknown languages, or, particularly in the case of movie titles, may even have their own internal linguistic structure.

The main advantage of CTS in SMARTKOM is therefore the availability of higher-level linguistic structure, which influences the prosodic structure of an utterance. Cinque (1993) gives a detailed account of how syntactic structure determines the default location of sentence stress. We have implemented an algorithm motivated by Cinque's findings. The prediction of prosodic structure including pitch accent and boundary types from linguistic structure is described in more detail in Schweitzer et al. (2002). Here we only give a brief description of the concept structure, the prediction of phrasing, and the implementation of Cinque's account for accent placement.

2.1 Concept Input

Concepts in SMARTKOM contain information on three linguistic levels. The highest level of annotation used for prosody prediction is the **sentence level**. Sentence mode (declarative, imperative, yes/no-question, or wh-question) is annotated on this level. This kind of information is mainly required for the prediction of boundary tones.

The next-lower level is **syntactic structure**. Syntactic trees in SMARTKOM are binary branching, and they may include traces resulting from movement of syntactic constituents. They are generated from smaller tree segments within the tree-adjoining grammar framework (Becker, 1998). Semantic and pragmatic information is integrated into the syntactic structure as follows. For each node of the syntactic tree, its argument status (subject, direct or indirect object, prepositional object, sentential object, or adjunct) and its information content (new vs. given) can be specified. Deixis is also specified on the syntactic level. Deictic elements occur when Smartakus executes pointing gestures referring to objects on the screen.

The lowest level of annotation is the **lexical level**. On this level, material that originates from database queries is inserted. The domain and language of this material are annotated if available. An example of the syntactic and lexical levels of a concept structure is given in Fig. 1.

2.2 Prediction of Prosodic Phrases

The first step in prosody generation is the prediction of prosodic phrase boundaries. There are two levels of phrases: Intonation phrases are terminated by major breaks ("big breaks", BB) and can be divided into several intermediate phrases, which in turn are separated by minor breaks (B).

Syntactic structure has been shown to be useful in the prediction of prosodic phrasing (Schweitzer and Haase, 2000). Particularly the insertion of prosodic phrase breaks between topicalized constituents in the *Vorfeld* (i.e., constituents preceding the finite verb in verb-second sentences) and the rest of the sentence has proved to be a common phenomenon in natural speech, if the material in the Vorfeld is long enough (Schweitzer and Haase, 2000). The Vorfeld corresponds to a syntactic constituent, a maximal projection, that is in the specifier position of another maximal projection (depending on the syntactic theory a verbal, inflectional, or complementizer projection). Prosodic breaks can also occur between constituents in the *Mittelfeld*. Another observation is that breaks are less likely between heads and complements than between heads and adjunct constituents. In any case, the longer the constituents are, the more likely the breaks are inserted. Usually, the inserted breaks are minor breaks; occasionally even major breaks occur.

These observations motivate the two rules in (1) and (2), which insert optional minor breaks. The [$\pm$B] feature indicates that a break can be inserted at the end of

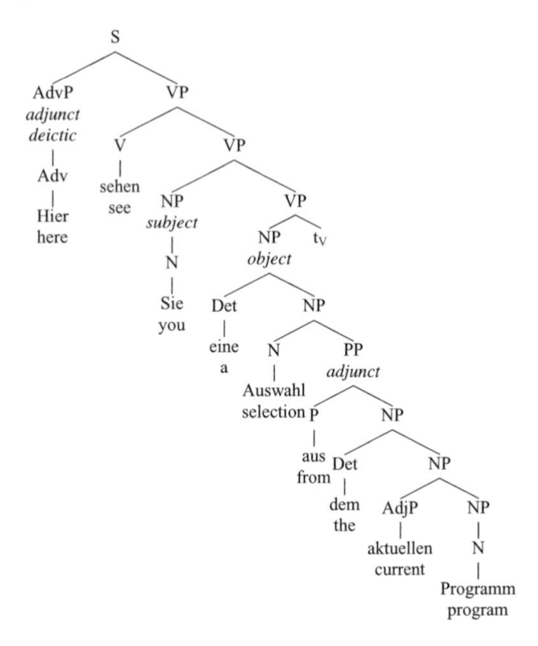

Fig. 1. Integration of additional information into the syntactic and lexical structure of the sentence *Hier Sehen Sie Eine Auswahl Aus Dem Aktuellen Programm*. In this example, deixis (*deictic*) and argument status (*adjunct*, *subject*, *object*) are added. These values are indicated in *italics*

the respective constituent. The rules each have two variants, (a) and (b), which are mirror images of each other.

(1) a. XP → $YP_{[\pm B]}$ XP

b. XP → $XP_{[\pm B]}$ YP

(2) a. XP → $X_{[\pm B]}$ YP (*adjunct*)

b. XP → $YP_{[\pm B]}$ (*adjunct*) X

The first rule states that maximal categories (the YPs in (1)) that are daughters of other maximal categories (the dominating XPs in (1)) can be separated from their sister node by a minor break. S constituents are also treated as maximal projections.

Since we do not distinguish X-bars from XPs, rule (1) applies to any maximal projection that is not the sister of a head. Examples of the application of (1) are the insertion of boundaries between topicalized constituents and the VP as well as between adjacent constituents within the VP. The second rule allows breaks to be inserted between the head of a phrase (the X in (2)) and its sister node, but only if the sister node is an adjunct. Thus, phrase boundaries between a head and its argument are excluded.

Deictic elements often trigger additional minor phrase breaks. A pilot study on material from 26 speakers showed that deictic expressions, i.e., expressions that were accompanied by pointing gestures, were usually marked by a phrase break or an emphatic pitch accent or both. This effect is modeled by inserting mandatory minor breaks preceding and following deictic expressions.

The result of the phrase break insertion for the sentence in Fig. 1 is shown in (3). Mandatory phrase breaks are (trivially) at the end of the utterance, and after the deictic AdvP *hier*, indicated by the [+BB] and [+B] features, respectively. Optional phrase breaks are inserted after the NP *Sie* according to rule (1a), and after the noun *Auswahl* according to rule (2a). These optional breaks are marked by the feature [±B] in (3).

(3) Hier [+B] sehen Sie [±B] eine Auswahl [±B]
aus dem aktuellen Programm [+BB]

In a second step, a **harmonization algorithm** selects candidates from the set of possible combinations of prosodic phrases. Candidates whose mean phrase length lies in a given optimal range and which show an even phrase length distribution are favored over other candidates. Thus, the observation that the insertion of breaks depends on the length of the resulting phrases is accounted for, and sequences of phrases that are unbalanced in terms of number of syllables per phrase are avoided if possible. The optimal range for the mean phrase length was found to be more than 4 to less than 11 syllables.

For the example in Fig. 1, the optimal candidate is shown in (4a). The other candidates are given in (4b) through (4d). Syllable number per phrase, mean phrase length, and variance are indicated in italics. Rule 4b is discarded because its mean phrase length is not in the optimal range. From the remaining three candidates, (4a) is chosen because it has the smallest variance.

(4) a. Hier [+B] sehen Sie eine Auswahl [+B] aus dem aktuellen Programm [+BB]
syllables: 1, 7, 8; mean: 5.33; variance: 9.55

b. Hier [+B] sehen Sie [+B] eine Auswahl [+B] aus dem aktuellen Programm [+BB]
syllables: 1, 3, 4, 8; mean: 4; variance: 6.5

c. Hier [+B] sehen Sie [+B] eine Auswahl aus dem aktuellen Programm [+BB]
syllables: 1, 3, 12; mean 5.33; variance: 22.89

d. Hier [+B] sehen Sie eine Auswahl aus dem aktuellen Programm [+BB]
syllables: 1, 15; mean 8; variance: 49

Finally, boundary tones are assigned to each predicted major phrase boundary. For sentence-internal phrase boundaries, a rising boundary tone is assigned to indicate continuation. In all other cases, the boundary tone depends on the sentence mode.

2.3 Accent prediction

The default location of sentence stress is determined by the syntactic structure according to Cinque (1993). We have adapted this procedure to predict the default accent location for each prosodic phrase. Additionally, semantic factors can cause deaccentuation. Pitch accent types depend on the information content of the accented word, on its position in the phrase, and on sentence mode.

According to Cinque (1993), the default accent is on the syntactically most deeply embedded element, as illustrated by the prepositional phrases in (5) (from Cinque (1993)). The underlined words are the most deeply embedded elements, and they are accented by default.

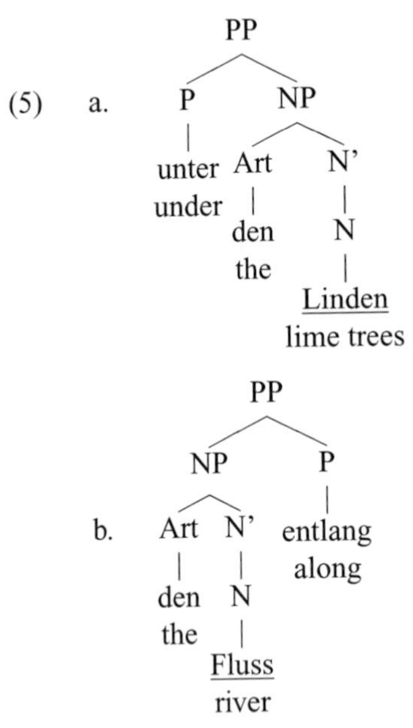

However, depth of embedding on the nonrecursive side is irrelevant, as shown in (6) (from Cinque (1993)): In neutral accentuation, the pitch accent falls not on the overall most deeply embedded element, *Italienern*, but on *Impfstoff*. This is because NPs are right-recursive. Depth of embedding, according to Cinque (1993), is only counted on a path along the X-bar axis (e.g., connecting XP and X', X' and X') and on the recursive side of each projection XP (e.g., connecting X' to a YP embedded on the left side, if XP is a left-recursive category; or connecting X' to a YP embedded on the right side, if XP is a right-recursive category). The main path of embedding is the path that reaches the top node. The overall most prominent element is the most deeply embedded element on the main path of embedding. In constituents on

(6)

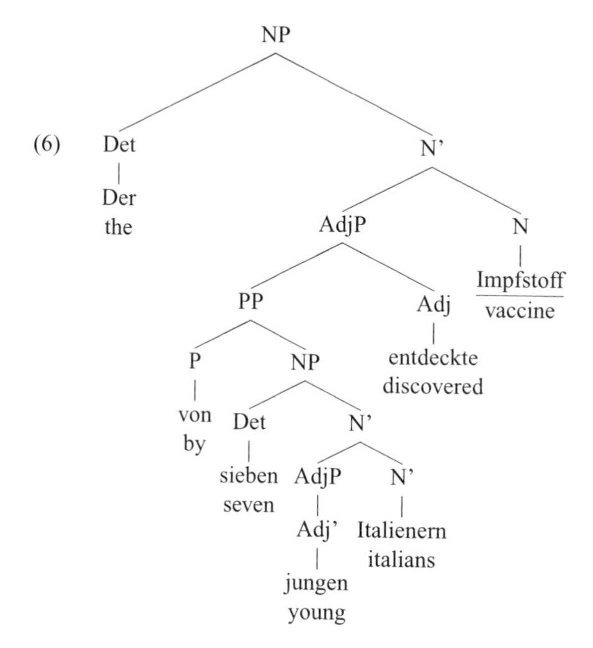

the nonrecursive side, depth of embedding determines the locally most prominent element in the constituent, but its depth of embedding is irrelevant for the location of the main stress.

This procedure had to be modified for two reasons. First, in the syntactic structures used in SMARTKOM, there are no X-bars. Thus, the main path is along an axis connecting XPs with embedded XPs, or connecting an XP with a maximal projection YP on the recursive side of the XP, if YP is a sister to the head X of XP. Second, large syntactic trees will usually be split into smaller units by the phrase prediction algorithm. In the phrases that do not contain the globally most prominent element according to the definition above, we still need to assign an accent to the locally most prominent element. The resulting procedure works as follows. In each phrase, the element with the smallest number of branches on the nonrecursive side is accented. If there are several elements with the same number of branches, the last one is accented. Depending on the information structure of an utterance, accentuation can deviate from the default accentuation: Words are deaccented if they are marked as "given" in the respective context, and narrow focus moves the accent from the default location to the focused constituent.

For each accented element, its accent category depends on its position in the phrase, its information content, and the sentence mode. We use a subset of the pitch accent inventory of the German ToBI labeling system as described in Mayer (1995), viz. L*H as a rising accent, H*L as a falling accent, and L*HL as an emphatic accent. For the diphone voice, the type of accent determines the template used for modeling the fundamental frequency contour (Möhler and Conkie, 1998). For the unit selection

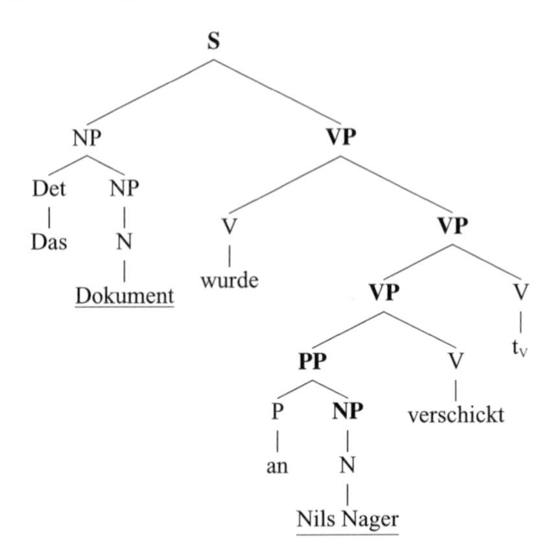

Fig. 2. Syntactic structure for the example "Das Dokument wurde an Nils Nager verschickt" (*The document was sent to Nils Nager*). The main path of embedding is marked by the nodes in *boldface*. The default accents for the two phrases are assigned to the underlined words

voice, it restricts the candidate set to candidates realized with a similar accent (see Sect. 4).

2.4 An Example

The complete prosody prediction algorithm is illustrated by the example in (7) and (8). An optional phrase break is inserted between the topicalized object *Das Dokument* and the finite verb *wurde*. The harmonization algorithm selects (8a) because (8b)'s mean phrase length exceeds the upper limit of 11 syllables per phrase and is therefore not considered in the selection step. Otherwise, it would have been preferred over (8a) because its variance is smaller.

(7) Das Dokument wurde an Nils Nager verschickt.
The document was to Nils Nager sent
The document was sent to Nils Nager.

(8) a. Das Dokument [+B] wurde an Nils Nager verschickt [+BB]
syllables: 4, 8; mean: 6.00; variance: 4.00
b. Das Dokument wurde an Nils Nager verschickt [+BB]
syllables: 12; mean: 12.00; variance: 0.00

The syntactic structure of (8) is shown in Fig. 2. The main path of embedding is indicated by the nodes in boldface. For the first phrase, the default accent is assigned to the noun *Dokument* because the path from the top of the tree to the noun contains only one branch (connecting S to the NP on its left) that is neither on the recursive side nor on the X-bar axis. The branches connecting NP to NP and NP to N are on the recursive side because NPs are right-branching. The path to the determiner *Das* contains two branches that are on the nonrecursive side: the branches connecting S to NP and NP to Det, respectively. In the second phrase, the name *Nils Nager* is on a path exclusively along the X-bar axis or along branches on the recursive side. It is therefore accented.

Since the sentence is a declarative sentence, it is terminated by a falling boundary tone. The accented element in the second phrase is assigned a falling accent for the same reason. The accent in the first phrase is predicted to be rising because the sentence continues across the intermediate phrase boundary between the two phrases. Thus, the prosodic structure for (8) is as shown in (9):

(9) Das Dokument wurde an Nils Nager verschickt.
L*H - H*L L%

3 Speaker Selection

Several constraints have to be met in the speaker selection process. On the one hand, users' expectations include not only intelligibility but also more subjective properties such as agreeableness, pleasantness, and naturalness. Adequacy of the voice for the target application may be even more important than subjective pleasantness. For instance, Smartakus is a small blue-colored cartoon-like character reminiscent of the letter "i". This visual appearance did not seem to go well with one particularly deep candidate voice, which was rated high by listeners only when presented independent of Smartakus. There are additional, more technical and practical, requirements, such as the experience of the speaker, which can decisively reduce the time needed for the recordings, but also foreign language skills, which are required for some nonnative diphones, as well as the speaker's availability over a longer period of time.

The subjectively perceived properties of a diphone voice are currently not predictable from the speaker's natural voice. The prediction is less difficult for unit selection voices because they preserve the characteristics of the original voice much better by reducing the number of concatenation points and the amount of signal processing. However, the number of concatenation points may be similar to that in diphone synthesis, in which case the subjective voice quality is almost as hard to predict as in the diphone case. To ensure that the selected speaker's voice is suitable for diphone synthesis in the sense that the resulting diphone voice is still judged to be agreeable, we built a test diphone voice for each speaker. To this end, a small diphone set was recorded that covered the diphones required for synthesizing three short sentences. The speaker with the best test voice was selected.

Since recording and building a diphone database is very time-consuming, even for the rather small set of diphones needed for the test voices, we split the speaker selection process into two phases. In the first phase, we asked speakers to record some SmartKom-specific material. This material included a short dialogue typical of a SMARTKOM domain, a list of (nonsense) diphone carrier words, and three short excerpts from movie reviews in German, English, and French. Some speakers sent in demo tapes, and some were recorded in an anechoic recording booth at our lab. Altogether, we collected demo material from 40 speakers, 29 female and 11 male. For each voice, some representative sentences were selected and rated for their subjective qualities in an informal evaluation procedure. Most participants in this rating procedure were colleagues from our institute.

In the second phase, the ten best speakers from the first phase, four male and six female, were invited to our lab to record the diphone set required for our three test sentences. The diphones were manually labeled and afterwards processed by the MBROLA[3] group at Mons. We carried out a formal evaluation with 57 participants; 20 participants were experienced, and 37 were "naive" with respect to speech technology. The three target sentences were synthesized for each speaker using different signal processing methods (MBROLA (Dutoit et al., 1996), PSOLA (Moulines and Charpentier, 1990) and Waveform Interpolation, (Morais et al., 2000)) and different prosody variants (the speaker's original prosody vs. prosody as predicted by our TTS system, with the pitch range adapted to the respective speaker's pitch range in the latter case). Some of the stimuli were presented as video clips showing Smartakus speaking, but without correct lip synchronization. Participants were asked to rate the stimuli for naturalness on a five-point scale from −2 to +2, where −2 corresponded to "not natural" and +2 corresponded to "very natural." Mean scores were calculated for every stimulus.

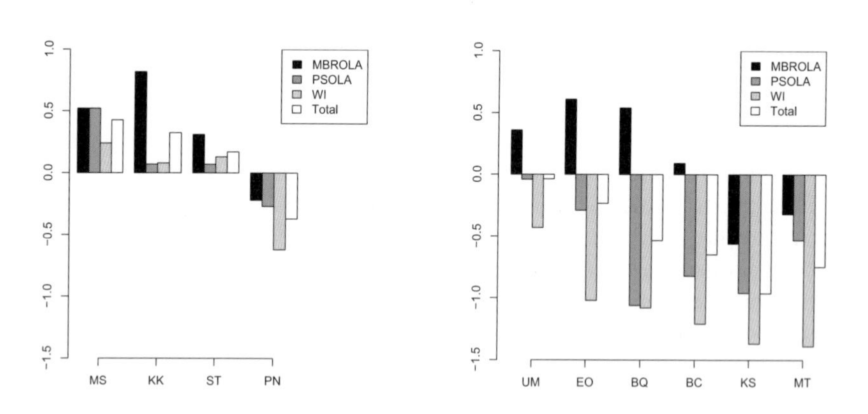

Fig. 3. Mean scores of audio stimuli for the male voices (*left panel*) and the female voices (*right panel*), broken down by signal processing method. Speakers are indicated on the x-axis, mean scores on the y-axis. Averaged over the different methods (*white bars*), the MS voice was rated the most natural, but when looking at MBROLA voices only (*black bars*), KK's voice was clearly better

[3] http://tcts.fpms.ac.be/synthesis/

The most important outcome of the evaluation procedure was that the subjective ranking of speakers was different for the two steps. For instance, the left panel in Fig. 3 shows that for the best four male speakers from the first step, the MBROLA diphone voice of KK, who was originally ranked third, was judged to be the most natural diphone voice, and the second most natural diphone voice was from MS, who was originally ranked fourth. Other signal processing methods yielded different rankings; in these cases, the MS diphone voice was judged to be the most natural. Similar effects are evident in the ranking of the female diphone voices in the right panel of Fig. 3. This confirmed our expectation that the subjective quality of the diphone voice does not correlate directly with the subjective quality of the original voice.

It is evident in Fig. 3 that MBROLA turned out to be the best signal processing method in all cases. Male voices were generally rated better than female voices, especially for signal processing methods other than MBROLA.

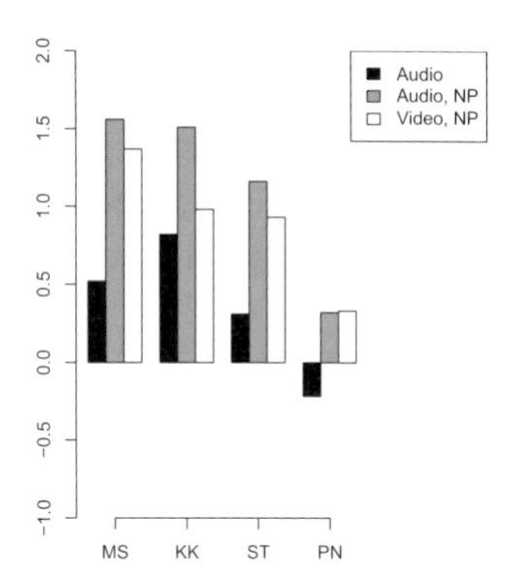

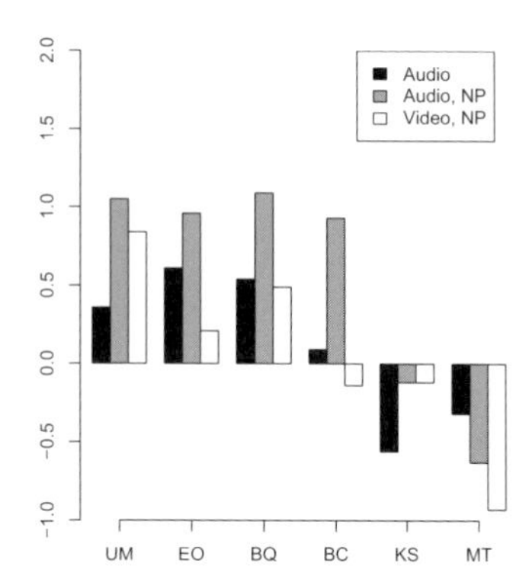

Fig. 4. Mean scores for MBROLA voices presented in audio mode, averaged over natural and rule-based prosody (*black bars*), with natural prosody (NP) only (*gray bars*), and for MBROLA voices presented in video mode with natural prosody only (*white bars*). All except one voice (MT) are rated better when the speaker's natural prosody is used in the audio-only condition. In the video condition, the ranking is different. MS is rated best in video mode. Generally, the male voices are preferred in this mode

In spite of Smartakus' relatively androgynous features, there was an even stronger preference for the male voices in the video clips. This is illustrated in Fig. 4. Only the results for the MBROLA voices are presented here, which were again rated better than the other voices. In the video condition, only the speakers' original prosody was used, which was transplanted onto the diphone voices. To assess the influence of the natural prosody in the ratings of the video stimuli, we included the ratings for audio stimuli with natural prosody in the diagram. The preference for MS in the video condition was not due to the natural prosody: While MS and KK were rated similarly good for those stimuli, MS was clearly preferred in the video condition. Thus, as alluded to above, the speaker rated best for the audio-only stimuli was rated much lower for the audio–video stimuli, presumably because of his low pitch, which did not seem to go well with the cartoonlike features of Smartakus.

Based on these results, MS was selected as the speaker for SMARTKOM. We recorded both a diphone and a unit selection database of this speaker. The diphone voice was used in the project during the development of the unit selection voice, and it was also used as a baseline synthesis voice in the evaluation of the unit selection voice.

4 Restricted Domain Unit Selection Synthesis

The SMARTKOM domains are restricted but not limited: Utterances are generated from a number of lexicalized partial syntactic trees (Becker, 1998, 2006), but open slots are filled with names, proper nouns, movie titles, etc., from dynamically changing external and internal databases. The vocabulary is therefore unlimited, although it is biased toward domain-specific material. The predominance of domain-specific material calls for a unit selection approach with a domain-specific speech database to ensure optimal speech synthesis quality for frequent phrases. However, since the vocabulary is theoretically unlimited, domain-independent material must be taken into account as well. This is especially important because the vocabulary shows typical "large number of rare events" (LNRE) characteristics (Möbius, 2003). Although each infrequent word on its own is very unlikely to occur, the probability of having an arbitrary infrequent word in an utterance is very high.

Domain-specific and domain-independent materials pose different requirements for the unit selection strategy. Domain-specific phrases may often be found in their entirety in the database. In this case, it may be unnecessary to even consider candidates made up of smaller noncoherent units. Domain-independent material, on the other hand, will usually have to be concatenated from much smaller units, such as single segments, demanding a carefully designed database with optimal coverage and a selection algorithm that can handle larger amounts of possible candidates. Therefore, a hybrid approach was implemented combining two existing strategies (Schweitzer et al., 2003). It is described in the following section. Details on the construction of the unit selection corpus are presented in Sect. 4.2.

4.1 Unit Selection Strategy

Current unit selection approaches mostly use segments (Hunt and Black, 1996; Black and Taylor, 1997; Breen and Jackson, 1998) or subsegmental units such as halfphones (Beutnagel et al., 1999; Conkie, 1999) or demiphones (Balestri et al., 1999) as the basic unit. For each unit in the target utterance, several candidates are selected from the speech database according to criteria such as segment identity and segmental and linguistic context. For each candidate, its *target cost* expresses how well it matches the specification of the target unit. For each pair of candidates, their *concatenation cost* measures the acoustic distortion that their concatenation would cause. Then the sequence of candidates is chosen that simultaneously minimizes target and concatenation costs. Since there is no distortion for originally adjacent units, longer stretches of successive units are favored over single nonadjacent units, reducing the number of concatenation points and rendering a more natural voice quality. We will call this a bottom-up approach because, starting from the segmental level, the selection of complete syllables, words, or phrases arises indirectly as a consequence of the lower concatenation costs for adjacent segments.

Such an approach faces two challenges. First, target costs and concatenation costs must be carefully balanced. Second, for frequent units the candidate sets can be very large, and the number of possible sequences of candidates grows dramatically with the number of candidates. For performance reasons, the candidate sets must be reduced, at the risk of excluding originally adjacent candidates.

One way to achieve the reduction of unit candidate sets is to cluster the units acoustically in an offline procedure and to restrict the candidate set to the units of the appropriate cluster (Black and Taylor, 1997). We will refer to this method as the acoustic clustering (AC) approach. The idea is to cluster all units in the database according to their linguistic properties in such a way that the acoustic similarity of units within the same cluster is maximized. In other words, the linguistic properties that divide the units into acoustically similar clusters are those properties that apparently have the strongest influence on the acoustic realization of the units in the cluster. During synthesis, the linguistic context determines the pertinent cluster. All other units are ignored, which reduces the number of candidates.

Some approaches (Taylor and Black, 1999; Stöber et al., 2000) use a different strategy. Candidates are searched top-down on different levels of the linguistic representation of the target utterance. If no candidates are found on one level, the search continues on the next-lower level. If appropriate candidates are found, lower levels are ignored for the part of the utterance that is covered by the candidates. For the phonological structure matching (PSM) algorithm (Taylor and Black, 1999), candidates can correspond to various nodes of the metrical tree of an utterance, ranging from phrase level to segment level, while Stöber et al. (2000) use only the word and segment levels. Both approaches are designed for limited domains and benefit from the fact that most longer units are represented in the database. The advantage of such a top-down approach is that it favors the selection of these longer units in a straightforward way. If candidates are found on levels higher than the segment level, this strategy can be faster than the bottom-up approaches because there are longer and therefore fewer unit candidates. Still, particularly on the segment level, candidate sets may be very large.

The LNRE characteristics of the SmartKom vocabulary with a limited number of very frequent domain-specific words and a large number of very infrequent words originating from dynamic databases suggested a hybrid strategy that integrates the two approaches described above. The PSM strategy ensures high-quality synthesis for frequent material by directly selecting entire words or phrases from the database. If no matching candidates are found above the segment level, which will typically be the case for domain-independent material, the AC approach serves to reduce the amount of candidate units.

Our implementation of the PSM algorithm differs from the original implementation (Taylor and Black, 1999) in some aspects. First, the original algorithm requires candidates to match the target specification with respect to tree structure and segment identities, but they may differ in stress pattern or intonation, phonetic or phrasal context, at the expense of higher target costs. This reflects the view that a prosodically suboptimal but coherent candidate is better than the concatenation of smaller non-coherent units from prosodically more appropriate contexts. We kept the matching condition more flexible by more generally defining two sets of features for each level of the linguistic hierarchy. *Primary features* are features in which candidates have to match the target specification (in addition to having the same structure), while they may differ in terms of *secondary features*. Mismatch of secondary features causes higher target costs, just as the mismatch of prosodic features increases the unit score in the original algorithm. The primary features typically are the unit identity and the classification of prosodic events occurring on the respective unit. Secondary features are mostly positional features expected to have a strong influence on the acoustic realization of the unit. More details can be found in Schweitzer et al. (2003).

Another, more important, difference from the original PSM algorithm is that candidate sets can optionally be reduced if their size exceeds a certain threshold. In this case, the candidate set is filtered stepwise for each secondary feature, thereby excluding candidates that do not agree on the respective feature, until the size of the candidate set is below the threshold. However, the PSM search is not performed below the syllable level because the initial candidate sets would be too large. Instead, the AC algorithm (Black and Taylor, 1997) takes over on the segment level, adding candidates for those parts of the target utterance that have not been covered yet.

As for the final selection of the optimal sequence of units, candidate units found by either search strategy are treated in the same way, i.e., they are subject to the same selection procedure. Thus, longer units are treated just as shorter units in that the optimal sequence of candidates is determined by a Viterbi algorithm, which simultaneously minimizes concatenation costs and target costs. Concatenation costs for two longer units are the concatenation costs for the two segments on either side of the concatenation point.

4.2 Text Material Design and Corpus Preparation

The requirements for the contents of the database are again different for domain-specific vs. domain-independent material. For the limited amount of domain-specific material, it is conceivable to include typical words in several different contexts (Stöber et al., 2000) or even to repeat identical contexts. In contrast, for the open-domain part a good coverage of the database in terms of diphones in different contexts is essential, as emphasized by van Santen and Buchsbaum (1997) and Möbius (2003).

We followed van Santen and Buchsbaum (1997) by applying a greedy algorithm to select from a large text corpus a set of utterances that maximizes coverage of units. The procedure was as follows. First, the linguistic text analysis component of the IMS German Festival TTS system[4] (Schweitzer and Haase, 2000) was used to determine for each sentence in a German newspaper corpus of 170,000 sentences the corresponding phone sequences as well as their prosodic properties. We built a vector for each segment including its phonemic identity, syllabic stress, word class, and prosodic and positional properties. Thus, we obtained a sequence of vectors for each sentence. Additionally, we determined the diphone sequence for each sentence. Sentences were then selected successively by the greedy algorithm according to the number of both new vectors and new diphone types that they covered. For German diphone types that did not occur at all, we constructed sentences that would contain them, added these sentences to the corpus, and repeated the selection process. This ensured that at least a full diphone coverage was obtained, and at the same time the number of phoneme/context vector types was increased.

We added 2643 SMARTKOM-specific words and sentences to the domain- independent corpus. They included excerpts from demo dialogues, but also domain-typical slot fillers such as people's names, place names, numbers, weekdays, etc. Movie titles, many of them in English, constituted the largest group of domain-specific material, partly to make up for the omission of English phones in the systematic design of the text material.

The speech database was recorded using the same professional speaker as for the diphone voice and amounted to about 160 minutes of speech. The automatically generated transcriptions were manually corrected according to what the speaker had said together with the corresponding orthographic notation. The hand-corrected transcriptions were then used for sentencewise forced alignment of the speech signal on the segment, syllable, and word levels. Pitch accents and boundary tones were automatically predicted from the orthographic notation and were subsequently corrected manually.

The corrected version of the database contains 2488 diphone types. 277 of the 2377 originally predicted types were not realized in the database, mostly because of incorrect predictions; instead, 388 additional types occurred. Similarly, the database had been predicted to cover 2731 out of 2932 phoneme/context vector types from the complete text corpus. 687 of these were not realized in the recorded database, whereas 791 new ones occurred, which yields 2835 types. Of these new vector types, only 10 belong to the 201 vectors that had been in the complete text corpus but not in the subset selected for the recordings.

These figures show that more than 90% of the diphone types were covered as expected, and many new types involving foreign phonemes were added. As for the coverage of phoneme/context vectors, the situation is more complex. Combinatorially, 19 440 phoneme/context vector types are possible. We estimate that no more than 4600 are theoretically possible because the context properties are not independent. For instance, boundary tones only occur on phrase-final syllables. Some consonants are phonotactically not allowed in syllable onsets, others not in the rhyme, and vowels are in the rhyme per definition. Also, pitch accents are always realized on

[4] http://www.ims.uni-stuttgart.de/phonetik/synthesis/index.html

syllables with syllabic stress, and function words usually have no pitch accent. However, only approximately 60% of these 4600 types were covered, even with a careful database design. One reason for this is that some of these types are so rare that they do not occur even in large corpora (Möbius, 2003). Apart from that, coverage of phoneme/context vectors was problematic because many of the predicted vectors were incorrect. This was partly due to foreign language material in the text corpus, which could not be adequately dealt with using the monolingual German lexicon; also, unknown words, mostly compounds, abbreviations, and acronyms, had often been predicted incorrectly. We expect that the prediction of context vectors can be significantly improved if foreign material is reliably marked as such in a preprocessing step. However, the prosodic contexts are difficult to predict, and often several alternative realizations are possible. Giving the speaker additional directions concerning intended prosodic realizations, on the other hand, may add too much load in supervising the recordings and moreover might result in unnatural realizations.

5 Lip and Gesture Synchronization

It has been shown that visual segmental information can enhance segmental speech perception (Massaro, 1998). Vice versa, inconsistencies between visual and acoustic information can significantly decrease intelligibility to the extreme that the segmental identity is compromised: MacDonald and McGurk (1978) demonstrated that acoustic [ba] is perceived as /da/ when presented with the visual information of [ga]. Thus, correct lip synchronization is an important issue in multimodal speech synthesis.

5.1 Lip Synchronization

In contrast to lip synchronization for more human-looking avatars, which may require modeling of various parameters such as the position of the jaw, the upper and lower lip, teeth, tongue tip, and tongue root, only two parameters are necessary for Smartakus because of his cartoonlike features: jaw opening and lip rounding. His teeth and tongue are never visible.

We used a simple mapping procedure to map phonemes to so-called *visemes*. Visemes are visually contrastive speech elements (Fisher, 1968). In our view, visemes are sets of feature-value pairs, where the features correspond to the different articulators and the values indicate their target positions. Movement results from interpolating between the target positions specified by the visemes. Each phoneme is represented by one or more visemes. Visemes can be underspecified regarding particular features. In this case, the value of the feature consists of a range of possible values. Underspecified visemes inherit the missing values from the context, which allows us to model coarticulation.

In SmartKom, only two features are specified for visemes: lip (or jaw) opening and lip rounding. Four degrees of lip opening (closed, almost closed, mid, open) and two degrees of lip rounding (unrounded, rounded) are differentiated. Visemes corresponding to **vowels** are fully specified for both opening and rounding. Figure 5

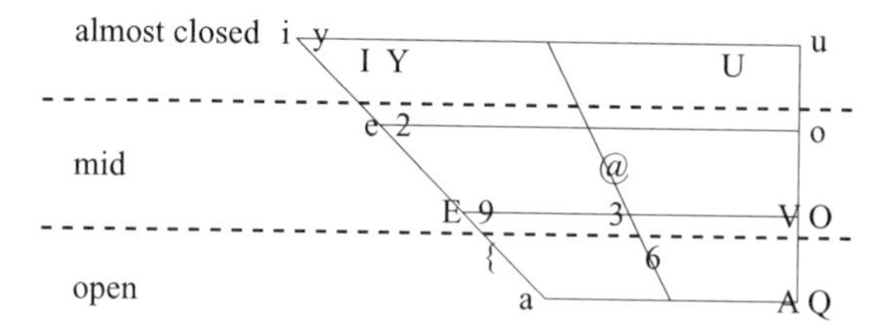

Fig. 5. IPA vowel space diagram of the vowels used in SmartKom synthesis, in SAMPA notation. The lip opening degree for each vowel is indicated on the *left*. The *dashed lines* indicate the boundaries between different degrees of opening

shows the vowels used in SmartKom arranged in the International Phonetic Association (IPA) vowel diagram, in SAMPA[5] notation. The position of each vowel in the IPA diagram reflects the tongue position in articulation. Thus, the vertical position of the vowels indicates tongue height, and the horizontal position indicates the front–back position of the tongue. To map vowel positions in the diagram to lip opening degrees, we stipulate that tongue height and lip opening correlate, but that the horizontal position is irrelevant for lip opening. The resulting mapping from the position in the diagram to the opening degree of the corresponding viseme is indicated on the left of Fig. 5. Schwa vowels (/6/, /@/) are an exception: They are usually realized in a reduced way, in which case the correlation between lip opening and tongue height seems to be less strong. Both schwas are therefore realized with almost closed lips. Rounding trivially follows the phonological specification of the respective vowels. Diphthongs are represented by a sequence of two visemes corresponding to the visemes representing the underlying vowels.

Visemes for **consonants**, on the other hand, may be underspecified. This is motivated by the hypothesis that consonants whose place of articulation is to the back can be articulated with an almost closed jaw or with an open jaw. Consonants with an anterior place of articulation, however, can only be articulated with a relatively closed jaw. Also, most consonants can be articulated with rounded or with unrounded lips. Thus, visemes corresponding to consonants are unspecified with regard to lip rounding, and the farther back their place of articulation is, the higher is the degree of underspecification for jaw opening. In Table. 1, the mapping from place of articulation to jaw opening degree is again demonstrated by an IPA chart containing the consonants used in SmartKom.

After mapping phonemes to visemes, the resulting sequences of partly underspecified visemes and the corresponding time intervals relative to the beginning of the speech signal are passed to the presentation manager (Poller and Tschernomas, 2006), which resolves underspecification and concatenates video sequences corresponding to the visemes.

[5] SAMPA is a common standard used for convenience instead of the IPA notation.

Table 1. IPA chart for the consonants used in SMARTKOM, in SAMPA notation (*upper part* of the table) and mapping to ranges of jaw opening degrees (*lower part*). Places of articulation are arranged in the chart from front (labial) to back (glottal). Possible jaw opening degrees are 0 (closed), 1 (almost closed), 2 (mid), and 3 (open). The range of values depends on the place of articulation of the corresponding phoneme. Velar, uvular, and glottal consonants exhibit the highest degree of underspecification for jaw opening: The range of jaw opening degrees is from 1 to 3

<table>
<tr><td></td><td colspan="9">Place of articulation</td></tr>
<tr><td></td><td>Labial</td><td>Labiodental</td><td>Dental</td><td>Alveolar</td><td>Postalveolar</td><td>Palatal</td><td>Velar</td><td>Uvular</td><td>Glottal</td></tr>
<tr><td>Plosives</td><td>p b</td><td>–</td><td colspan="3">t d</td><td>–</td><td>k g</td><td>–</td><td>?</td></tr>
<tr><td>Nasals</td><td>m</td><td>–</td><td colspan="3">n</td><td>–</td><td>N</td><td>–</td><td>–</td></tr>
<tr><td>Fricatives</td><td>–</td><td>f v</td><td>T D</td><td>s z</td><td>S Z</td><td>C</td><td>x</td><td>R</td><td>h</td></tr>
<tr><td>Approximants</td><td>–</td><td>–</td><td colspan="3">l r</td><td>j</td><td>w</td><td>–</td><td>–</td></tr>
<tr><td></td><td colspan="9">Jaw opening</td></tr>
<tr><td>Minimal</td><td>0</td><td>1</td><td colspan="4">1</td><td colspan="3">1</td></tr>
<tr><td>Maximal</td><td>0</td><td>1</td><td colspan="4">2</td><td colspan="3">3</td></tr>
</table>

5.2 Gesture–Speech Alignment

Smartakus may execute gestures while he is speaking. In this case, temporal alignment of speech and gesture is required. Pointing gestures also influence prosody, as mentioned in Sect. 2.

Building on the sketch model (de Ruiter, 2000), speech is synthesized independently of the temporal structure of the accompanying gesture. Instead, the gesture is executed in temporal alignment with the speech signal. According to de Ruiter (2000), gestures can be divided in three phases, viz. the preparation phase, the stroke phase, and the retraction phase. The stroke phase is the core phase, which accompanies corresponding speech material. Preparation and retraction phases of gestures can be adjusted to align the stroke phase with the relevant speech material. In SMARTKOM, most of the gestures occurring during speech are pointing gestures, which accompany deictic elements in the linguistic structure. In this case, the timing information for the deictic material is passed to the presentation manager to enable alignment of the stroke phase with the corresponding deictic element.

6 Evaluation of the Speech Synthesis Module

Due to the complexity of multimodal systems, it is difficult to evaluate single components because they are not designed to perform in a stand-alone mode, isolated from other system components with which they interact. Also, the performance of the system as a whole, not the performance of its modules, is decisive when it comes to user acceptance or usability. Consequently, the SMARTKOM system has been evaluated extensively as a whole (Schiel, 2006).

However, in addition to an end-to-end evaluation of the complete system, the evaluation of its speech synthesis component is necessary to give more detailed, possibly diagnostic, insights into potential synthesis specific problems. This can be difficult since the boundaries between system components are often not clear-cut from a functional point of view. In SMARTKOM, language generation and synthesis are strongly linked. Without language generation, simulating concept input for CTS synthesis is tedious. But if concept input is generated automatically for synthesis evaluation purposes, the language generation component is implicitly evaluated together with the synthesis module. A second problem is that the appropriateness of the synthesis voice for Smartakus cannot be evaluated without the animation component.

To detect possible synthesis-specific problems, we carried out evaluations of the synthesis module, detached as far as possible from the SMARTKOM system, at two times. The first evaluation took place early in the project and served to verify that the diphone synthesis voice produced satisfactory intelligibility; the second evaluation was carried out in the last project phase to assess the quality of the new unit selection voice, particularly in comparison with the diphone voice. Table 2 shows an overview of the tasks performed in the evaluation procedures.

Table 2. Overview of the tasks performed by participants in the evaluation procedures. The general type of task is indicated in the *left column*. The table lists text material, viz. normal text (open domain), semantically unpredictable sentences (SUS), or SMARTKOM specific material (SK), and the voices used to generate the stimuli, viz. diphone voice or unit selection voice (US)

<table>
<tr><th rowspan="3"></th><th colspan="2" rowspan="2">First evaluation</th><th colspan="4">Second evaluation</th></tr>
<tr><th colspan="2">Pilot study</th><th colspan="2">Full experiment</th></tr>
<tr><th>Material</th><th>Voice</th><th>Material</th><th>Voice</th><th>Material</th><th>Voice</th></tr>
<tr><td rowspan="3">Dictation</td><td rowspan="2">SUS</td><td>Natural</td><td rowspan="3">–</td><td rowspan="3">–</td><td rowspan="3">SUS</td><td>Diphones</td></tr>
<tr><td>Diphones</td><td rowspan="2">US</td></tr>
<tr><td>SK</td><td>Diphones</td></tr>
<tr><td rowspan="2">Listening comprehension</td><td rowspan="2">–</td><td rowspan="2">–</td><td rowspan="2">–</td><td rowspan="2">–</td><td rowspan="2">Open domain</td><td>Diphones</td></tr>
<tr><td>US</td></tr>
<tr><td rowspan="2">Subjective impression</td><td rowspan="2">SK</td><td rowspan="2">Diphones</td><td rowspan="2">SK</td><td>Diphones</td><td rowspan="2">SK</td><td rowspan="2">US</td></tr>
<tr><td>US</td></tr>
</table>

6.1 First Evaluation

The first evaluation involved a total of 58 participants, which can be classified in two groups. The first group comprised 39 students from the University of Ulm. These subjects are referred to as "naive" because they reported having no prior experience with speech synthesis or language processing. The second group consisted of employees of DaimlerChrysler at Ulm, who were experienced with regard to speech technology. All participants completed three dictation tasks: one with SMARTKOM-specific utterances rendered by the diphone voice, one with semantically unpredictable sentences (SUS (Benoît et al., 1996) recorded from a speaker, and one using SUS stimuli synthesized by the diphone voice.

The SMARTKOM-specific dictation task was intended to verify that the intelligibility of the diphone voice was satisfactory for the use in SMARTKOM. The participants transcribed nine system turns in a continuous dialogue between the system and a user. Ninety three percent of these system turns were transcribed without any errors; 4% involved obvious typing errors; and in 2% of the transcriptions there were errors that can probably be attributed to memory problems rather than to intelligibility. These figures show that the diphone voice offers excellent intelligibility for normal speech material.

The SUS dictation tasks are perceptually more demanding because the linguistic context does not provide any cues in cases of locally insufficient intelligibility. The tasks thus aimed at testing the intelligibility of the diphone voice under more challenging conditions. The sentences were generated automatically using five different templates, which are listed in Table 3. The material to fill the lexical slots in the templates came from lists of words selected from CELEX (Baayen et al., 1995) according to their morphological and syntactic properties. The lists were randomized before generating the SUS stimuli. All lexical items were used at least once, but in varying combinations.

Table 3. Overview of syntactic templates used for the generation of SUS stimuli. The table shows the lexical slots in the templates corresponding to the constituents in each of the templates. Although not explicitly stated here, noun phrases were also congruent in gender, and the complements of transitive verbs and prepositions were in the appropriate case

Template	Constituent	Lexical slots
S V O	subject	determiner (sg.) + noun (sg.)
	verb	transitive verb (3rd person sg.)
	object	plural noun
S V PP	subject	determiner (sg.) + noun (sg.)
	verb	intransitive verb (3rd person sg.)
	adjunct PP	preposition + determiner (acc. sg.) + noun (acc. sg.)
PP V S O	adjunct PP	preposition + determiner (dat. sg) + noun (dat. sg.)
	verb	transitive verb (3rd person sg.)
	subject	determiner (nom. sg.) + noun (nom. sg.)
	object	determiner (acc. sg.) + noun (acc. sg.)
V S O!	verb	transitive verb (imperative pl.)
	subject	“Sie”
	object	determiner (acc. sg.) + noun (acc. sg.)
V S O?	verb	transitive verb (3rd sg.)
	subject	determiner (nom. sg) + noun (nom. sg)
	object	determiner (pl.) + noun (pl.)

The SUS task using natural stimuli immediately preceded the task with the diphone stimuli. It served to estimate the upper bound of scores in such a task. The subjects transcribed 15 stimuli in each of the 2 tasks. For the natural stimuli, the sentence error rate was 4.9%. Of these, 0.6% were obvious typing errors. The error rate for the synthesized stimuli was 33.9%. Again, 0.6% were typing errors. The error analysis for the diphone stimuli showed three relatively frequent error types. One concerned the confusion of short and long vowels. This can probably be attributed to the duration model used for determining segmental durations, which had been trained on a speech corpus from a different speaker. We replaced this model with a speaker-specific model trained on the unit selection voice data later in the project. Another problem was that sometimes the subjects did not correctly recognize word boundaries. We expect that in these cases listeners should also benefit from the improved duration model. The other two types of errors concerned voiced plosives preceding vowels in word onsets, and voiced and voiceless plosives preceding /R/ in the same position. We claim that the latter is a typical problem in diphone synthesis: The two /R/-diphones concatenated in these cases are two different positional variants of /R/, viz. a postconsonantal variant, and an intervocalic variant.

After performing the dictation tasks, participants were asked for their subjective impression of the diphone voice. They rated the voice on a five-point scale ranging from −2 to +2 for each of the two questions *"How did you like the voice?"* (−2 and +2 corresponding to "not at all" and "very much," respectively), and *"Did you find the voice easy or hard to understand?"* (−2 and +2 corresponding to "hard" and "easy", respectively). Subjects also answered "yes" or "no" to the question *"Would you accept the voice in an information system?"*. The results strongly indicate that non-naive subjects generally rated the voice better than naive subjects. The mean scores for the first two questions broken down by experience with speech technology were +0.53 and +1.37 for non-naive participants, and −0.21 and +0.67 for naive participants, respectively. Of the non-naive subjects, 95% said they would accept the voice in an information system, whereas only 72% of the naive subjects expressed the same opinion. In summary, the first evaluation confirmed that the diphone voice yielded satisfactory results.

6.2 Second Evaluation

The second evaluation focused on the unit selection voice. Here the diphone voice served as a baseline for the dictation and listening comprehension tasks. The actual evaluation was preceded by a pilot study on the acceptability of the unit selection voice versus the diphone voice specifically for typical SmartKom utterances. The subjects in this pilot study were students from Stuttgart and their parents. The younger student group and the older parent group each consisted of 25 participants. Subjects listened to 25 SmartKom-specific dialogue turns in randomized order, both rendered in the unit selection voice and in the diphone voice. Afterwards, they were asked to answer the questions *"How do you judge the intelligibility of the synthesis voice?"* and *How do you judge the suitability of this voice for an information system?"* on a five-point scale ranging from −2 ("very bad") to +2 ("very good"). There was a similar effect observable between the younger and the older group as in the first evaluation between the non-naive and the naive group. The younger group was more tolerant to diphone synthesis regarding intelligibility: The mean scores for the diphone voice were +0.83 for the younger group and +0.51 for the older one. The unit selection voice was rated significantly better by both groups; the mean score was +1.76 in both cases. The results for the question regarding the suitability of the voices in an information system show that the unit selection voice is strongly preferred. Mean scores were clearly below zero for the diphone voice (−1.21 and −1.33

for the younger and the older group, respectively), and clearly above zero for the unit selection voice (+1.79 and +1.23 for the younger and the older group, respectively).

In the following evaluation, 77 subjects participated, none of which had taken part in the earlier evaluations. Three tasks were completed in this evaluation. Participants first transcribed SUS stimuli. The stimuli were taken from the first evaluation, but they were synthesized using both the diphone and the unit selection voices. The results are comparable to the earlier results: The sentence error rate was 27% including typing errors for the diphone voice (earlier: 33%). This shows that the diphone voice has gained in intelligibility compared to the first evaluation. For the unit selection voice, however, the error rate was 71%. This is due to the fact that the SUS stimuli contained only open domain material. The unit selection voice was designed for a restricted domain with prevailing SmartKom-specific material (Sect. 4). In this respect, completely open domains are a worst-case scenario, in which the synthesis quality must be expected to be inferior to that of SmartKom-specific material. Additionally, at the time of conducting the evaluation, the speech database was still in the process of being manually corrected. Informal results obtained at the end of the project, i.e., two months after the formal evaluation and after extensive manual correction of prosodic and segmental corpus annotations, indicate that the subjective synthesis quality especially for open domain material has improved since the completion of the evaluation.

After completing the SUS dictation task, participants were presented three video clips showing the SmartKom display during a user's interaction with Smartakus. The user's voice had been recorded by a speaker. The system's voice in the video clips was the unit selection voice, synchronized with Smartakus's lip movements and gestures. Subjects were asked to answer three questions by adjusting a sliding bar between two extremes. The three questions were *"How do you judge the intelligibility of the voice?"*, with possible answers ranging from "not intelligible" to "good"; *"How natural did you find the voice?"*, with answers between "not natural at all" and "completely natural"; and *"How did you like the voice?"*, with answers between "not at all" and "very well." The results for the three answers were 71% for intelligibility, 52% for naturalness, and 63% for pleasantness. These figures show that in the SmartKom-specific contexts, the unit selection voice is very well accepted and judged to be satisfactorily intelligible. This confirms the results obtained in the pilot study for audio-only stimuli.

In the last task, the listening comprehension test, the subjects listened to four short paragraphs of open domain texts. After each paragraph, they were asked three questions concerning information given in the text. Two texts were rendered using the diphone voice, and two using the unit selection voice. The results were again better for the diphone voice, with 93% of the answers correct, while 83% were correct for the unit selection voice. In this context, both voices were rated lower than in the SmartKom-specific task. The scores for intelligibility, naturalness, and pleasantness were 53%, 34%, and 42% for the diphone voice, and 23%, 22%, and 26% for the unit selection voice, respectively. Again, we expect much better results after the manual correction of the speech database.

7 Conclusion

To summarize, the superiority of the unit selection voice is evident for the SMARTKOM domain. This was confirmed by the pilot study and the SMARTKOM-specific part of the second evaluation. The quality of the diphone voice improved between the first and the second evaluation. We attribute this effect mainly to the new duration model obtained from the unit selection data of our speaker. The ongoing manual correction of the unit selection database is evidently effective. Subjectively, the synthesis quality has improved since the completion of the second evaluation. However, this will have to be confirmed in more formal tests.

Future work will focus on the extension of our unit selection approach from the restricted SMARTKOM domain to open domains in general. The experience gained in working with the SMARTKOM unit selection voice suggests that accuracy of the database annotation is crucial for optimal synthesis quality. Also, the strategy to deal with large numbers of unit candidates as they often occur in open domain sentences without excluding potentially good candidates will need some more attention in the future.

Acknowledgments

The significant contributions of Martin Ernst (DaimlerChrysler, Ulm) and Gerhard Kremer, Wojciech Przystas, Kati Schweitzer, and Mateusz Wiacek (IMS) to the synthesis evaluations are gratefully acknowledged.

References

S.P. Abney. Chunks and Dependencies: Bringing Processing Evidence to Bear on Syntax. In: *Computational Linguistics and the Foundations of Linguistic Theory*, Stanford, CA, 1995. CSLI.

H. Baayen, R. Piepenbrock, and L. Gulikers. *The CELEX Lexical Database – Release 2 (CD-ROM)*. Centre for Lexical Information, Max Planck Institute for Psycholinguistics, Nijmegen; Linguistic Data Consortium, University of Pennsylvania, 1995.

M. Balestri, A. Pacchiotti, S. Quazza, P.L. Salza, and S. Sandri. Choose the Best to Modify the Least: A New Generation Concatenative Synthesis System. In: *Proc. EUROSPEECH-99*, vol. 5, pp. 2291–2294, Budapest, Hungary, 1999.

T. Becker. Fully Lexicalized Head-Driven Syntactic Generation. In: *Proc. 9th Int. Workshop on Natural Language Generation*, Niagara-on-the-Lake, Canada, August 1998.

T. Becker. Natural Language Generation With Fully Specified Templates, 2006. In this volume.

C. Benoît, M. Grice, and V. Hazan. The SUS Test: A Method for the Assessment of Text-To-Speech Synthesis Intelligibility Using Semantically Unpredictable Sentences. *Speech Communication*, 18:381–392, 1996.

M. Beutnagel, M. Mohri, and M. Riley. Rapid Unit Selection From a Large Speech Corpus for Concatenative Speech Synthesis. In: *Proc. EUROSPEECH-99*, vol. 2, pp. 607–610, Budapest, Hungary, 1999.

A.W. Black and P. Taylor. Automatically Clustering Similar Units for Unit Selection in Speech Synthesis. In: *Proc. European Conference on Speech Communication and Technology*, vol. 2, pp. 601–604, Rhodos, Greece, 1997.

A.P. Breen and P. Jackson. Non-Uniform Unit Selection and the Similarity Metric Within BT's Laureate TTS System. In: *Proc. 3rd Int. Workshop on Speech Synthesis*, pp. 373–376, Jenolan Caves, Australia, 1998.

D. Büring. *The Meaning of Topic and Focus – The 59th Street Bridge Accent.* Routledge, London, UK, 1997.

G. Cinque. A Null Theory of Phrase and Compound Stress. *Linguistic Inquiry*, 24 (2):239–297, 1993.

A. Conkie. Robust Unit Selection System for Speech Synthesis. In: *Collected Papers of the 137th Meeting of the Acoustical Society of America and the 2nd Convention of the European Acoustics Association: Forum Acusticum*, Berlin, Germany, 1999.

P.W. Culicover and M.S. Rochemont. Stress and Focus in English. *Language*, 59(1): 123–165, 1983.

J.P. de Ruiter. The Production of Gesture and Speech. In: D. McNeill (ed.), *Language and Gesture*, pp. 284–311, Cambridge, UK, 2000. Cambridge University Press.

T. Dutoit, V. Pagel, N. Pierret, F. Bataille, and O. van der Vrecken. The MBROLA Project: Towards a Set of High Quality Speech Synthesizers Free for Use for Non Commercial Purposes. In: *Proc. Int. Conference on Spoken Language Processing*, vol. 3, pp. 1393–1396, Philadelphia, PA, 1996.

C.G. Fisher. Confusions Among Visually Perceived Consonants. *Journal of Speech and Hearing Research*, 11:796–804, 1968.

D.J. Hirst. Detaching Intonational Phrases From Syntactic Structure. *Linguistic Inquiry*, 24:781–788, 1993.

A.J. Hunt and A.W. Black. Unit Selection in a Concatenative Speech Synthesis System Using a Large Speech Database. In: *Proc. IEEE Int. Conf. on Acoustics and Speech Signal Processing*, vol. 1, pp. 373–376, Munich, Germany, 1996.

J. MacDonald and H. McGurk. Visual Influences on Speech Perception Process. *Perception and Psychophysics*, 24:253–257, 1978.

D.W. Massaro. *Perceiving Talking Faces: From Speech Perception to a Behavioral Principle.* MIT Press, Cambridge, MA, 1998.

J. Mayer. Transcribing German Intonation – The Stuttgart System. Technical report, University of Stuttgart, 1995.

J. Mayer. Prosodische Merkmale von Diskursrelationen. *Linguistische Berichte*, 177:65–86, 1999.

B. Möbius. Rare Events and Closed Domains: Two Delicate Concepts in Speech Synthesis. *Int. Journal of Speech Technology*, 6(1):57–71, 2003.

G. Möhler and A. Conkie. Parametric Modeling of Intonation Using Vector Quantization. In: *Proc. 3rd Int. Workshop on Speech Synthesis*, pp. 311–316, Jenolan Caves, Australia, 1998.

E. Morais, P. Taylor, and F. Violaro. Concatenative Text-To-Speech Synthesis Based on Prototype Waveform Interpolation (A Time-Frequency Approach). In: *Proc. Int. Conf. on Spoken Language Processing*, pp. 387–390, Beijing, China, 2000.

E. Moulines and F. Charpentier. Pitch-Synchronous Waveform Processing Techniques for Text-To-Speech Synthesis Using Diphones. *Speech Communication*, 9: 453–467, 1990.

P. Poller and V. Tschernomas. Multimodal Fission and Media Design, 2006. In this volume.

F. Schiel. Evaluation of Multimodal Dialogue Systems, 2006. In this volume.

A. Schweitzer, N. Braunschweiler, T. Klankert, B. Möbius, and B. Säuberlich. Restricted Unlimited Domain Synthesis. In: *Proc. European Conference on Speech Communication and Technology*, pp. 1321–1324, Geneva, Switzerland, 2003.

A. Schweitzer, N. Braunschweiler, and E. Morais. Prosody Generation in the SmartKom Project. In: *Proc. Speech Prosody 2002 Conference*, pp. 639–642, Aix-en-Provence, France, 2002.

A. Schweitzer and M. Haase. Zwei Ansätze zur syntaxgesteuerten Prosodiegenerierung. In: *Proc. 5th Conf. on Natural Language Processing – Konvens 2000*, pp. 197–202, Ilmenau, Germany, 2000.

E. Selkirk. *Phonology and Syntax – The Relation Between Sound and Structure*. MIT Press, Cambridge, MA, 1984.

K. Stöber, P. Wagner, J. Helbig, S. Köster, D. Stall, M. Thomae, J. Blauert, W. Hess, R. Hoffmann, and H. Mangold. Speech Synthesis Using Multilevel Selection and Concatenation of Units From Large Speech Corpora. In: W. Wahlster (ed.), *Verbmobil: Foundations of Speech-to-Speech Translation*, pp. 519–534, Berlin Heidelberg New York, 2000. Springer.

P. Taylor and A.W. Black. Speech Synthesis by Phonological Structure Matching. In: *Proc. EUROSPEECH-99*, vol. 2, pp. 623–626, Budapest, Hungary, 1999.

J.P.H. van Santen and A.L. Buchsbaum. Methods for Optimal Text Selection. In: *Proc. European Conference on Speech Communication and Technology*, vol. 2, pp. 553–556, Rhodos, Greece, 1997.

Part V

Scenarios and Applications

Building Multimodal Dialogue Applications: System Integration in SmartKom

Gerd Herzog and Alassane Ndiaye

DFKI GmbH, Saarbrücken, Germany
{herzog,ndiaye}@dfki.de

Summary. We report on the experience gained in building large-scale research prototypes of fully integrated multimodal dialogue systems in the context of the SmartKom project. The development of such systems requires a flexible software architecture and adequate software support to cope with the challenge of system integration. A practical result of our experimental work is an advanced integration platform that enables flexible reuse and extension of existing software modules and that is able to deal with a heterogeneous software environment. Starting from the foundations of our general framework, we give an overview of the SmartKom testbed, and we describe the practical organization of the development process within the project.

1 Introduction

The realization of fully fledged prototype systems for various application scenarios constituted an integral part of the research and development activities within the SmartKom project (Reithinger et al., 2003; Wahlster, 2003). Building such large-scale multimodal dialogue applications poses a significant technical challenge and requires a principled approach for succesful system integration.

Figure 1 provides an overview concerning the technical components of the SmartKom demonstrator system that covers all application scenarios, including SmartKom-Public, Home, and Mobile as well as the English system. The instrumented vehicle, which is not permanently available for demonstrations, can also be connected to this demonstrator setup. Alternatively, standard audio components and a separate display can be employed to simulate the in-car equipment. Furthermore, a GPS simulation can be used for SmartKom-Mobile to demonstrate incremental route guidance. The overall system incorporates more than 40 software components from 11 different project partners. Customized SmartKom installations exist at all partner sites. For example, two laptops and a handheld computer are sufficient for a compact instantiation of SmartKom-Mobile.

The difficulty of system integration is high and is often underestimated. The distributed nature of a large-scale research activity like SmartKom imposes severe

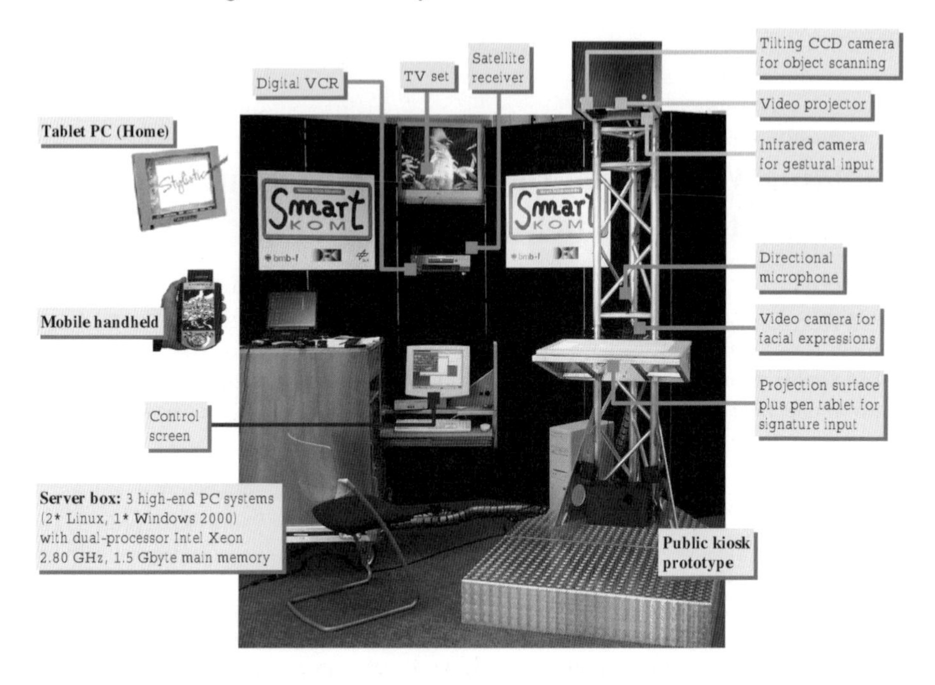

Fig. 1. SMARTKOM reference installation for demonstrations

constraints on the development of an integrated system. This kind of joint effort is characterised by a lot of variance in code development across teams. Each workgroup is accustomed to its preferred software environment and development practices. Given the need for optimal use of available project resources, it is also not feasible to start an implementation from scratch. The resulting heterogeneity of contributions from different project partners impedes the realization of a monolithic system. Thus adequate technical and organizational means are required to assemble heterogeneous components into an integrated and fully operational system.

The following section presents some basic considerations that govern our technical approach for system integration. Section 3 then provides a compact overview of the SMARTKOM testbed, our specific software integration platform that has been employed to realise the multimodal dialogue system prototypes. In addition to technical issues, organizational aspects are of particular importance for a succesful research and development project. Section 4 describes how the development and integration process has been organized within the SMARTKOM project.

2 A Framework for Software Integration

Typical aspects of large-scale system development have to be considered during the construction of advanced intelligent user interfaces. The following key factors influence the technical approach to be taken:

- strong need to reuse and extend existing software
- pressure to build initial prototypes rapidly
- significant number of distinct components from different sources
- heterogeneous software environment, including various programming languages and different operating systems
- high variability of target configurations for the integrated dialogue system applications

A distributed system constitutes the natural choice to realize an open, flexible, and scalable software architecture, able to integrate heterogeneous software modules implemented in diverse programming languages and running on different operating systems. A key point of our specific integration framework is the idea of having a component-based system, composed out of executable modules. Our approach proposes a modularization of the dialogue system into distinct and independent software modules to allow maximum decoupling. A main advantage of such a component architecture (Hopkins, 2000) is the fact that component integration and deployment are independent of the development life cycle, and there is no need to recompile or relink the entire application when updating with a new implementation of a component.

The kind of component-based software engineering proposed here requires an integration platform that provides the necessary software infrastructure to assemble the large-grained components into a multimodal dialogue system. Such an integration platform can be characterized as a software development kit and consists of the following main parts:

- a run-time platform for the execution of a distributed dialogue application, including also means for configuring and deploying the individual parts of the complete multimodal dialogue system
- application programming interfaces (API) for linking into the run-time environment and for interprocess communication
- tools and utilities to support the whole development process, including also installation and software distribution

In general, different options are available for the interconnection of individual components (Bernstein, 1989; Emmerich, 2000; Geihs, 2001). Distributed processing with remote procedure calls or remote method invocation follows the client–server paradigm and using these techniques, each component has to declare and implement a specific API to make its encapsulated functionality transparently available for other system modules. Asynchronous message-based communication, however, better supports the need for scalability, flexibility, and decoupling.

Two main schemes of message-oriented middleware can be distinguished. Basic *point-to-point* messaging employs unicast routing and realizes the notion of a direct connection between message sender and a known receiver. The more general *publish/subscribe* approach is based on multicast addressing. Instead of addressing one or several receivers directly, the sender publishes a notification on a named message queue, so that the message can be forwarded to a list of subscribers. This kind of distributed event notification makes the communication framework very flexible, as it focuses on the data to be exchanged and it decouples data producers and data consumers. The well-known concept of a blackboard architecture (Erman et al., 1980) and distributed shared memory approaches that are based on the notion of tuple spaces (Carriero and Gelernter, 1989) follow similar ideas.

In order to ensure syntactic and semantic interoperability definite contracts among components are required. Using message-based communication, the interface of a specific component can be characterized by the admissible message contents that it is able to process and to produce. The so-called *extensible markup language*, XML, has meanwhile evolved into a standard tool for the flexible definition of application-specific data formats for information exchange. Thus we propose to operationalize interface specifications in the form of an XML language. XML-based languages define an external notation for the representation of structured data and simplify the interchange of complex data between the separate components of an integrated multimodal dialogue system.

For the technical realization of an integration platform, we postulate basic design principles that form the conceptual foundation of our integration framework:

- Asynchronous message-passing is more flexible than component-specific remote APIs and better fits parallel processing in a distributed system.
- Blackboard-like publish/subscribe messaging is a generalisation of point-to-point communication that supports decoupling of components and allows better modularization of data interfaces.
- Well-designed XML languages support the definition of declarative data models for information interchange and semantically rich interface specifications.
- The proposed framework does not rely on centralised control. Instead of that, individual components take care if coordination is needed. Required synchronization information is modeled explicitly within data interfaces.
- There should be no hidden interaction between components. All data have to be exchanged transparently via the provided communication system to ensure traceability of processing behaviour.

3 The SmartKom Testbed

Taking into account the design aspects outlined in the previous section, it becomes obvious that available off-the-shelf middleware is not enough to provide a flexible architecture framework and a powerful integration platform for large-scale dialogue systems. These considerations motivated our investment into the development of a suitable system integration platform.

SmartKom relies on the so-called *multiplatform testbed* (Herzog et al., 2003), a software solution that originates from our practical work in various research projects. The testbed, which is built on top of open-source software (Wu and Lin, 2001), provides a fully-fledged integration platform for dialogue systems and has been improved continuously during the last several years (Bub and Schwinn, 1999; Herzog et al., 2004; Klüter et al., 2000).

3.1 Middleware Solution

Our specific realization of an integration platform follows the architecture framework and the design principles described in the previous section. Its main characteristics are:

- component architecture
  - modularization into distinct and independent software modules to allow maximum decoupling
  - encapsulation of heterogeneous, large-grained software components into a uniform module structure
- event-based architectural style
  - very flexible publish/subscribe messaging
  - implicit invocation (event-driven) to support decoupling of components
  - highly adaptable content format specifications for each data pool, i.e., named message queue

Instead of using custom programming interfaces, the interaction between distributed components within the testbed framework is based on the exchange of structured data through messages. The SMARTKOM testbed includes a message-oriented middleware implementation that is based on the *parallel virtual machine* (PVM) software described in Geist et al. (1994). In order to provide publish/subscribe messaging with data pools on top of PVM, a further software layer called *pool communication architecture* (PCA) had to be added (Klüter et al., 2000).

Compared with commercial-grade middleware products, our low-cost solution does not consider issues like fault tolerance, load balancing, or replication. It does, however, provide a scalable and efficient platform for the kind of real-time interaction needed within a multimodal dialogue system. The PVM/PCA-based messaging system is able to transfer arbitrary data contents and provides excellent performance characteristics. Within SMARTKOM-Public, for example, it is even possible to perform a telephone conversation. Message throughput on standard PCs with Intel Pentium III 500 MHz CPUs is offhand sufficient to establish a reliable bidirectional audio connection, where uncompressed audio data are being transferred as XML messages in real time.

The so-called *module manager* provides a thin API layer for component developers with language bindings for the programming languages that are used to implement specific dialogue components. It comprises the operations required to access the communication system and to realise an elementary component protocol needed for basic coordination of all participating distributed components. The module manager implementation of the SMARTKOM testbed includes, in particular, APIs for C, C++, Java, and Prolog.

3.2 Technical Modules for System Execution

In addition to the functional components of the dialogue system, the run-time environment also includes special testbed modules in support of system operation.

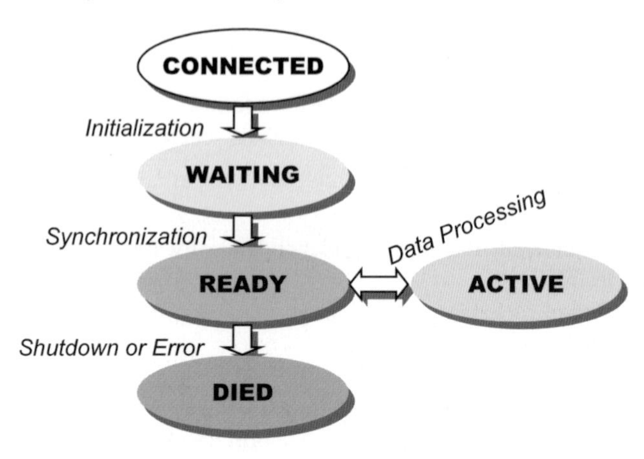

Fig. 2. Basic component states for module coordination

The *testbed manager* component is responsible for system initialization and activates all distributed components pertaining to a given dialogue system configuration. It also cooperates with the functional modules to carry out the elementary component protocol, which is needed for proper system start-up, controlled termination of processes and restart of single components, or a complete soft reset of the entire dialogue system. A state-transition network forms the basis for this protocol, which is also directly supported by the API functions of the module manager interface. As illustrated in Fig. 2, only a few basic component states need to be distinguished.

The testbed manager supports flexible run-time configuration of the application system. Module configurations can be declared using an XML-based specification. The modular concept allows start-up of individual components or subsystems for debugging purposes, e.g., to test a critical path from speech or gesture input to speech and graphic output.

The *testbed control graphical user interface (GUI)* shown in Fig. 3 constitutes a separate component that provides a graphical user interface for the administration of a running system. The adaptable testbed GUI provides a simplified view of the system architecture and offers the necessary means to monitor system activity, to interact with the testbed manager, and to manually modify configuration settings of individual components while testing the integrated system.

Both testbed modules and the functional components interact using the underlying communication system. Specific data pools are employed to exchange testbed-related control data vs. application data of the multimodal dialogue system. From the point of view of the component developer, the internal system pools are not directly visible, as they are hidden inside the module manager API implementation.

The SMARTKOM testbed provides automatic logging of module output messages and all communicated data. A further testbed module, the logging component, is employed to save a complete protocol of all exchanged messages—including system pools—for later inspection.

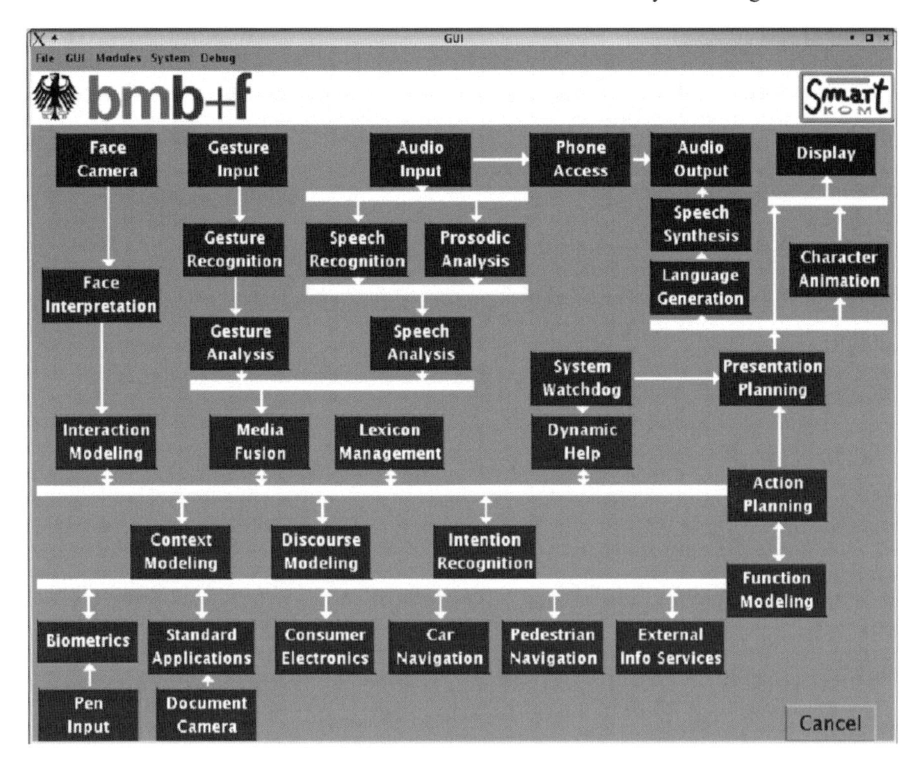

Fig. 3. Graphical user interface for testbed administration

3.3 M3L Support

The SMARTKOM system employs *Multimodal Markup Language* (M3L) as external data format for information exchange between functional components. M3L, an XML-based language, was developed within the SMARTKOM project to cover all data interfaces inside this complex multimodal dialogue system.

Parts of the language specification are generated automatically from an underlying conceptual taxonomy that provides the foundation for the representation of domain knowledge inside the dialogue system. The offline tool described in Gurevych et al. (2003) transforms an ontology written in OIL format (Fensel et al., 2001) into an M3L-compatible XML schema definition. Our tool simplifies updates of the M3L specification as the ontology evolves and ensures a consistent mapping from structural and semantic knowledge encoded in the ontology to directly interpretable communication interfaces.

From the point of view of component development, XML offers various techniques for the processing of transferred content structures. The standard document object model (DOM) API makes the M3L data available as a generic tree structure—the *document object model*—in terms of elements and attributes, whereas *simple API for XML* (SAX) provides an event-based API for XML parsers.

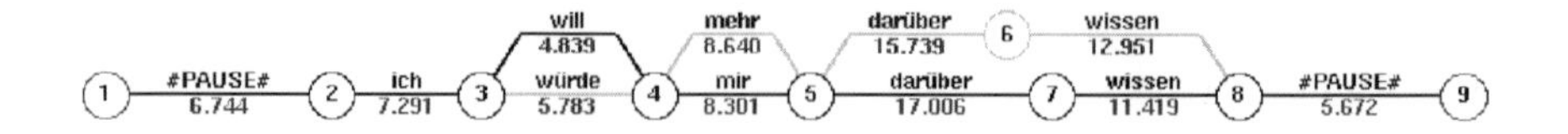

Fig. 4. Word hypothesis graph for "I would like to know more about this"

Another interesting option is to employ XSLT stylesheets to flexibly transform between the external XML format used for communication and a given internal markup language of the specific component. The use of XSLT makes it easier to adapt a component to interface modifications and simplifies its reuse in another dialogue system.

Instead of working on basic XML structures like elements and attributes, XML data binding can be used for a direct mapping between program internal data structures and application-specific M3L markup. In this approach, the M3L language specification is exploited to automatically generate a corresponding object model in a given programming language.

All the techniques decribed here have been applied in different implementations of the various functional components of the SMARTKOM system. In addition, a specific M3L API has been developed and included into the SMARTKOM testbed. This M3L API offers for all target platforms a light-weight programming interface to simplify the processing of XML structures within the implementation of a component. It incorporates a standard XML parser and uses the DOM interface to implement the M3L-specific API layer. Furthermore, the SMARTKOM testbed also provides a generic, XML-enabled toolbox and tailored XSLT stylesheets for the offline and online inspection of M3L data.

3.4 Additional Testbed Tools

A rich set of utilities can be used to manipulate log data easily, e.g., to select and identify relevant subsets from the complete data store resulting from a system run. Flexible replay of selected pool data provides a simple, yet elegant and powerful mechanism for the simulation of small or complex parts of the dialogue system in order to test and debug components during the development process. Using a simple tool, a system tester is also able to generate trace messages on the fly to inject annotations into the message log.

Another important development tool is a generic XML-enabled data viewer for the online and offline inspection of pool data. Figure 4 presents an example of a word hypothesis graph corresponding to the natural language part of a multimodal user input. Such a graphical visualization of speech recognition results during system execution aids the system integrator while testing the dialogue system.

The flexible data viewer toolbox can also be used in offline mode. A practical example is given in Fig. 5, which shows a comprehensible display format that documents an interaction with the SMARTKOM system.

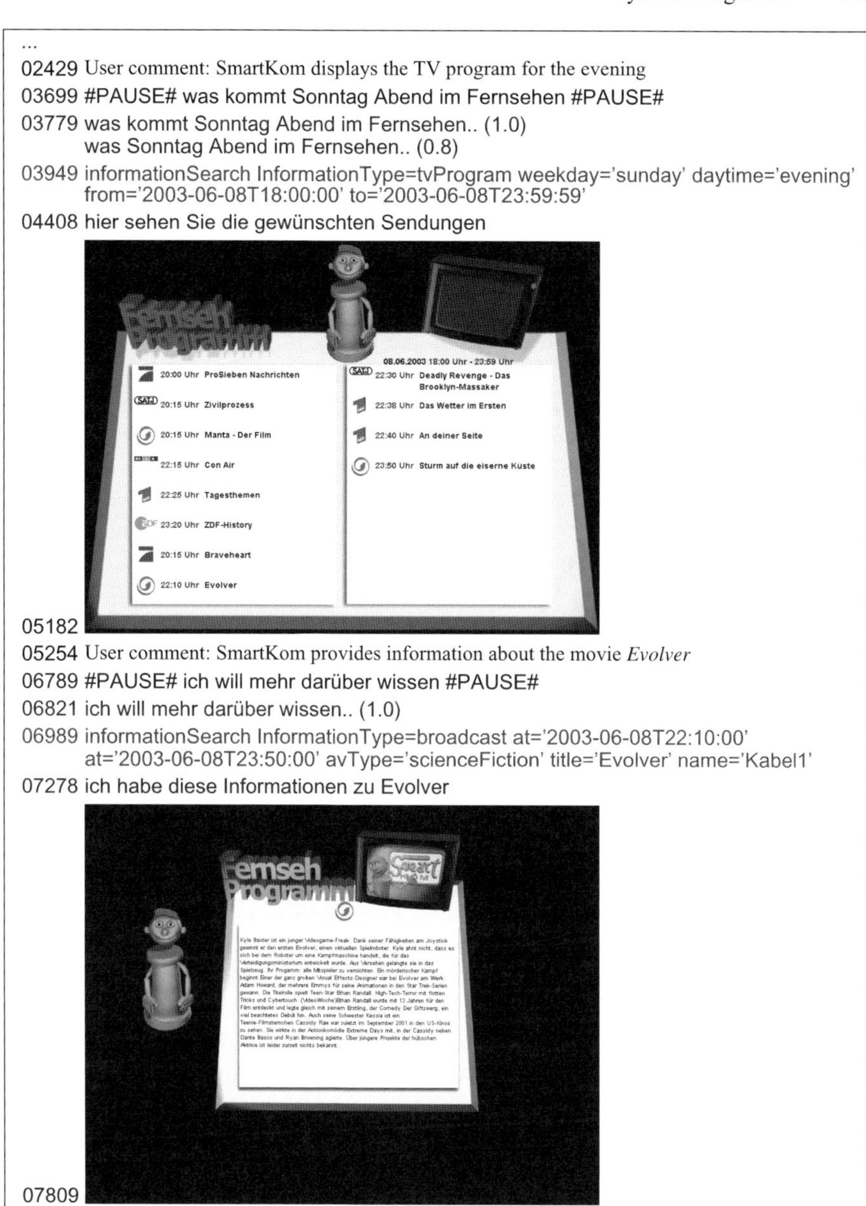

Fig. 5. Excerpt from an automatically generated dialogue protocol

This kind of dialogue protocol provides a compact summary of important processing results and is generated fully automatically from the message log. The individual entries consist of the specific message number followed by the summarized content of selected data pools. Each item is prepended with the corresponding message number in order to enable more detailed inspection of relevant data if potential processing errors need to be resolved. Further offline tools include a standardized build and installation procedure for components as well as utilities for the preparation of software distributions and incremental updates during system integration.

4 System Integration Process

In addition to the software infrastructure, the practical organization of the project constitutes a key factor for the successful realization of an integrated multimodal dialogue system. Distributed teams and heterogeneous software environments within a large research project add complexity that must be resolved. In order to cope with system integration challenges, it is beneficial to clearly distinguish two major roles in the development process. A system integrator takes care of the overall software infrastructure and assembles executable components into an integrated application system, whereas a module developer should concentrate on a specific functional component.

For the SMARTKOM project, a dedicated system integration group composed of professional software engineers has been established to ensure that the resulting software is robust and maintainable, that an adequate architectural framework and testbed are provided to the participating researchers, that software engineering standards are respected, and that modules developed in different programming languages by a distributed team fit together properly. These aspects are too important to leave them to the individual scientists, who regard them as a side issue because they have to focus on their own research topics. An important achievement of the SMARTKOM project is the consistent integration of a very large number of components created by diverse groups of researchers from disparate disciplines.

Given the underlying architecture framework, the approach taken in SMARTKOM bears many similarities to component-based software engineering, where the notion of building an application by writing code has been replaced with building a system by assembling and integrating supplied software components (Bass et al., 2003). In contrast to conventional software development, where system integration is often the tail end of an implementation effort, component integration is the centrepiece of the approach.

Figure 6 provides a sketch of our development process, which is integration-centric as opposed to development-centric. In this kind of view, implementation has given way to integration as the focus of system construction.

After the design phase, which leads to a specification of the concrete component architecture and functional interfaces, the realization starts with the provision of a tailored testbed software package. The SMARTKOM testbed follows the *breadboard paradigm*, which provides an initial instantiation of the overall architecture using placeholders that represent the functional components (Bub and Schwinn, 1999; Klüter et al., 2000). During system integration, the dummy components are replaced

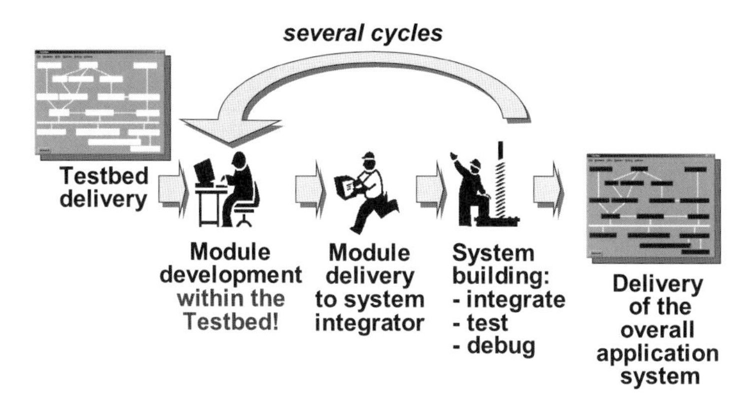

Fig. 6. Stages of system integration

successively by more substantial modules with increased functionality, i.e., components are like black boxes which can be plugged into a breadboard. Nonexistent modules can be simulated through the replay of appropriate test data. This technique enables rapid prototyping, given that even at an early stage an environment is available for testing and evaluating individual modules in the context of the entire application.

The one constant aspect of component-based software development is change. Constituent components are constantly evolving, leading to an iterative process where the different phases have to repeated several times until the multimodal dialogue application can be finalized. This includes also the customization of the software infrastructure, which is an integral element of a project life cycle.

Stepwise improvement and implementation of the design of architecture details and interfaces necessitate an intensive discussion process. It has to include all participants involved in the realization of system components in order to reach a common understanding of the intended system behaviour. The moderation and coordination of these activities constituted an important task of the system integration group in SMARTKOM.

The functional architecture of the multimodal dialogue system provides a basis for modularization into distinct software components. Significant parts of the initial system architecture were already predetermined by the project structure. Functional components are mapped to modules, realized as independent software processes, to obtain a specific software architecture. Figure 7 depicts the components and communication links inside the SMARTKOM system, which includes about 40 functional modules and 100 data pools for information exchange.

The specification of the content format for each data pool defines the common language that the dialogue system components use to interoperate. The careful design of information flow and accurate specification of content formats act as essential elements of our approach. We operationalize interface specifications in the form of an XML language. Our experience with M3L shows that the design of such an XML language for the external representation of complex data is not a simple task. All

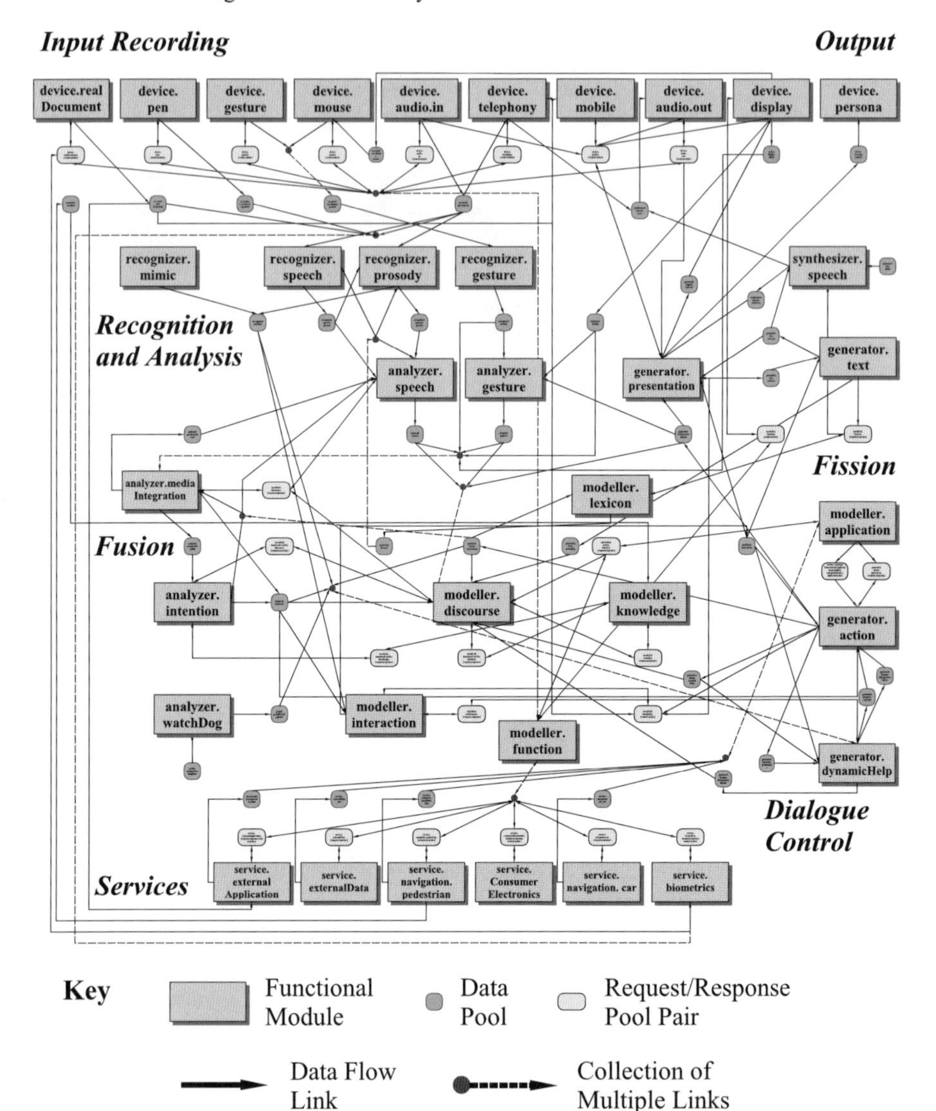

Fig. 7. Detailed software architecture of the SMARTKOM system

design decisions have to be made carefully. For example, it is better to minimize the use of attributes. They are limited to unstructured data and may occur at most once within a single element. Preferring elements over attributes better supports the evolution of a specification since the content model of an element can easily be redefined to be structured and the maximum number of occurrences can simply be increased to more than one. A further principle for a well-designed XML language requires that the element structure reflects all details of the inherent structure of the represented data, i.e., textual content for an element should be restricted to well-defined elementary types. Another important guideline is to apply strict naming rules so that it becomes easier to grasp the intended meaning of specific XML structures.

The latest version of M3L consists of more than 1,000 different XML elements. In order to make the specification process manageable and to provide a thematic organization, the M3L language definition has been decomposed into about 40 schema specifications. This kind of modularization allowed to distribute responsibility for the different parts of the specification, assigning each XML schema to one selected researcher with specific expertise in that area.

5 Conclusion

This contribution has presented a discussion on system integration in building multimodal dialogue applications with an emphasis on the SmartKom integration platform. It has shown the analogy of our approach to component-based software engineering.

The capabilities of the multiplatform software framework are being enhanced on a continuing basis since the first design in Verbmobil, the predecessor of the SmartKom project. The SmartKom testbed provides a powerful environment and advanced new tools, like, for example, the added support for XML-based data representations. Multiplatform has been made available as open source software and is currently being used at several industrial and academic sites.

The paper also gave a look into the underlying system integration process and underlines the role of nontechnical factors for effective inter- and intrainstitutional collaboration.

Acknowledgments

First of all, we would like to thank our former colleagues in the system integration group. Heinz Kirchmann, Andreas Klüter, Hans-Jörg Kroner, Stefan Merten, Martin Neumann, and Rainer Peukert all contributed as well to the work described here.

Given the central position of our team within the project structure, there have been intensive contacts with nearly every project participant. We enjoyed this joint work and gratefully acknowledge the very successful collaboration in the SmartKom consortium.

References

L. Bass, P. Clements, and R. Kazman. *Software Architecture in Practice*. Addison-Wesley, Boston, MA, 2nd edn., 2003.

P.A. Bernstein. Middleware: A Model for Distributed System Services. *Communications of the ACM*, 39(2):86–98, 1989.

T. Bub and J. Schwinn. The Verbmobil Prototype System — A Software Engineering Perspective. *Natural Language Engineering*, 5(1):95–112, 1999.

N. Carriero and D. Gelernter. Linda in Context. *Communications of the ACM*, 32(4): 444–458, 1989.

W. Emmerich. Software Engineering and Middleware: A Roadmap. In: *Proc. Conf. on the Future of Software Engineering*, pp. 117–129, Limerick, Ireland, 2000. ACM.

L.D. Erman, F. Hayes-Roth, V.R. Lesser, and D.R. Reddy. The Hearsay-II Speech-Understanding System: Integrating Knowledge to Resolve Uncertainty. *ACM Computing Surveys*, 12(2):213–253, 1980.

D. Fensel, F. van Harmelen, I. Horrocks, D.L. McGuinness, and P.F. Patel-Schneider. OIL: An Ontology Infrastructure for the Semantic Web. *IEEE Intelligent Systems*, 16(2):38–45, 2001.

K. Geihs. Middleware Challenges Ahead. *IEEE Computer*, 34(6):24–31, 2001.

A. Geist, A. Beguelin, J. Dongorra, W. Jiang, R. Manchek, and V. Sunderman. *PVM: Parallel Virtual Machine. A User's Guide and Tutorial for Networked Parallel Computing*. MIT Press, Cambridge, MA, 1994.

I. Gurevych, S. Merten, and R. Porzel. Automatic Creation of Interface Specifications from Ontologies. In: H. Cunningham and J. Patrick (eds.), *Proc. HLT-NAACL 2003 Workshop on Software Engineering and Architecture of Language Technology Systems (SEALTS)*, pp. 59–66, Edmonton, Canada, 2003. Association for Computational Linguistics.

G. Herzog, H. Kirchmann, S. Merten, A. Ndiaye, P. Poller, and T. Becker. MULTIPLATFORM Testbed: An Integration Platform for Multimodal Dialog Systems. In: H. Cunningham and J. Patrick (eds.), *Proc. HLT-NAACL 2003 Workshop on Software Engineering and Architecture of Language Technology Systems (SEALTS)*, pp. 75–82, Edmonton, Canada, 2003. Association for Computational Linguistics.

G. Herzog, A. Ndiaye, S. Merten, H. Kirchmann, T. Becker, and P. Poller. Large-Scale Software Integration for Spoken Language and Multimodal Dialog Systems. *Natural Language Engineering*, 10(3–4):283–305, 2004. Special Issue on Software Architecture for Language Engineering.

J. Hopkins. Component Primer. *Communications of the ACM*, 43(10):27–30, 2000.

A. Klüter, A. Ndiaye, and H. Kirchmann. Verbmobil From a Software Engineering Point of View: System Design and Software Integration. In: W. Wahlster (ed.), *Verbmobil: Foundations of Speech-to-Speech Translation*, pp. 635–658, Berlin Heidelberg New York, 2000. Springer.

N. Reithinger, G. Herzog, and A. Ndiaye. Situated Multimodal Interaction in SmartKom. *Computers & Graphics*, 27(6):899–903, 2003.

W. Wahlster. SmartKom: Symmetric Multimodality in an Adaptive and Reusable Dialogue Shell. In: R. Krahl and D. Günther (eds.), *Proc. Human Computer Interaction Status Conference 2003*, pp. 47–62, Berlin, Germany, June 2003. DLR.

M.W. Wu and Y.D. Lin. Open Source Software Development: An Overview. *IEEE Computer*, 34(6):33–38, 2001.

SmartKom-English: From Robust Recognition to Felicitous Interaction

David Gelbart[1], John Bryant[1], Andreas Stolcke[1], Robert Porzel[2], Manja Baudis[2] and Nelson Morgan[1]

[1] International Computer Science Institute (ICSI), Berkeley, USA
texttt{gelbart,jbryant,stolcke,morgan}@icsi.berkeley.edu
[2] European Media Laboratory GmbH [EML], Heidelberg, Germany
{robert.porzel,manja.baudis}@eml-d.villa-bosch.de

Summary. This chapter describes the English-language SMARTKOM-Mobile system and related research. We explain the work required to support a second language in SMARTKOM and the design of the English speech recognizer. We then discuss research carried out on signal processing methods for robust speech recognition and on language analysis using the Embodied Construction Grammar formalism. Finally, the results of human-subject experiments using a novel *Wizard and Operator* model are analyzed with an eye to creating more felicitous interaction in dialogue systems.

1 Introduction

The SMARTKOM-Mobile application provides navigation and tourism information using either a handheld personal digital assistant (PDA) interface or an in-car interface. Some images from the application's display are shown in Fig. 1. A user communicates with SMARTKOM-Mobile using pointing gestures and natural speech, and the system responds with speech (from an animated agent, displayed on-screen) and the display of images and text. The natural and reliable conversational interaction that this calls for provided motivation for a range of research. In this section and in Sect. 2 we describe the work required to port SMARTKOM-Mobile to the English language and the design of the English speech recognizer. In the following sections, we describe research on robust speech recognition, language analysis, and human–computer interaction carried out by the SMARTKOM-English team.

The development of an English-language SMARTKOM-Mobile verified the language portability of the SMARTKOM architecture and facilitated the demonstration of SMARTKOM at international conferences. Staff and visiting researchers at ICSI and staff at DFKI took the lead roles in the creation of the English-language system, and important contributions came from several other SMARTKOM partner sites.

The English speech recognizer was developed completely independently from the German one; it is a hybrid connectionist system descended from the one used in

Old Bridge
The Old Bridge, otherwise known as the
Karl-Theodor Bridge, has become a central
part of Heidelbergs townscape. For 600 years
it was the only solid link across the Neckar
River. The bridge has been destroyed a
number of times both through forces of
nature and the hand of man, the last time in

Fig. 1. SmartKom-Mobile screen shots showing pedestrian navigation (*left*) and tourist site information (*right*)

the BERP dialogue system project (Jurafsky et al., 1994). All other modules in the English SmartKom-Mobile system were based on, or identical to, modules in the German-language SmartKom-Mobile.

The modular architecture of SmartKom greatly eased porting to English by encapsulating language dependencies in specific modules. Most SmartKom modules required no modification to support English. Of the modules requiring modification, only speech recognition and speech synthesis required significant changes to software source code. The speech analyzer (which parses the recognized speech) and text generator (which creates the system's output sentences) required only a change in their template (grammar) files and otherwise used the same software engines for both German and English. The lexicon module had to be aware of the current language in order to provide the correct word pronunciations. Some displayed text provided by the pedestrian and vehicle navigation modules (such as tourist site information and map labels) was translated to English. If dynamic help had been included in the English system, some additional displayed text would have required translation. The speech analyzer outputs a language-independent semantic representation of the user input, and so modules which tracked dialogue state and user intention did not need to be language-aware.

2 The SmartKom-English Speech Recognizer

2.1 Overview

The recognizer uses the hidden Markov model (HMM) approach to speech recognition illustrated in Fig. 2. This approach models speech as a sequence of observations sampled from different possible probability distributions, with a distinct distribution corresponding to each member of a finite set of possible hidden states. In our recognizer the hidden states represent phones, and the observations are assumed to be acoustic realizations of those phones. Each observation represents a single frame of time. The time step from the start of one frame to the start of the next frame is 16 ms and the length of each frame is 32 ms; the resulting overlap between frames is useful since the frame boundaries are not necessarily aligned with phone boundaries. The observations being modeled are not the original audio but rather are the output of a feature extraction process intended to reduce dimensionality and discard irrelevant variation. We used perceptual linear prediction (PLP) feature extraction (Hermansky, 1990), which captures the envelope of the frame power spectrum but discards some spectral detail. The probability of a particular observation for each possible hidden state is determined by an acoustic modeling stage, which we carry out using a multilayer perceptron (MLP), following the hybrid connectionist approach of Bourlard and Morgan (1993). The probabilities determined by the acoustic modeling stage for all frames are used by a decoding stage that searches for the most likely sentence, taking into account a dictionary of word pronunciations and a language model (tables of word transition probabilities).

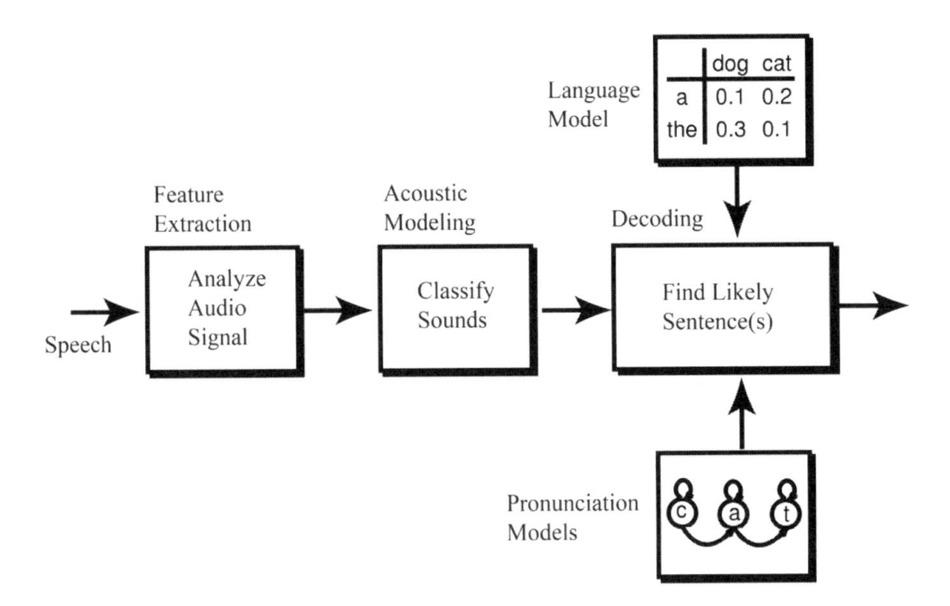

Fig. 2. Speech recognizer architecture

The English speech recognizer was built as a chain of small tools used in a pipeline, communicating with each other using Unix pipes. This simple, modular design makes adding or upgrading speech recognizer components easy. A wrapper program starts the pipeline and handles communication with other SmartKom modules. Feature extraction is performed by ICSI's RASTA tool. Utterance-level mean and variance normalization of the features and MLP output calculation is performed by ICSI's FFWD-NORM tool. Decoding was performed by the NOWAY tool (Renals and Hochberg, 1996); changes made to the NOWAY tool to support the needs of the SmartKom project are described below. Source code for all these tools is available free for research use.

2.2 Language Modeling and Decoding

The speech recognizer uses a trigram language model with backoff to bigrams and unigrams. The SRI language modeling toolkit (Stolcke, 2002) was used to estimate the language model from training data that consisted of sample dialogues, partly edited by hand to achieve good coverage of what were perceived to be natural user inputs. Since no naturally collected English data was available, we relied on English translations of German dialogues, taking care to produce idiomatic, rather than literal translations.

To meet the needs of SmartKom, we modified the NOWAY decoder to use the C++ libraries associated with the SRI LM toolkit to access language models. This modification allowed NOWAY to be used with class-based language models. Class-based LMs can include class labels as part of N-grams, which are then expanded by a list of class member words. Class-based models have two key advantages for SmartKom. First, by using classes the LM generalizes better to novel word sequences, which is especially important given the scarcity of training data. For example, a DIRECTION class was used to stand for possible map directions (e.g., left, right, up, down, east, west, north, and south) in training sentences. To achieve generalization, the word classes are defined by hand based on task knowledge, and the appropriate words are replaced by class labels in the training data.

The second key function of word classes is that they allow new class members to be added on the fly while SmartKom is running, without reestimating the entire language model. In the English Mobile application this occurs with parking garage names, which are retrieved from the car navigation module. Parking garage names can occur any time the class name GARAGE occurs in the language model, thereby covering sentences such as

- Can you tell me more about GARAGE
- I would like to know more about GARAGE
- I'd like to know more about GARAGE

3 Signal Processing for Robust Speech Recognition

3.1 Introduction

Compared to recordings made using a close-talking microphone placed near the user's mouth (e.g., a headset microphone), recordings from more distant microphones have higher levels of background noise relative to the speech level (since the speech level is lower) and are subject to reverberation and other effects due to the longer and sometimes indirect paths taken by the traveling sound waves. While these degradations affect human speech recognition performance as well, current automatic speech recognition systems are much more sensitive to them. However, close-talking microphones are often inconvenient, and improvements in the recognition accuracy that can be achieved without them are very likely to increase adoption of speech recognition technology.

The SMARTKOM-Mobile application can be used inside a car (using an installed display and microphones) or on foot (using the microphone and display on a hand-held computer or PDA).[3] Of these two circumstances, ICSI research focused on the in-car case, making use of the SpeechDatCar (Moreno et al., 2000) corpus. The SpeechDatCar corpus is available in several languages; in this article we will only describe results for SpeechDatCar-German. This contains in-car recordings of German connected digit strings made simultaneously with close-talking and hands-free microphones, in various noise conditions: "Stop Motor Running," "Town Traffic," "Low-Speed Rough Road," and "High-Speed Good Road." Our speech recognition experiments with this corpus were performed for three cases: the well-matched case used all microphone types and noise conditions in both training and test data, the medium-matched case used only the hands-free microphone and tests using only the "High-Speed Good Road" noise condition (which is excluded from the training data in this case), and the highly mismatched case used close-talking microphone training data and hands-free microphone test data.

3.2 Noise Reduction and Deconvolution

Figure 3 shows a model of how acoustic degradation is caused by reverberation and background noise. Reverberation and other acoustic effects related to the transmission of speech from talker to microphone, together with the frequency response of the microphone itself, are modeled as a linear time-invariant system with impulse response $c(n)$ and frequency response $C(\omega)$. Background noise (including noise internal to the microphone itself) is additive after this system. We investigated the effectiveness of some signal processing techniques based on this model.

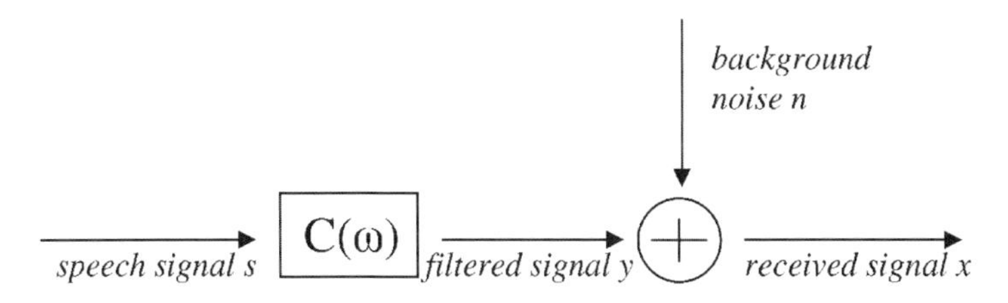

Fig. 3. Model of acoustic degradation

[3] In in-car applications, non–close-talking microphones are sometimes referred to as *hands-free* microphones, as in hands-free mobile phone operation.

3.2.1 Noise Reduction

We used a noise reduction implementation developed for an Aurora (Hirsch and Pearce, 2000) front-end proposal, a joint effort between ICSI, OGI, and Qualcomm engineers, described in Adami et al. (2002). The algorithm performs Wiener filtering with modifications such as a noise overestimation factor, smoothing of the filter response, and a spectral floor. It calculates an instantaneous noise power spectral estimate $|\acute{N}(m,k)|^2$ (where k is the frequency bin and m is the frame index) by averaging the noisy power spectra $|X(m,k)|^2$ over an initial period before speech starts as well as later frames, which are judged to be nonspeech because their energy level falls below a threshold. This estimate is used to calculate a filter

$$|H(m,k)| = max\left(\frac{|X(m,k)|^2 - \alpha|\acute{N}(m,k)|^2}{|X(m,k)|^2}, \beta\right),$$

where α is an SNR-dependent oversubtraction factor and the spectral floor parameter β is used to avoid negative and very small filter values. This filter $|H(m,k)|$ varies from frame to frame. To reduce artifacts in the noise-reduced output, the filter is smoothed over time and frequency. Then the smoothed filter is applied to the noisy power spectra X to obtain an estimate of the noise-free power spectra Y:

$$|\acute{Y}(m,k)|^2 = max(|X(m,k)|^2 * |H(m,k)|^2, \beta_{final} * |\acute{N}(m,k)|^2),$$

where β_{final} is a second spectral floor parameter, which specifies the floor as a fraction of the estimated noise power.

This Wiener filtering approach is based on the assumption that the noise power spectrum is steady. This is probably a fairly good assumption for engine and wind noise during unchanging driving conditions.

Table 1 shows word error rates (WER) on SpeechDatCar-German using the Aurora baseline speech recognizer described in Hirsch and Pearce (2000), with and without noise reduction preprocessing. The noise reduction is very effective in the medium-mismatch condition. In the high mismatch condition it is counterproductive; perhaps the application of noise reduction to the close-talking microphone training data makes that data an even worse match for the hands-free microphone test data. For these experiments, the noise estimation was performed independently for each utterance, and we used overlap-add resynthesis to create noise-reduced output waveforms. This allowed the noise reduction to be used with existing feature extraction code without modifying that code. This noise reduction approach was added to the SMARTKOM-English system as a new pipeline stage preceding the RASTA feature extraction tool.

3.2.2 Deconvolution by Mean Subtraction

Deconvolution by mean subtraction is commonly employed in speech recognition systems, most often via the cepstral mean subtraction (CMS) algorithm. The reasoning behind it is as follows. Consider a discrete-time speech signal $s(n)$ (the origin

Table 1. SpeechDatCar-German word error rates (WER) with and without noise reduction. The well-matched test data contain 5009 words, the medium-matched test data contain 1366 words, and the highly mismatched test data contain 2162 words.

WER (%)	Without noise reduction	With noise reduction
Well-matched	8.0	7.0
Medium-matched	20.6	14.9
Highly mismatched	15.4	17.5

of speech as a continuous-time signal is ignored here for simplicity) sent over a linear time-invariant channel with impulse response $c(n)$, producing a filtered signal $y(n)$. Then the product property holds for the spectra of s and c: $Y(\omega) = S(\omega)C(\omega)$. Assume that the filtered signal is processed using a windowed discrete Fourier transform, giving $Y(m,\omega)$, where m is the frame index determining around which sample the window function was centered. If the window function is long and smooth enough relative to $c(n)$, then the product property approximately still holds (Avendano, 1997): $X(m,\omega) \approx S(m,\omega)C(\omega)$. Note that the channel is assumed not to vary over time. Taking the logs of the magnitudes of both sides of the previous equation, we find that $log|X(m,\omega)| \approx log|S(m,\omega)| + log|C(\omega)|$. Therefore, by subtracting the mean over m (i.e., over time) of $log|X(m,\omega)|$ from $log|X(m,\omega)|$, we remove $log|C(\omega)|$ and the mean over time of $log|S(m,\omega)|$. If $s(n)$ is speech, and the mean is calculated over a large enough number of frames, then we expect the mean of $log|S(m,\omega)|$ to contain little linguistic information, and therefore its removal need not be detrimental to speech recognition performance. Thus the subtraction can be used to compensate for the magnitude response of the channel (for various reasons, there is usually no attempt to compensate the phase response). Speech recognition feature extraction is usually essentially based on windowed discrete Fourier transforms. However, following the transform it is typical to do further processing like Mel/Bark-scale filter bank integration and cepstral transformation, and most often the mean subtraction is done after that processing. The cepstral transformation, being linear, does not affect the reasoning above, but the filter bank integration implies an additional assumption that the channel frequency response is close to constant across the frequency bins that are being integrated.

Table 2 shows WER on SpeechDatCar-German with and without mean subtraction methods added to the Aurora baseline speech recognizer described in Hirsch and Pearce (2000), which uses Mel-frequency cepstral coefficient (MFCC) feature extraction and by default does not perform mean subtraction.

Cepstral mean subtraction (CMS) removes the mean across frames from the MFCCs. The table shows CMS was most helpful in the well-matched and highly mismatched cases, where both close-talking and hands-free recordings are used. This is not surprising because there is a channel mismatch between close-talking and hands-free recordings. The reasoning behind mean subtraction assumes that all frames contain speech, while in fact the data is a mix of speech and pauses. Calculating the mean only over frames judged by a multilayer perceptron classifier to contain speech resulted in a significant performance improvement, which is consistent with the results for speech recognition over telephone connections in Mokbel et al. (1996). We also tried the log-DFT mean normalization (LDMN) proposed in Neumeyer et al. (1994). In this method the mean subtraction occurs midway through the MFCC computation, before the filter bank integration, so the assumption of channel constancy across bins being integrated is not required. At ICSI, we have sometimes found this to be more effective than CMS, but on this task it does not perform better (in fact, there is a slight, though not statistically significant, drop in performance).

Table 2. SpeechDatCar-German word error rates (WER)

WER (%)	Baseline	Noise reduction alone	Noise red. and CMS	Noise red. and CMS; mean taken over speech frames	Noise red. and LDMN; mean taken over speech frames
Well-matched	8.0	7.0	6.7	6.1	6.1
Medium-matched	20.6	14.9	15.7	14.6	15.2
Highly mismatched	15.4	17.5	14.7	11.3	11.0

For further results, including other mean subtraction methods and other data sets, see Gelbart (2004).

3.3 Gabor Filtering

The noise reduction and mean subtraction approaches described above are intended to increase the robustness of existing feature extraction methods by adding additional processing. Another approach, which can be complementary, is to create new feature extraction methods which have desirable properties. We collaborated with Michael Kleinschmidt of the Universität Oldenburg on his project on Gabor filter feature extraction for automatic speech recognition (Kleinschmidt and Gelbart, 2002; Kleinschmidt, 2002). Gabor filters are a family of two-dimensional filters, which Kleinschmidt proposed to use for feature extraction by convolving Gabor filters with a time-frequency representation such as a mel-band spectrogram (see Fig. 4). Depending on what Gabor filters are used, this can behave similarly to short-term spectral envelope-based feature extraction approaches like the popular MFCC and PLP methods, or to the TRAPS (Jain and Hermansky, 2003) approach of long-term analysis in narrow frequency bands, or it can look for patterns at an oblique angle to the time and frequency axes. In Kleinschmidt's approach the Gabor filters used are chosen by a data-driven selection procedure which searches for Gabor filters that appear likely to give good classification performance.

The second column of Table 3 gives WER on SpeechDatCar-German using the back end of the Aurora baseline speech recognizer (Hirsch and Pearce, 2000) with the QIO-NoTRAPS feature extraction module developed by Qualcomm, ICSI, and OGI (Adami et al., 2002), which calculates robust MFCC features using techniques such as noise reduction by Wiener filtering and mean subtraction. When features derived from Gabor analysis were concatenated to the robust MFCC features to form a longer feature vector, the error rate decreased, as shown in the third column. We found that for good performance with the Gabor filters it was necessary to pass them through a stage of nonlinear discriminant analysis by MLP (the back end used a different acoustic modeling approach). For source code for Gabor feature extraction, and additional information about this approach, please refer to our website.[4]

Table 3. SpeechDatCar-German word error rates (WER)

WER (%)	QIO-NoTRAPS	With Gabor analysis
Well-matched	5.8	5.4
Medium-matched	12.1	11.7
Highly mismatched	12.0	11.6

4 Robust, Semantically Rich Language Analysis

The approach to analysis (parsing and semantic analysis) of recognized speech normally used in the SMARTKOM system was Ralf Engel's SPINmodule (Engel, 2002). The SMARTKOM project also funded investigation into an alternative approach, aimed at robust and semantically rich language analysis. The alternative approach, described here, makes use of a linguistically sophisticated language formalism called Embodied Construction Grammar (ECG (Bergen and Chang, 2002; Chang et al., 2002). ECG is a construction-based grammar formalism (Goldberg, 1995) that uses embodied primitives like frames (Fillmore, 1982), image schemas (Lakoff, 1987) and executing schemas (Narayanan, 1997) as its semantic representation. In addition to supporting these cognitive primitives, ECG is an extension of unification-based formalisms like HPSG (Pollard and Sag, 1994), and as a consequence, it is precise enough for computational models of language analysis.

ECG is an extremely expressive grammar formalism, and as such new algorithms needed to be designed that could take advantage of the wealth of semantic information contained in an ECG grammar, and yet still efficiently and robustly process each utterance. These algorithms are implemented within the so-called constructional analyzer (Bryant, 2003).

The first key innovation employed in the design of the constructional analyzer is how it combines linguistic knowledge with process. Instead of treating each construction as a passive piece of grammatical knowledge, the analyzer compiles each construction into an active unit called a construction recognizer. Each construction recognizer is responsible for applying both the form and the semantic constraints associated with its construction. The recognizer then generates an instance of the construction if an acceptable set of constituents is found.

[4] http://www.icsi.berkeley.edu/Speech/papers/icslp02-gabor/

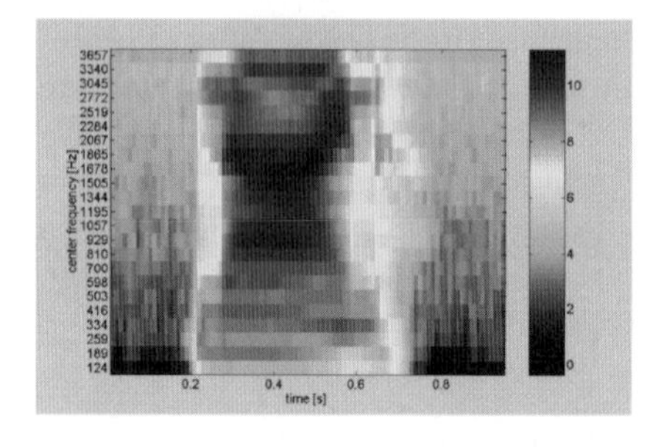

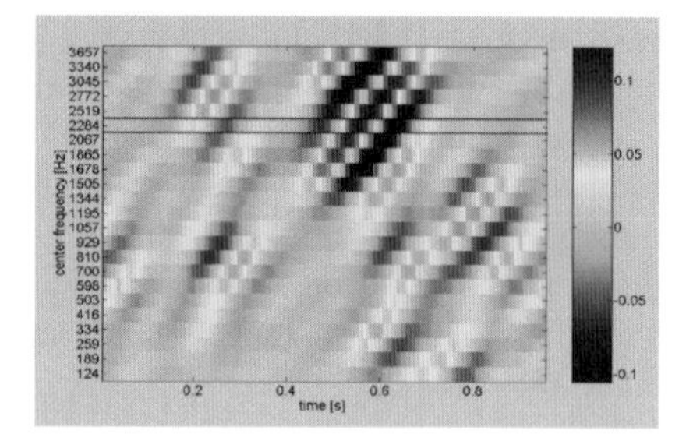

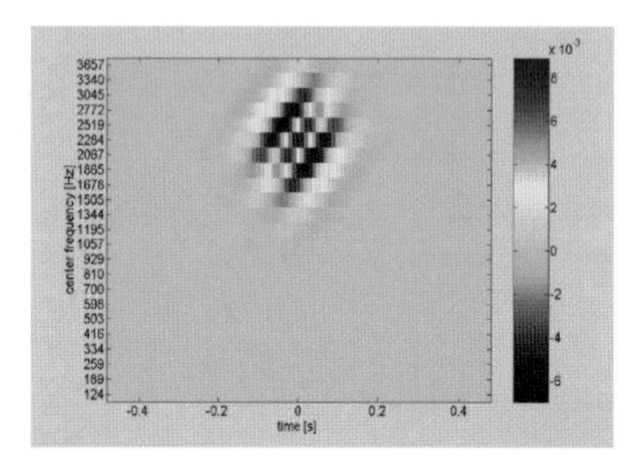

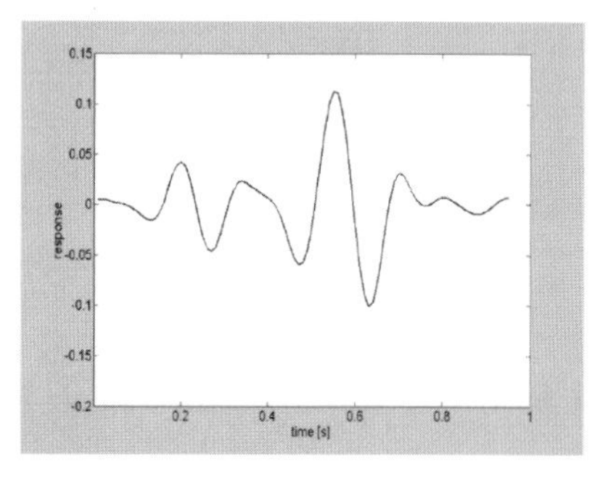

Fig. 4. *Upper left*: A log mel-band spectrogram for the spoken word "nine". *Lower left*: The real part of a Gabor filter. The Gabor filters are complex-valued. Rather than doing acoustic modeling with complex-valued features, Kleinschmidt chose to use the real part, imaginary part, or magnitude of the results of correlation as features. *Upper right*: The real part of the log mel-band spectrogram after correlation with the Gabor filter. *Lower right*: The values of the filtered log mel-band spectrogram at the center frequency of the Gabor filter (it is those that would be used as features for speech recognition)

The analyzer itself must manage the interaction between each of the recognizers, while looking for semantically and formally complete analyses of the whole utterance. The analyzer also has the responsibility for robustly responding when an analysis of the whole utterance cannot be found because of an unforeseen syntactic pattern. This scenario leads us to the second key conceptual innovation employed in the constructional analyzer: leveraging semantics for robust analysis.

Given the rich semantics found in an ECG grammar, the strategy for robust behavior is twofold. First, the analyzer needs a way to infer likely identifications between frames. Such a strategy is specified by the Parsimony Principle (Kay, 1987). It states that the ideal reader unifies compatible frames whenever possible. Applying this principle to a collection of frames and schemas simulates the identifications that were likely to result had a complete analysis been found.

Second, the analyzer needs a mechanism for choosing between the competing analyses that result from application of the Parsimony Principle. The heuristic used for this task is called semantic density (Bryant, 2003). Semantic density defines the completeness of an analysis as the ratio of filled frame roles to total frame roles.

Analyses that are more semantically dense specify more of the frame roles in the utterance than those that are less semantically dense, and the analyzer thus prefers denser analyses.

The constructional analyzer was tested by integrating it with the SMARTKOM domain and context model (Porzel et al., 2006), taking semantic information from that module's ontology. Not only did it successfully analyze questions from the tourist domain, but it also capitalized on the structure found with the ontology to perform linguistic type-coercion known as construal (Porzel and Bryant, 2003).

5 Wizard and Operator Study of Felicitous Human Computer Interaction[5]

5.1 Introduction

End-to-end evaluations of conversational dialogue systems with naive users are currently uncovering severe usability problems that result in low task completion rates. Preliminary analyses suggest that these problems are related to the system's dialogue management and turn-taking behavior. We present the results of experiments designed to take a detailed look at the effects of that behavior. Based on the resulting findings, we spell out a set of criteria which lie orthogonal to dialogue quality, but nevertheless constitute an integral part of a more comprehensive view on dialogue *felicity* as a function of dialogue quality and efficiency.

Research on dialogue systems in the past has focused on engineering the various processing stages involved in dialogical human–computer interaction (HCI), e.g., robust automatic speech recognition, intention recognition, natural language generation, or speech synthesis (Allen et al., 1996; Cox et al., 2000; Bailly et al., 2003). Alongside these efforts, the characteristics of computer-directed language have also been examined as a general phenomenon (Zoeppritz, 1985; Wooffitt et al., 1997; Fraser, 1993; Darves and Oviatt, 2002). The flip side, i.e., computer–human interaction (CHI), has received very little attention as a research question by itself.

The intuitive usability of such conversational dialogue systems can be demonstrated by usability experiments with real users that employ the PROMISE evaluation framework (Beringer et al., 2002), which offers some multimodal extensions over the PARADISE framework (Walker et al., 2000). The work described herein constitutes a starting point for a scientific examination of the "whys" and "wherefores" of the challenging results stemming from such end-to-end evaluations of conversational dialogue systems.

One of the potential reasons for the problems thwarting task completion stems from the problem of *turn overtaking*, which occurs when users rephrase questions or make a second remark to the system while it is still processing the first one. After such occurrences a dialogue becomes asynchronous, meaning that the system responds to the second-last user utterance while in the user's mind that response con-

[5] Robert Porzel and Manja Baudis were the principal authors of this section.

cerns the last. Given the state of the art regarding the dialogue handling capabilities of HCI systems, this inevitably causes dialogues to fail completely.

5.2 Wizard and Operator Study

Here, we describe a new experimental paradigm and the first corresponding experiments tailored toward examining the effects of the computer's communicative behavior on its human partner. More specifically, we will analyze the differences in human–human interaction (HHI) and HCI/CHI turn-taking and dialogue management strategies, which constitutes a promising starting point for an examination of the effects of the computer's communicative behavior on the felicity and intuitiveness of dialogue systems. The overall goal of analyzing these effects is for systems to become usable by exhibiting a more felicitous communicative behavior. After reporting on the results of the experiments in Sect. 5.3, we highlight a set of hypotheses that can be drawn from them and finally point toward future experiments that need to be conducted to verify these hypotheses in Sect. 5.4.

For conducting the experiments we developed a new paradigm for collecting telephone-based dialogue data, called *Wizard and Operator Test* (WOT), which contains elements of both Wizard-of-Oz (WOZ) experiments (Francony et al., 1992) as well as Hidden Operator Tests (HOT (Rapp and Strube, 2002)). This procedure also represents a simplification of classical end-to-end experiments, as it is much like WOZ and HOT experiments conductable without the technically very complex use of a real conversational system. As postexperimental interviews showed, this did not limit the feeling of *authenticity* regarding the simulated conversational system by the human subjects. The WOT setup is described in detail by Porzel and Baudis (2004) and Gurevych and Porzel (2006). It consists of two major phases that begin after the subject has been given a set of tasks to be solved with the telephone-based dialogue system. In the first phase the human assistant is acting as a wizard who is simulating the dialogue system by operating a speech synthesis interface. In the second phase, which starts immediately after a system breakdown has been simulated by means of beeping noises transmitted via the telephone, the human assistant is acting as a *human* operator asking the subject to continue. In our experiments, subjects used the simulated dialogue system to gather information related to tourism in the city of Heidelberg. Simulating a telephone-based dialogue system (rather than a local multimodal dialogue system such as the SMARTKOM-Mobile demonstrator) allowed a natural-seeming switchover from computer–human interaction to human–human interaction.

The experiments were conducted in the English language at ICSI in California. A total of 25 sessions were recorded. At the beginning of the WOT, a person acting as test manager told the subject that they were testing a novel, telephone-based dialogue system that supplies tourist information on the city of Heidelberg. In order to avoid the usual paraphrases of tasks worded too specifically, the manager gave the subjects an overall list of 20 very general tourist activities, such as *visit museum* or *eat out*, from which each subject had to pick 6 tasks that were to be solved in the experiment. The manager then removed the original list, dialed the system's number on the phone, and exited from the room after handing over the telephone receiver. The subject was always greeted by the system's standard opening ply: *Welcome to the Heidelberger tourist information system. How I can help you?* After three tasks were finished (some successful, some not) the assistant simulated the system's breakdown and came onto the telephone line saying *Excuse me, something seems to have happened with our system, may I assist you from here on*, and finishing the remaining three tasks with the subjects.

5.3 Experimental Results

The PARADISE framework (Walker et al., 1997, 2000) proposes distinct measurements for dialogue quality, dialogue efficiency, and task success metrics. The remaining criterion, i.e., user satisfaction, is based on questionnaires and interviews with subjects and cannot be extracted (sub)automatically from log-files. The measurements described here mainly revolve around dialogue efficency metrics in the sense of Walker et al. (2000). As we will show below, our findings show that a felicitous dialogue is not only a function of dialogue quality, but critically hinges on a minimal threshold of efficiency and overall dialogue management as well. While these criteria lie orthogonal to the Walker et al. (2000) criteria for measuring dialogue quality (such as recognition rates), we regard them to constitute an integral part of an aggregate view on dialogue quality and efficiency, here referred to as *dialogue felicity*. For examining dialogue felicity we will provide detailed analyses of efficiency metrics per se as well as additional metrics for examining the number and effect of pauses, the employment of feedback and turn-taking signals, and the amount of overlaps.

First of all, we apply the classic Walker et al. (2000) metric for measuring dialogue efficiency, by calculating the number of turns over dialogue length on the collected data. (The average length of a dialogue was 6 minutes. The subjects featured approximately uniform distributions of gender, age (12–71), and computer expertise.) As the discrepancy between the dialogue efficiency in phase 1 (HHI) versus phase 2 (HCI) of the experiment might be accounted for by latency times alone, we calculated the same metric with and without pauses. For these analyses, pauses are very conservatively defined as silences during the conversation that exceeded one second.

The overall comparison, given by Porzel and Baudis (2004), shows that naturally latency times severely decrease dialogue efficiency, but also that they alone do not account for the difference in efficiency between human–human and human–computer interaction. This means that even if latency times were to vanish completely, yielding actual real-time performance, we would still observe less efficient dialogues in HCI. While it is obvious that the existing latency times increase the number and length of pauses of the computer interactions as compared to the human operator's interactions, there are no such obvious reasons why the number and length of pauses in the human subjects' interactions should differ in the two phases. However, they do differ substantially.

Next to this *pause effect*, which contributes greatly to dialogue efficiency metrics by increasing dialogue length, we have to take a closer look at the individual turns and their nature. While some turns carry propositional information and constitute utterances proper, a significant number solely consist of specific particles used to exchange signals between the communicative partners, or combinations of such communicative signals with propositional information. We differentiate between dialogue-structuring signals and feedback signals in the sense of Yngve (1970). Dialogue-structuring signals — such as hesitations like *hmm* or *ah* as well as expressions like *well*, *yes*, *so* — mark the intent to begin or end an utterance, or to make corrections or insertions. Feedback signals, while sometimes phonetically alike — such as *right*, *yes* or *hmm* — do not express the intent to take over or give up the speaking role, but serve as a means to stay in contact, which is why they are sometimes referred to as *contact signals*. All dialogues were annotated manually for dialogue structuring and feedback particles.

The data show that feedback particles almost vanish from the human–computer dialogues — a finding that corresponds to those described above. This linguistic behavior, in turn, constitutes an adaptation to the employment of such particles by the respective interlocutor. Striking, however, is that the human subjects still attempted to send dialogue structuring signals to the computer, which — unfortunately — would have been ignored by today's "conversational" dialogue systems. (In the data the subject's employment of dialogue structuring particles in HCI even slightly surpassed that of HHI.)

Most overlaps in human–human conversation occur during turn changes, with the remainder being feedback signals that are uttered during the other interlocutor's turn (Jefferson, 1983). In the collected data the HHI dialogues featured significantly more overlap than the HCI ones, which is partly due to the respective presence and absence of feedback signals as well as to the fact that in HCI turn-taking is accompanied by pauses rather than immediate overlapping handovers.

Lastly, our experiments yielded negative findings concerning the type-token ratio and syntax. This means that there was no statistically significant difference in the linguistic behavior with respect to these factors. We regard this finding to strengthen our conclusions, that emulating human syntactic and semantic behavior does not suffice to guarantee effective and therefore felicitous human–computer interaction.

5.4 Analysis of the Results

The results presented above enable a closer look at dialogue efficiency as one of the key factors influencing overall dialogue felicity. As our experiments show, the difference between the human–human efficiency and that of the human–computer dialogues is not solely due to the computer's response times. There is a significant amount of *white noise*, for example, as users wait after the computer has finished responding. We see these behaviors as a result of a mismanaged dialogue. In many cases users are simple unsure whether the system's turn has ended or not and consequently wait much longer than necessary.

The situation is equally bad at the other end of the turn-taking spectrum, i.e., after a user has handed over the turn to the computer, there is no signal or acknowledgment that the computer has taken the baton and is running with it — regardless of whether the user's utterance is understood or not. Insecurities regarding the main question, i.e., *whose turn is it anyways*, become very notable when users try to establish contact, e.g., by saying *hello* —pause— *hello*. This kind of behavior certainly does not happen in HHI, even when we find long silences.

Examining why silences in human–human interaction are unproblematic, we find that such silences are being announced, e.g., by the human operator employing linguistic signals, such as *just a moment please* or *well, I'll have to have a look in our database* in order to communicate that he is holding on to the turn and finishing his round.

To push the relay analogy even further, we can look at the differences in overlap as another indication of crucial dialogue inefficiency. Since most overlaps occur at the turn boundaries, thereby ensuring a smooth (and fast) handover, their absence constitutes another indication why we are far from having winning systems. As the primary effects of the human-directed language exhibited by today's conversational dialogue systems, the experiments showed that dialogue efficiency decreased significantly even beyond the effects caused by latency times. Additionally, human interlocutors ceased in the production of feedback signals, but still attempted to use his or her turn signals for marking turn boundaries, which, however, go ignored by the system. Last, an increase in pausing is observable, caused by waiting and uncertainty effects, which are also manifested by missing overlaps at turn boundaries.

Generally, we can conclude that a felicitous dialogue needs some amount of extrapropositional exchange between the interlocutors. The complete absence of such dialogue controlling mechanisms by the nonhuman interlocutors alone literally causes the dialogical situation to get out of control, as observable in turn-taking and turn-overtaking phenomena. As evaluations show, this way of behaving does not serve the intended end, i.e., efficient, intuitive, and felicitous human–computer interaction.

Acknowledgments

The English Mobile team at ICSI was Johno Bryant, David Gelbart, Eric Lussier, Bhaskara Marthi, Robert Porzel (visiting from EML), Thilo Pfau, Andreas Stolcke, and Chuck Wooters. Contributions to the development, integration, and testing of the English Mobile system also came from Tilman Becker, Ralf Engel, Gerd Herzog, Norbert Reithinger, Heinz Kirchmann, Markus Loeckelt, Stefan Merten, Christian Pietsch, and Hans-Joerg Kroner at DKFI; Antje Schweizter at IMS; Silke Goronzy, Juergen Schimanowski, and Marion Freese at Sony; Hidir Aras at EML; and Andre Berton at DaimlerChrysler. Funding for ICSI and EML participation in the SMARTKOM project was provided by the German Federal Ministry for Education and Research (BMBF). Some of the work described in this chapter received additional support from other sources, including the Canadian Natural Sciences and Engineering Research Council, the Klaus Tschira Foundation, and Qualcomm.

References

A. Adami, L. Burget, S. Dupont, H. Garudadri, F. Grezl, H. Hermansky, P. Jain, S. Kajarekar, N. Morgan, and S. Sivadas. Qualcomm-ICSI-OGI Features for ASR. In: *Proc. ICSLP-2002*, Denver, CO, 2002.

J.F. Allen, B. Miller, E. Ringger, and T. Sikorski. A Robust System for Natural Spoken Dialogue. In: *Proc. 34th Annual Meeting of the Association for Computational Linguistics*, pp. 62–70, Santa Cruz, CA , June 1996.

C. Avendano. *Temporal Processing of Speech in a Time-Feature Space*. PhD thesis, Oregon Graduate Institute, 1997.

G. Bailly, N. Campbell, and B. Mobius. ISCA Special Session: Hot Topics in Speech Synthesis. In: *Proc. EUROSPEECH-03*, pp. 37–40, Geneva, Switzerland, 2003.

B. Bergen and N. Chang. Embodied Construction Grammar in Simulation Based Language Understanding. Technical Report TR-02-004, ICSI, 2002.

N. Beringer, U. Kartal, K. Louka, F. Schiel, and U. Türk. PROMISE: A Procedure for Multimodal Interactive System Evaluation. In: *Proc. Workshop "Multimodal Resources and Multimodal Systems Evaluation"*, pp. 77–80, Las Palmas, Spain, 2002.

H. Bourlard and N. Morgan. *Connectionist Speech Recognition: A Hybrid Approach*. Kluwer Academic, Dordrecht, The Netherlands, 1993.

J. Bryant. Constructional Analysis. Master's thesis, University of California Berkeley, 2003.

N. Chang, S. Narayanan, and M. Petruck. From Frames to Inference. In: *Proc. Scalable Natural Language Understanding(SCANALU)*, Heidelberg, Germany, 2002.

R.V. Cox, C.A. Kamm, L.R. Rabiner, J. Schroeter, and J.G. Wilpon. Speech and Language Processing for Next-Millenium Communications Services. *Proc. IEEE-2000*, 88(8):1314–1337, 2000.

C. Darves and S. Oviatt. Adaptation of Users' Spoken Dialogue Patterns in a Conversational Interface. In: *Proc. ICSLP-2002*, Denver, CO, 2002.

R. Engel. SPIN: Language Understanding for Spoken Dialogue Systems Using a Production System Approach. In: *Proc. ICSLP-2002*, pp. 2717–2720, Denver, CO, 2002.

C. Fillmore. Frame Semantics. In: Linguistics Society of Korea (ed.), *Linguistics in the Morning Calm*, Seoul, Korea, 1982. Hanshin.

J.M. Francony, E. Kuijpers, and Y. Polity. Towards a Methodology for Wizard of Oz Experiments. In: *Proc. 3rd Conf. on Applied Natural Language Processing - ANLP-92*, Trento, Italy, 1992.

N. Fraser. Sublanguage, Register and Natural Language Interfaces. *Interacting with Computers*, 5, 1993.

D. Gelbart. Mean Subtraction for Automatic Speech Recognition in Reverberation. Technical Report TR-04-003, ICSI, 2004.

A. Goldberg. *Constructions: A Construction Grammar Approach to Argument Structure*. University of Chicago Press, Chicago, IL, 1995.

I. Gurevych and R. Porzel. Empirical Studies for Intuitive Interaction, 2006. In this volume.

H. Hermansky. Perceptual Linear Predictive (PLP) Analysis of Speech. *Journal of the Acoustical Society of America*, 87(4), 1990.

H.G. Hirsch and D. Pearce. The AURORA Experimental Framework for the Performance Evaluations of Speech Recognition Systems Under Noisy Conditions. In: *Proc. ISCA ITRW ASR2000*, Paris, France, 2000.

P. Jain and H. Hermansky. Beyond a Single Critical-Band in TRAP Based ASR. In: *Proc. EUROSPEECH-03*, Geneva, Switzerland, 2003.

G. Jefferson. Two Explorations of the Organisation of Overlapping Talk in Conversation. *Tilburg Papers in Language and Literature*, 28, 1983.

D. Jurafsky, C. Wooters, G. Tajchman, J. Segal, A. Stolcke, E. Fosler, and N. Morgan. The Berkeley Restaurant Project. In: *Proc. ICSLP-94*, Yokohama, Japan, 1994.

P. Kay. Three Porperties of the Ideal Reader. In: R.O. Freedle and R.P. Durán (eds.), *Cognitive and Linguistic Analyses of Test Performance*, pp. 208–224, Norwood, NJ, 1987. Ablex.

M. Kleinschmidt. *Robust Speech Recognition Based on Spectro-Temporal Processing*. PhD thesis, Carl von Ossietzky-Universität, Oldenburg, Germany, 2002.

M. Kleinschmidt and D. Gelbart. Improving Word Accuracy With Gabor Feature Extraction. In: *Proc. ICSLP-2002*, Denver, CO, 2002.

G. Lakoff. *Women, Fire, and Dangerous Things*. University of Chicago Press, Chicago, IL, 1987.

C. Mokbel, D. Jouvet, and J. Monné. Deconvolution of Telephone Line Effects for Speech Recognition. *Speech Communication*, 19(3), 1996.

A. Moreno, B. Lindberg, C. Draxler, G. Richard, K. Choukri, S. Euler, and J. Allen. SpeechDat-Car: A Large Speech Database for Automotive Environments. In: *Proc. 2nd Int. Conf. on Language Resources and Evaluation (LREC 2000)*, Athens, Greece, 2000.

S. Narayanan. *Knowledge-Based Action Representations for Metaphor and Aspect*. PhD thesis, University of California, Berkeley, CA, 1997.

L. Neumeyer, V. Digalakis, and M. Weintraub. Training Issues and Channel Equalization Techniques for the Construction of Telephone Acoustic Models Using a High-Quality Speech Corpus. *IEEE Transactions on Speech and Audio Processing*, 2(4), 1994.

C. Pollard and I. Sag. *Head-Driven Phrase Structure Grammar*. University of Chicago Press and CSLI Publications, Chicago, IL, 1994.

R. Porzel and M. Baudis. The Tao of CHI: Towards Felicitous Human-Computer Interaction. In: *Proc. Human Language Technology Conference / North American Chapter of the Association for Computational Linguistics Annual Meeting 2004*, Boston, MA, 2004.

R. Porzel and J. Bryant. Employing the Embodied Construction Grammar Formalism for Knowledge Represenation: The Case of Construal Resolution. In: *Proc. 8th Int. Cognitive Linguistics Conference*, Logrono, Spain, 2003.

R. Porzel, I. Gurevych, and R. Malaka. In Context: Integrating Domain- and Situation-Specific Knowledge, 2006. In this volume.

S. Rapp and M. Strube. An Iterative Data Collection Approach for Multimodal Dialogue Systems. In: *Proc. 3rd Int. Conf. on Language Resources and Evaluation (LREC 2002)*, pp. 661–665, Las Palmas, Spain, 2002.

S. Renals and M. Hochberg. Efficient Evaluation of the LVCSR Search Space Using the NOWAY Decoder. In: *Proc. Int. Conf. on Acoustics, Speech, and Signal Processing (ICASSP-96)*, Atlanta, GA, 1996.

A. Stolcke. SRILM — An Extensible Language Modeling Toolkit. In: *Proc. ICSLP-2002*, Denver, CO, 2002.

M.A. Walker, C.A. Kamm, and D.J. Litman. Towards Developing General Model of Usability with PARADISE. *Natural Language Engineering*, 6, 2000.

M.A. Walker, D.J. Litman, C.A. Kamm, and A. Abella. PARADISE: A Framework for Evaluating Spoken Dialogue Agents. In: *Proc. 35th ACL*, Madrid, Spain, 1997.

R. Wooffitt, N. Gilbert, N. Fraser, and S. McGlashan. *Humans, Computers and Wizards: Conversation Analysis and Human (Simulated) Computer Interaction.* Brunner-Routledge, London, UK, 1997.

V. Yngve. On Getting a Word in Edgewise. In: *Papers From the 6th Regional Meeting of the Chicago Linguistic Society*, Chicago, IL, 1970.

M. Zoeppritz. Computer Talk? Technical Report 85.05, IBM Scientific Center Heidelberg, 1985.

SmartKom-Public

Axel Horndasch[1], Horst Rapp[2], and Hans Röttger[3]

[1] Sympalog Voice Solutions GmbH, Erlangen, Germany
horndasch@sympalog.de
[2] MediaInterface Dresden GmbH, Dresden, Germany
rapp@mediainterface.de
[3] Siemens AG, CT IC 5, München, Germany
hans.roettger@siemens.com

Summary. SMARTKOM-Public is the result of consistent development of traditional public telephone booths for members of a modern information society in the form of a multimodal communications booth for intuitive broad-bandwidth communication.

1 Introduction

The SMARTKOM-Public communications booth is provided with a wide palette of modern communication appliances. SMARTKOM-Public offers a higher degree of privacy, much higher communication bandwidth and optimal quality with regard to input and output results than is available to mobile systems for technical reasons. High rates of data flow and high-resolution screens lead, in contrast to a personal digital assistant (PDA), to increased comfort and better ergonomics. This communication space offers the simultaneous and comfortable use of language, pictorial and data services. A document camera is integrated for the transmission and processing of documents. An interesting further possibility concerns the saving of the user's data in the communications booth. Upon biometric identification, users are able to access their personal data. The idea of a virtual communications workstation that is capable of traveling with the users and adapting to their needs becomes reality. As the product at hand deals with a communication space for the public, along with supplying services, intuitive usability is of primary importance for acceptance. This is particularly so as even people who do not have access to a home computer with similar services will also make use of SMARTKOM-Public. The human–machine interface is designed to replicate normal human means of communication. This includes the broadest range of free speech dialogue possible and the use of natural gestures. Consideration must be made that conventional uses, and not only those based on multimodal communication, can be performed from the communication booth. Therefore our project started with a user study. The results are presented in the next section, followed by a description of the implemented functionality and some detailed information on selected realization topics in subsequent sections.

2 User Study

2.1 Introduction

SMARTKOM-Public demonstrates fundamental techniques of a user-friendly, multimodal human–machine communication in public areas. In order to have a close connection between the project specification and the behavior and the wishes of the potential users, the project was accompanied by a user study from the beginning. This study was to give answers to following questions:

- What are the expectations of potential users concerning a multimodal communication cell?
- Which kinds of human–machine communication does the user accept?
- Which communication services have to be provided?
- Which fears and application thresholds are connected with these new techniques?

2.2 Milestones

The data of the user study were obtained by questionnaires in two phases. In a first phase (9/99–5/00) a questionnaires from the technical point of view was elaborated. The questionnaires were adjusted among the SMARTKOM partners. Around 100 persons took part in the first part – mainly „insiders," persons who were directly involved in SMARTKOM or who had a very good knowledge about the project.

In the second phase (6/00–12/00) a more structured questionnaire was developed in collaboration with the Institute for Communication, Information and Education (KIB) of the University of Applied Sciences Zittau/Görlitz. The questions were asked in more detail, and the answers had to be given on a 5-point scale. It took the participants around 30–60 minutes to fill in the questionnaire. The questionnaire was provided in an Internet version, and owing to different marketing strategies, like mailings actions and a quiz, around 400 persons took part in the second phase.

2.3 Questions and Results

2.3.1 Demographic Data ("Who?")

- Gender: 50 % female, 50% male
- Age: 40% below 30 years
  50% between 30 and 50 years
  10% above 50 years

The mean age was 34 years.

- Education: ca. 33% school
  ca. 33% colleges
  ca. 33% universities

- Concerning their professions, there was a broad variety among the participants with a certain concentration of employees.
- Their computer skills were assessed as follows:
  60% good
  30% average
  10% bad

Thus it can be concluded that the study achieved a well-balanced and representative mixture of potential users concerning sex, age, education and skills.

2.3.2 Application Domains ("What?")

With these questions the preferred application domains of the potential users for SmartKom services were investigated. The participants were asked to give their opinions on four domains:

- information (timetables, weather, holiday planning, cultural and sport events, tourist and cultural locations, stock values, navigation and complex information)
- reservations (hotel, cultural events)
- orders and booking (tickets, products, travels)
- financial services (bank and stock business)

The potential users gave the following assessments to the application domains:

- Information concerning timetables and routes found a high degree of acceptance.
- Information about locations, events and complex information was also accepted.
- Information about weather, travel and especially about stock values found only a low acceptance.
- Reservations in cinemas and hotels were accepted.
- From the application domains directly connected with financial transactions, ordering of tickets was the only accepted area.
- Ordering of products and the booking of holiday trips were refused, more or less.
- For financial services the SmartKom-Public communication cell was not found to be suitable.

2.3.3 Media Services ("Which?")

Within these questions the technological equipment of a multimodal communication cell had to be estimated. The following media could be chosen:

- telephone
- video telephone
- fax
- Internet
- Access to personal mailboxes

The answers coincided in a high degree:

- All the media offered were accepted.
- The telephone as the most usual and best-known means of communication found the highest acceptance.

2.3.4 User Groups and Places of Installation ("Where?")

From these questions it should be concluded which places of installation would be preferred by different user groups. The participants of the questionnaire were offered the following situations:

- business traveler in an unknown city
- vacationer in an unknown city
- inhabitant in their home city
- employee in an office

2.3.5 Interaction and Technology ("By what?")

With these questions the modalities and possibilities of interaction were assessed. The participants had to give their preferences concerning:

- input options (speech, gestures, free input by pencil, mimic, document input)
- output options (text and pictures on a screen, speech, video telephone, printer)
- others (use of personal devices, chipcards for identification and payment)

The participants gave the following adjustments:

- High acceptance rates for input by speech and hand gestures.
- Input by pencil and document camera was also accepted.
- Mimic as input modality was refused. (Here it can be presumed that the importance of mimic for the dialogue was not understood sufficiently. Further, the mimic input requires a permanent video recording of the face, which was not accepted.)
- Among the output modalities speech and visual output were preferred.
- The printer was estimated as very useful (for faxes, roadmaps, receipts for reservations, etc.)
- The chipcard reader and the plug-in possibility for personal devices were also accepted.

2.3.6 Verification ("How?")

For purposes of verification the following items had to be adjusted:

- personal identification number (PIN)
- signature
- voice
- hand contour

- finger print
- face

The results were as follows:

- The well-known methods with chipcards or PIN were accepted.
- Among the biometrical methods, the fingerprint found the highest acceptance.
- Signature and face recognition were refused in many cases.

2.3.7 General Requirements for a Communication Cell

Different questions about general requirements for a communications cell gave the following views:

- There were high expectations concerning the intelligence of the system. More than 70% of the participants expected a successful dialogue with only a few clearing steps.
- The communication cell and the services were regarded more as an office employee. Only a third of the participants saw the technical system in the foreground.
- Discretion and anonymity were expected by the majority of potential users. Alternatively to the loudspeaker, a telephone was demanded. The video recording for the mimic recognition was refused in most of the cases. Around 75% of the participants would not make use of the services if they were connected with personal identification.
- There was a high consensus about the fact that better technological possibilities should be offered at constant prices. The only additional costs accepted were taxes comparable to information services and other special services like fax.

2.3.8 Correlation Analysis

In order to investigate the relations between the judgments of the potential users, their answer were analyzed in a paired correlation matrix. For example, there were high correlations in the answers for questions like "information for holiday travel" and "booking of holiday travel". This seems logical — if somebody wants to book travel, of course they want to get information about it beforehand. From the variety of such logical relations, two main conclusions can be obtained:

- The correlation matrix gives indications for producers of future devices on which applications are compatible with each other and can be integrated into one product.
- The answers were well chosen, which underlines the fact of representative nature of the study.

2.3.9 User Group Analysis

From the analysis according to the gender it can be concluded that female and male participants tended to give the same answers. Women showed less interest in technological details, and their expectations concerning the intelligence of the system were not so high. On the other hand, they were more restrictive in the face recognition problems. The user groups "below 30 years" and "between 30 and 50 years" tended to give the same answers in tendency. The user group "above 50 years" had different opinions on some questions. Elderly people want make less use of the multimodal communication cell in general. They were less interested in the technical possibilities, and they are less restrictive in security problems. The comparison between "insiders" and "others" showed that the answers also were given with the same tendency. The "insiders" however were slightly more optimistic and in agreement with the scenario.

2.3.10 User Suggestions

The questionnaires asked the participants to notify individual suggestions, opinions, apprehension or ideas. Fortunately, this possibility was used in a high degree. In summary, a catalogue of useful ideas containing details of technical question and general requirements could be obtained. Some of the most interesting statements follow:

Applications

- information and booking for actual events in the city (exhibitions, sport, cinema, conferences, etc.)
- restaurant guide with detailed information concerning the menu and quality assessments
- service for results of sport events
- social and political information about the city/the country
- virtual city tours, information about sights
- regional and global news service
- entertainment for waiting times, Internet games
- city map and city information with guide to locations of cultural interest, administration, service
- information about working hours of offices
- substitution of ways to the office (for registration of cars, new passports, etc.)
- flat and property service
- job and qualification announcements

Locations

- villages and small towns for elderly and less mobiles persons
- central places and transportation hubs for business and tourism
- inner city and important/big buildings (post and other offices, shopping malls...)
- medical centers, parking areas, truck stops, schools and universities

General Requirements

- interaction results should not be obvious to any but current user
- safety from vandalism
- anonymity, only PIN accepted
- guarantee of data security
- ease of use for physically challenged people
- service free of charge in the tourist information centers
- special services (fax, etc.) at normal charges
- clearly structured with a good overview in each application domain
- simple and intuitive handling

2.4 Summary

In the study, a representative variety of potential users took part, thus the results are valuable. The judgments of the application confirmed the SMARTKOM Demo scenarios. The most important locations are tourist and transportation centers. The different input and output modalities were accepted; only the mimic recognition was refused. User verification by PIN and fingerprint was more accepted than by signature or face recognition. The intelligence of the system was expected to be very high in general. The requirements concerning discretion, anonymity and data security were emphasized with a high degree of importance. The criterion "ease of use for physically challenged people" played an important role. The high number of logical correlations proves the representative nature of the study. They give indications for producers of future devices on which applications are compatible with each other and can be integrated into one product. From the point of view of user groups there were significant differences between female and male users. Elderly people showed less interest in SMARTKOM itself and its technological possibilities. The "insiders" of the SMARTKOM scenario showed a more optimistic behavior compared to the "general user." With the verbal notices a valuable collection of ideas concerning application domains and technological improvements is available.

3 Functionality

The SMARTKOM-Public demonstrator provides a wide range of applications considering the results of the user study. The focus is on communication and information services. Since it was not possible to implement all user-suggested services, some exemplary applications were selected:

- phone call as the most common and best-known means of communication, here supported by an electronic address book
- fax and e-mail services to realize the transmission of written documents
- a cinema information and booking system combined with a city map
- multimodal biometric authentification to access personal data

Fig. 1. The former German federal president using SMARTKOM-public e-mail

Altogether, SMARTKOM-Public offers more than 20 functionalities that can be combined in many ways.

These functionalities are accessed through an easy to use multimodal user interface that combines speech, gesture and mimic input with symmetric output. The optical output is projected on an interaction surface, where it can be referenced by natural pointing gestures. This part of the system is based on the Siemens Virtual Touchscreen (SiViT) system (Maggioni and Röttger, 1999). A second dedicated camera scans documents placed on the interaction area for further processing. As demanded by potential users, all applications can be accessed with only a few clearance steps. Although in the user study potential users expressed objections to mimic input, it was integrated into the demonstrator for scientific reasons and to evaluate possible applications.

Another compromise was made on the physical design of the communication booth. On one hand, potential users stressed their need for discretion and anonymity, demanding a traditional closed telephone booth design. On the other hand, SMARTKOM-Public is a showcase to demonstrate new technologies to the interested public, which requires an open design. We decided to use an open information terminal design for the second reason. For the same reason we used hands-free communication instead of a traditional phone handset.

The next sections describe the basic SMARTKOM-Public applications in detail. For more information on the gesture and mimic input, see Racky et al. (2006) and Frank et al. (2006).

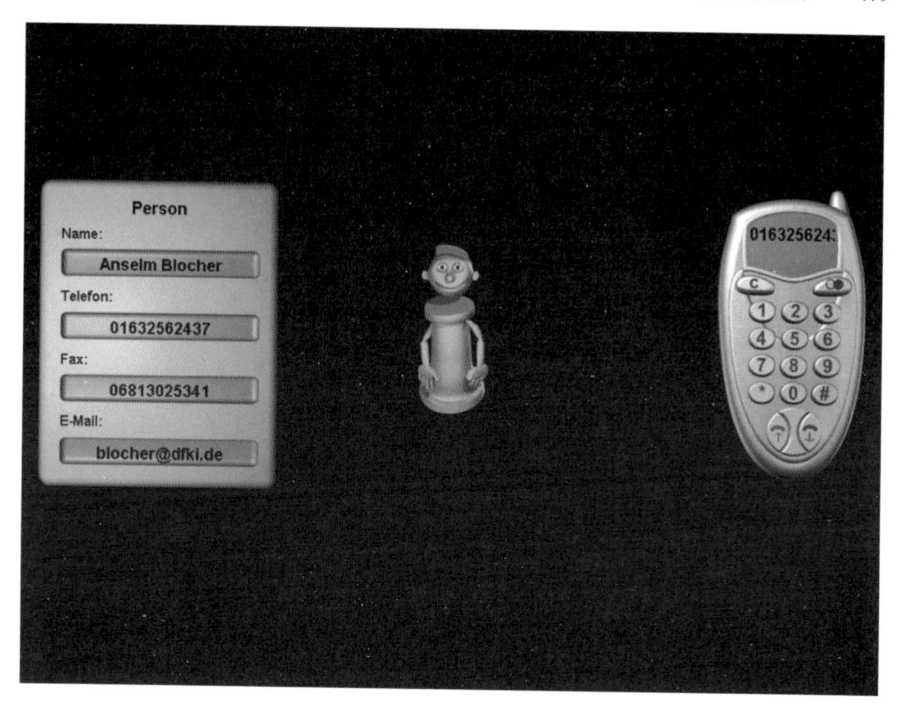

Fig. 2. Phone application in SMARTKOM-Public

3.1 Phone and Address book

Just like a traditional public telephone booth SMARTKOM-Public allows users to make standard telephone calls. The life-like character "Smartakus," visible on the projected output, serves as an interface agent to meet users need. Thus SMARTKOM-Public's interaction style breaks radically with the traditional desktop metaphor. It is based on the situated delegation-oriented dialogue paradigm (SDDP (Wahlster et al., 2001). The user simply tells Smartakus to establish a phone connection. This can be done in two different ways:

- dialing manually on a projected phone keyboard
- dialing by a phone book reference

In contrast to other scientific demonstrators, SMARTKOM-Public establishes real phone connections based on the European ISDN standard and thus has to handle signalization events correctly. It also uses hands-free communication based on the SMARTKOM audio modules and provides an address book based on the standard application module (see Sect. 4 for technical details).

Let us close the phone section with a look at two sample phone dialogues:

Example 1: Dialing manually on a projected phone keyboard

USER:	I'd like to make a call.
SMARTAKUS:	(presents phone keys) Please dial a number.
USR:	(dials number)
SMARTAKUS:	(establishes connection)

Example 2: Dialing by a phone book reference

USER:	I'd like to call Mr. Blocher.
SMARTKOM	(presents entry for Mr. Blocher) I've found an entry for Mr. Blocher. Is this correct?
USR:	Okay.
SMARTAKUS:	(establishes connection)

3.2 Facsimiles and E-Mail

Users can send facsimiles (fax) and e-mails with SMARTKOM-Public as well. Smartakus receives the users' requests and guides them through the necessary steps.

Transmission of pictorial information is mainly used for documents related to business or original artifacts that cannot be described in words. Therefore SMARTKOM-Public provides the possibility to capture prepared information with a high-resolution document camera and to send it as a picture to the receiver. For example, it is possible to send an image of your just-purchased new tie — do not try this with a standard fax machine! If the user decides to send an e-mail, the original is captured as a true-color image and sent to the selected e-mail address as an e-mail attachment. For a fax the color image is converted into a black and white version and sent as a real fax to the selected ISDN phone number.

The image transmission functionality is a complex task combining several modules into one service (see Sect. 4 for technical details).

Here is a typical fax dialogue to illustrate the interaction steps:

USER:	I'd like to send a fax to Mr. Wahlster.
SMARTAKUS:	(presents scan area) Please place your document here.
USER:	(places the document)
SMARTKOM:	(captures the document) Please remove the document.
USER:	(removes document)
SMARTKOM:	The document was captured. I've found an entry for Mr. Wahlster. Is this correct?

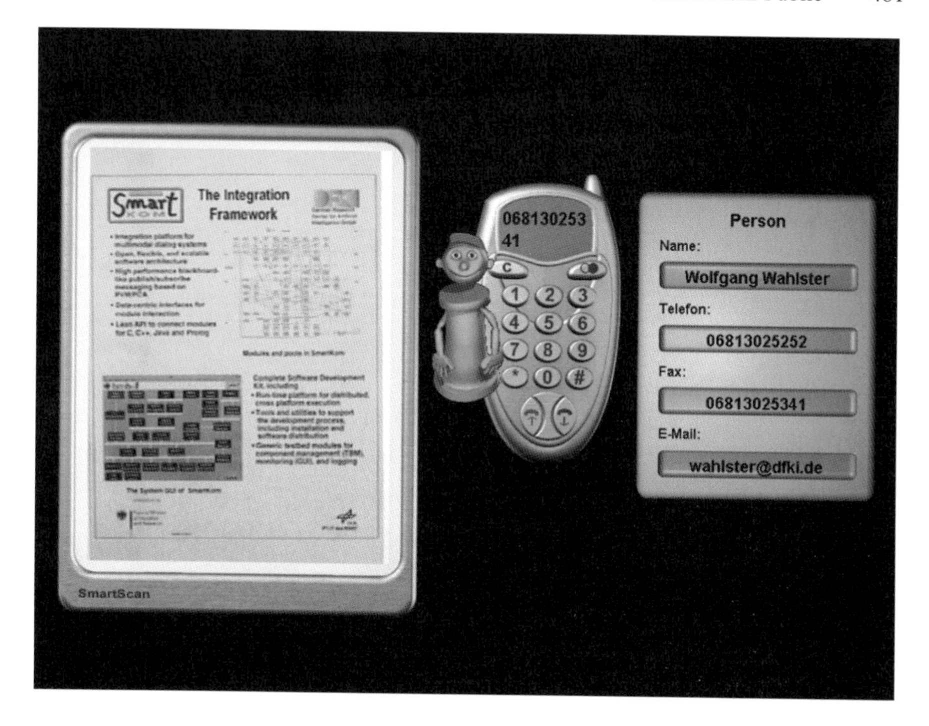

Fig. 3. Fax application in SMARTKOM-Public

USER: Yes.
SMARTAKUS: (sends fax to Mr. Wahlster)

3.3 Cinema

The cinema application was chosen to demonstrate the advantages of the SDDP in a complex information service environment. It was also favored in the user study. The cinema application provides a wide range of choices:

- the current cinema program
- the movie schedule in specific theaters
- information on specific movies
- information on movie actors
- locations of the theaters and directions to those

As in all SMARTKOM-Public applications, the presented information is real and always up to date. A module for external information services scans the Internet for current data. The vocabulary for the speech recognition is updated dynamically as well. Thus it is even possible to ask questions on completely unknown actors, for

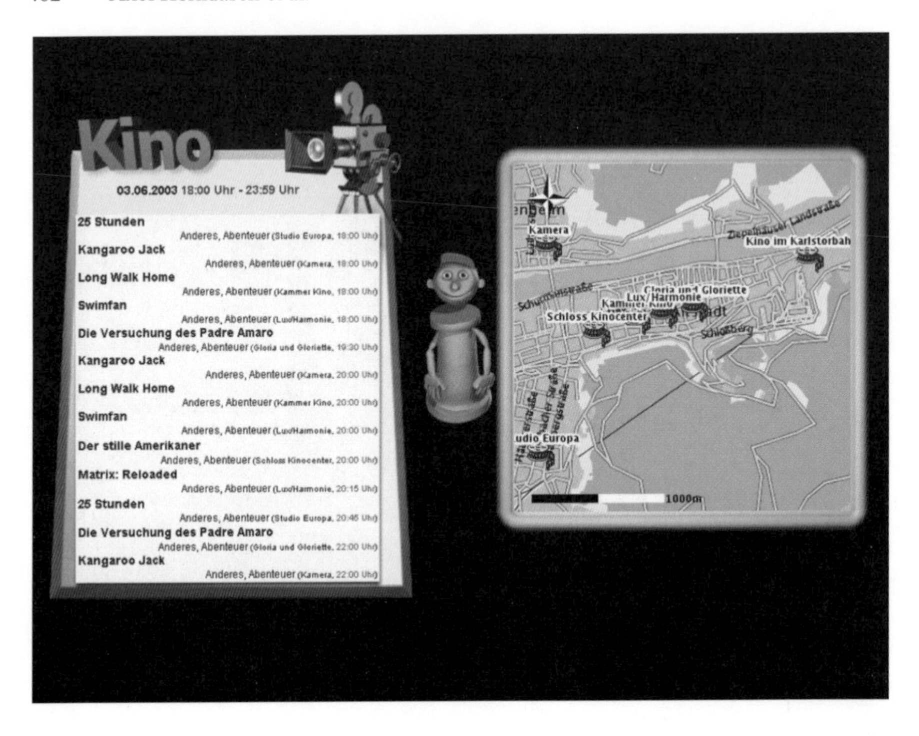

Fig. 4. Cinema application in SmartKom-Public

example, "Is there a movie with Leonardo DiCaprio?". Here is more serious sample dialogue:

USER:	I'd like to go to the movies tonight.
SMARTAKUS:	(presents a list of movies and a map) Here is a selection of tonight's movies. I've marked the cinemas on the map.
USER:	(points to the cinema "Kamera" on the map) What's on here?
SMARTAKUS:	(presents the program of the cinema "Kamera") These are tonight's movies at the "Kamera".
USER:	(points to *Matrix: Reloaded*) Tell me more about this movie.
SMARTAKUS:	(presents additional information) Here is more information on *Matrix: Reloaded.*
USER:	I'd like two tickets for this film at 8 pm.
SMARTAKUS:	(presents seat plan of the "Kamera") Where do you want to sit?
USER:	(encircles specific seats) Here.

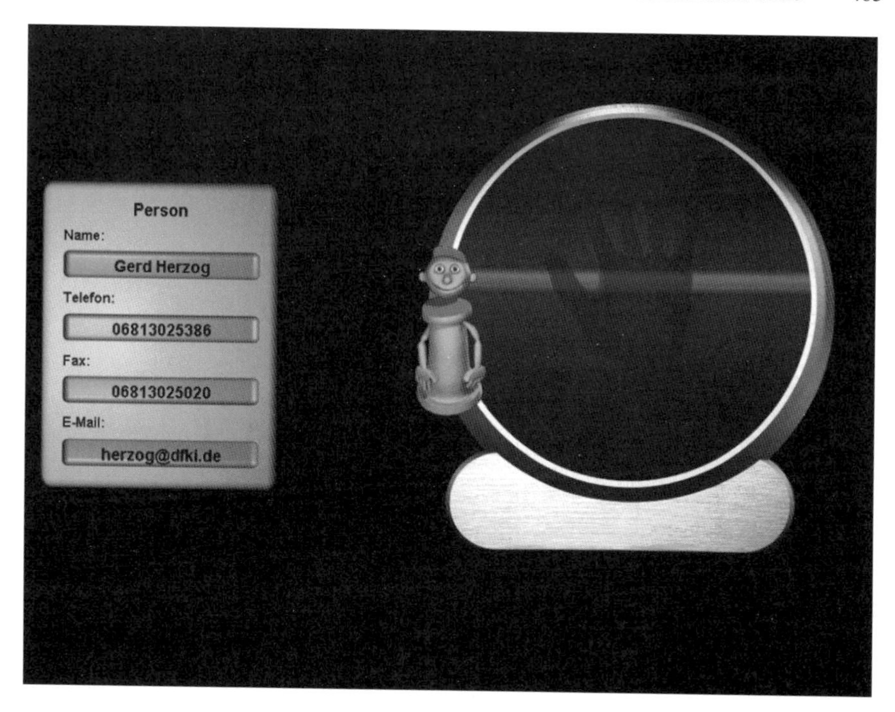

Fig. 5. Multimodal biometrics in SmartKom-Public

SMARTAKUS:	(highlights the selected seats on the plan) Is this correct?
USER:	Exactly.
SMARTAKUS:	I reserved the tickets. Your reservation code is 42.

3.4 Biometrical Authentification

As mentioned in the Introduction, SmartKom-Public provides personalized services. Therefore it is essential to determine the identity of the user. This is done by multiple biometrics (see Grashey and Schuster (2006) in this volume). To guarantee comfort and safety, three different biometrics have been integrated: speaker verification, signature verification and hand contour verification. Authentification can be actively initiated by the user with the phrase "My name is" or can be demanded as the situation calls for:

SMARTAKUS:	This action requires an authentification. What's your name?

USER:	My name is "Herzog".
SMARTAKUS:	I've found an entry for Gerd Herzog. Hand contour is the default biometry. Please place your hand with spread fingers on the marked position.
USER:	(puts his hand on the interaction area)
SMARTAKUS:	Authentification accomplished.

4 Technical Realization

As is the case for all SmartKom demonstrators, SmartKom-Public is based on the common SmartKom system architecture handling the dialogue with the user. The SmartKom system architecture consists of several independent modules connected via communication pools. Most of these modules have been described in the previous chapters. The following sections describe SmartKom-Public–specific modules:

- the communication module supported by the audio modules, realizing phone calls and data transfers on standard ISDN lines
- the module for standard applications, realizing address book, e-mail and calendar functionalities
- the virtual module for sending documents, as a combination of document camera and communication module.

4.1 Communication (TK-Module)

The TK-Module in SmartKom realizes the telephone and fax services on a Microsoft Windows PC. It works on the hardware base "Fritz!Card PCI" manufactured by AVM, which provides an ISDN S0 connection with two channels.

The software of the TK-Module is realized in a layer structure. The hardware driver is controlled by CAPI2.0. This is enclosed in a library that provides the functionalities typical for telecommunication. The most important features are the support of n-channel telephone calls and fax, the support of ISDN functions and the conversion between the audio formats PCM (in SmartKom) and A Law (in ISDN). The highest layer consists of a console application. It controls the internal workflow and allows the parameter input.

In the "telephone" operation mode the TK-Module works together with the modules Audioln and AudioOut, which are controlled by the "function modeling." During a telephone call AudioOut is occupied exclusively and thus it cannot submit other system outputs.

Each task for the TK-Module contains the telephone number to be dialed and a name of the connection. This name is also used to identify all the data streams of the connection.

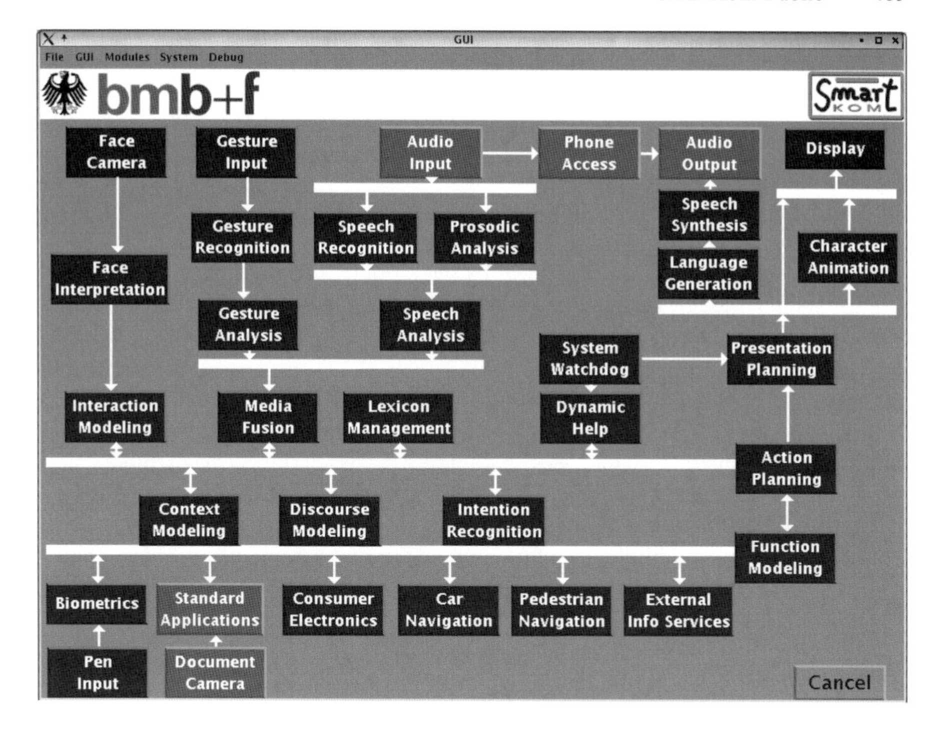

Fig. 6. SMARTKOM-Public module overview

In the "fax delivery" operation mode the TK-Module works together with the "realDocument" module controlled by the "function modeling." The TK-Module receives the documents to be sent in the structured fax file (SFF) format together with the number to be dialed and the name of the connection.

4.2 Audio Modules AudioIn and AudioOut

The AudioIn and AudioOut modules are responsible for the input and output of audio data into and out of SMARTKOM, not only to realize phone calls but also for speech input and output. The modules are realized in different objects:

WIN-32 Console Application: This application realizes the communication of the audio modules with the SMARTKOM system using the SMARTKOM pool architecture supervised by the module manager. It controls the internal workflow and allows the parameter input. It is also responsible for the time synchronization of the audio data.

DLL Object: This multithread capable dynamic link library contains all functions for the communication of SMARTKOM with different audio sources and audio consumers. The most important features are the support of *n* channels, the mixer functionality and the handling of different PCM audio formats (sample rate, etc.). Audio data can be submitted and delivered local area networks (LAN) and also by wireless LAN. The connection to the local audio devices is realized by the Multimedia-Microsoft-Win32-API. The remote connection is based on the RTP protocol. The devices for audio input and output can be switched in runtime by parameters.

The audio modules are controlled by the “function modeling” module. The tasks from “function modeling” contain the actual parameters, the purpose and a unique identification of the data stream. Both modules support the synchronous mechanism of SMARTKOM. The actions controlled by the modules can be activated by external events, and the modules themselves can send events related to certain system states.

AudioIn Module: The AudioIn module receives the audio data from the input devices and puts it into chunks of defined formats and sizes. The format definitions are made by the function modeling module. So, for instance, in the operation modes “telephone” and “speaker verification” the data chunks contain 2-KB audio samples with 8-kHz sample rate — thus speech data of 256 ms. In the operation mode “normal” the chunks contain typically 8-KB data sampled with 16 kHz.

The audio data are sent in regular intervals in XML documents. The documents contain time stamps with the absolute time of the first sample of each data chunk. Thus the information of different modalities can be synchronized within the whole system.

In addition, the AudioIn module provides the functionalities “silence detection” and “echo compensation”. In the case of silence, the transmission of audio data is going on but with empty data chunks. The echo compensation provided by partner DaimlerChrysler removes the SMARTKOM system messages via loudspeakers from the microphone signals.

AudioOut Module: The AudioOut module provides the output samples for the output devices at the given sample rates. In order to be able to synchronize the audio stream with other output streams like video data, the AudioOut module can start a certain audio output at an absolute time given in the audio document.

4.3 Standard Applications

4.3.1 Introduction

The interface for users of SMARTKOM is intuitive and offers many possibilities, as described in the previous sections. Every interaction with the system can lead to numerous complex activities involving devices and services especially designed and programmed for SMARTKOM. The ability to use standard IT applications within SMARTKOM is, however, also essential.

Groupware applications to manage e-mail, address books or calendars are typical examples of such services. Standard software offering these features has been integrated into SMARTKOM instead of duplicating their functionality within the system. Currently, the supported program suites are Lotus Domino Server and Lotus Notes, which include an application programming interface (API) capable of controlling all necessary groupware components. Other common software products can, however, also take their place.

4.3.2 Lotus Domino Server and Lotus Notes — A Short Overview

Lotus stores all data in objects which can be accessed using an API that exists for different programming languages. These databases, also known as *Notes Storage Facility* or *NSF files*, contain views and/or documents, which are the equivalent of, for example, an address book, an e-mail folder or a calendar. From a technical point of view the Java API that was used to integrate Lotus offers two possible ways of reading or manipulating Lotus data:

- The *local access pattern*, which allows the programmer to open NSF files without the server running in the background. The disadvantage is that the code has to run on the machine on which Lotus is installed.
- The *remote access pattern*, which hides a lot of tedious implementation details. Its drawback is that all methods are invoked through CORBA's *Internet Inter-ORB Protocol (IIOP)* interface, which means that processing is done remotely. Even if no traffic between different computers on the network is created, the bits have to travel through the network adapter of the local machine.

With performance being one of the main issues during the development of SMARTKOM, the local access pattern represented the natural choice for the implementation.

Apart from the possibility of retrieving data from a database or modifying it through its programming interface, Lotus offers standard e-mail protocols such as POP3, IMAP and SMTP, enabling users of the SMARTKOM system to communicate with the outside world.

4.3.3 Lotus Applications Modeled in SmartKom

To satisfy personal communication needs, the following Lotus applications were made available to SMARTKOM users:

- searching the public address book
- sending e-mails
- using the personal calendar

During the course of the multimodal dialogue with SMARTKOM, the system internally generates a plan to meet the demands of the user. The standard applications are employed to perform parts of the overall task through the abstract layer, which models the functionality of the different devices and services. For example, if the user wishes to send e-mail, the retrieval of the address as well as the task of sending the e-mail itself are building blocks of the entire process.

4.3.4 Storing SmartKom User Profiles

All the applications mentioned above make it necessary to log on to the Lotus Domino Server with a user name and a password. Because it is not acceptable to request this information after the user has been authenticated, it has to be stored somewhere within the SmartKom system. Since all users who want to use the standard applications have to be registered on the Lotus Domino Server anyway, it makes sense to store the information relating to the person in the public address book. This also makes it possible to store other pieces of information about users, for example, their preferred method of authentication. Within SmartKom these parts of a user profile are only available to the administrator of the Lotus Domino Server through a metasearch in the public address book. When SmartKom is initialized, the login information of this metauser is read from the global configuration file. Returning to the example of an e-mail being sent by a SmartKom user, it now becomes clear that the standard applications are involved when restoring the user profile, authenticating the person and fulfilling the request of sending the e-mail.

4.3.5 The Import Tool for User Profiles and Address Book Entries

Once Lotus Domino Server and Lotus Notes have been installed, user profiles and contacts which should be accessible to SmartKom have to be created on the server. This task is handled by an import tool with a graphical user interface, which lets the SmartKom administrator add and edit the relevant data. The offline tool, which supports comma-separated value (CSV) files as input, connects to the server and makes all the desired entries in the address book.

4.3.6 The Interface to Standard Applications

Since all communication within SmartKom is carried out using the Multimodal Markup Language (M3L), which is based on XML, it was necessary to design an M3L interface and implement a middleware to pass on the functionality of the groupware applications to the SmartKom system. This interface accepts messages with the following content as input:

- an identifier for the data stream
- login information of the user
- the data necessary to complete the request

The data for the request are organized so they can be easily mapped to the data structures of the groupware product.

The output for all requests consists in the simplest case of the following control information:

- the identifier of the data stream that was processed
- the request status

Normally the request status is *finished* or *fail*. In the case that the interaction exceeds a certain time limit, it is set to *pending*. A further description of the output generated requires a consideration of two different kinds of request. First, there are messages that do not expect an answer from the server, apart from status information. Creating, changing or deleting calendar entries are requests of that type, as is the sending of an e-mail. Under special circumstances it may also be necessary to return unique identification numbers for the data items processed on the server. These numbers represent groupware documents and are referred to as *note ID* or *universal ID* in Lotus (N.B. this is not the same as the identifier of the data stream). An example for the use of such an ID is a user utterance such as "please move the appointment to 7 pm," which leads to the modification of an entry in the calendar that was created earlier during a SMARTKOM session but is still held in the system. The second kind of request is characterized by the fact that it results in the generation of data output on top of the metainformation. The retrieval of contacts from the address book or entries from a personal calendar are requests of that type. These searches are modeled much like an SQL query: restrictions (the WHERE-clause) such as the last name of a person, and the desired answer format (the SELECT-clause) for the resulting data have to be specified. A typical request for example would be "I need John Doe's telephone number" (for the M3L code of this request, see the next section). For the calendar it is also possible to ask for appointments that take place during a certain period of time. This supports the processing of speech input such as "show me all my appointments between 4 and 6 pm".

4.3.7 Address Book Search: An Example

The following example shows a valid M3L representation of a request to the standard applications module for the telephone number of *John Doe* by a user that is registered on the Lotus Domino Server with the name *TestUser*. The format for the result of the query in the address book contains the telephone number and the first and last name of the persons found. Please note that for better legibility the header information has been deleted from all example messages in this section.

```
<externalApplicationRequest> \ a request to
<dataStream> \ standard applications
<name> \
abook_1\
</name> \
<purpose> \
queryAddressBook\
</purpose> \
</dataStream> \
<smartkomUserID> \ user information
TestUser\
</smartkomUserID> \
<action> \
<addressBookSearch> \ the type of the request
<personAnswerFormat> \ the desired answer format
<name> \ (SELECT part of query)
<firstName> \
</firstName> \
<lastName> \
```

</lastName> \
</name> \
<contact> \
<telephoneNumber> \
</telephoneNumber> \
</contact> \
</personAnswerFormat> \
<searchPerson> \ *the restrictions*
<name> \ *(WHERE part of query)*
<firstName> \
John\ *look for 'John Doe'*
</firstName> \
<lastName> \
Doe\
</lastName> \
</name> \
</searchPerson> \
</addressBookSearch> \
</action> \
</externalApplicationRequest> \

The answer containing the data matching the request is the following (the request status information, which is also produced as output, has been omitted for the sake of clarity):

<externalApplication> \ *data from standard applications*
<dataStream> \
<name> \
abook_1\
</name> \
<purpose> \
queryAddressBook\
</purpose> \
</dataStream> \
<addressBook> \ *the type of the data returned*
<smartkomUserID> \ *user information*
TestUser\
</smartkomUserID> \
<person> \ *all address book entries*
<name> \ *matching the restrictions*
<firstName> \
John\
</firstName> \
<lastName> \
Doe\
</lastName> \
</name> \

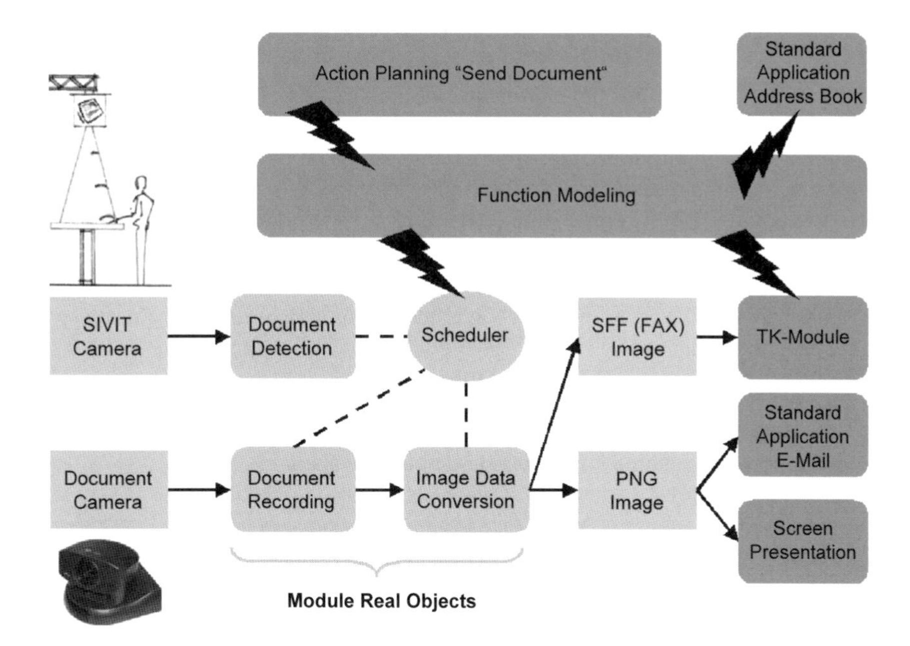

Fig. 7. SMARTKOM-Public fax and e-mail functionalities

<contact> \
<telephoneNumber> \
16508451000\
</telephoneNumber> \
</contact> \
</person> \
</addressBook> \
</externalApplication> \

4.3.8 Summary: Standard Applications in SmartKom

All standard applications that are available to SMARTKOM users are provided by a middleware that offers an M3L interface and models certain functionalities of a groupware product such as the Lotus Domino suite. Single tasks, like the retrieval of an e-mail address, are small building blocks of plans that are generated based on a multimodal dialogue between the SMARTKOM user and the system.

The standard applications are used to store pieces of information about SMARTKOM users, for example, their preferred method of authentication, whether voice, signature or hand. The different tasks generate a flow of information that is separated into a data and a control flow conforming to the architecture of SMARTKOM. Unique IDs enable back-referencing to data objects already held in the system.

4.4 Composed Applications: Fax and E-Mail

Among the most complex applications in the SMARTKOM system are the fax and e-mail functionalities. A lot of SMARTKOM modules are involved to produce the desired results. Although the internal operating sequence is complex, the user gets the impression of an easy to use functionality. But now we go into the details. Let us start with the previous Fax example user utterance "I'd like to send a fax to Mr. Wahlster." Let us also assume that the action planning module of SMARTKOM already understood the sense of the user's utterance and sent the order "send document to Wahlster fax" to the function modeling module, which coordinates the next steps. What is happening internally?

1. The ISDN fax number of Mr. Wahlster has to be determined.
   (Standard Application Address Book)
2. The Screen presentation "Document scanning" is shown.
   (Module Display)
3. SMARTKOM waits for a document on the interaction area.
   (Module Real Objects analyzes SiViT gesture camera input)
4. The high-resolution image is captured by the document camera.
   (Module Real Objects performs Panoramic Imaging)
5. The image is converted into a display format and rendered on the screen.
   (Module Real Objects and Module Display)
6. The image is converted into a SFF black and white Fax image.
   (Module Real Objects)
7. Image generated in step 7 is sent to the Fax number determined in step 1.
   (TK-Module).

If the user wants to send an e-mail the steps are slightly different: instead of the ISDN fax number the e-mail address is determined, and instead of the black and white image a PNG color image is generated and sent as an attachment to the selected e-mail address.

References

C. Frank, J. Adelhardt, A. Batliner, E. Nöth, R.P. Shi, V. Zeißler, and H. Niemann. The Facial Expression Module, 2006. In this volume.

S. Grashey and M. Schuster. Multiple Biometrics, 2006. In this volume.

C. Maggioni and H. Röttger. Virtual Touchscreen – A Novel User Interface Made of Light — Principles, Metaphors and Experiences. In: *Proc. 8th Intl. Conf. on Human-Computer Interaction*, pp. 301–305, Munich, Germany, 1999.

J. Racky, M. Lützeler, and H. Röttger. The Sense of Vision: Gestures and Real Objects, 2006. In this volume.

W. Wahlster, N. Reithinger, and A. Blocher. SmartKom: Multimodal Communication with a Life-like Character. In: *Proc. EUROSPEECH-01*, vol. 3, pp. 1547–1550, Aalborg, Denmark, September 2001.

SmartKom-Home: The Interface to Home Entertainment

Thomas Portele[1], Silke Goronzy[2], Martin Emele[2], Andreas Kellner[1], Sunna Torge[2], and Jürgen te Vrugt[1]

[1] Philips Research Laboratories GmbH, Aachen, Germany
{Thomas.Portele,Andreas.Kellner,Juergen.te.Vrugt}@philips.com
[2] Sony International (Europe) GmbH
Sony Corporate Laboratories Europe, Advanced Software Laboratory, Stuttgart, Germany
{goronzy,emele,torge}@sony.de

Summary. SMARTKOM-Home demonstrates the use and benefit of an intelligent multimodal interface when controlling entertainment devices like a TV, a recorder, and a jukebox, and when accessing entertainment services like an electronic program guide combining speech and a handheld display with touch input. One important point is emphasizing the functional aspect, i.e., the user's needs, conveyed to the system in a natural way by speech and gesture, are satisfied. The user does not need to know device-specific features or service idiosyncrasies. The function modeling component in SMARTKOM-Home has the necessary knowledge to transform the abstract user request into device commands and service queries.

1 Introduction

The central goal of the SMARTKOM-Home scenario is to support intuitive access to home entertainment functionality using the SMARTKOM paradigms. The SMARTKOM-Home system was jointly defined and monitored by Philips Research in Aachen and Sony International (Europe) GmbH in Stuttgart.

We start with a description of the features of the SMARTKOM-Home system in Sect. 2. The hardware used is described in more detail in Sect. 2.1; the applications are discussed in Sect. 2.2. Hardware and applications pose some particular constraints on the dialogue system. On one hand, devices the user usually uses at home such as TV, video cassette recorder (VCR), MP3 jukebox, etc., need to be controlled by SMARTKOM. As a consequence, the many functionalities of these devices and services need to be modeled within the system to be able to control them and offer their full functionality to the user. In our system this is handled by the function modeling component, which is described in Sect. 2.3. Furthermore, services such as an electronic program guide (EPG) are characterized by their highly dynamic content, i.e., the daily changing TV program and information for current shows. In order to handle dynamic content appropriately, various modules in the system need

User	Hello!
System	Welcome to SMARTKOM. My name is Smartakus. How may I help you?
User	What's on TV this noon?
System	This is the TV program at noon.

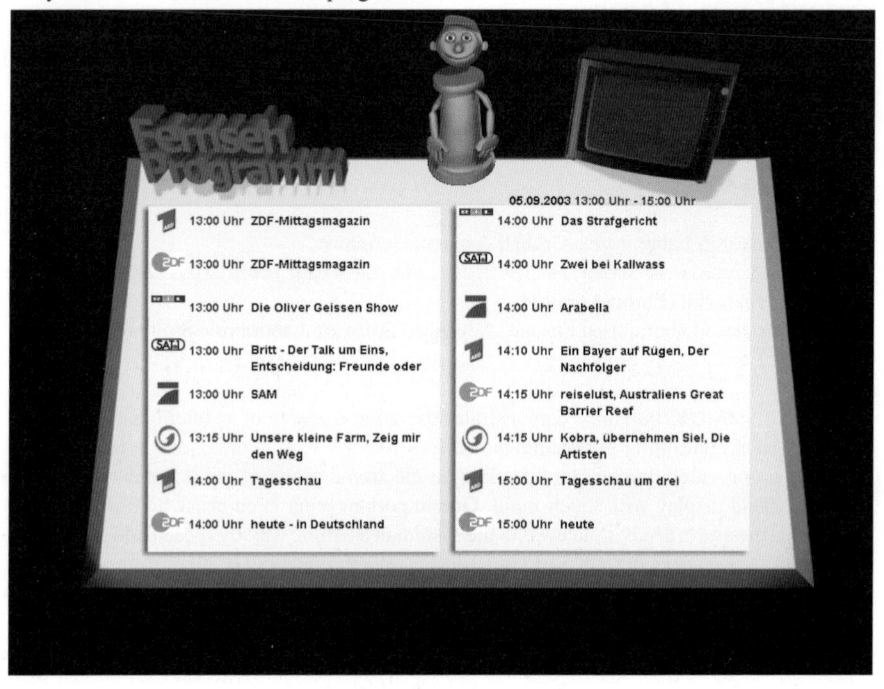

Fig. 1. Typical interaction with SMARTKOM-Home (translated from German), part 1. After initial greeting, the results of the electronic program guide are shown

to be aware of it. Therefore, a dynamic lexicon, described in Sect. 2.4 serves as a central knowledge source. Also, the interaction modes are different from the other scenarios. In the home scenario we explicitly distinguish between lean-forward and lean-backward modes (Sect. 2.5). Feedback strategies as an important part of the interaction are also discussed. In Section 3 we mention the evaluations that were conducted to evaluate the SMARTKOM-Home demonstrator.

2 Features

The environment of a SMARTKOM-Home system is, not surprisingly, the living room of the user. The global application is accessing entertainment functionality. Key features are multimodal interaction using speech and a tablet PC, switching between a multimodal lean-forward and a speech-only lean-back mode, and an animated inter-

User — What else is on this channel (points to a program from provider ARD)?
System — This is the TV program on ARD in the afternoon (shows only program items from ARD).
User — Which genres do you know?
System — These are the genres you can use.

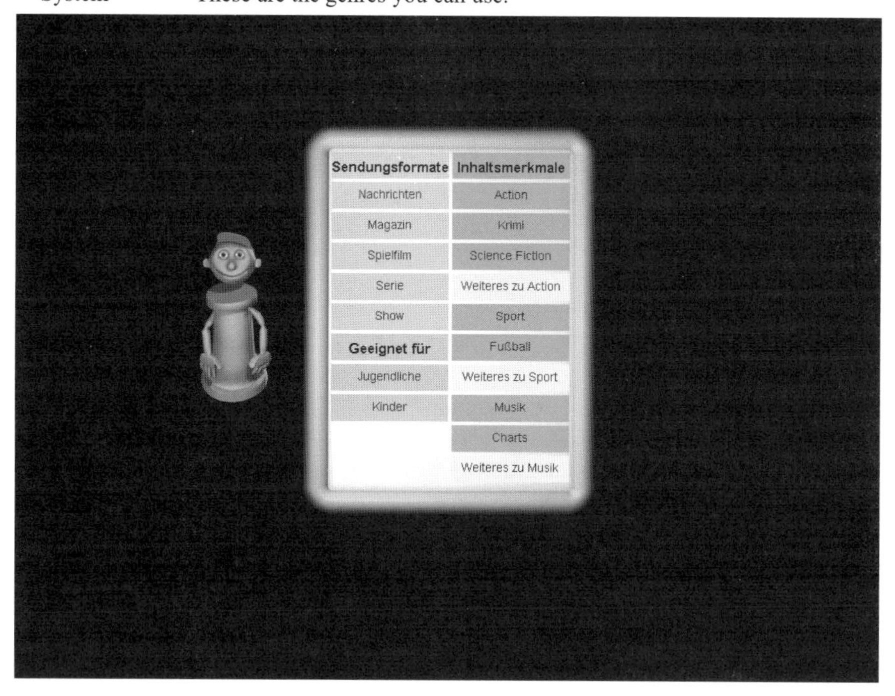

Fig. 2. Typical interaction with SMARTKOM-Home (translated from German), part 2. Dynamic help delivering information about the genres

action agent called "Smartakus". One important point is emphasizing the functional aspect. The user does not need to know device-specific features or service idiosyncrasies. Figures 1–3 show a typical interaction (translated from German). After an initial greeting, a TV show is selected using dynamic help. The video recorder is programmed to record that show, and the TV is switched on to a currently running program. The system is set to lean-back mode while the user watches TV.

2.1 Hardware

The application hardware is a high-end digital Sony VCR DHR-1000VC and a standard Philips TV set. This represents a sample set of devices. However, the layout of the modules that control these devices is designed to support other devices as well, such as hi-fi components, etc. In choosing the interaction device we had to consider that it should reflect the necessities of the home environment. A platform that is

User: Is there sports on any channel?
System: These are the sports programs this afternoon (shows all sports program items).
User: Record that (points to a tennis show).
System: The show “US Open Semifinals” today at eight p.m. on RTL will be recorded.
User: Switch the TV to ARD.
System: (switches the TV on and selects channel ARD). The TV is now showing the station ARD.
User: Turn off the display.
System: (turns display off) The display is now deactivated.

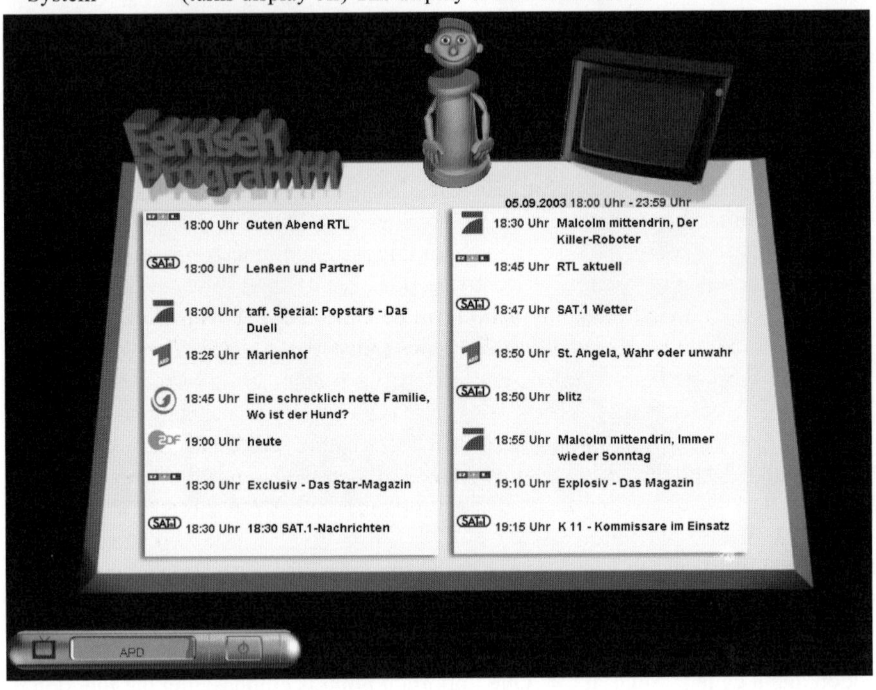

Fig. 3. Typical interaction with SmartKom-Home (translated from German), part 3. A TV show is selected. The video recorder is programmed to record that show, and the TV is switched on to a currently running program. The system is set to lean-back mode while the user watches TV.

mounted at a fixed place in the room will not be accepted by people used to remote controls, but the constraints of a mobile scenario on device size do not apply either. A portable easy-to-use device with a sufficiently large interaction display to allow showing several information items (EPG info and device status) and the SmartKom persona Smartakus simultaneously was judged to be the optimal interaction device. At the time of the definition project phase (early 2000), the Fujitsu Stylistic 3500 tablet PC (a full-fledged PC with a touch screen, see Fig. 4) was the best choice

— nowadays several similar Web pads are available. The device is sufficiently powerful to run the SMARTKOM modules for audio I/O and display generation locally.

2.2 Applications

The main applications are the control of consumer electronics (CE) devices and access to an EPG. Furthermore, the personal calendar application is used to store and retrieve shows to be recorded. For each application a description for the SMARTKOM action planning module was created. The action planning module is responsible for coordinating/triggering the different modules when necessary. Links between these descriptions allow switching from one application description (also termed *action plan*) to another, e.g., if the user indicates in the EPG application that she wants to record a certain program with the VCR. This process is transparent for the user, but facilitates development and integration of new applications. An important difference between device control and EPG is the way of interaction. EPG is performed in a *browsing mode*, where information is always displayed, and missing information is substituted by meaningful default values (e.g., "today, now" for time, "all" for channels and genres). These values are changed and refined during the browsing process (Kellner and Portele, 2002). On the other hand, device control needs to have all values determined before any action can be carried out. This is in line with traditional mixed-initiative slot-filling dialogues (Aust et al., 1995). The SMARTKOM backbone supports both styles of interaction. Each application has a visual representation (see screenshots in Figs. 1–3). The EPG application usually takes the full screen (Figs. 1 and 3), while TV and VCR information are displayed in the bottom (see Fig. 3). When delivering user assistance in a metadialogue, the screen changes again (Fig. 2). The EPG application supports preferences for genres and channels. These preferences, however, are not used as a filter to eliminate hypothetically unwanted information, but help to resort the results — preferred items are displayed (or spoken in lean-back mode) first. The preference values are extracted by the dynamic help component (Streit, 2003) according to emotive user reactions. For instance, if information about a thriller is displayed and the user likes it (she says something like "fine" or smiles), the preference value for the genre "thriller" is increased.

2.3 The Function Model

Since we consider SMARTKOM-Home as a uniform interface to various services and devices, all of these are modeled by the function model. This was motivated by the fact that users will have complex wishes involving several devices and/or services. However, the user should not have to care about which device needs to do what at a certain point in time. Therefore the function model formally describes the functionalities of all services and devices in the system. The action planner triggers the requests to the function model. A planning component within the model generates a plan that includes all devices/services needed to fulfill the user wish. Furthermore, this plan includes the necessary steps to be taken for each of the applications. A user wish could, e.g., be "Record the thriller with Julia Roberts tonight". This involves querying the EPG services for the information needed (in this case channel and starting time) to then program the VCR accordingly. The function model is described in more detail in Torge et al. (2002a,b).

2.4 The Dynamic Lexicon

Most of the applications in the home scenario are characterized by their highly dynamic content. This holds in particular for the EPG, since the TV program changes daily or even more frequently. Many modules are affected by this dynamic content. The dynamic lexicon handles this dynamic content and serves as a knowledge source for those modules that need lexical information. Concretely, these are the speech recognizer, the prosody recognizer, the speech analysis, and the speech synthesis for obvious reasons: Of course the user should be able to speak any name of a show, actor etc., genre, time channel and so on when browsing the EPG. These new words must be recognizable and interpretable, and also the speech synthesis on the output side uses the new words when presenting the EPG search results. It is the task of the lexicon to keep track of words that need to be added/removed to/from the lexicon when the user request requires this. Words can also be removed from the lexicon when not needed any longer. This is important for keeping the speech recognizer's vocabulary as small as possible to maintain high speech recognition rates.

Also, the pronunciations for new words are necessary for the modules named above. These are automatically generated by a decision tree–based grapheme-to-phoneme conversion. Pronunciations are mainly generated for German. However, some lexical entries, such as movie titles or actor names, are often not German but rather English. As a consequence, English pronunciations are generated simultaneously for those entries.

2.5 Interaction

The interaction with SMARTKOM-Home is governed by the principle that the user should have control over the system. The system is designed more like a butler than like an autonomous agent. The user must be able to understand *what* the system is doing, and *why* it is doing it, and it is one important task of the system to support this.

The home scenario with the tablet PC demands some specific interaction features. Most important, users should be able to switch between two modes:

- The *lean-forward* mode supports using the display for touch input and visual output. This is the "normal" mode for focused interaction with the system.
- The *lean-back* mode uses only the acoustic channel (i.e., speech) for input and output. This mode is assumed to be used when the user is either not willing to leave the sofa and reach for the device, or when only short commands like channel switching are necessary. An example corresponding to the second system turn of Fig. 1 would be read as: *"Some of the broadcasted programs are the news show "Tagesschau" on ARD at 12:30, the sitcom "Alles unter einem Dach" at*

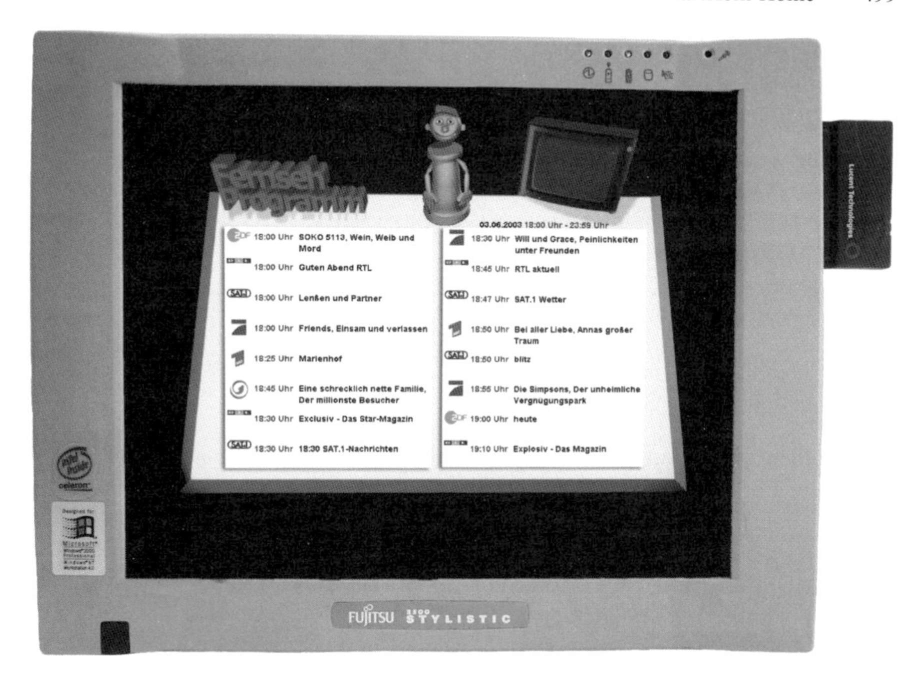

Fig. 4. Tablet PC as interaction device of SMARTKOM-Home

12:30, and the news show "Tagesschau" on ZDF, also at 12:30. Only a limited number of item (currently three) is read.

Switching between the modes must be initiated by the user. The switch to lean-back mode is given via one of several speech commands (e.g., *"Go to sleep!"*). The lean-forward mode is turned on either by a speech command (e.g., *"Wake up!"*) or by touching the display. The system can suggest a mode switch to lean-forward mode if a list with more than three items has to be conveyed to the user.

Feedback is an important issue in a complex dialogue system (Brennan and Hulteen, 1995). Three levels of feedback are distinguished in SMARTKOM-Home, and each level has its own modality in the lean-forward mode:

- **System State:** The state of the system is conveyed by the Smartakus persona. When a user command can be accepted, it makes a listening gesture. Processing stages (recognition, analysis, processing, presentation) are symbolized by different persona gestures. Thus, the user knows when the system is working (and even how long it may take), and when it is ready to accept commands.
- **System Belief:** The results of the analysis are presented by speech, e.g., EPG constraints ("Here you see this evening's sports shows."), or device states ("The TV is now switched to ARD."). Thus, the user can detect misinterpretations. The visual presentation also contains information about the system belief, e.g., the current constraints in the EPG application.

- **System Answer:** The results of the user query (EPG information, device states) are displayed on the screen.

In the lean-back mode, only spoken feedback is possible. The system state is not conveyed to the user, but the other two items are (at least partially for result lists) spoken.

Another issue for complex systems is user assistance. Three cases are considered in SMARTKOM-Home:

- **Help:** The user can ask for help. Supported questions contain formulations like "What can I say?" for general help, "Which channels/genres/... do you know?" for the possible EPG constraints, and "Which commands do you know?" for the device control. The possible alternatives are then displayed (see central screenshot in Fig. 1).
- **Problem detection:** The system can detect problems during the interaction. These problems can be caused by failure to recognize or interpret user input. The system then gives assistance by describing current input possibilities (similar to the response to "What can I say?"), or by asking for missing information.
- **Constraining and Relaxing:** The EPG database access can lead to no or too many results. The system then displays the available information and gives hints to constrain or relax the query — it will not do that automatically, because the user should remain in control. The same holds for failures during VCR operation (e.g., missing recording media).

Modality-switching, feedback, and user assistance are designed according to the butler metaphor: the user has control, the system gives assistance.

3 Integration and Evaluation

The complete system contains 28 modules communicating via a multiblackboard architecture (Klüter et al., 2000). The interface format is an XML dialect which allows using online validation and several XML tools. The modules were developed by ten partners and span the whole range from device handlers for audio and display, recognizers for gestures and speech, analyzers, dialogue, function, lexicon and presentation managers, service and hardware interface modules, to generation and synthesis modules. The complexity of the system is therefore substantial, and intensive testing and checking against the specifications was performed by the integration group at the DFKI and the scenario managers (Philips and Sony). To avoid a mixture of incompatible module versions, fixed integration dates were defined by the system group, and several versions of SMARTKOM-Home existed until the final version was rolled out.

3.1 Informal Evaluation of Prerelease Systems

In the initial phase of the project, the SMARTKOM-Home scenario was specified by the scenario managers. Among other things, the scenario specification included:

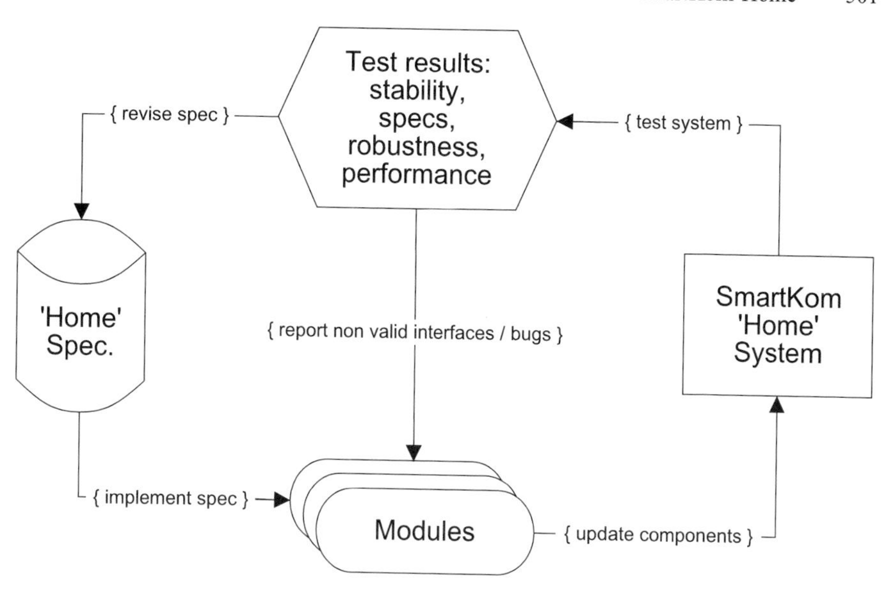

Fig. 5. Workflow of informal evaluations of prerelease SmartKom-Home systems

- **Interaction patterns:** These patterns define the modalities that can be used to interact with the system in various situations. Basic patterns that describe the typical task-specific dialogue behavior have been formulated. For example, for the user input speech and pointing gestures are used to interact with the system in the lean-forward mode. In the lean-back mode, only speech is enabled as input modality. In this mode, audio is the only modality used for system output, while in lean-forward mode the screen of the portable touch-screen device is the main output device.
- **Applications:** A number of applications to be controlled by the SmartKom-Home system have been identified. The unified control of different applications by SmartKom supports accessing functionality enabled by the interplay of two or more applications in an easy and intuitive way.

Besides the abstract specification, examples of the interactions between user and system have been provided. These contained formulations for input and output and sketches for the screen output. These examples have been used by the module designers to create first versions of their modules and to generate the required knowledge sources (e.g., the grammar templates).

To build a system for end-user tests, the workflow depicted in Fig. 5 was applied by the sites responsible for the SmartKom-Home scenario.

Starting with an initial SmartKom-Home system, this system was tested by developers according to the scenario specifications regarding possible interaction patterns and applications. These interactions were recorded, and the results were evaluated against the scenario specification. The robustness of the interaction was also tested: The interaction examples provided by the scenario managers and enriched by the system group defined a minimal baseline the system must handle. To obtain a robust system, alternative approaches to solve a task were also investigated. Other issues during the test were the system stability and performance in terms of processing speed.

After a test phase, the specification of the scenario was revised. The updated scenario description was delivered to the project partners to adapt the modules. Interface format problems and obvious bugs were directly reported to the responsible partners.

The system group integrated components that had been updated (i.e., modules or knowledge sources) due to the system tests or due to feature updates into a new prerelease system. Based on this new system, the tests started again.

The main difficulties in this workflow arose from the complexity and the distribution of the development over several sites. The complexity stems from the large number of modules that had to work together and the large set of functionalities specified for the system. The distributed development lead to a significant delay in reactions on problem reports and communication overhead.

The inherently dynamic nature of a dialogue system made it inevitable to perform online tests. Subsets of the complete systems have been tested in offline tests to ensure stability of the modules. Continuous testing and checking of new versions against a given specification proved to be one of the key factors for the success of the project.

3.2 Evaluation Using the PROMISE Framework

A formal evaluation of an earlier version (SMARTKOM-Home 2.1) had been carried out by project partners from the university of Munich using their PROMISE framework (Beringer et al., 2002). The evaluation established several problematic aspects such as unacceptable system response times that could be resolved by module optimization and new hardware. Furthermore, the evaluation resulted in an improved feedback mechanism and optimized module behavior. The current prototype has also been subject to a similar evaluation, and an improved performance has been observed.

4 Conclusion

SMARTKOM-Home is a multimodal assistant for home environments. SMARTKOM-Home serves as a uniform interface to all service and devices by combining applications for EPG and device control in a transparent way. The interaction style adheres to the principle: the user has control, the system gives assistance. We described particular requirements for this scenario and how the involved modules take this into account. This included a description of the hardware, the function model that jointly controls all services and devices, and the dynamic lexicon, which serves as central knowledge source for all modules that need to deal with dynamic content such as the entries in the EPG data. Also the interaction modes — lean-forward and lean-backward mode — that are specific for the home scenario were discussed together with strategies for feedback and user assistance. The system was iteratively improved by conducting user evaluations, which confirmed the importance of feedback and user assistance.

References

H. Aust, M. Oerder, F. Seide, and V. Steinbiss. The Philips Automatic Train Timetable Information System. *Speech Communication*, 17:249–262, 1995.

N. Beringer, U. Kartal, K. Louka, F. Schiel, and U. Türk. PROMISE: A Procedure for Multimodal Interactive System Evaluation. In: *Proc. Workshop "Multimodal Resources and Multimodal Systems Evaluation"*, pp. 77–80, Las Palmas, Spain, 2002.

E. Brennan and E.A. Hulteen. Interaction and Feedback in a Spoken Language System: A Theoretical Framework. *Knowledge-Based Systems*, 8:143–151, 1995.

A. Kellner and T. Portele. Spice - A Aultimodal Conversational User Interface to an Electronic Program Guide. In: *Proc. ISCA Tutorial and Research Workshop on Multi-Modal Dialogue in Mobile Environments*, Kolster Irsee, Germany, 2002.

A. Klüter, A. Ndiaye, and H. Kirchmann. Verbmobil From a Software Engineering Point of View: System Design and Software Integration. In: W. Wahlster (ed.), *Verbmobil: Foundations of Speech-to-Speech Translation*, pp. 635–658, Berlin Heidelberg New York, 2000. Springer.

M. Streit. Dynamic Help in the SmartKom System. In: *Proc. ISCA Workshop on Error Handling in Spoken Dialogue Systems 2003*, Chateau-d'Oex, Switzerland, 2003.

S. Torge, S. Rapp, and R. Kompe. The Planning Component of an Intelligent Human Machine Interface in Changing Environments. In: *Proc. Workshop on Multi-Modal Dialogue in Mobile Environments*, Bad Irsee, Germany, 2002a.

S. Torge, S. Rapp, and R. Kompe. Serving Complex User Wishes With an Enhanced Spoken Dialogue System. In: *Proc. Int. Conf. on Speech and Language Processing*, Denver, Colorado, 2002b.

SmartKom-Mobile: Intelligent Interaction with a Mobile System

Rainer Malaka, Jochen Häußler, Hidir Aras, Matthias Merdes, Dennis Pfisterer, Matthias Jöst, and Robert Porzel

European Media Laboratory GmbH, Heidelberg, Germany
{rainer.malaka, jochen.haeussler, hidir.aras, matthias.merdes, dennis.pfisterer, matthias.joest, robert.porzel}@eml.villa-bosch.de

Summary. This paper presents SMARTKOM-Mobile, the mobile version of the SMARTKOM system. SMARTKOM-Mobile brings together highly advanced user interaction and mobile computing in a novel way and allows for ubiquitous access to multidomain information. SMARTKOM-Mobile is device-independent and realizes multimodal interaction in cars and on mobile devices such as PDAs. With its siblings, SMARTKOM-Home and SMARTKOM-Public, it provides intelligent user interfaces for an extremely broad range of scenarios and environments.

1 Introduction

Novel ensembles of services and interaction paradigms have become available for stationary systems and applications. As a result of higher bandwidths for mobile networks, increasing computing power in mobile devices, and improvements in display and I/O technologies, it is now possible to make these services and interaction paradigms available on mobile systems. The need to adapt to tiny displays or anachronistic protocols like WAP is rapidly diminishing (Rapp et al., 2000), thereby causing a transition from thin clients toward thick(er) clients. Interaction via speech input can be performed with high-quality audio streams instead of GSM-based ones (Haiber et al., 2000). To some extent acoustic as well as graphical rendering is possible on the mobile device itself or can be executed on a server, and the ensuing results can be streamed from the server to the client. Some of the most advanced techniques of human–computer interaction can now be ported to the mobile world, including natural language interaction, multimodal interfaces, embodied conversational agents, semantic Web integration, and more (Console et al., 2002; Huang et al., 2000; Malaka, 2000; Malaka and Zipf, 2000).

Many other research systems fail in one of the aspects outlined above. They are either too narrowly limited to some "toy world" applications (Boualem et al., 2002), or they have rather limited user interface capabilities (Gurevych et al., 2003a). SMARTKOM, however, features a complex system with free speech, multiple domains, symmetric multimodality, and a broad range of realistic tasks.

In general, mobile systems and their prototypes can be classified in terms of their realization of one or more of the aforementioned levels of complexity. For example, we find individual prototypes and systems that allow for very sophisticated multi-modal input, such as the MATCH system (Johnston et al., 2002) or the QuickSet system (Oviatt, 1996), which, however, are either limited to a single end-device or a single application domain.

Others such as described by Krüger et al. (2004) enable seamless device changes and connectivity but, in turn, are lacking interaction and service flexibility. In this paper, we describe how—for the first time—a system was developed in which all levels of complexity are fused into an operational prototype.

We will show how such a system was adapted and customized from the core SMARTKOM system for mobile use, as sketched in Fig. 1. This task is by no means trivial, as we want to keep the full flexibility along all the aforementioned levels of complexity and also support mobility-specific aspects in a scalable and intuitive way. In particular, we show how the processing of geographical information and mobile situation-dependent interaction was integrated in realizing the SMARTKOM-Mobile system.

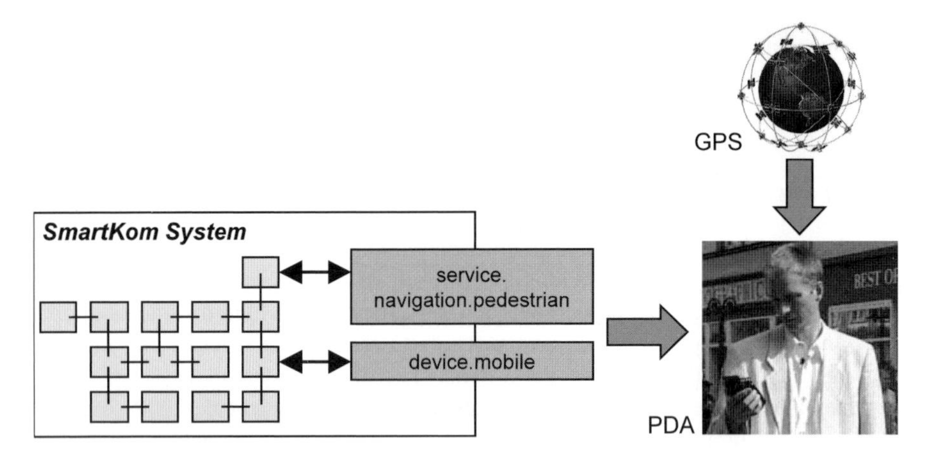

Fig. 1. General architecture layout of SMARTKOM-Mobile. The SMARTKOM system is connected through two modules to the mobile infrastructure: "service.navigation.pedestrian" and "device.mobile". Localization information comes from a GPS module

However, in this fusion of technologies a multitude of mobility-specific aspects remain to be solved. In particular mobile applications, such as route planning or spatial directions, are inherently different from desktop applications. Mobile architectures must be suited for the various user interface components involved and must support multiple end-devices. For mobile systems context changes more dynamically and dramatically than for desktop ones and thus plays a key role. An intelligent mobile interface must therefore take the user's context into account. This context can vary from driving or walking to performing several tasks concurrently.

This paper introduces SMARTKOM-Mobile. It is the mobile version of the SMARTKOM system (Wahlster et al., 2001), a novel multimodal and multidomain dialogue system. SMARTKOM-Mobile enables the user to interact with the system while driving a car as well as while walking around, the pedestrian case, using different mobile devices, interfaces, and modalities. In one case, the built-in screen of the navigation system and the audio system of the car serve as SMARTKOM end-devices, while in the pedestrian case the user cooperates with SMARTKOM through a handheld computer. We show how such a complex interactive system can be adapted to create a *full-fledged* mobile system. For comparing mobile systems, we propose at least three distinct dimensions of complexity:

1. single system and device *versus* heterogeneous and seamless system/device integration
2. single static application *versus* multiple internal and external service integration
3. single-mode, reactive, and controlled interaction *versus* intuitive, proactive, and conversational multimodal interaction.

A system that does not fall short along any of these dimensions of complexity can, in our opinion, be considered a full-fledged mobile system.

In this paper we will introduce the ontology-driven interactions with and within SMARTKOM-Mobile, the mobile services, and the mobile system architecture. Thereafter, we will describe how the system is implemented and present first evaluation results.

2 Mobile Semantic Integration

One of the key factors in realizing a robust multidomain system is the use of an extensible ontology that consists of a domain-independent top-level and process ontology that is combined with multiple specialized domain ontologies. These domain ontologies capture the domain-specific knowledge necessary for representing the domains covered by the individual SMARTKOM scenarios. This ontology is used in two ways:

1. for inference tasks involved in understanding, analyzing, and responding to user requests (Gurevych et al., 2003c)
2. for generating interface specifications for all the SMARTKOM core components (Gurevych et al., 2003b).

This approach allows a consistent extensible and unified knowledge repository. In contrast to many other systems, it comes with an advanced interface technology together with a broad range of application tasks.

For the integration of multidomain services and the mobile interaction described above into a multimodal dialogue system, it is necessary to represent all information and tasks in an abstract and amodal way. For this, a system-wide ontology is used to represent information on all respective domains (Gurevych et al., 2003c). The ontology is organized in a top-level ontology that follows the modeling principles outlined in Rößler et al. (2001), a domain-independent process model that follows the frame semantic analyses described in Baker et al. (1998), and a number of subontologies for the domain- and task-specific elements. The resulting joint ontology uses domain- and task-independent concepts for general top-level concepts such as *types* or *roles* and processes such as *MotionDirectedTransliterated* for representing both the user's wish to be guided to a specific place as well as for the system's incremental responses. The subontologies refine top-level concepts down to route segments, spatial relations, electronic devices, or the Heidelberg Castle.

The ontology is used for a number of tasks within the system, e.g.:

1. Scoring coherence of speech recognition hypothesis (Engel, 2002): The ontology allows for the identification of the most coherent speech recognition hypothesis, thus increasing recognition reliability even under adverse acoustic conditions.
2. Multimodal fusion (Wahlster, 2002): Ontological knowledge is the basis for multimodal disambiguation, e.g., in the case of pointing gestures on a displayed map, the ontology-driven representation of the accompanying speech input helps to identify the intent of the pen input.
3. Basis for interface specification (Herzog et al., 2003): The interface specification for all multimodal processing modules is formulated on top of the amodal representation of relevant objects, processes, and categories.
4. Speech interpretation (Engel, 2002): For interpreting and intentional analysis of utterances the ontology provides the necessary knowledge base for making inferences and allows for underspecification.
5. Dialogue and discourse management (Pfleger et al., 2002): Complex tasks are also modeled as processes within the ontology. They are then, in turn, used to manage dialogue context.

For modeling the additional domains necessary for the mobile scenario, we added concepts for space, tourism, and navigation to the ontology. This domain-specific ontological modeling was based on the existing ontology of Deep Map and ontologies that have already been proposed for geographical objects (Malaka and Zipf, 2000; Zipf and Krüger, 2001). For the SmartKom-specific mobile interactions, no additional processes had to be modeled since user interactions such as navigation requests, tour planning, sight information, or map interaction could already be captured by the domain-independent process model of the SmartKom core system.

The semantic bridge from the SmartKom world to external agent ontologies in Deep Map or that of Web services supplying topical information is realized in the SmartKom module "navigation. pedestrian," which translates between internal and external worlds. The abstract modeling of the mobile SmartKom domain allows for a seamless integration of a variety of internal and external services into the SmartKom-Mobile system.

As with the other domains Public and Home, language interpretation, multimodal disambiguation, action planning, multimodal presentations, and dialogue management are based on a homogeneous ontological model. Since the ontology is used systemwide, it also allows for mobility-related requests in the other scenarios, i.e., users at home can also plan a route to a city, request information on sights, or interact with a map. Additionally, the ontology, which is formulated in DAML/OIL, can be used as the basis for message contents in SMARTKOM, and, in fact, the SMARTKOM tool OIL2XSD translates concepts from the ontology into M3L schemata (Gurevych et al., 2003b). This allows for a consistent way of defining message formats that can then be further interpreted and analyzed with the ontology.

3 Mobile Services

In order to build a flexible, user-, and context-adaptive mobile system, one has to take into account a model of the user, the environment, and the interaction between them on the basis of a common ontology. The SMARTKOM-Mobile system realizes such a system by using an agent-based approach that implements intelligent agents that are capable of dealing with positioning, spatial cognition, and contextual information. The agent system incorporates these aspects by using an overall ontology that encodes the environment model as well as the interaction with the user. This way the system tries to react to user requests in an autonomous, independent, and context-aware way (Porzel and Gurevych, 2002), while taking into account the user preferences and certain conditions of a given situation or location.

In SMARTKOM-Mobile, we integrated such a multiagent system for intelligent location-based services for the tourism domain. The Deep Map multiagent system provides a number of location-based services for the city of Heidelberg (Malaka and Zipf, 2000) realized as intelligent agents. Deep Map was developed using a FIPA-based agent platform, Resource-Aware Java Agents (RAJA), that is built on top of FIPA-OS (Ding et al., 2001).

Figure 2 gives an overview of the integration of Deep Map into the SMARTKOM-Mobile system via the navigation.pedestrian module. It also illustrates the interaction between both systems and the involved agents. Both systems have their own respective communication language. In SMARTKOM all content in intercomponent communication is encoded as M3L messages. In Deep Map, messages are encoded in an agent communication language (ACL), where content objects are represented as Deep Map Objects (DMO). Both M3L and DMO are defined by their respective ontologies. The SMARTKOM Deep Map bridge translates between both systems, while the "navigation.pedestrian" control agent (NCA) manages the individual pedestrian navigation sessions. The other agents within the Deep Map system are:

- SpaCE is an agent for a number of spatial cognitive tasks such as interactive knowledge acquisition, localization, route management, and more (Kray and Porzel, 2000). The agent is able to divide a route into segments that are to be verbalized in navigation instructions. This segmentation has to be computed according to features of the route such as landmarks, and turning points, and not just distance (Johnston et al., 2002). SpaCE makes use of a set of low-level services provided by the other agents.
- The GPS agent, for example, is needed by SpaCE to get the current user position or to be informed when the user enters or leaves a certain region or location (region or time triggering).

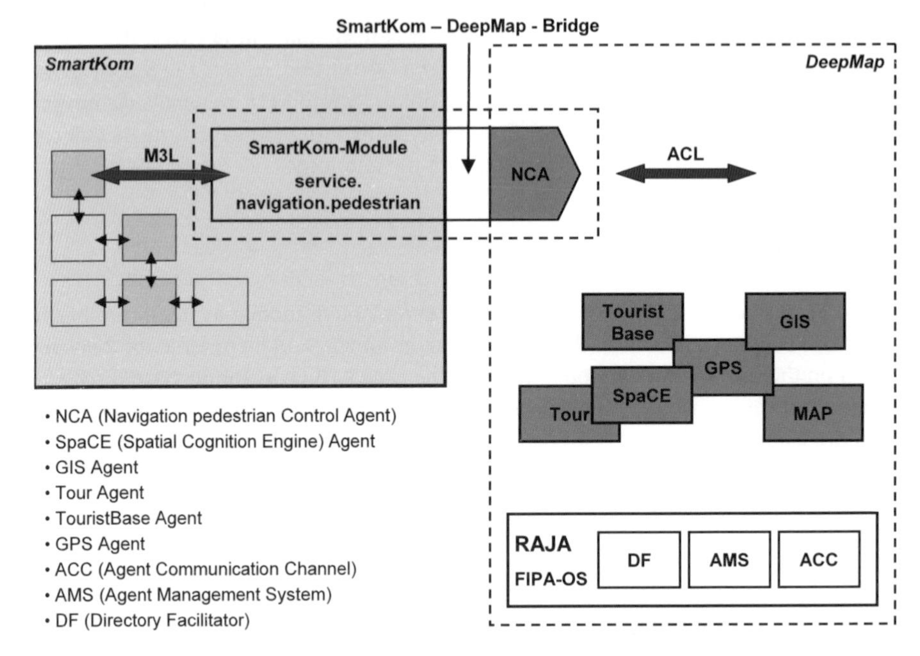

Fig. 2. SMARTKOM Deep Map multiagent system integration. The multiagent system RAJA/FIAP-OS consists of the core components DF, AMS and ACC, which are responsible for agent communication, management, and service discovery, respectively

- The tour agent is a flexible tour planner that optimizes a tour for a given number of sights, time, and other parameters, while calculating user-adapted tours (Joest and Stille, 2002).
- The GIS agent allows access to geospatial information and allows for geospatial querying, distance measurement, and visibility analysis for real-world objects.
- The map agent generates maps for the pedestrian environment and thus allows free map interaction.

Together, these agents constitute a multilayer GIS that contains the topology of the street network, the geometry of buildings, and other spatial objects as well as georeferenced objects *cum* touristic information that can be subject to user interactions.

As described above, Deep Map provides SMARTKOM with the essential knowledge and reasoning capabilities for realizing mobile and location-based services, which are fused with the SMARTKOM system and services. In the following we will show what kinds of services are realized and how the involved ontologies support the creation of flexible and intelligent mobile user interfaces.

The SMARTKOM-Mobile system differs from the other two scenarios through its support of mobile context-dependent domains and applications. The two main applications in SMARTKOM-Mobile are route planning and spatial navigation. These applications are specific for a mobile environment and are very suitable for developing and testing multimodal interaction.

Since both applications have a spatial context, interactive maps are offered to the user not only to give her an orientation but also as a means to talk about and refer to spatial objects. Therefore a central base application for ‘map interaction’ was realized (Häußler and Zipf, 2003). The interactions with maps can be divided into:

1. Informative map interactions: The map can be used to identify or select objects and to request information about them. The map itself is not manipulated.
2. Manipulative map interactions: Map interactions with the intention to manipulate the map itself. It can be distinguished between map navigation (pan and zoom), where only the extent of the visible area is changed, and interactions where the content of the map is changed, e.g., by displaying additional objects or hiding others.

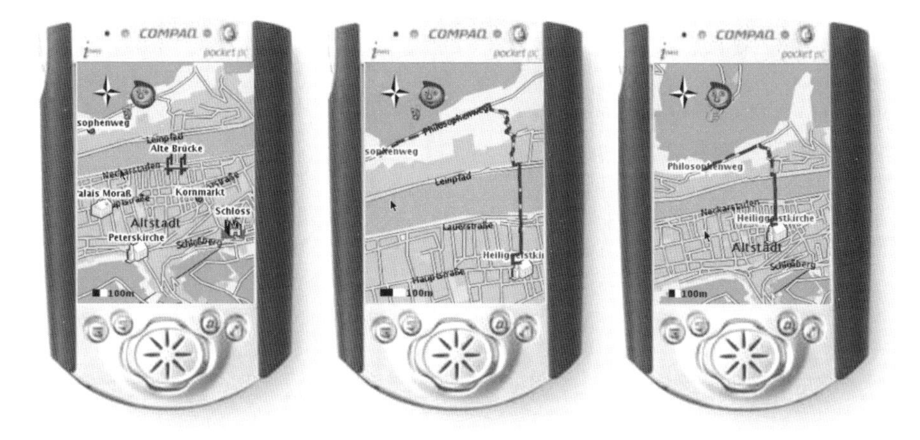

Fig. 3. The use of dynamic maps in the application “route planning” *(from left to right)*: Avatar “Smartakus” presents a street map of Heidelberg. After the user has selected the start and destination points for her sightseeing tour, the computed path is presented on the map. For a better orientation, the user may interact with the map again by giving commands like “zoom out”

To make the described map interactions possible, the SMARTKOM system has to know about the content of the currently displayed map. Thus the “no presentation without representation” paradigm (Wahlster et al., 1991) needs to be fully realized for this application. An image as provided by usual map services is insufficient — a semantic description of every generated map is required. We have modeled the description of a map in M3L. The map agent answers every map request from the SMARTKOM system with an image plus semantic description of this map. This description contains general information like map size and geocoordinates of the map’s bounding box as well as information about the displayed objects like buildings, etc. Therefore other SMARTKOM modules get all the information that is necessary to make these objects themselves part of the further dialogue: names, identifiers and types of the objects, their position, and the display style of an object (e.g., green route, blue route). An example for the use of dynamic maps is shown in Fig. 3.

The next application for the SMARTKOM-Mobile system is the pedestrian navigation system with incremental tour guidance support. For route planning in the car, a commercial system was integrated. The pedestrian system differs in many ways. For actual navigational instructions, the route is segmented into "cognitively"-reasonable chunks (Johnston et al., 2002). The GPS can now track the user, and, depending on the relative location of the user to the actual route segment, the system will proactively start interacting with the user. It should be noted that a route instruction is a very unique user interaction within this system. Most tasks in SMARTKOM can be accomplished in only a few dialogue turns and take only a few minutes. A whole navigation or walk through a city can take up to several hours. Moreover, system initiative can be triggered by external/context changes.

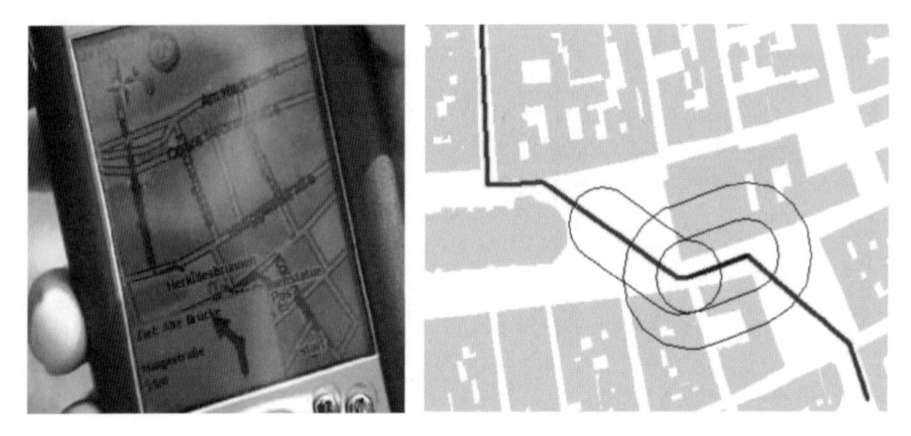

Fig. 4. Route segmentation and instructions. *Left:* SMARTKOM display with dynamically generated map, interface agent, and arrows indicating route direction. *Right:* Segmented route and trigger regions. Two regions around the active route indicate the valid and the egression region. The next route segment's valid region is also shown

The basic interaction of the navigation system, however, is always related to route segments. The system assumes that the user wants to follow the actual route segment and computes a corridor around this segment which is the "valid region" for the user's location (Fig. 4). As long as the GPS indicates that the user stays within this region, no action is initiated. When the user leaves the valid region of a segment and enters the next valid region of the next segment, the system will request a navigational instruction from Deep Map. The SpaCE agent generates a preverbal message that includes the route segment, landmarks, and necessary user actions such as to turn right or left, for example. While SpaCE controls the spatial context of the user and detects changes concerning the environment, the NCA controls the whole pedestrian navigation session and provides SMARTKOM-Mobile with the needed M3L response messages, i.e., maps, preverbal navigation instructions, or textual descriptions of objects, about which the user is informed. The NCA translates between the agent system's ACL representation and SMARTKOM's multimodal markup language.

In the described example, the preverbal message is sent to SMARTKOM and translated into a natural language generation request that, in turn, results in an utterance of the embodied character representing SMARTKOM (Müller et al., 2002). While the SMARTKOM system allows for English and German as interaction languages, the preverbal messages have also been employed for the generation of Japanese instructions in the Deep Map framework (Kray and Porzel, 2000). With this instruction, an update of the map with an arrow indicating the direction of the navigational instruction — relative to the direction from which the user is coming — is generated. Together with the user's position and sights in the vicinity, this is sent to the presentation management components of SMARTKOM.

For each route segment an "egression region" is calculated that surrounds the valid region (Fig. 4). If the user leaves the valid region and enters an egression region, Deep Map proactively notifies SMARTKOM and suggests either a redirection or a recalculation of the tour. If a simple redirection can lead the pedestrian easily back to the route, SMARTKOM will generate an instruction to go back to the precalculated tour. Again, this is done on the basis of preverbal messages from the navigation.pedestrian subsystem. If the user is already too far off the route, a dynamic recalculation of the tour is done and new route instructions lead the user to her goal.

A third proactive behavior occurs when a user enters the vicinity of an interesting sight. SMARTKOM then generates a message that there is a sight and offers to present more information. The user may also decide to request information on objects directly by asking SMARTKOM. This can be done by language requests only, e.g., "Give me more information on St. Peter's Church," or by multimodal interaction using pen input on the map with a pointing gesture and an utterance like, "I want more information on this."

Since all map information is semantically available (and not just rendered as pixel images), it can be used in SMARTKOM and deictic gestures on the display can be resolved (Gurevych et al., 2003c; Häußler and Zipf, 2003). The names of locations labeled on a map are extracted from a generic description of the map by the action planner (dialogue manager) and used for updating the dynamic lexicon (Rapp et al., 2000). This enables the system to recognize and resolve user input with spoken references to locations and to synthesize speech output, including these novel location names. Thus, the mobile SMARTKOM system can make full use of the multimodal processing capabilities of SMARTKOM.

4 Mobile Infrastructure

The mobile SMARTKOM system has to face a number of challenges. They can be classified into those related to resources, those related to services, and those related to interaction. The resource challenges arise as a result of SMARTKOM's complexity.

As described above, more than 40 software components are combined in a multi-blackboard architecture. The prototype system needs at least two high-end PCs for acceptable performance (Herzog et al., 2003; Wahlster, 2003). However, this does not constitute a major restriction for future application of this technology, considering the rapid developments regarding the computing power in mobile devices. Additionally, new wireless broadband networks such as third-generation cellular connections (e.g., UMTS) or wireless local area networks (WLAN) make it possible to distribute software components that are based on such a multiblackboard architecture onto multiple machines and to connect the end-devices through a wireless network.

The goal of the SMARTKOM-Mobile research framework, however, goes beyond porting the system to just one mobile setup and using a mobile terminal instead of a PC screen. The mobile SMARTKOM system should instead implement ubiquitous information access to the SMARTKOM services through any device. The user, in turn, should be able to decide which device she wants to interact with. In our experiments and implementations, we prototypically focused on the information technologies (IT) infrastructure typically found in modern cars — such as the display of a car navigation system, microphone array for free speech input, and audio system — as one setup, and a handheld computer such as a personal digital assistant (PDA) as a second setup. The management of these multiple devices is implemented in a SMARTKOM module called "device.mobile," sketched in Fig. 5 and described below.

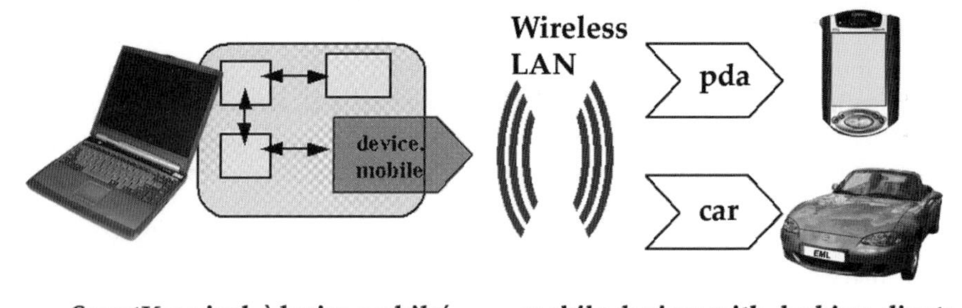

Fig. 5. Setup of the "device.mobile" docking mechanism for SMARTKOM-Mobile

For the integration of end-devices, it is not sufficient to think of one particular device but rather of a personal IT infrastructure owned by the user that can be used in multiple combinations. In order to realize such ubiquitous computing in SMARTKOM, we want to support the user with her own devices. In particular, in situations where multiple devices are available, the user has the final decision regarding which device to use. For instance, in the car a user can either use the built-in screen for visual system output and the microphones for speech input, or she may decide to use her PDA. This could be useful, for instance, because of more convenient peninput with the PDA when the car is parked. It should, however, be noted that it would in principle be possible to serve multiple devices concurrently.

In order to allow maximal flexibility, we realized the device.mobile module as a device broker (Fig. 5). It has two components. One is the SMARTKOM module that resides on the SMARTKOM system. On each potential client, a docking client component establishes the connection to the SMARTKOM system via a WLAN. In principle, it can support any number and type of mobile device. Within the mobile scenario support for four different client devices has been developed, namely PDA (Compaq IPAQ), test and simulation clients, and a client for the onboard computer of a test car. With these inherently different client types it is neccessary to abstract from the specific hardware by a client-specific connection software.

The device.mobile module is designed to support the user in her mobile activities with as little disruption as possible. This includes the seamless transition from one mobile device to another. For example, the user might travel in a car using the built-in display, speakers, and microphone to communicate with the SMARTKOM system in order to have a trip planned. When the user arrives at her destination she can pick up her PDA, leave the car, and the SMARTKOM display will "follow" her to the new mobile device. In order for this to be possible, the physical connections between the SMARTKOM system and the end user device have to be managed and the specific mobile device's capabilities must be communicated to the SMARTKOM system.

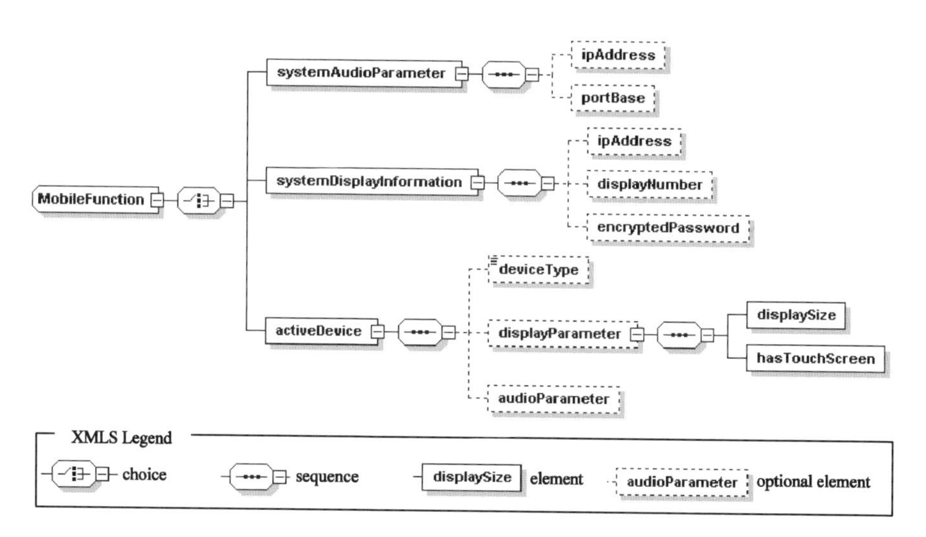

Fig. 6. Model of mobile devices with their specific capabilities expressed in M3L schema

It is therefore necessary to model the display and audio input and output capabilities and the parameters for establishing and maintaining the display and audio connections. As with other domains within the SMARTKOM system, this model is expressed as part of M3L. The relevant portion of the schema is displayed in Fig. 6.

4.1 Modeling Multimodality for Mobile Devices

Apart from the management of physical connections for display and audio data, it is necessary to model the input and output capabilities of the devices on a more general and abstract level. For a multimodal dialogue system such as SMARTKOM, the explicit modeling of supported modalities and their combinations is very important. When the system is aware of the modalities both for input and output of the current active device it can adjust the presentation and its communicative goals to the existing infrastructure. This is especially important since mobile devices can change frequently, requiring constant adaptation of the dialogue to the new situation. Through this mechanism of device abstraction, the system can adapt the output and rendering as well as input and dialogue behavior. This constitutes a necessary step in realizing a framework for a flexible multimodal human–machine interaction in mobile environments (Bühler et al., 2002). Device- and content-dependent interfaces can then be build upon such a setup (Elting et al., 2002).

Whenever the mobile device of the SMARTKOM system changes, the device management device.mobile module checks the multimodal capabilities of the new device and communicates these to the suitable SMARTKOM communication pool in order to enable other modules such as the presentation planning or display modules to use this information. The supported modalities are speech, gesture, mimic, audio, graphics, and persona. Of course, not all modalities are physically identical for input and output devices. For example, the "persona" which represents the life-like character Smartakus is only a virtual entity serving as output modality as compared to the physical entity of the user on the input side. Characteristics of special mobile devices as well as typical usage situations may impose further restrictions on modalities. An example is the situation in a moving car where the user should not be distracted too much and also should not user her hands for pen input.

4.2 User-Driven and System-Driven Activation of Mobile Devices

In order for a multimodal mobile system to be truly flexible it must be able to cope with a number of situations that arise from the special properties of mobile devices and the interaction with these devices. We discovered the following requirements for the handling of mobile devices within the SMARTKOM system:

1. Mobile devices must be discovered automatically.
2. The system must keep track of all mobile devices.
3. If a mobile device gets lost the system must be aware of this.
4. The user should be able to explicitly switch from the currently active device to one of the waiting devices (user-driven activation).
5. If the currently active device becomes unavailable, e.g., due to connection problems, one of the waiting devices should become the new active device (system-driven activation).

The device.mobile module was developed to meet these requirements. The correct implementation of this behavior could be successfully demonstrated. This is espe-

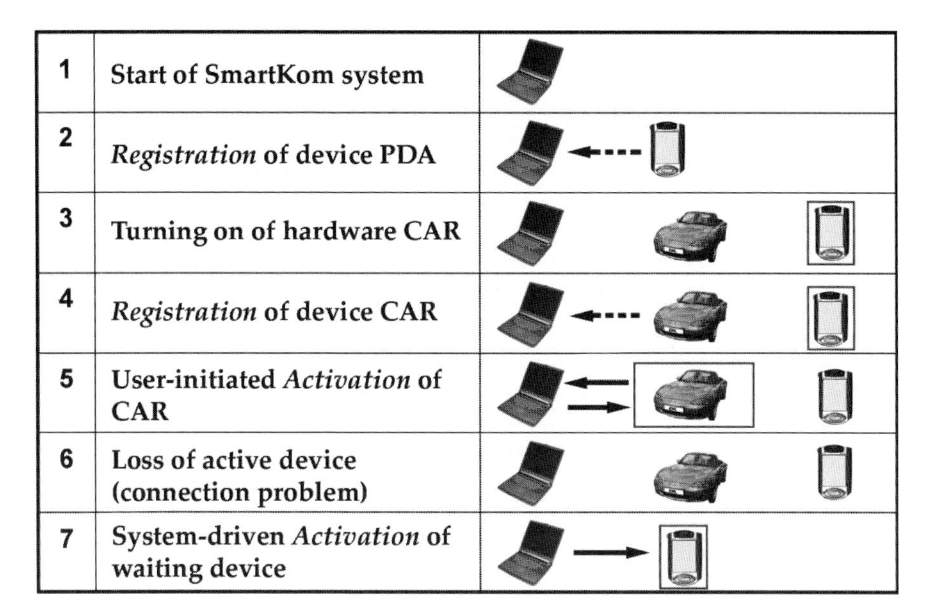

1
Start of SmartKom system
2
Registration of device PDA
3
Turning on of hardware CAR
4
Registration of device CAR
5
User-initiated Activation of CAR
6
Loss of active device (connection problem)
7
System-driven Activation of waiting device

Fig. 7. Typical mobile device scenario with user- and system-driven device change

cially important for supporting the seamless transition between pedestrian and automotive use of the SMARTKOM system. A typical scenario including user- and system-driven change of device is depicted in Fig. 7.

This general setup of the SMARTKOM controls also enables future additions for using multiple devices at the same time and to mix input and output channels, e.g., using audio input from the free speech microphone array in the car while using the display of a handheld PDA.

5 Implementation and Evaluation

The SMARTKOM-Mobile system has been implemented and tested in a prototypical setting. We realized several variants of the system. For testing with mobile pedestrians in a city, SMARTKOM runs on two laptops. As a handheld device, we used a Compaq iPAQ H3660. The GPS is a Garmin eTrex system. We connected the PDA using WLAN (ad hoc mode) to one laptop running LINUX with two network adapters, one for connecting the Windows laptop and one for the WLAN connection. For the setup in the car, the PCs are located in the trunk. The pedestrian can either take the laptops in a backpack or another person carries them. An alternative is setting up enough WLAN access points in the test area.

Using this configuration, we tested the systems in a number of trials through Heidelberg. In the current configuration, performance is still one of the biggest problems. However, this can be remedied using more powerful high-end computers. Due to the fact that interaction turns in the pedestrian scenarios are rather infrequent, the performance was generally perceived as sufficient.

The first trials with the system also showed that in noisy environments, the PDA speakers are only usable to a limited extent because the speaker volume is insufficient. However, this problem was remedied by using earphones. In many narrow alleys in the old town of Heidelberg, the positioning using GPS is problematic because of missing contact to a sufficient number of satellites. However, such problems can be fixed using more advanced dead-reckoning algorithms.

In general, the integration phase of SMARTKOM-Mobile showed that it is a big challenge to integrate such a system, and often the AI-loaded components are not the bottleneck but rather technical components such as GPS or WLAN that are not available in all situations. However, we think it is worthwhile to continue the effort of bringing the system into the streets because realistic evaluation environments are needed for further research and development.

As the SMARTKOM system combines the intuitive interaction paradigm of multimodal user input and output with aspects of mobile information systems, it is crucial to evaluate under which user contexts such multimodal interaction is more suitable than existing alternatives. Recently we conducted an evaluation of the SMARTKOM-Mobile system under laboratory conditions in order to address this question (among others).

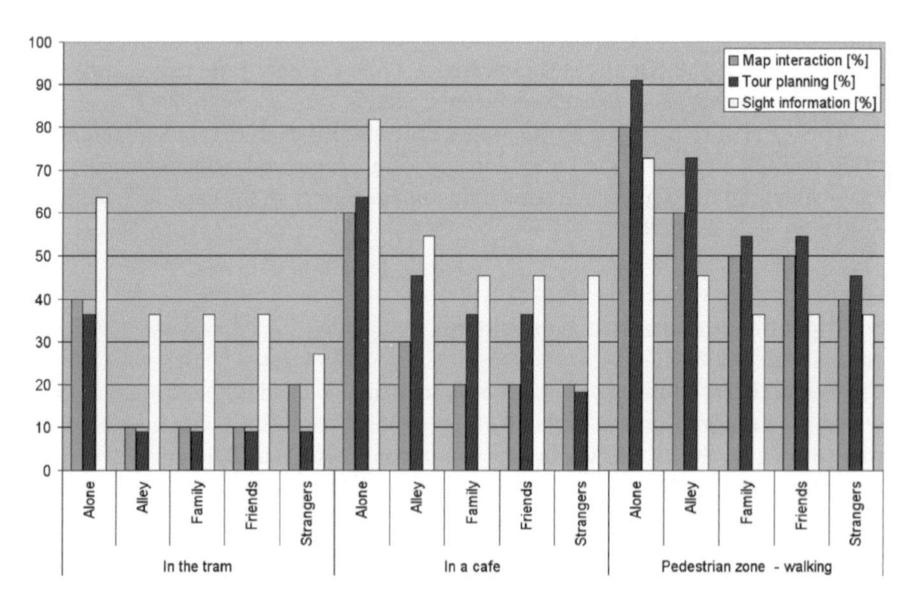

Fig. 8. Rating of multimodal interaction in different tasks. This interaction was compared to the best method a user knows to perform the specific task, e.g., printed map or tour guide

During the study the participants had to perform three different tasks. At first they had to interact with a map of the city of Heidelberg, including panning and zooming. Afterwards they had to request information about famous sights, and last they had to request a tour from a start location to an end location. These entire tasks had to be performed via multimodal interaction including speech input and pointing gestures. After each task the participants had to name the best method they know to perform the specific task, for example, using a printed map or a Web-mapping solution. Furthermore, they had to compare the performed method with the method they mentioned in different situations, e.g., alone or with friends, walking, or sitting in a cafe.

The results (see Fig. 8) showed that multimodal interaction is rated better than any named method if the user is alone (except for the situation when the user is using public transportation). Also, multimodal interaction is generally more accepted for more complex tasks like tour planning than for easier ones such as map interaction or sight information.

6 Conclusion

In this paper, we presented the mobile SMARTKOM system. SMARTKOM-Mobile represents a new generation of systems for mobile multimodal user interaction, combining three so far disjoint levels of complexity. It realizes these new levels of complexity by integrating many tasks from multiple domains into a single dialogue system and allowing users to interact in a variety of scenarios using heterogeneous devices. Our mobile architecture fuses mobile services and devices together with novel interaction paradigms that enable a user to conduct an uninterrupted multimodal dialogue even while changing devices, e.g., from cars to PDAs.

SMARTKOM-Mobile can be used with different display, audio, and deictic devices. For this, the content is adapted to each situation and device configuration. Therefore, the system needs to adapt to the dynamic mobile situation of the user, e.g., position, topic, activity. Consequently, the integration of a multiagent system was developed in order to deal with location and context-awareness. This system provides the user with flexible location-based services. Through the semantic modeling of geographical content, it is possible that users can also communicate about their spatial environment with the system.

The next steps will be dedicated to user evaluations. The system enables us to evaluate all tasks either in the old town of Heidelberg with mobile users or (feasible for specific questions) in laboratory conditions. In contrast to many other systems, we are now able to evaluate a complex system with real-world data.

Besides those questions relating to human factors of the system, a number of technological questions remain. For instance, network access is still a big bottleneck in public spaces. On the one hand, commercial networks such as UMTS are still not available; on the other hand, private networks like WLAN have become so popular that we encountered "frequency pollution" in some spots.

With the fusion of the described systems, services, and semantics we set the basis for ubiquitous interaction with a personal IT infrastructure. The system allows interactions with potentially arbitrary devices in the user's environment. If we extend this vision to "ambient intelligence" and multiuser scenarios, the system will need a kind of presentation planning wherein SMARTKOM cooperates with the users through the environment.

The focus of this paper was to present the architecture and concepts of the SMARTKOM-Mobile system. Many aspects could, therefore, not be discussed in detail. One of the biggest advances of SMARTKOM-Mobile is not a single new technology or algorithm but the fusion of a multitude of intelligent user interface methods, services, and devices into a single mobile system. In order to do so, many research questions had to be solved, but also a great deal of engineering had to be performed. The result is a highly complex research prototype that allows for user studies which so far have only been possible in Wizard-of-Oz studies.

References

C.F. Baker, C.J. Fillmore, and J.B. Lowe. The Berkeley FrameNet Project. In: *Proc. COLING-ACL'98*, pp. 86–90, Montreal, Canada, 1998.

M. Boualem, L. Almeida, I. Amdal, N. Beires, L. Boves, E. den Os, P. Filoche, R. Gomes, J. Eikeset Knudsen, K. Kvale, J. Rugelbak, C. Tallec, and N. Warakagoda. Eurescom MUST Project: Multimodal, Multi-lingual Information Services for Small Mobile Terminals. In: *Proc. 9th Conf. of Traitement Automatique des Langues Naturel*, Nancy, France, 2002.

D. Bühler, J. Häußler, S. Krüger, and W. Minker. Flexible Multimodal Human-Machine Interaction in Mobile Environments. In: *ECAI Workshop Notes of AIMS 2002*, Lyon, France, 2002.

L. Console, S. Gioria, I. Lombardi, V. Surano, and I. Torre. Adaptation and Personalization on Board Cars: A Framework and its Application to Tourist Services. In: *Adaptive Hypermedia and Adaptive Web-Based Systems*, LNCS, pp. 112–121, Berlin Heidelberg New York, 2002. Springer.

Y. Ding, C. Kray, R. Malaka, and M. Schillo. RAJA — A Resource-Adaptive Java Agent Infrastructure. In: *Proc. 5th Int. Conf. on Autonomous Agents*, pp. 332–339, Montreal, Canada, 2001.

C. Elting, J. Zwickel, and R. Malaka. Device-Dependant Modality Selection for User-Interfaces — An Empirical Study. In: *Proc. Int. Conf. on Intelligent User Interfaces (IUI) 2003*, Miami, FL, 2002.

R. Engel. SPIN: Language Understanding for Spoken Dialogue Systems Using a Production System Approach. In: *Proc. ICSLP-2002*, pp. 2717–2720, Denver, CO, 2002.

I. Gurevych, R. Malaka, R. Porzel, and H.P. Zorn. Semantic Coherence Scoring Using an Ontology. In: *Proc. Human Language Technology Conference / North American Chapter of the Association for Computational Linguistics Annual Meeting 2003*, pp. 88–95, Edmonton, Canada, 2003a.

I. Gurevych, S. Merten, and R. Porzel. Automatic Creation of Interface Specifications from Ontologies. In: H. Cunningham and J. Patrick (eds.), *Proc. HLT-NAACL*

2003 Workshop on Software Engineering and Architecture of Language Technology Systems (SEALTS), pp. 59–66, Edmonton, Canada, 2003b. Association for Computational Linguistics.

I. Gurevych, R. Porzel, E. Slinko, N. Pfleger, J. Alexandersson, and S. Merten. Less Is More: Using a Single Knowledge Representation in Dialogue Systems. In: *Proc. HLT-NAACL'03 Workshop on Text Meaning*, pp. 14–21, Edmonton, Canada, May 2003c.

U. Haiber, H. Mangold, T. Pfau, P. Regel-Brietzmann, G. Ruske, and V. Schleß. Robust Recognition of Spontaneous Speech. In: W. Wahlster (ed.), *Verbmobil: Foundations of Speech-to-Speech Translation*, pp. 46–62, Berlin Heidelberg New York, 2000. Springer.

J. Häußler and A. Zipf. Multimodale Karteninteraktion zur Navigationsunterstützung für Fußgänger und Autofahrer. In: *Proc. Symposium für Angewandte Geographische Informationstechnologie — AGIT 2003*, Salzburg, Austria, 2003.

G. Herzog, H. Kirchmann, S. Merten, A. Ndiaye, and P. Poller. MULTIPLATFORM Testbed: An Integration Platform for Multimodal Dialog Systems. In: *Proc. HLT-NAACL 2003 Workshop on Software Engineering and Architecture of Language Technology Systems (SEALTS)*, pp. 76–83, Edmonton, Canada, May 2003.

X. Huang, A. Acero, C. Chelba, L. Deng, D. Duchene, J. Goodman, D. Hon, H. Jacoby, L. Jiang, R. Loynd, M. Mahajan, P. Mau, S. Meredith, S. Mughal, S. Neto, M. Plumpe, K. Wang, and Y. Wang. MIPAD: A Next Generation PDA Prototype. In: *Proc. ICSLP-2000*, Beijing, China, 2000.

M. Joest and W. Stille. A User-Aware Tour Proposal Framework Using a Hybrid Optimization Approach. In: *Proc. 10th ACM Int. Symposium on Advances in Geographic Information Systems*, pp. 81–88, Washington, DC, 2002.

M. Johnston, S. Bangalore, G. Vasireddy, A. Stent, P. Ehlen, M. Walker, S. Whittaker, and P. Maloor. MATCH: An Architecture for Multimodal Dialogue Systems. In: *Proc. 10th ACM Int. Symposium on Advances in Geographic Information Systems*, pp. 376–383, Washington, DC, 2002.

C. Kray and R. Porzel. Spatial Cognition and Natural Language Interfaces in Mobile Personal Assistants. In: *Proc. Workshop Artificial Intelligence in Mobile Systems at the 14th ECAI (AIMS 2000)*, Berlin, Germany, 2000.

A. Krüger, A. Butz, C. Müller, C. Stahl, R. Wasinger, K. Steinberg, and A. Dirschl. The Connected User Interface: Realizing a Personal Situated Navigation Service. In: *Proc. Int. Conf. on Intelligent User Interfaces (IUI) 2004*, Funchal, Madeira, Portugal, 2004.

R. Malaka. Artificial Intelligence Goes Mobile. In: *Proc. Workshop Artificial Intelligence in Mobile Systems at the 14th ECAI (AIMS 2000)*, pp. 5–6, Berlin, Germany, 2000.

R. Malaka and A. Zipf. Deep Map: Challenging IT Research in the Framework of a Tourist Information System. In: D.R. Fesenmaier, S. Klein, and D. Buhalis (eds.), *Proc. 7th. Int. Conf. on Information and Communication Technologies in Tourism (ENTER 2000)*, pp. 15–27, Berlin Heidelberg New York, 26–28 April 2000. Springer.

J. Müller, P. Poller, and V. Tschernomas. Situated Delegation-Oriented Multimodal Presentation. In: *Proc. AAAI-Workshop on Intelligent Situation-Aware Media and Presentations (ISAMP), AAAI-2002*, Edmonton, Canada, 2002.

S.L. Oviatt. Multimodal Interfaces for Dynamic Interactive Maps. In: ACM (ed.), *Proc. CHI-96*, pp. 95–102, New York, 1996.

N. Pfleger, J. Alexandersson, and T. Becker. Scoring Functions for Overlay and Their Application in Discourse Processing. In: *Proc. KONVENS 2002*, pp. 139–146, Saarbruecken, Germany, September–October 2002.

R. Porzel and I. Gurevych. Towards Context-Adaptive Utterance Interpretation. In: *Proc. the 3rd SIGdial Workshop on Discourse and Dialogue*, pp. 90–95, Philadelphia, PA, 2002.

S. Rapp, S. Torge, S. Goronzy, and R. Kompe. Dynamic Speech Interfaces. In: *Proc. Workshop Artificial Intelligence in Mobile Systems, ECAI 2000*, Berlin, Germany, 2000.

H. Rößler, J. Sienel, W. Wajda, J. Hoffmann, and M. Kostrzewa. Multimodal Interaction for Mobile Environments. In: *Proc. Int. Workshop on Information Presentation and Natural Multimodal Dialogue*, Verona, Italy, 2001.

W. Wahlster. SmartKom: Fusion and Fission of Speech, Gestures, and Facial Expressions. In: *Proc. 1st Int. Workshop on Man-Machine Symbiotic Systems*, pp. 213–225, Kyoto, Japan, 2002.

W. Wahlster. SmartKom: Symmetric Multimodality in an Adaptive and Reusable Dialogue Shell. In: R. Krahl and D. Günther (eds.), *Proc. Human Computer Interaction Status Conference 2003*, pp. 47–62, Berlin, Germany, June 2003. DLR.

W. Wahlster, E. André, B. Bandyopadhyay, W. Graf, and T. Rist. WIP: The Coordinated Generation of Multimodal Presentations From a Common Representation. In: O. Stock, J. Slack, and A. Ortony (eds.), *Computational Theories of Communication and Their Applications*, Berlin Heidelberg New York, 1991. Springer. Also available as Technical Report DFKI-RR-91-08, DFKI, Saarbruecken, Germany.

W. Wahlster, N. Reithinger, and A. Blocher. SmartKom: Multimodal Communication with a Life-like Character. In: *Proc. EUROSPEECH-01*, vol. 3, pp. 1547–1550, Aalborg, Denmark, September 2001.

A. Zipf and S. Krüger. TGML — Extending GML Through Temporal Constructs — A Proposal for a Spatiotemporal Framework in XML. In: *Proc. 9th ACM Int. Symposium on Advances in Geographic Information Systems*, Atlanta, GA, 2001.

SmartKom-Mobile Car: User Interaction with Mobile Services in a Car Environment

André Berton, Dirk Bühler, and Wolfgang Minker

DaimlerChrysler AG, Research and Technology, Ulm, Germany
{andre.berton,Dirk.Buehler,wolfgang.minker}@daimlerchrysler.com

Summary. People tend to spend an increasing amount of time in their cars and therefore desire high comfort, safety, and efficiency in that environment. A large variety of electronic devices has been made available to meet these requirements in the vehicle. These electronic devices should allow for speech interaction in order to minimize driver distraction and to maximize driver comfort.

This contribution studies the user requirements for potential assistant functionalities operated by speech in the car. The architecture is of the dialogue system is defined based on the user requirements study. Speech dialogues were designed according to state-of-the-art principles of human machine interaction for the functionalities desired by the users. Results and ideas for future work conclude this contribution.

1 Introduction

People spend an increasing amount of time in mobile environments, notably in their cars, and want to make that time as enjoyable and productive as possible. Therefore, a large variety of electronic devices has been made available in the vehicle for comfort, ease of driving, entertainment, and communications. All these systems require a rather complex human–machine interaction.

Since driver distraction can become a significant problem, highly efficient human–machine interfaces are required. In order to meet both comfort and safety requirements, new technologies need to be introduced into the car, enabling drivers to interact with onboard systems and services in an easy, risk-free way. Undeniably, spoken language dialogue systems considerably improve safety and user-friendliness of human–machine interfaces (Gärtner et al., 2001; Green, 2000; Minker et al., 2003).

More recently, multimodal recognition and knowledge processing techniques have rapidly grown, enabling their use in mobile environments. A user requirement study has revealed that certain driving assistant functionalities, such as navigation and information services, are highly appreciated by potential mobile users. Multimodal operations may not only be performed in the car environment, but also by pedestrians. A theoretical discussion will analyze the interaction modalities (speech, gesture, mimics), which are suited for both environments. After a presentation of the conceptual framework, we will discuss the technical realization of the mobile services and propose a well-adapted system architecture. In order to adapt the interaction to human habits and to render it as natural and efficient as possible, we will analyze the most important principles and guidelines in dialogue design. All these definitions, including the different mobile environments, the interaction modalities, the system architecture, and the dialogue design principles, will form the basis for specifying and implementing the services required by the potential users of these multimodal interfaces. The conclusion and outlook section of this chapter summarizes the results and proposes some ideas for future work in this area.

2 User Requirement Studies

A user study on mobile speech-operated services was performed at DaimlerChrysler Research and Technology. It was aimed at defining applications (services) for the SmartKom-Mobile environments, which may be of interest for potential users and should be realizable with state-of-the-art technology. Further important issues have been discussed:

1. Which prospective human–machine interaction modalities will enable the use of new approaches to access and process information?
2. Which applications are expected to emerge first on the market?
3. Which new possibilities to exchange and process information may be exploited in the different applications and environments?
4. Which information and interaction modality will be preferred by the potential user?
5. How may the different SmartKom applications be combined?
6. Which databases and background intelligence processing are required for a particular application?
7. How may background knowledge be exploited to design interaction modalities that are as natural as possible?
8. How may communication and information processing be efficiently combined?
9. Which modular concepts need to be considered in order to realize the different applications in one generic SmartKom device?

The user requirement studies should allow us to combine the effective pursuit of technical concepts with an ingenious strategy of their introduction in the market. Therefore the study focused on three main points: technological trends, business models, and applications. In order to make reliable predictions about emerging technological trends and user behavior (the years 2005–2010 were considered), it is necessary to define target groups. Such groups maybe classified by age/life phase, income, and education. The following groups were identified as potential users for the SmartKom services:

1. Early adaptors and fun seekers: young (16-24 years), enthusiastic about technologies

2. High flyers: focused on career and conscious about communication styles
3. Family managers: modern middle class, practically-oriented consumers
4. Conservative, highly educated elderly people

The volume market (late followers), which will, among others, also be composed of these four groups, will develop with a certain delay. Using a user acceptance questionnaire we found that among a large number of services, the following were the most desired ones: mobile phone, emergency calls, dynamic route planning, parking place reservation, and health check.

The most important predictions of this study will be grouped in four main areas: future interaction modalities, financial aspects, data security, and personalization. The findings for each of these areas are summarized in the following.

2.1 Emerging Interaction Modalities (Speech, Gesture, Mimics)

1. Preserving privacy will play an important role in the use of mobile devices. The operation of the devices by speech in public places may be affected by this issue. Mobile phone conversations in the subway are, for instance, outlawed in Japan. Speech input with close microphone seems to be less suited for public use.
2. Cars will be important carriers for speech technology due to safety aspects in operating complex services while driving.
3. The input modalities gesture and mimics are considered science fiction, since a camera observing the driver might lead to acceptance problems.
4. The tolerance of errors will be pretty low in all groups. While input errors using the keyboard are usually considered as personal mistakes, the system will be blamed for input errors using speech operation in most cases. Younger user groups will be more tolerant than older ones.

2.2 Financial Aspects

1. A strong downward pressure of prices is expected in telematics services.
2. Financing models:
   a) Subscription fees: like I-Mode in Japan, about 3 euro per month.
   b) OnStar model by General Motors (telematics services): monthly charges of US$ 15, 1 million customers.
   c) The type of the mobile device plays a significant role.

2.3 Data Security:

1. High user requirements (comparable with elk test for cars).
2. Mistrust toward stored information will be increased if the user requirements are ignored.
3. A process of getting used to operating the services is expected.
4. Trusted suppliers will be certified.
5. Permanent biometric access control will oblige the security needs.

2.4 Personalization

1. A critical attitude of elderly users is expected.
2. Users will require some influence on the level of personalization.
3. Personalization may help to overcome weaknesses of the system.
4. Users might not be willing to explicitly indicate their preferences.
5. Even slightly negative experiences with the system will create a negative user attitude against the device and the technology.

3 Concept of Functionality for Mobile Use

Today's users of mobile devices expect to have access to data and services in the car. The large differences in development and life cycles of cars and communication devices have so far prevented the establishment of a standard for the majority of mobile devices. State-of-the-art solutions either only serve some customer groups (like adaptors for mobile phones) or they do not offer complete functionality when used in the car.

The car manufacturers develop proprietary infotainment systems with high effort. The components of this type of system are designed and developed particularly for in-car use. These integrated solutions have so far overcome the external solutions due to a more comfortable operation. But they are restricted to a very small number of services, like navigation, and operation of the audio system and the telephone. Meanwhile, all car manufacturers offer solutions based on a central display and several operation elements (buttons). In-car telephony has pushed the integration of particular audio hardware, like microphones, amplifiers, and noise cancellation techniques.

The future scenario represented in SMARTKOM-Mobile aims at combining the advantages of a hardware particularly adapted to car environments and the flexibility and large variety of services available for mobile devices. The communication system logically and physically runs on the mobile device, while the car provides a particular hardware for input and output operations. The driver benefits from the larger and better situated in-car display and the improved audio input and output functionalities provided by the adapted microphones and the sound system in the car. The GPS data received from the car may potentially be used in combination with car data like velocity and steering angle.

In the user requirement studies we found that the different SMARTKOM services need to be combined in many situations in order to solve a task in a scenario (like calendar and route planning). A higher-level knowledge source is required to combine the services and to coordinate the processing order. Repeating input prompts not only irritate the user but also leads to more complicated interaction models even for simple tasks. For example, it is not necessary to ask a user who has already reserved several rooms in the same hotel for his preferences in subsequent hotel reservation dialogues. Repeating prompts result in a lower user acceptance and hence should be avoided if possible. This information may be achieved by storing the user profile and using it in the intelligent background processing. Such a procedure allows for efficient human–computer interaction and an improved user-friendliness.

The goal of the development in the mobile scenario is a mobile communication assistant. The general framework in such a mobile environment significantly differs from the home and public environments. Furthermore, the mobile environment in the car and for a pedestrian/traveler yield different characteristics. The user is in a particular situation while driving:

1. The main task of the driver is to concentrate on the traffic and to drive in an adapted manner. The SMARTKOM services need to be realized in such a way that the driver may operate them while driving using speech, but incidentally. The SMARTKOM services aim at assisting the driver without causing safety risks through driver distraction. The interaction should incorporate only little graphical output presented in a simple manner. Input by means of gesture or pen will not be considered, due to the their higher level of driver distraction. Therefore, speech plays a very important role as input and output modality in car applications.
2. The services should only interact with the driver in situations without safety risks. The dialogue needs to be aware of the situation if possible. Possibilities to interrupt and to continue the dialogue should exist.
3. The environment of the driver is characterized by noises of the car, the street, and the weather conditions, which interfere with speech input and output.
4. The user will operate the service in the car. The functionalities of the system will be offered whilst driving and when parking. When the car stops, additional modalities, like pen and gesture input may be considered..

3.1 Transition Between Mobile Devices

The conceptual framework of the mobile functionality considers a scenario in which the user may switch between the two mobile environments "car" and "pedestrian". The user enters the car with an active mobile device (PDA) and fits the device in the designated cradle. The mobile device maybe operated by pen if the car is not moving. For moving cars, however, only the car-specific input and output modalities should be considered. The user pulls the push-to-activate (PTA) lever at the steering wheel, which causes the transition from the mobile device to the car-specific hardware, namely the central car display and the audio input and output devices. Now the driver may operate the mobile services using speech dialogues by pulling the PTA lever. If the user arrives at destination he may deactivate the car-specific hardware by pressing the PTA lever in the car. This deactivation transfers the activity to the previously active in-car device-specific hardware (the mobile device in our case). Alternatively to transfer the activation to the mobile device would be to press the PTA button at the mobile device.

4 System Architecture

The two mobile environments, car and pedestrian, use their particular communication devices, e.g., the car-specific hardware and the PDA. Since it is impossible to run the entire SMARTKOM system on state-of-the-art PDAs and car hardware due to limitations in memory and processing power, both systems have to communicate with the SMARTKOM main processing system, which runs on a computer with high processing power.

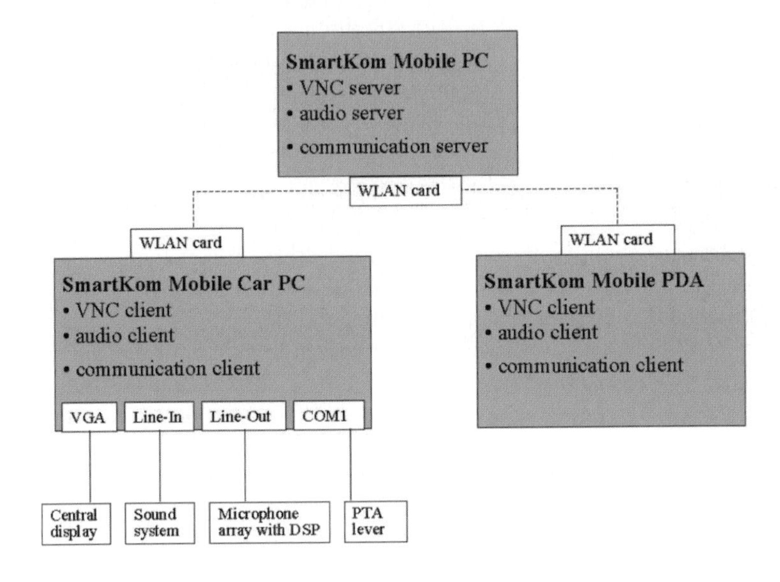

Fig. 1. Architecture of the SMARTKOM-Mobile system

The consortium agreed on the usage of an iPAQ H3600 Series PDA by Compaq as mobile device and a standard laptop as the in-car device. Both devices will run PocketLinux and communicate using wireless LAN[1](WLAN) and the standard protocol TCP/IP. The PDA will only be used as input/output device by incorporating its display, the pen, the internal microphone, the speaker, and the PTA lever. The in-car PC is connected to the car- specific hardware: the central display, a microphone array, the sound system, and the PTA lever at the steering wheel. Both mobile devices only serve for transmitting the audio and graphic data and as input/output devices. The processing of the data in a dialogue system is handled on the SMARTKOM PC. The concept and implementation for the audio server and clients were realized at MediaInterface GmbH, Dresden. The audio clients can be adapted to the environmental

[1] The limited range of WLAN was accepted in the consortium, since communication technology is not the main focus of the project and UMTS technology will enable sufficient bandwidth and high availability within the next few years.

noises, which greatly differ between the two mobile scenarios. Such an adaptation leads to a more precise voice activity detection in the speech recognition front end. The communication server and client were developed at the European Media Lab, Heidelberg.

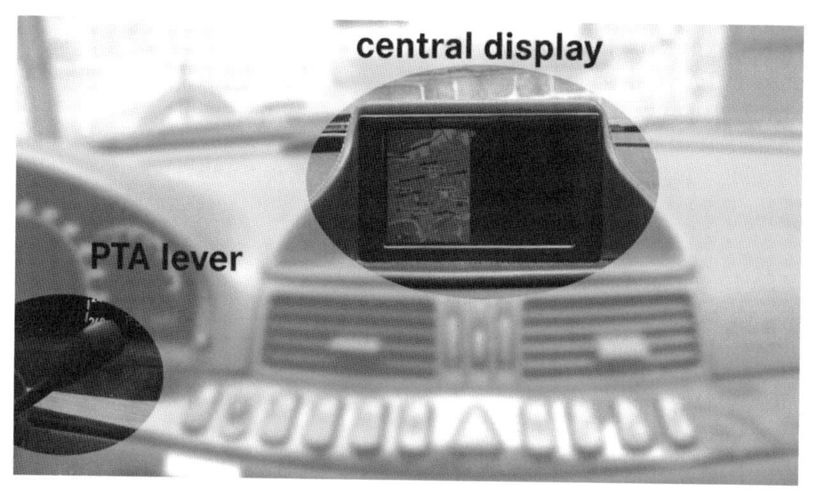

Fig. 2. Integration of the display and the push-to-activate (*PTA*) lever in the car

The SmartKom-Mobile environment in the car consists of the speech input and output and the graphical output modalities. A PTA lever is incorporated to initialize the speech input. The display will not have a touch screen functionality since grease residues on touch-screen displays would not appear valuable. The following components have been installed in the car and connected to the SmartKom system using the mobile car PC.

4.1 Central Display for Visual Output

A 7-inch coated backlight-amplified central TFT display from Toshiba with a resolution of 800 × 480 pixel has been integrated in the car. Since the resolution of the display differs significantly from the resolution of the iPAQ (240 × 320 pixel), a different presentation mode was developed in cooperation with the DFKI. The car display is segmented in two presentation zones. On the left, the iPAQ display is reproduced. On the right, detailed information about sights or parking places may be displayed. If the user does not ask for detailed information, the right display area contains a large representation of the virtual animated agent Smartakus like in the SmartKom-Public scenario. The cockpit also contains buttons to switch the display on and off and to adjust its brightness.

4.2 Sound System for Audio Output

The audio output was realized by streaming audio data via the mobile car PC to the stereo speakers in the front doors. An amplifier improves sound quality.

4.3 Microphone Array for Speech Input

An array of four microphones has been installed in the car for speech input. It has been calibrated toward the head of the driver, so that simultaneous speech input from the codriver can be compensated to a large extent. A particular signal processor (DSP) allows for reducing noise and combining the four microphone signals to a monosignal, passed to the speech recognition component.

4.4 PTA Lever for Input Initialization

The PTA lever on the right side of the steering wheel is connected to the mobile car PC via a microcontroller and a serial connection. The lever is pulled either to activate the car environment and its mobile device or to start speech input. Pushing the lever deactivates the mobile device in the car and switches the activation to the mobile device that was active before the one in the car. Most users decide, however, to activate the desired device instead of deactivating the undesired one, since all available devices are registered at the SmartKom system.

5 Dialogue Design

When designing multimodal dialogues for mobile environments the situation of the user needs to be considered, since some modalities maybe reasonable and desired, while others may distract the user, violate privacy, or disturb other people. The transition between different modalities needs to be modeled accordingly, i.e., the pedestrian user is able to use a combination of speech and pen input on the PDA, while the car driver has to rely on speech input only.

Dialogues have to be designed with respect to the users, technical constraints, budget, and time. The user requirement studies allowed us to specify the desired functionalities and modalities in detail. Wizard-of-Oz experiments needed to be performed, due to the high complexity of the system. These experiments, carried out at Ludwig-Maximilians-University, Munich in cooperation with the partners who designed the mobile scenario, i.e., European Media Lab and DaimlerChrysler, influenced the design process strongly. The dialogues were realized as naturally as possible using mixed-initiative, natural speech, barge-in, and brief prompting. With one aim of the project being to give the user a high level of control of the interaction, suitable repair mechanisms for potential errors had to ensure that the dialogue strategy appears clear and consistent to the user. Another consideration was to keep the internal organization of dialogue structures as simple as possible by breaking complex dialogues into smaller, simpler modules, if possible.

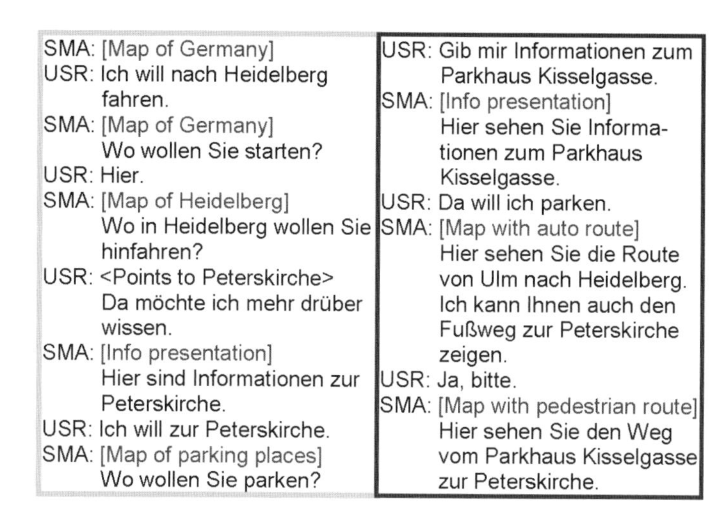

SMA: [Map of Germany]
USR: Ich will nach Heidelberg fahren.
SMA: [Map of Germany] Wo wollen Sie starten?
USR: Hier.
SMA: [Map of Heidelberg] Wo in Heidelberg wollen Sie hinfahren?
USR: <Points to Peterskirche> Da möchte ich mehr drüber wissen.
SMA: [Info presentation] Hier sind Informationen zur Peterskirche.
USR: Ich will zur Peterskirche.
SMA: [Map of parking places] Wo wollen Sie parken?
USR: Gib mir Informationen zum Parkhaus Kisselgasse.
SMA: [Info presentation] Hier sehen Sie Informationen zum Parkhaus Kisselgasse.
USR: Da will ich parken.
SMA: [Map with auto route] Hier sehen Sie die Route von Ulm nach Heidelberg. Ich kann Ihnen auch den Fußweg zur Peterskirche zeigen.
USR: Ja, bitte.
SMA: [Map with pedestrian route] Hier sehen Sie den Weg vom Parkhaus Kisselgasse zur Peterskirche.

Fig. 3. A sample route planning dialogue using different modalities

6 Mobile Services

The consortium agreed to focus on services that are highly desired according to the user requirement studies and that are not already available in state-of-the-art cars and mobile devices. Therefore it was decided to implement route planning and parking place reservation as car-specific services and an integrated route planning, which is a mixture of car and pedestrian navigation service using a device transition from car to PDA. Standard speech-operated functionalities like air conditioning and audio operation were not considered since they are already available in state-of-the-art cars. The module that implements the mobile services designed for the car was dubbed *service.navigation.car*.

This module contains the following three core functionalities: route planning, mapping, and parking place reservation. The route planning consists of calculating a route between geographical locations according to the type of the route (shortest or quickest). The mapping retrieves a map image for a geographical region and allows for panning and zooming the map image. The parking place reservation consists of a database of parking places and garages. Each database entry is annotated with its geographic location to be shown in the map. The entry also contains detailed information about opening hours, the number of available parking spaces, and the parking fee. The functionality also allows for the calculation of the parking places that are closest to the current position of the car.

The simple route planning service for cars incorporates car route planning and mapping functionalities, while the complex integrated route planning service also incorporates the car route search to the destination city, the pedestrian path to the destination within that city, and the search for a parking place close to the desired destination. The car route is active until the parking place has been reached, when activity is switched to the PDA and pedestrian route guidance is activated. The user may also ask for detailed information about parking places and places of interest in the destination city; the latter are available from the module for pedestrian navigation. A complete integrated route planning dialogue is shown in Fig. 3.

The M3L protocol employed is defined in the *geographical.xsd* XML schema, which guarantees compatibility with the *service.navigation.pedestrian* module for the pedestrian navigation. The protocol is based on request–response exchanges that are initiated by the function modeling module.

The map request function returns a high-quality bitmap image of the desired geographical region. The effective extent of the geographical region may be specified exactly or, alternatively, be inferred from the geographical extent of the vendor objects to be shown in the map. The vendor objects include calculated routes and points of interest (city locations).

The module *service.navigation.car* can be configured to work with different map data sets. For the SMARTKOM project two data sets have been devised: a full map of Germany, and a smaller map of southern Germany. The reason for using different map data sets relies in the extensive amount of disk space (about 500 megabytes for the full map) that is needed for the data.

By default, streets and cities are shown in the map image. When using the full map data, a geographical relief is also shown as a background image. The binary bitmap data is encoded as a JPEG file. The compression level is configurable and allows us to significantly reduce the amount of transmitted data.

Technically, the first version of the module *service.navigation.car* was implemented as a Windows Java application that interacted with the Navigon 2000 product via an ActiveX interface. This interface, however, proved to be rather cumbersome to use, and so for the SMARTKOM 2.0 milestone an individual solution with direct access to the underlying libraries was implemented in C++. This also made it possible to provide the complete module and data sets within the SMARTKOM distribution.

The existing functionality is sufficient for the current scope of the car application. However, some limitations seem to be apparent:

1. The service does not return high-level information about text labels. This information could be important for interpreting user gestures, which tend to focus on the text label rather than on the city center, which is also represented in the map. Also, for the presentation of the map (e.g., the placement of Smartakus), it would be advantageous to highlight potentially important text in the maps.
2. It is unclear if the font sizes used (that seem to be pixel-based) are adequate for different resolutions (in dots per inch). A bitmap approach would be rather inelegant for presentation, since it prevents us from incremental presentations of routes and other objects.

7 Outlook: Toward Flexible Modality Control

We argue that a truly multimodal interaction paradigm must not place restrictions on the user or on the environment as to which modalities to use. Instead, a multimodal dialogue system should be flexible enough to be used even as a monomodal system, if it is required. The situations in which restricted multimodal interaction is useful are manifold: Unrestricted multimodal output may lead to privacy problems. Requiring multimodal input, such as pointing gestures, from the user who may be occupied with something other than interacting with the system may also lead to inconvenience, or, much worse, to security problems, most apparently in the SmartKom-Mobile scenario.

The technical realization of a flexible modality control will be an enormously complex task. However, we are convinced that SmartKom and M3L would provide a suitable groundwork for doing so.

In this section we concentrate on the SmartKom-Mobile scenario and try to identify its particular requirements for modality control from an application-oriented perspective.

Based on experience in developing and testing speech-based user interfaces in mobile environments [1] we have identified five major combinations of modalities, *interaction modes*, that seem characteristic to the SmartKom-Mobile scenario: (see Fig. 4):

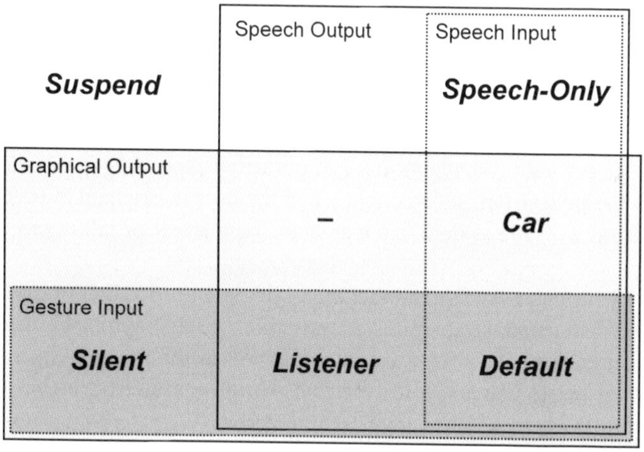

Fig. 4. Interaction modes as combinations of communication modalities in SmartKom-Mobile. Transitions are described in the text

1. **Default:** Mainly used in the pedestrian environment and when privacy or disturbing others is not an issue. All modalities should be enabled in this mode. Stemming from safety considerations (discussed below), this mode is available in the driver environment only when the car is not moving.
2. **Listener:** In this mode the user accepts spoken system output but he does not use speech as an input modality (i.e., the user is the listener here). Spoken output is useful for quickly providing concise summaries or extended background

information to the user whose focus of attention is placed on tasks other than interacting with the system. The user may employ earphones in order to protect his privacy. The listener mode may also prove useful for avoiding confusion of the system by off-talk. In this mode, the system does not try to excite spoken input from the user. It should therefore generate prompts like "*Show me ...*" rather than "*Tell me ...*".

3. **Silent:** This mode is useful when spoken language human–machine dialogue is problematic, for instance, in certain public places or in public transportation. The interface should be similar to traditional graphical user interfaces.
4. **Car:** This mode is a restricted version of the default mode for use mainly in the driver environment. Speech is the dominant communication modality, while graphical displays are used only for presenting additional (nonvital) information, such as maps for presenting a route while driving or the locations of parking options. Consequently, the display does not need to be touch-sensitive.
5. **Speech-Only:** This mode is mainly used in the driver environment notably when the car is moving. Any additional safety risk (such as driver distraction from the driving task by gesture input or graphical output) must be avoided. System interaction is restricted to the audio channel, i.e., the use of a speech-only dialogue is mandatory.

In the driver environment an emergency requires the dialogue with the system to be temporarily suspended. This *Suspend* situation (Fig. 4) occurs when the driving situation becomes dangerous and the entire user attention is required. Although the system suspends any interactivity, it should be prepared to resume the dialogue, especially after very short interruptions. Temporarily suspending interaction may also be useful in the pedestrian environment when the user is engaged in task others than communicating with the system, such as crossing a street or talking to another person.

Having identified these five interaction modes, we need to determine under which circumstances intermodal transitions are possible. We distinguish between *user-* and *system-driven* transitions. A user-driven transition is the result of an explicit command to use or not to use a specific modality. In turn, system-driven transitions are triggered by the system without any user interaction.

The user should be able to switch between modalities anytime, although some modalities may temporarily be disabled for safety reasons. A user-driven modality transition may be performed by the following actions:

1. The user may suspend/resume the operation with the system. In the pedestrian environment this operation could be performed by activating a button on the portable device. In the driver environment this could be realized by a push-to-activate button or lever.
2. In the pedestrian environment the user may switch off speech input and output using a button on the portable device.
3. In the pedestrian environment, the user may toggle display operation. A request for turning on the display could be recognized by the touch-sensitive screen of the portable device.

4. Reopening the speech channel by uttering spoken commands re-enables spoken language dialogue, e.g., when the system interaction is suspended or when graphical output is exclusively used.
5. Requesting graphical output by spoken commands (like “Show me the map”) enables graphical output, unless this modality is disabled for safety reasons (see speech-only mode).

In addition to the user, the system may initiate a modality transition in one of the following ways (Table 1):

Table 1. Possible modality transitions in SmartKom-Mobile

Signal	System reaction
User-driven transitions	
Push-to-activate	Suspend/resume operation
Command/button	Turn on/off speech
Command/button	Turn on/off graphics
System-driven transitions	
Install device in the car	Turn off gesture input
Take device out of the car	Switch to default mode
Start car	Turn off graphical displays
Stop car	Turn on graphical displays
Detect emergency	Suspend operation
Misrecognition/noise	Turn off speech input

1. As outlined above, specialized software running on the car PC may detect an emergency in the driver environment. To this end, the software has access to the necessary pieces of information about the state of the car (e.g., driving speed, state of brakes, etc.) through the CAN bus.
2. Analogously, starting the car is detected by the car PC and is interpreted as a transition into the speech-only mode. Likewise, stopping the car leads to the default mode.
3. In the pedestrian environment, i.e., when graphical output is available, repeatedly failing to understand spoken input from the user should lead the system to infer that the current situation is not suitable for speech recognition (e.g., though strong background noise). The system therefore switches to the listener mode in order to not rely exclusively on speech input.

8 Conclusion

A large variety of electronic devices serve for comfort, ease of driving, entertainment, and better communication in modern cars. These devices as well as future services require a rather complex human–machine interaction. Our work focused on new interaction modalities, which may allow for fewer driver distraction and higher comfort for operating the devices and future mobile services.

Studying user requirements for future mobile services we found that the most important issue will be which interaction modalities (speech, gesture, mimics) may be applied for which services, how to finance the services, how to guarantee data security and preserve privacy, and how to personalize the services. The services most desired by the subjects, who were asked to fill in the questionaire of our study, were the mobile phone, emergency calls, dynamic route planning, parking place reservation, and health check.

In a theoretical discussion we analyzed the interaction modalities (speech, gesture, mimics) that are suited for operating the services. This discussion led to the conceptual framework for the mobile services. We argued that in the future scenario represented by SmartKom-Mobile, the advantages of hardware particularly adapted to car environments and the flexibility and large variety of services available for mobile devices should be combined in such way that the communication system logically and physically runs on the mobile device, while the car provides a particular hardware for input and output operations. The technical realization of the mobile services carefully integrated the hardware available in future cars into future mobile services. Such hardware may be a microphone array, a sound system, and a large display.

In order to adapt the interaction to human habits and to render it as natural and efficient as possible, we considered the most important principles and guidelines in dialogue design, and applied them to specify the new services, their interaction modalities, the transition between modalities, and the transition between different mobile devices.

The consortium agreed to focus on services that are highly desired according to the user requirement studies and that are not already available in state-of-the-art cars and mobile devices. Therefore the following mobile services were specified and implemented in SmartKom: route planning and parking place reservation as car-specific services and an integrated route planning, which is a mixture of car and pedestrian navigation service using a device transition from car to PDA.

In an outlook, we have presented a framework for a flexible modality control oriented at the needs of the SmartKom-Mobile application scenario. The main goals are to illustrate the necessity of being able to handle modality transitions, if an efficient and unobtrusive multimodal interaction is to be achieved in different usage situations. An implementation will, of course, face interesting challenges. Even if the events signaling a modality transition can be detected effectively and a modality-independent output mechanism is in place, the question of how to achieve a consistent interaction experience will arise. Nonetheless, a modality control framework is clearly necessary, not only in the face of new interaction devices with different capabilities, but also for the sake of a personalization of the interaction (considering, for instance, speech- or sight-impaired users).

References

U. Gärtner, W. König, and T. Wittig. Evaluation of Manual vs. Speech Input When Using a Driver Information System in Real Traffic. In: *Int. Driving Symposium on Human Factors in Driver Assessment, Training and Vehicle Design*, Aspen, CO, 2001.

P. Green. Crashes Induced by Driver Information Systems and What Can Be Done to Reduce Them. In: *Proc. Convergence 2000*, Warrendale, PA, 2000. Society of Automotive Engineers.

W. Minker, U. Haiber, P. Heisterkamp, and S. Scheible. Intelligent Dialog Overcomes Speech Technology Limitations: The SENECa Example. In: *Proc. Int. Conf. on Intelligent User Interfaces (IUI) 2003*, Miami, FL, 2003.

Part VI

Data Collection and Evaluation

Wizard-of-Oz Recordings

Florian Schiel and Ulli Türk

Bavarian Archive for Speech Signals
c/o Institut für Phonetik und Sprachliche Kommunikation
Ludwig-Maximilians-Universität Münchenn, Germany
{schiel,tuerk}@bas.uni-muenchen.de

Summary. This chapter gives a concise overview of the empirical Wizard-of-Oz recordings done within the SMARTKOM project. We define the abstract specifications of the intended simulated communicative situations, describe the necessary technical setup (including numerous useful practical hints), and finally outline the specifications of the resulting multimodal corpus, which may be obtained from the Bavarian Archive for Speech Signals (BAS).

1 Introduction

The SMARTKOM data collection was aimed to provide empirical data about human behavior in human–machine communication (HMC) for several purposes: to train statistical models for pattern recognition as well as user modeling, to gain insight on how humans may react or interact in HMC situations and to elicit situations where the human displays a certain degree of an emotional state. [1] To achieve these goals a so-called Wizard-of-Oz recording paradigm (WOZ) was used throughout the investigations. That is, users were not aware that they were interacting not with a machine but rather a simulated machine.

The successful simulation of a complex system with several input and output modalities such as SMARTKOM is a very complex task that requires a large range of expertise ranging from purely technical to psycholocigal knowledge. However, in this contribution we will concentrate on two main issues:

- How to define a task for a human user that enables her to interact with a simulated system without any explicit restrictions. On the one hand, we would like to give the user as much freedom as possible to encourage her to show new ways to interact multimodally with a system like SMARTKOM. On the other hand, we have to observe the implicit restrictions of a limited technical system. A WOZ simulation would be rather implausible if the system is able to process any kind of human multimodal input data.

[1] Please note that the description found here is rather abbreviated due to space limitations. More detailed descriptions on the topics outlined here are found in the references given at the end of this chapter.

- How to implement a technical recording setup that not only allows the successful simulation of the intended system but simultaneously allows the recording of all relevant communication channels between human and machine. The simulation must run smoothly, without any artifacts that might tip the user of the simulation and allow the controlling wizards to react appropriately within the assumed restrictions of the simulated system. The data must not only be gathered in real time but must also be synchronized across several data types and sampling frames.

Many other issues such as the design of a graphical interface that does not exist yet, the design of task flow plans for the simulation, the evaluation of the simulation, the training of the controlling agents (the so-called wizards) behind the system, the annotation, labeling and categorization of the recorded data, and others cannot be addressed here. Some of them are discussed with more detail in the literature given at the end of this contribution.

This contribution may serve two purposes for the interested reader: First, he might get some important background knowledge about the SMARTKOM Multimodal Corpus that may be obtained at the Bavarian Archive for Speech Signals (BAS, http://www.bas.uni-muenchen.de/Bas). Second, he might learn about the specific problems of such a large-scale WOZ recording and get some hints and tricks on how to circumvent known problem areas if he intents to do his own investigations.

The next section deals with is the first problem addressed above, likewise section 3 gives a concise description about the technical setup that was used in the SMARTKOM WOZ recordings. Section 4 summarizes the main features of the resulting multimodal corpus.

2 Recording Plan

2.1 Overview and Definitions

The SMARTKOM WOZ recordings consist of a large number (to satisfy the statistical requirements) of recording sessions with a single user interacting unsupervised with the system. The user gets a short instruction and a task to solve with the help of SMARTKOM. She is convinced that the system is real in the sense that it is an automatic technical system. In reality she is interacting with two so-called wizards steering the technical "surface" of the system.

The *recording plan* is the script of the experiment. It contains all information regarding the experiment, ranging from the instruction to the user to the rules of behavior for the wizards, and gives an overview on how all these single components of the experiment shall be combined. Of course, the recording plan implicitly defines the expected data that will be recorded during the real experiment. However, there is always room for surprises, and there should be.

The fully detailed recording plan is documented in Steininger and Kartal (2000) and implicitly also in the corpus documentation and corpus metadata. Here we will restrict the presentation to the basic design principles and performance rules. [2]

2.2 Experimental Design

2.2.1 Typical Session

A typical SK recording session[3] consists of the following phases:

- Questionnaire for user
- Instructions for user
- Session 1 (4.5 min)
- Second instruction
- Session 2 (4.5 min)
- Interview or questionnaire
- Gathering of user metadata
- Signing of nondisclosure agreement and payment of incentive

The user fills out the first part of the user questionnaire. After that the supervisor leads the user to the recording room, instructs the user briefly on how to use the system, "starts" the system and leaves the room.

The first session begins; it lasts 4.5 min. The supervisor is in radio contact with the two hidden wizards and may decide to interrupt the session if something seriously goes wrong.

The settings for each of the two sessions are defined according to an overall session plan, which defines in which technical scenario (Public, Home or Mobile) the task takes place, which primary and secondary task is to be solved and some other conditions as listed in Sect. 2.3. Sessions 1 and 2 usually have different definitions. In most cases the tasks are different; in some cases they might be based on each other (that is, a task from session 1 is continued in session 2).

After the second session, the second part of the questionnaire is filled out by the user. Then the user metadata are registered in the database (later exported into the "spr" file of the corpus), and the session definitions are stored into the database as well (later exported into the "rpr" file of the corpus).

Finally, the user signs an agreement that the recorded data may be used for the intended scientific investigations and receives her incentive. The user is strongly urged not to reveal the nature of the recordings to others because this might lead to severe recruiting problems (only "naive" users can be used for WOZ recordings).

---

[2] The recording plan as presented here is based on a concept that was proposed in Dec. 1999 to the SK consortium, augmented by contributions of the partners in Jan. 2000, at the second main SK workshop and again after the MoKo workshop of 2001.

[3] In the future simply referred to as "session."

2.2.2 Description of the Scenarios

There are three basic technical scenarios in SMARTKOM. We will give a very brief description of the setups as they occur to the user. Further technical details are found in Sect. 3.

Public and Mobile

The user is standing in front of a table (Fig. 3). On the table a graphic tableau (Wacom A2) is mounted. The graphic tableau is not tilted as usual but is flat and covered with a purely white foil so that it is not recognizable as a graphic tableau. In some sessions the user has a pen available, in some not. Approx. 1.5 m above the table a cube about 40-cm long is attached to a double column behind the table. From there the graphical output is displayed onto the white surface of the table. Between the columns in front of the user a camera is aimed at the general direction of the head; a second camera is situated about 2-m away to the left-hand side. The sound output of the system appears to come from under or behind the table; other sounds like artificial background noise may be played in on a speaker set behind the user. In some sessions the user is required to wear a headset, in some a collar microphone, in some no microphone at all. The environment is neutral office equipment with some sound damping elements on ceiling, floor and walls. The visual background behind on the right-hand side of the speaker is grey–white; the background behind the speaker is colored in different abstract patterns and can be changed from session to session.

In the Mobile scenario the size of the projected output is reduced to indicate the small display in a car or on a PDA. The user wears a headset, because SMARTKOM-Mobile will not have any far-field microphones.

Home

The user is sitting on one of two low, comfortable living room chairs. One of them faces a low table of the size and height of a coffee table on which the display is projected; the other chair faces, in the opposite direction, a TV set and a VCR. On top of the VCR there is a video tape ready for use. The user wears a collar microphone that is connected to the ceiling so that the user can move freely in the room. Behind the coffee table is a large bookshelf along the wall; other living room accessories are scattered across the room.

2.2.3 Dealing with the User

The recruitment, briefing, questioning and instruction of the user is the task of the experimental supervisor. The supervisor has to take care that:

- The user takes the WOZ simulation for real.
- The user is not reluctant to deal with the machine.
- The user is relaxed.
- The user acts naturally and does not comment all her actions.

- The user uses multimodal input channels and does not concentrate on one channel only.
- The user might also show something about her user state.

During the recruitment the supervisor creates the illusion that the user will be involved in a market study. This diverts the attention of the user from the technical details to the "usability" of the system and thus improves the plausibility of the simulation. Between the sessions the supervisor might ask the user if she used gestures or not and encourages her to do so in the next session etc.

As for the instruction itself, it turned out that written instructions are not very effective (especially with elderly users). Verbal instructions by the supervisor according to an (internal) manual and a short handout with very short keywords that the user may reread just before the session starts were better. Of course, both the manual and the handout have to be adapted to the selected session definition. It seems to be important that the core instruction (the "task") can be formulated in one sentence only. Otherwise the user starts with repetitions of parts of the instruction to remind herself of the different points during the recording, which is clearly not what is wanted.

The supervisor must verify that the user is ready for the session and fulfills all the points listed above. If not, he may add some more coaching before he decides to start the session; in extreme cases he might reject the user altogether.

Fig. 1. The simulated GUI in scenario Public and domain Tourism before the dialogue starts as seen by the controlling wizard (navigation bar is visible to the *right*)

2.2.4 Graphical User Interface (GUI)

The graphical user interface (GUI) consists of the SMARTKOM logo, a help field, the persona or agent, and the main display area where the information is presented. It turned out that users almost never pointed to the help area. In most cases help was ordered verbally. For the simulation the GUI is implemented as simply HTML pages enriched with animated flash sequences. To avoid association with the standard PC GUI all familiar elements like mouse pointer, slider and menus were removed. The controlling wizard sees the same display as the user but has a visible mouse pointer as well as a navigation bar to the right where he can control the changes of pages or other elements of the GUI. The navigation bar is, of course, not visible to the user. The information content may be short texts, keywords for the next command, pictures/graphics or maps.

The system status is displayed to the user via different appearances of the persona. The persona in our simulation does not really interact with the user; that is, we do not have an animated speaking face. The persona Smartakus — which was the name the users were asked to use to address the system — showed only four different system states:

- inactive (“sleeping”)
- active and listening
- active and processing
- system error

Fig. 2. The WOZ version of the persona “Smartakus” in the state “processing”

In the Public scenario the persona starts in the “inactive” state and goes to “active and listening” when pointed to or when addressed with the name Smartakus. In the Home scenario the persona goes into “active and listening” state when the portable display is switched on. In the Mobile scenario no persona was shown on the display because of the reduced size of the display; the system is always in the “active and listening” state here.

The simulated system will only react to input in the state "active and listening"; all inputs at other times — even partly — are ignored. A more detailed description to the simulated GUI can be found in Beringer et al. (2000).

2.2.5 Speech Output

The voice output is spoken by one of the wizards and then electronically distorted to achieve a machinelike voice. We do not use a real synthetic voice because it turned out that typos in the synthesis raise more suspicion than speaking errors. The wizard can open the channel to the recording room by means of a foot pedal. A red warning light in the control room indicates that the channel is open and that no other noise should be produced. Alternatively, he may activate prerecorded standard sentences, but this requires a full set of prerecorded samples per wizard team.[4] The voice output is softly restricted to a list of sentences or sentences types, respectively. The wizard should preferably stick to this list of possible outputs. The lists of preferred output sentences is context-dependent. That is, for each major display the wizard has a special list of sentences available (of course, some situation-dependent sentences are always possible, like error messages or help messages).

2.2.6 Simulated Contents

The simulated system must, of course, provide information to the user to enable her to solve the task successfully. Since all scenarios had to take place in the German city of Heidelberg, most of the contents were taken from there.

The date (if shown on the display) is always that of the day of recording. The TV program is based on the EPG database provided by the partner Philips and updated by this partner on a regular basis. The cinema program of Heidelberg, Heidelberg maps as well as restaurant and tourist information were not changed during the experiments. This might lead to a reduced vocabulary in terms of content words. However, the effort to keep a fully updated simulation all the time for more than three years was simply to high.

2.2.7 Guidelines to the Wizards

Two wizards control the simulation from a hidden room. One is responsible for the voice output; the other for the GUI. The user may interact verbally or via gestures with the system. All inputs are interpreted by a wizard; the user has no real controlling abilities. Both wizards can watch the user's face and the upper body from the left as well as the display area. The speech output wizard needs a considerable amount of training to always speak with a constant rate and a more or less normalized intonation. Also, he has to avoid any other noises such as coughs, laughing, clearing his throat, etc.
General rules for the wizards are:

[4] In prime times we had up to five teams working in shifts.

- Try to simulate the intended SMARTKOM system as well as possible (this includes misinterpretations as well as technical errors). The "intelligence" level of the simulated system is significantly higher than would be expected with today's (2000) state-of-the-art technology.
- Try to achieve a natural and varying dialogue with the user.
- Only clear gestures that are in relation with the actual graphical display are recognized; gestures outside the interaction area (for instance, in front of the body) are ignored.
- The user may use any kind of (correct) sentence or command to formulate a question, granted that the sentence is not to long (more than 25 words) and not to complex (e.g., contains subclauses).
- Even long and complex utterances are accepted by the wizards, if the very first part contains a simply statement or question, e.g., in the spoken input sentence:

  "I'd like to go to the cinema tonight ... can you help me ... at best I'd like to go to a cinema in the center ... oh, yeah, and it should have THX ..." [5]

  the wizards will ignore everything except "go to the cinema tonight".
- If the dialogue stalls (for instance, if the user does not know how to go on), the system will offer simple assistance. For instance, it will indicate how to give input to the system:

  "You may use free speech or indicate things on the screen"

  or which options exist at a certain dialogue stage:

  "If you want, I could make some suggestions."
- If the user consequently uses only one modality for input, the wizards try to steer her into a multimodal interaction.
- Technical instructions, e.g., "Please try to repeat or use other words for your questions" are embedded in the dialogue; there exists no training sequence at the beginning of the dialogue.
- If the dialogue runs too smoothly, the wizards will insert artificial errors into the processing.
- If a user thinks she has already solved the task before the recording time of 4.5 min. has ended or the user "sticks" at a problem and does not go on with her task, the wizards will cause an error to continue the interaction.
- If a user shows signs of anger, the wizards will try to enforce that emotion by using wrong answers persistently, etc.
- If the user tries to leave the possible domains, the wizards will block this by issuing a standard sentence like, "This function is not implemented yet. May I help you with something else?"
- The wizard has to avoid repetitive outputs, because it is almost impossible to reproduce the same sentence with the same rhythm and speaking manner. It is very risky that the user may detect such differences and deduce that this cannot possible be a machine speaking.

[5] All examples in this contribution were adapted to English.

2.3 Session Plan

For each recording there exists a so-called session plan that defines the varying parameters (called "dimensions") of the individual session. The distribution of dimensions across the individual sessions is controlled by the global session plan, which in turn reflects the requirements of the project partners.

The dimensions of each session concern the technical setup as well as the task the user has to solve. There are nine basic dimensions that may vary from session to session:

- Scenario (Public, Home, Mobile): This dimension defines a large number of technical settings and changes to the recording setup. Therefore the change of this dimension is something that can be done only a few times during the recording period. See Sect. 3 for more details about this.
- Primary domain (cinema, tourism, navigation, TV program, VCR, fax, task planner): The primary domain reflects the task explicitly given to the user, e.g., "With the aid of Smartakus, find a good cinema in Heidelberg for tonight." The primary domain also includes a role model for the user and a situation in which the user is at the moment of the recording. Both have to be explained thoroughly to the user before the experiment starts.
  For example, "You are on a business trip in Heidelberg; you had an appointment in the morning, and you are free for the afternoon and evening. You would like to see a movie, possibly together with a friend or colleague." The primary task also includes a defined number of possible complications that may or may not be implemented by the wizards during the recording. This is done to give the wizards some guidelines and to prevent too exotic problems that might come to the minds of the wizards.
- Secondary domain (restaurant guide, telephone, task planner, select music, e-mail, TV program, VCR): The secondary task is not explicitly given to the user, but to the wizards. The wizards will try to steer the dialogue that starts with the primary task to the secondary task. For instance, after selecting a movie and getting reservations in the cinema, the system asks the user if she would like to look for a nice restaurant before or after the movie.
- Politeness: The system addresses the user either with "Du" or with the more polite "Sie".
- Biometric authentification: This means that the user has to authentificate herself before using certain protected functions of the system (for instance, to use the task planner). This dimension was never implemented in the SMARTKOM recordings because the simulation effort was too high.
- Evoked emotions: In some sessions the wizards are instructed to evoke anger in the user.
- Pen: There are three types of sessions:
  - User has no pen; she will always use her hand or finger to gesture.
  - User is told to use only the pen to gesture.
  - User is told that she might use the pen or might use her finger.

- Background acoustic: For each technical scenario one of three different types of corresponding background noise is played back during the recording. The sound level is kept constant throughout the recording. Two conditions have a rather high noise level (S/N ratio approx. 20 dB, but not verified); one condition is of rather a low level such as in a quiet phone booth, a quiet home or a standing vehicle (S/N ratio approx. 35 dB).
- Background visual: The visual background for the lateral camera was constant (either gray for Public and Mobile, or a bookcase for Home). The background of the frontal camera, however, was changed between three different patterns:
  - the "real" background consisting of white acoustical damper elements mounted on the ceiling ("None")
  - colored cubical patterns with sharp edges ("Kubismus")
  - colored flowing patterns without any sharp edges ("Bunt")

The exact distribution of these dimensions across the recordings are described in Steininger and Kartal (2000) or in the metadata of the corpus, respectively (see Sect. 4.4).

3 Technical Setup

3.1 Overview

The SMARTKOM system should be able to recognize natural speech as well as gestures on a flat interaction area. Additionally, facial expressions are analyzed with regard to visible emotions or so-called user states. The output of the system is presented with a GUI, which is projected on the interaction area, and with synthesized speech (see Fig. 3 for a schematic view of the recording setup).

In WOZ experiments subjects are recorded in sessions of 4.5-min length while they are interacting with a simulated version of the system. During these sessions all capture devices of the system (and some more) are used for collecting data:

- Audio is captured using a directional microphone, a microphone array with four channels and (alternating) a headset or a stereo clip microphone.
- Video is recorded by two standard DV cameras (one for the facial expression, one for the side view of the subject) and by an infrared camera that is part of the gesture recognizer SiViT (Siemens).
- The graphical output is recorded in a low-frame-rate video (used only for labeling)
- Gesture coordinates captured by the SiViT system and the graphical tablet.

Figure 4 shows the processing stages during and after the recording together with resulting changes on the data file set. Except for the last two stages, which are covered by one team, each block represents a different working team. In the following we give a detailed description for each stage, focusing especially on the capture of synchronous multimodal data with standard equipment.

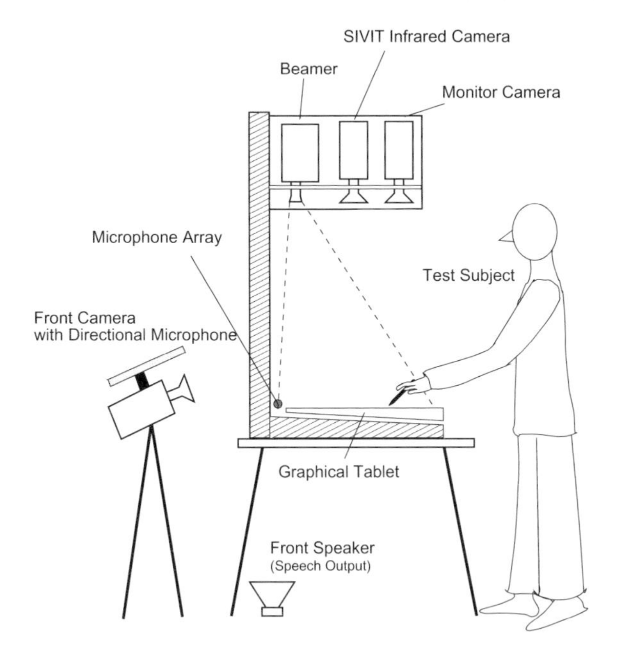

Fig. 3. SMARTKOM setup used for data collection in the Public scenario

3.2 The QuickTime File Format

QuickTime[6] allows the integration of several kinds of media like text, video, audio, images and vector graphics in one multimedia file format. All SmartKom files are compatible with this technology. Together with these data files, we provide a so-called framework QuickTime file in the corpus. The user can play back the recorded session, select the tracks to be presented, choose the style of the presentation and navigate to individual turns in the recording. In addition, we use QuickTime's capabilities to render new data files from the existing set of files (see, e.g., Sects. 3.4.1 or 3.6).

3.3 Data acquisition

During the recording of a WOZ experiment all data are stored on a set of currently five Windows NT workstations. Each computer is dedicated to record a single data stream because continuous capturing of audio and, especially, of video demands a great amount of computing power. The data from two video streams (front view and side view) are recorded separately via a FireWire bus between camera and computer and are stored as Quicktime files encoded in DV. The video signal of the infrared camera is digitized by an external analog–DV converter and then recorded in the same way. Video data encoded with DV offer a high picture quality and show

[6] http://www.apple.com/quicktime/

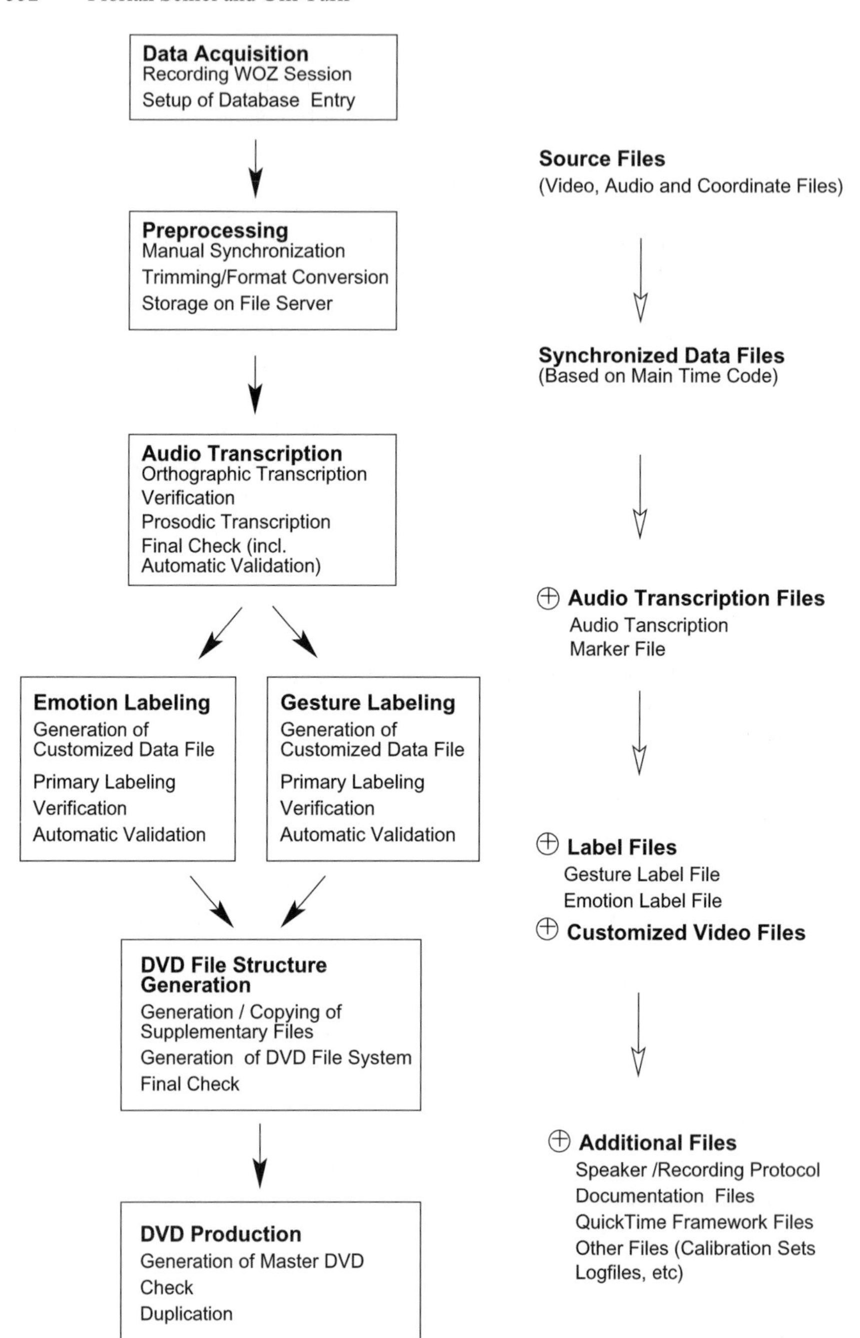

Fig. 4. Processing stages in SMARTKOM data collection (*left*) and corresponding changes in the data file set (*right*)

very few artifacts that could be otherwise problems for image processing algorithms. However, the advantage in quality comes together with large amounts of data (see Sect. 3.9 for details on storage requirements). For capturing the audio data we use a ten-track audio card. The recorded audio files are stored in Windows WAVE format, using a resolution of 16 bit and a sampling frequency of 48 kHz.

Capturing of the graphical system output is done via a screen capturing tool. It allows recording to a video file in AVI format at low frame rates; in our setup capturing at 4 fps is sufficient because the interaction speed is low. The remaining data tracks, the coordinate files of the SiViT gesture recognizer and of the graphical tablet are recorded with specially designed tools. The file format is a text file containing a list of coordinates together with the time stamps. In addition to the actual recording, this stage also comprises the generation of new entry for the session in the database. Information about the speaker as well as his or her behavior during the recording or details about the setup are stored here. In the following stages gradually more and more information is added, e.g., about the current state of the session in the processing pipeline. The database is presented in more detail in Sect. 3.9.

3.4 Preprocessing

This stage comprises all steps that are necessary to deliver a set of synchronized data files to the following processing stages. Because the recording is done on several computers, each running on its own internal clock, it is thus necessary to align all data tracks to a single master clock and to define a common start and end point.

3.4.1 Generation of "visible" coordinate log files

In a first step the coordinate data files must be made accessible for the editor in order to perform the temporal alignment. Our idea was to use the sprite track feature of Quicktime, where small bitmaps (called sprites) can be blended over another video track. The parameter of the sprites like, for example, the position or orientation can be changed dynamically. Figure 5 shows an overlay of three video tracks:

- the infrared video from the SiViT camera, showing a gray-scale picture of subject's gestures
- the video track of the graphical system output
- the visualization of the SiViT coordinates

The coordinate data is displayed by a moving dot with a cometlike tail, which gives not only a graphical representation of the current position but also of the history of coordinates at previous time steps.

3.4.2 Manual synchronizing

The actual synchronization process is done manually on a video editing workstation running Adobe Premiere.[7] We investigated several automatic or semiautomatic approaches using defined synchronizing signals that appear in all captured modalities,

[7] http://www.adobe.com/products/premiere/main.html

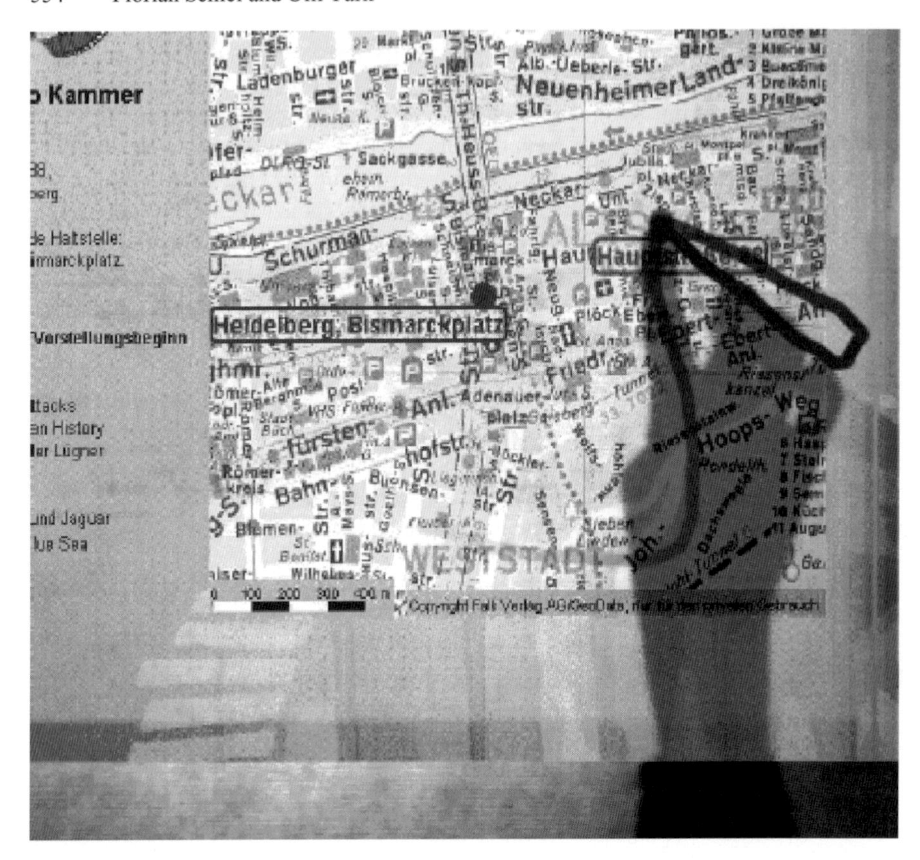

Fig. 5. Coordinate log file of the SiViT gesture recognizer visualized by cometlike data plot

but it turned out that all of these attempts resulted in a considerable number of errors so that a manual inspection of the results was necessary anyway. Therefore we decided to skip these techniques and do the synchronization manually. [8]

The process is done in two steps. First, the audio files, which are recorded separately on the audio capture computer, are aligned to the two video camera tracks. This is done by visually aligning the waveforms of the audio tracks to the waveforms of the audio tracks recorded by the cameras.

Second, the remaining video tracks are aligned by matching characteristic changes in the video tracks, e.g., a change of graphical system output or a stroke of a hand gesture. The circumstance that the test subject starts each session with a finger tap on the Web persona helps with this task. At this point we have aligned all data tracks at the beginning of the recording.

[8] Another advantage of manual synchronization is that a human operator actually looks at and hears the captured data. That way faulty data caused by malfunctioning devices, etc., can be detected early.

In a last step the start and end point of the recording is defined and all tracks are cut to equal length. At the same time, some file format conversions are done, e.g., downsampling the audio data files to 16kHz. The coordinate files are updated: the temporal shift and trimming that were performed on the visual coordinate tracks must also be considered in the raw data files.

3.4.3 Verification of the Synchronization Process

A big challenge in the SMARTKOM data collection was the ability to record different data sources and to combine them to one temporally consistent data entity. Because we had a limited budget, we had to use standard components for the recording studio. While audio hardware with synchronizing features is available, video systems that allow capturing of multiple streams can only be found in the high-end (and also high-price) market. The problem of how the other data sources like the gesture coordinate data in our project can be integrated in such a setup still remains.

Another approach would have been to synchronize the time base of the recording computers via network using, e.g., the network time protocol (NTP).[9] This would have required the design of new capturing tools in order to profit from the distributed master clock. Besides the fact that this would have caused a great delay for the start of the SMARTKOM data collection, this approach failed because of missing APIs for the video hardware.

Our current synchronization process is based on the assumption that the timing deviations of the different capturing processes stay in an acceptable range. We try to achieve this by a setup that is as homogeneous as possible. For example, the three video capture computers are equipped and configured identically.

We performed several test recordings in order to specify the timing deviations in our setup. These test recordings were limited to a length of 9 min as the size of the video files would otherwise hit the 2-GB limit for the file size given by the operating system. All data tracks were recorded and manually aligned as described in the previous section. We found an increasing timing deviation between the video tracks and the remaining tracks, which reached the maximum of one full video frame (40 ms in PAL) after 9 min. If we assume a linearly increasing drift, we end with a timing deviation of 20 ms in a typical SMARTKOM session of 4.5 min. Deviations in this range are below the range human beings normally can perceive. Dialogue systems like SMARTKOM operate with much higher timing windows when combining events from different modalities. We therefore consider the timing deviation as acceptable for this multimodal corpus.

3.4.4 Data Storage on Server

Finally, the aligned data files are stored on a Linux file server. The working teams of the following process stages access these files over the network in order to prevent repeated copying of large files. The actual path to the data is noted in the database sheet.

[9] http://www.eecis.udel.edu/ ntp/

3.5 Audio Transcription

Audio transcription is a time-consuming manual process. It includes orthographic transcription (proper names, hesitations, noise and pronunciation variants are also tagged), a verification step and prosodic transcription (primary and secondary stress, intonational movement at phrase boundaries). A detailed description of the SMARTKOM transcription system can be found in Steininger et al. (2006)

During the orthographic transcription a marker file is also generated that describes the start and end point of each turn of the HMC (segmentation). This timing information may be used as structural element. For example, it is used to display the transcription turnwise in the video stream in a subtitle style. The final transcription file is checked manually and validated automatically against the transcription conventions.

3.6 User State and Gesture Labeling

These work steps require customized video files that are rendered from the source data files on the file server. Figure 6 shows a frame of a video track used for gesture labeling. It contains four different views on the video data in one picture (front camera, side camera, graphical system output, infrared camera merged with the system output) and an audio track. In addition, the audio transcription of the current turn is inserted on the top. The file format is AVI with Cinepak encoding as this the only video format which can be used together with the labeling tool *Interact.*[10] For the emotion labeling a simplified version (only the front camera together with the audio TRL and an audio track) is produced. Both customized video tracks are created by a Java application using the Quicktime API for Java (Maremaa and Stewart, 1999).

The label files generated in these stages are stored in ASCII text files. They are later integrated in the framework Quicktime file and in the BAS Partitur Format files (Schiel et al., 2002) in the same manner as the audio TRL file. For more information on user state and gesture labeling see Steininger et al. (2006)

3.7 Final Data Preparation

In this stage all source files and annotation, resp. label files, of a session are collected, final-checked and prepared for distribution on DVD-R media. A specially designed tool provides help with the required working steps.

Supplementary files for the sessions are generated, for example, a speaker protocol file, which gives information about the characteristics of the subject or a recording protocol file describing parameters of a recording. Final checks of the source files are done. They include semiautomated checks of cross-references in the data files (e.g., same speaker ID used in speaker protocol and recording protocol) as well as manual consistency checks.

[10] http://www.mangold.de/

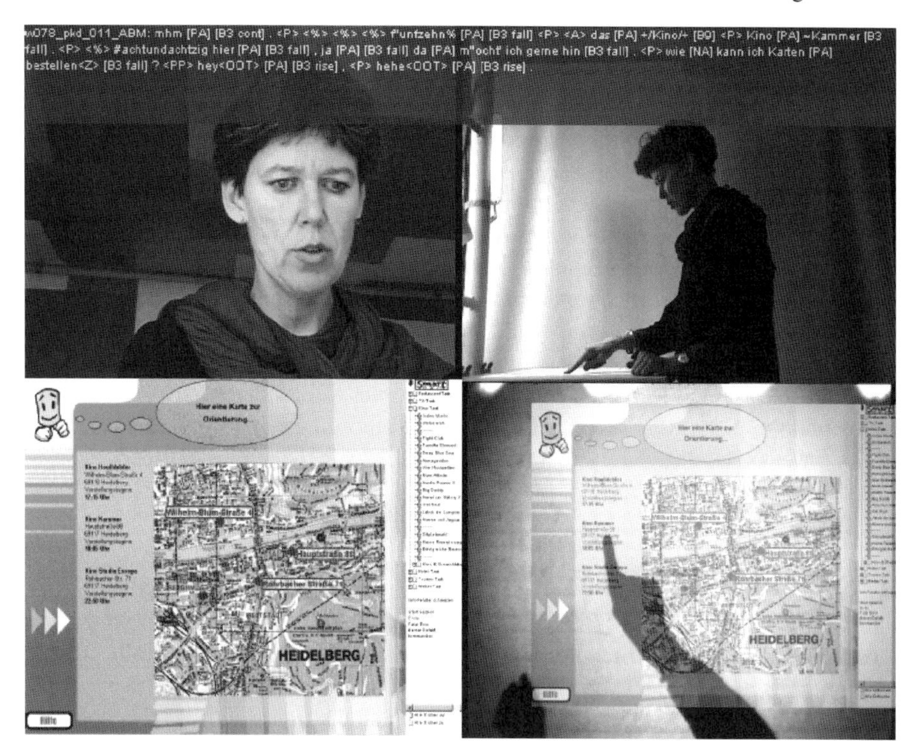

Fig. 6. Video stream used for gesture labeling

The framework Quicktime file is generated in the next step. It offers the possibility to choose among different views on the data files and to use these specialized views for playback of the recorded session. In the last step the DVD file structure is generated and additional files, for example, technical documents or useful tools for viewing the data, are copied.

3.8 DVD Master Production

Finally, all data are moved to the DVD-R station, an Apple Macintosh running DVD Toast by Adaptec. A master DVD is burned and verified; it then serves as source for duplication. As soon as the DVD is verified successfully and the master data on the server is archived into a *Tivoli* storage system (which is located in a different building to maximize security), all data are removed from the server. [11]

3.9 Data Logistics

The database mentioned earlier is a key element in organizing the recordings and the further processing of the data. Due to the fact that each recording makes up a data

---

[11] See Sect. 4.2 for details about the distribution of data.

amount of 4 GB (temporarily even up to 6.5 GB) and because up to 60 recordings are processed in parallel, the whole processing must be well planned.

The database reflects the processing scheme as indicated in Fig. 4 and gives up-to-date information about the status of each recording during its processing via a Web interface. Members of the team may check in certain tasks they perform, lock certain files while working on them and report the successful completion of a task. Furthermore, special observations or found bugs and faulty data can be reported here, session and speaker specifics entered and linked to session IDs. Finally, the database can be used as an automatic reporting generator not only for progress statistics but also for XML-rendered metadata, which are later included into the final distribution of the corpus.

In the final stage of SMARTKOM the file server offered approx. 2200 GB disc space and the processing time needed from the recording to burning the DVD was about 6 weeks. Unfortunately, these bottlenecks restrict the number of recorded test subjects to an average of 4 per week.

4 SmartKom Corpus

The final corpus was distributed continuously during the course of the project to the project partners. The Ministry of Education and Sciences (BMBF) as the principal funding source of the SMARTKOM project assigned the task of further maintenance and distribution of the corpus to the BAS. As in the former Verbmobil project, all data produced during the project phase should be freely available to interested scientists and engineers one year after the project internal release date. There are no restrictions with regard to the usage of the corpus except the redistribution to third parties. The corpus will also be available through the European Language Resources Association (ELRA).

The SMARTKOM Corpus consists of four parts:

- BAS SMARTKOM-Public (SK-P)
- BAS SMARTKOM-Home (SK-H)
- BAS SMARTKOM-Mobile (SK-M)
- BAS SMARTKOM Biometry (SK-B)

In the following we will give a short description of the first three parts containing the WOZ recordings and annotations as described above. [12]

4.1 Corpus Overview

The BAS edition of the SMARTKOM multimodal data collection contains the same data as the original project edition newly structured and validated against basic BAS guidelines. The following deviations to the original edition should be noted here:

[12] The SMARTKOM Biometry corpus will not be described here since it consists of an entirely different set of data.

- transliteration, annotations, lexicon and BAS Partitur files added
- additional documentation of formats
- revalidation against BAS standards
- partly extended to replace incomplete sessions

The size of the full corpus is approx. 1.4 TByte. It contains the recorded data of 233 users (92 male, 141 female). The basic structural unit is the recording session; the total number of sessions is 428 (P: 162, M: 139, H: 127). Each recording session contains all recorded modalities together with the annotations and metadata. The size of a single recording should be less than 4.5 GB to allow distribution on standard DVD-R. Sessions are stored on numbered DVDs (since the original corpus was distributed on DVD-R). Therefore each directory named "DVD-##.V" is the root directory to a "DVD" even if more than one DVD is stored on a medium.[13] If the corpus was distributed on a larger medium than DVD-R, there usually exist structures above the DVD directories that summarize contents in a more practical way. Please refer to the README on the top level of the medium for that.
Each session is named by a seven-character string:

```
TNNN_SD  where:   1st char
                  T = type of recording  b : Biometric data
                                         w : Wizard-of-Oz
                                         d : Demo session
                                         p : Test session
                                         e : evaluation
                                             session
```

Wizard-of-Oz is the standard case where the person is not aware that he/she is dealing with a human-controlled system.

A *demo session* is a fake recording to demonstrate the capabilities of the system; the person acts after a script.

A *test session* is a real recording but the person is cooperative and knows about the WOZ situation.

A *evaluation session* is the recording during the real usage of the prototype system either communicating with a naive person or with an expert.

Only the *Wizard-of-Oz* sessions are part of the BAS SMARTKOM edition; the remaining data are kept available at the BAS for interested parties.

```
2nd to 4th char
NNN = session number starting with 001
```

Session numbers start with 001 within each type. The numbering is not continuous; there might be missing numbers due to faulty recordings that were omitted before distribution.

```
5th char
'_'
```

[13] "##" = DVD number; "V" = DVD version number.

```
6th char
S = technical scenario p : Public
                       m : Mobile
                       h : Home

7th char
D = primary task       k : cinema
                       t : touristic
                           planning
                       f : TV guide
                       r : restaurant
                       n : navigation
                       v : VCR programing
                       m : music jukebox
                       a : phone
                       x : fax
```

In addition to the seven-character session name above, there exist the following name extensions to denote more specific parts of the session:

```
8th char : recording
           channel:      a : clip-on microphone, channel 1
                             Sennheiser KA 100 P XLR
                             placement: collar
                         b : clip-on microphone, channel 2
                             Sennheiser KA 100 P XLR
                             placement: collar
                         h : headset microphone
                             Sennheiser NB2
                         1-4 : microphone array 4 channels
                             Sennheiser KA 100 SP
                             placement: upper end of
                             interaction area
                         d : directional microphone
                             Sennheiser ME 66
                             placement: on face camera
                             distance to mouth: 60 cm
                         w : system output
                         p : playback backround noise front
                         q : playback background noise back
                         t : tableau coordinates
                         s : SiViT coordinates
                         i : infrared video of
                             interaction area
                         m : front capture camera (face)
```

```
l : left lateral capture
o : system display capture
```

Please refer to Türk (2000) for a detailed description of the technical setup.

```
9th char : '_'

10th - 12th char : turn numbering, starting with '001'

13th char : '_'

14th - 16th char : speaker ID
```

Extensions denoting turns or speaker IDs are only used to denote events in a turn of a session, e.g., within a labeling scheme. The SMARTKOM system always has the "speaker ID": "SMA".

The following extensions are used throughout the corpus:

```
.qt   QuickTime file
.trl  transliteration
.spr  speaker protocol file
.rpr  recording session protocol
.par  BAS Partitur Format file
.avi  video file AVI
.mov  video file DV
.ges  gestic labeling file
.ush  user state labeling file ('holistic')
.usm  user state labeling file ('mimic')
.trp  user state labeling file ('prosody')
```

For example, a session recording "w003_pk" consists of a set of files like what follows (when all channels are present):

```
Type                          Name

QuickTime frame *             w003_pk.qt
Front capture *               w003_pkm.mov
Lateral capture *             w003_pkl.mov
Infrared video                w003_pki.mov
System display output         w003_pko.avi
Audio clip 1 #                w003_pka.wav
Audio clip 2 #                w003_pkb.wav
Audio headset #               w003_pkh.wav
Audio Directional #           w003_pkd.wav
Audio array 1                 w003_pk1.wav
Audio array 2                 w003_pk2.wav
Audio array 3                 w003_pk3.wav
Audio array 4                 w003_pk4.wav
```

```
Audio system output *                w003_pkw.wav
Audio background front               w003_pkp.wav
Audio background back                w003_pkq.wav
Tableau coordinates                  w003_pkt.qt
SiViT coordinates                    w003_pks.qt
Recording session protocol *         w003_pk.rpr
Speaker protocol file *              AAB.spr
Transliteration *                    w003_pk.trl
Bas Partitur Files (BPF) *           w003_pk.par
User State Holistic $                w003_pk.ush
User State Mimic $                   w003_pk.usm
User State Prosody $                 w003_pk.trp
Gesture Labeling *                   w003_pk.ges
Turn Segmentation *                  w003_pk.mar
```

The obligatory files are marked with a "*". At least one of the audio channels a, b, h or d (marked with a "#") must be recorded. If the front video recording is present, the annotation files ush, usm and trp (marked with "$") must be present. If one of these files is missing, this is regarded as an error and listed in the 'Known Errors' section of the documentation.

4.2 Distribution

Because of the large size of the corpus (at least large at the time of production), the distribution of the whole corpus or even of the individual subcorpora Public, Home and Mobile cannot be done on regular DVD-5 media. Nevertheless, the BAS offers the distribution of selected sessions or of special extracts of modalities on DVD-5.

A full corpus will be distributed on IDE hard disc drives of 120 or 160 GB size each. The file system is FAT 32 to allow for easy mounting on all PC platforms. The total SMARTKOM corpus requires 11–12 hard disc drives. The data integrity of these drives is guaranteed for approx. 5 years. The end user should therefore copy the received data as soon as possible on a file server (with a backup facility) or — as an inexpensive alternative — insert the original drives into a server system.[14] The user may also copy parts of the corpus to DVD-5 at his/her facilities; to simplify this we structured all BAS SMARTKOM corpora into chunks smaller than 4.5 GB.

All symbolic information to the corpus, that is, all metadata, documentation and annotations as described below, are available on the BAS FTP server for free. There you will always find the latest version of these data. Since it is very likely that during the usage of the corpus some errors occur, the BAS strongly encourages the users of the SMARTKOM corpus to report these errors back to the BAS. Bug-fix messages will then be automatically sent to all known users, and the fixed data can be downloaded from the BAS servers.

[14] The data integrity lasts much longer if using the drives in a life server system rather than just storing the drives in an archive.

4.3 Signal Formats

Most of the file formats used in this corpus are widely used standards. The top QuickTime (QT) file has the extension ".qt" and summarizes all recorded data into one synchronized data stream. Individual streams are stored in standard format files (e.g., *.wav, *.mov, *.avi), which can be also accessed independently from the top QT file. The current QT player allows only one video track to be activated. You may switch from different views during playback. The same is true for the nine audio tracks. As an additional feature, the QT file contains specially compiled versions of the transliteration to display the spoken text in a subtitle frame. A chapter track allows users to jump to selected turns of the recorded dialogue. Each distributed medium of the corpus contains the publicly available QT library and player for Mac OS and MS platforms.

4.4 Metadata

For each recording session and each SMARTKOM user there exist metadata description files.

4.4.1 Recording Protocols

The recording protocols list all relevant parameters to a recording session and are stored in the file "TNNN_SD.rpr", where "TNNN_SD" denotes the session ID. The recording protocol is stored in XML and is mostly self-explanatory (English tag set). A DTD for the format can be found in the online documentation.

For example (excerpt only):

```
<?xml version="1.0" encoding="UTF-8" standalone="yes"?>
<!DOCTYPE RPR SYSTEM "rpr.dtd">
<RPR>

  <HEAD>
    <version number="2" subnumber="0" />
  </HEAD>

  <DVD no="163" />

  <SPEAKER id="ADB" />
  <SESSION-PARAMETERS>
    <session_id value="w160_mt" />
    <recorded_domains>
      <domain_planned use-case="Touristik" />
      <domain_recorded use-case="Touristik" />
    </recorded_domains>
    <atmosphere place="mobile" number="1" volume="soft" />
    <background pattern="Bunt1" />
    <pen mode="pen" />
```

```
    <emotions evoked="no" />
    <content_variation version="1" />
    <recording_date year="2001" month="08" day="23" />
    <recording_location value="LMU" />
    <recording_setup wizard="wizard3" />
    <experimenter name="Silke_Steininger" />
    <wizard_speech_output name="Ulrich_Reubold" distortion="soft
     distortion" />
    <wizard_navigation name="Ulrich_Reubold" />
    <session_sequence_no position="2" />
  </SESSION-PARAMETERS>

  <DATA-TRACKS>
    <VIDEO>
      <Data fieldname="(m)_mimic" device="Panasonic NV DX 100 EG"
       present="yes" />
```

4.4.2 Speaker Profile

Speaker profiles are stored in the file “AAA.spr”, where “AAA” denotes the three-character speaker ID. The speaker ID to a given session can be either obtained from the corresponding recording protocol, from the standardized turn marker or from the header of the corresponding BAS Partitur Format file. The speaker profile is stored in XML and is mostly self-explanatory. A DTD for the format can be found in the online documentation.

For example:

```
<?xml version="1.0" encoding="UTF-8" standalone="yes"?>
<!DOCTYPE SPR SYSTEM "spr.dtd">
<SPR>

  <HEAD>
    <version number="2" subnumber="0" />
  </HEAD>

  <SPEAKER>
    <Personal_Data speaker-id="AAJ" sex="F"
     date_of_birth="13.07.1940"
     height="164  cm" weight="60  kg" handed="right" />

    <School degree_state="BY (Bayern)"
     degree="Volksschule/Hauptschule"
     profession="Rentnerin/Rechtsanwaltsgehilfin" />

    <Languages mothertongue="DEU" mothertongue_mother="DEU"
      mothertongue_father="DEU" dialect="I3 (S\"udbairisch)"
      bilingual="no">
```

```
      <foreign_languages>
        <language value="ITA" />
      </foreign_languages>
      <bilingual_languages>
        <language value="none" />
      </bilingual_languages>
    </Languages>

    <Culture german_nationality="yes"
     cultural_environment="DEU" />

    <Experience speech_singing_training="no"
      computer_experience="yes"
      speech_dialogue_experience="no" />

  </SPEAKER>

  <RECORDING-SPECIFIC>
     <glasses exists="yes" />
     <smoker exists="yes" />
     <beard exists="no" />
     <piercing exists="no" />
     <jewels exists="yes" />
  </RECORDING-SPECIFIC>

  <COMMENTS>
VPN war zweimal da! Zweite Session: w335_mt und w336_mt
  </COMMENTS>

</SPR>
```


4.5 Annotation Formats

The SMARTKOM corpus contains several annotation layers: transliteration, 2D gesture, user state holistic, user state face and prosodic segmentation. Please refer to the contribution by Steininger et al. (2006) for a more detailed discussion of the problem on how to annotate multimodal data. In the following we will merely give a short formal description of the formats used for the different annotation types.

4.5.1 Transliteration

Input: audio channel

This file type contains the orthographic transliteration of spontaneous spoken language. That includes, for instance, information like interruptions within phrases, revision and repetition of utterances, reduction or hesitation. There are symbols for the following categories:

1. lexical units
2. syntactical–semantical formation
3. nonverbal articulatorical utterances
4. noise
5. pauses
6. acoustical interaction (speech overlap)
7. comments
8. special comments
9. prosodies

The audible events are recorded orthographically only. There is no phonological transliteration, no phonetical transcription and no temporal correlation with the signal.[15]

The format is based on the well-known Verbmobil II transliteration format (TR2) extended by markers for prosodic event, for barge-in, for off-talk as well as some methodological adaptations. Therefore this type of transliteration is denoted by "TRS" (transliteration SMARTKOM) in the BAS Partitur Format to distinguish it from other formats.

A detailed description of the conventions can be found at our website.[16] The transliteration is stored in a file with extension *.trl with one turn per line.

4.5.2 2D Gestures

Input: audio channel, facial video, lateral video, system output overlapped with infrared video of interaction area

This file type contains information on gestures that occurred in the session. All gestures that occur within the borders of the SiViT camera field are labeled. Additionally, emotional gestures that occur elsewhere are labeled. The following label information is assigned to each found gesture:

- begin of the gesture (onset) in samples
- duration of the gesture in samples
- category of the gesture
- label of the gesture
- hand and finger used
- reference word
- reference zone on display
- reference object
- begin of the stroke in samples
- duration of the stroke in samples
- optional: comments about the gesture

The data fields are tab-separated and stored in a BPF tier file with extension *.ges.

[15] However, a temporal correlation can be obtained by using the segmental information in the MAR files (see below) or by using the information in the TRN tier, respectively.

[16] http://www.phonetik.uni-muenchen.de/Forschung/SmartKom/Konengl/engltrans

4.5.3 Turn Segmentation

Input: audio channel

Since all recording contain the data of a complete dialogue of approx. 4.5-min length, it is useful to segment this data stream into shorter units called turns. The turn is defined in the transliteration conventions to SMARTKOM (see our website[17]). This information is stored in so-called "marker files" (*.mar).

Marker files contain the information when a turn of the dialogue starts and ends in the signal files as used in the transliteration. Each line contains three entries separated by white spaces:

```
Start of turn            End of turn            Turn marker
```

Start and end are given in samples (16 kHz) from the beginning of the file (0). The turn marker reads as follows: XXX_A_B, where "XXX" is the speaker ID; "A" is the turn number and "B" is the channel number (channel 1 is human; channel 2 is machine). For example:

```
47232 187264 SMA_1_2
205312 249472 ULT_2_1
...
2170240 2190080 SMA_19_2
```

4.5.4 Prosodic Labeling for User State

Input: audio channel

The metalinguistic prosodic annotation is part of the so-called user state annotation. User states are positive or negative attitudes of users towards a certain system in human-computer interaction. The labeled prosodic events are part of the input to the user state classifier that works exclusively on the audio signal.

The files *.trp contain the orthographic transliteration of spontaneous spoken language as described above modified in the following way:

1. Elimination of the following tags:
   - linguistic prosodic features (labels enclosed in '[]')
   - noise (tags beginning with '#' or '<#>' or bracketing with those)
   - pauses ('<P>')
   - overlapped speech (word modifiers '...1@' or '@1...')
   - special comments ('<;...>')
   - tildes for marking names ('~..')
   - words hard to understand ('...%')
2. Added the annotation of the following metalinguistic prosodic features
   - PAUSE_PHRASE: irregular pause on a phrasal level

[17] http://www.phonetik.uni-muenchen.de/Forschung/SmartKom/Konengl/engltrans

- PAUSE_WORD: irregular pause on a word unit level
- PAUSE_SYLL: irregular pause on a syllable level
- LENGTH_SYLL: lengthening of a syllable
- EMPHASIS: emphatic accentuation
- STRONG_EMPH: stronger emphatic accentuation
- CLEAR_ART: clear articulated speech
- HYPER_ART: hyperarticulated speech
- LAUGHTER: speech overlapped by laughter, sigh or the like

More than one label per word is allowed and a feature occurring at several adjacent words is marked at each word by a separate label.

3. Added information about durations in ms, surrounded by curved brackets: {`%snd:SOUNDFILENAME_begin_end`}, where `%snd` stands for "sound" and `_begin_end` are given in msec from the beginning of the sound file `SOUNDFILENAME`. This segmental information is used in the following way, depending on the annotated label:
   - PAUSE_PHRASE, PAUSE_WORD: duration of the word before the pause plus the pause duration
   - PAUSE_SYLL: duration of the word with a pause between its syllables
   - LENGTH_SYLL: duration of the word with a lengthened syllable
   - EMPHASIS: duration of the word with an emphatic accentuation
   - STRONG_EMPH: duration of the word with a stronger emphatic accentuation
   - CLEAR_ART: duration of the clear articulated word
   - HYPER_ART: duration of the hyperarticulated word
   - LAUGHTER: duration of the overlapped word

The various labels are placed after the word that reveals the annotated phenomena; of course, pause label are placed between the relevant units. See Steininger et al. (2000) for more detailed information about this annotation.

4.5.5 Holistic Labeling of User State

Input: facial video, audio channel

The file type *.ush contains an segmentation and labeling of user-states (interesting emotional and cognitive states) that occurred in the session. The whole session is segmented that is there are no gaps within the segmentation.

For each segment begin (second column) and duration (third column) are given in samples (16 kHz) starting from the begin of recording (0). Each segment is assigned to one of the seven labels (third column) below (label names in square brackets):

- neutral [Neutral]
- joy/gratification (being successful) [Freude/Erfolg]
- anger/irritation [Ärger/Mißerfolg]
- helplessness [Ratlosigkeit]
- pondering/reflecting [Überlegen/Nachdenken]

- surprise [Überraschung/Verwunderung]
- unidentifiable episodes [Restklasse]

The labels are assigned with respect to the impression of the labeler. Not only the facial expression but also the voice quality or other contextual information is considered. Only the use of words with emotional content, but without an emotional expression is *not* considered as an indicator of a respective emotion/user state.

The intensity of a user state is labeled in the forth column for the classes 2–6 by:

- strong [stark]
- weak [schwach]

In parallel to the user-state labels the *.ush files also contain a segmentation of occlusions when the hand or pen occlude parts of the face. This segmentation is not without gaps and is overlapped to the segmentation of user states: [18]

- hand over face [Hand im Gesicht]
- hand over face/mouth [Hand im Gesicht/Mund]
- hand over face/nose [Hand im Gesicht/Nase]
- hand over face/eyes [Hand im Gesicht/Augen]
- pen over face [Stift im Gesicht]
- pen over face/mouth [Stift im Gesicht/Mund]
- pen over face/nose [Stift im Gesicht/Nase]
- pen over face/eyes [Stift im Gesicht/Augen]
- face partly not in the range of recording [Teilweise nicht im Bild]
- other object in face [Objekt im Gesicht]

For more detailed information please refer to Steininger et al. (2002a), Steininger et al. (2002b) or Steininger et al. (2000).

4.5.6 Mimic Labeling of User State

Input: facial video, USM annotation without labels

The file type *.usm contains information on user states that were segmented and labeled *without the audio channel.*

The file structure and the labels used are identical to the *.ush annotation described before. Only the labeling procedure differs: A different labeler watches the video face recording again and assigns new names to the segments. The information that is available for the segmenter from the previous USH labeling process are:

- the neutral segments
- the segments with occlusions of the face
- the time information but not the labels of all other user state events

[18] In the BAS Partitur Format (Schiel et al., 2002) that summarizes all annotations in a common format this segmentation has a different tier name (OCC) than the user state segmentation (USH).

The segmenter is advised to ignore the neutral and occlusion segments and to label the remaining user-state segments anew. He may also assign the borders of the segments anew (if applicable).

Acknowledgments

The work described here was carried out by a rather large group of scientists and students at the Ludwig-Maximilians-Universität München stemming from a variety of scientific fields such as electrical engineering, psychology, phonetics, computer linguistics, computer science, phonology, choreography, and music. Specifically, we would like to thank Silke Steininger, Nicole Beringer, Susan Rabold, Olga Dioubina, Daniel Sonntag, Ute Kartal, Hannes Mögele, and approx. 40 students for their valuable contributions to the SmartKom Corpus.

References

N. Beringer, V. Penide-Lopez, K. Louka, M. Neumayer, U. Türk, and C. Grieger. Wizard-of-Oz Display. Technical Document SmartKom 12, Bavarian Archive for Speech Signals (BAS), 2000.

T. Maremaa and W. Stewart. *QuickTime for Java: A Developer Reference*. Morgan Kaufmann, San Francisco, CA, 1999.

F. Schiel, S. Steininger, N. Beringer, U. Türk, and S. Rabold. Integration of Multi-Modal Data and Annotations Into A Simple Extendable Form: The Extension of the BAS Partitur Format. In: *Proc. Workshop "Multimodal Resources and Multimodal Systems Evaluation"*, pp. 39–44, Las Palmas, Spain, 2002.

S. Steininger, O. Dioubina, R. Siepmann, C. Beiras-Cunqueiro, and A. Glesner. Labeling von User-State im Mensch-Maschine Dialog — User-State-Kodierkonventionen SmartKom. Technical Document SmartKom 17, Bavarian Archive for Speech Signals (BAS), 2000.

S. Steininger and U. Kartal. Aufnahmeplan Wizard-of-Oz Aufnahmen SmartKom. Technical Document SmartKom 5, Bavarian Archive for Speech Signals (BAS), 2000.

S. Steininger, F. Schiel, O. Dioubina, and S. Rabold. Development of User-State Conventions for the Multimodal Corpus in SmartKom. In: *Proc. Workshop "Multimodal Resources and Multimodal Systems Evaluation"*, pp. 33–37, Las Palmas, Spain, 2002a.

S. Steininger, F. Schiel, and A. Glesner. Labeling Procedures for the Multi-Modal Data Collection of SmartKom. In: *Proc. 3rd Int. Conf. on Language Resources and Evaluation (LREC 2002)*, pp. 371–377, Las Palmas, Spain, 2002b.

S. Steininger, F. Schiel, and S. Rabold. Annotation of Multimodal Data, 2006. In this volume.

U. Türk. Technisches Setup bei den SmartKom-Aufnahmen. Technical Document SmartKom 7, Bavarian Archive for Speech Signals (BAS), 2000.

Annotation of Multimodal Data

Silke Steininger, Florian Schiel, and Susen Rabold

Bavarian Archive for Speech Signals
c/o Institut für Phonetik und Sprachliche Kommunikation
Ludwig-Maximilians-Universität Münchenn, Germany
{steins,schiel,rabold}@bas.uni-muenchen.de

Summary. Do users show emotions and gestures if they interact with a rather intelligent multimodal dialogue system? And if they do, what do the "emotions" and the gestures look like? Are there any features that can be exploited for their automatic detection? And finally, which language do they use when interacting with a multimodal system — does it differ from the usage of language with a monomodal dialogue system that can only understand speech?

To answer these questions, data had to be collected, labeled and analyzed. This chapter deals with the second step, the transliteration and the labeling.
The three main labeling steps are covered: orthographic transliteration, labeling of user states, labeling of gestures. Each step will be described with theoretical and developmental background, an overview of the label categories, and some practical advice for readers who are themselves in the process of looking for or assembling a coding system. Readers who are interested in using the presented labeling schemes should refer to the cited literature — not all details necessary for actually using the different systems are presented here for reasons of space. For information on the corpus itself, please refer to Schiel and Türk (2006).

1 Transliteration of Speech

The transliteration of spontaneous speech is a tricky problem. Spontaneously speaking people tend to produce a lot of events that are not actually speech: Looking at a transcript one sometimes gets the impression that a speaker actually produces more nonlanguage events than actual language events. One can assume that the ratio of well-pronounced speech to reduced or unconventional speech and nonspeech events is different for different situations of spontaneous speech. For example, a person giving a talk pays much more attention to proper pronunciation and comprehensibility and tries to suppress coughs, hesitations and similar events. A person who is talking rapidly to a friend, perhaps joking, perhaps even eating, simultaneously represents the other extreme: One can expect a lot of neologisms, reductions, interruptions, etc. The same will be the case for emotional speech. The situation in SmartKom is a case in-between the two of these extremes: The speaker will probably only concentrate to produce well-pronounced speech in a small part of the dialogue and show lots of the events that are characteristic for everyday, spontaneous speech. A transcription system has to be able to capture all the peculiarities of such casual speech. Fortunately, spontaneous speech has been transliterated for many years, therefore much experience about the to-be-expected problems has been accumulated and a wide variety of transcription systems is available.

We decided to use the Verbmobil conventions to be consistent with the Verbmobil corpus and because our requirements were very similar to the ones in Verbmobil. Thus we adapted the Verbmobil conventions, most notably was the addition of prosodic labels. For the complete conventions please refer to our website [1] or the German version in Beringer et al. (2000).

2 Labeling of the User States

2.1 Overview User-State Labeling

This section covers the most interesting points of the user-state labeling in SMARTKOM. The labeling process comprised three separate procedures: holistic labeling of the data, labeling of the facial expression without audio information and prosodic annotation. The first two steps were specially designed for the requirements in SMARTKOM. The third step was added by request of one of the developer groups and comprised a coding system that was used in Verbmobil. First, we explain which requirements we wanted the coding systems to fulfill and how these were translated into the coding conventions for holistic and facial expression labeling. After that the basic principles and categories of these two steps are outlined. Problems and observations that could prove useful to future emotion-labeling endeavors are discussed afterwards. A brief description of the prosodic annotation concludes the section.

2.2 Requirements of the User-State Labeling

The labeling of emotions or user states[2] in SMARTKOM had to serve two main goals:

1. the training of recognizers
2. the gathering of information on how users interact with a multimodal dialogue system and which user states occur during such an interaction

The questions that have to be solved to detect user states automatically are: Which features of the face and of the voice contribute to an emotional impression? Which of these features can be detected automatically?

If we already knew the answers to these questions, it would make sense to define coding conventions that mark these features in the data (for example, with a formal or structural coding system that labels morphological features), compare FACS by

[1] http://www.phonetik.uni-muenchen.de/Forschung/SmartKom/Konengl/engltrans

[2] The name "emotion labeling" was changed to "user-state labeling" because the targeted episodes in the data comprised not only emotional but also cognitive states.

Ekman and Friesen (1978). But since we are far from answering these questions conclusively, we had to work the other way round.

Our "other way round" technique amounted to the following: We decided to define the user states with regard to the subjective impression that a human communication partner would have, if he would be in place of the SMARTKOM system. A human in a conversation with another human is able to judge which emotion or user state his communication partner shows. Of course, he does not know which emotion is "truly" present in his communication partner. But this is not necessary for him to use the emotions he perceives as clues for managing the dialogue: going on if his partner seems to be in a good mood, stopping and intervening if his partner shows signs of a negative mood and offering help if his partner seems confused. A misjudgment would not be seen as problematic (in most cases), but ignoring an expressed (negative) emotion in many cases would lead to irritation. If a communication partner therefore can detect user-states that are relevant for the communication process — then a labeler should be able to do the same — and the user states that he detects would be the ones that are relevant for the system. It would not be too bad if the detected user state was not present (it would only be bad if this happened too often). Missing a user state (with a function in the dialogue), on the other hand, would be less desirable. Our first requirement therefore was:

- The label system should target user states that were perceived subjectively and were regarded as important for the dialogue. This is a functional[3] approach; we look for the user states that serve a certain function in the dialogue.

In addition to the task of marking the relevant user states to fulfill the two goals mentioned above, the coding system had to satisfy some additional, mostly formal requirements. They overlap to a large part with the requirements for transliteration and gesture labeling:

- The labels should be selective.
- The coding system should be fast and easy to use.
- The resulting label file should facilitate automatic processing (a consistent file structure, consistent coding, nonambiguous symbols, ASCII, parsability) and preferably should be easy to read.
- The main categories and most of the modifiers should be realized as codes and not as annotations, in order to heighten consistency. Annotations (free comments and descriptions that do not follow a strict rule) are more flexible, but codes (predefined labels from a fixed set) increase the conformity between labelers.

These considerations were the starting point for the development of the coding systems for holistic and facial expression labeling, which will be outlined in the next subsection.

[3] "Functional code" or "functional unit" is sometimes defined differently by different authors. We use the term in accordance with Faßnacht (1979) for a unit that is defined with regard to its effect or its context.

2.3 Development of the Coding System for the User States

We had no data on emotions that could be expected in a human–machine dialogue, therefore we picked those that seemed interesting or relevant (for us or for the developers): "anger/irritation," "boredom/lack of interest," "joy/gratification (being successful)," "surprise/amazement," "neutral/anything else." A selection of sessions was labeled with these categories. After the labeling the categories were discussed. "Boredom/lack of interest" was excluded because it could not be distinguished from "neutral." "Neutral" and "anything else" were separated into two different categories because many sequences were found where the users definitely did not show a neutral expression, but no meaningful label could be given. "Anything else" was renamed an "unidentifiable episode." Two new categories were included to describe user states that occurred quite often in the data and are important in the context of human-computer interaction: "helplessness" and "pondering/reflecting." In a second step with the new (final) set of categories all sessions from the pretest were labeled again, and after that all other sessions. Since the goal was to find detectable features, for each category some morphological characteristics were listed. Please note that these descriptions were assembled relatively late in the process of labeling the data. We started in each category only with the "simple" instruction to the labeler: "Mark a sequence as 'joy/gratification' if you have the impression that the subject is in a positive mood, enjoys himself, is visibly content, amused or something similar." Only after a considerable amount of data was labeled were the formal criterions were extracted and written down. The goal was to enhance the consistency of the labeling system but at the same time to be true to the goal to label the subjective impressions, not existing formal categories.

With the holistic labeling system we were relatively sure to be able to label all relevant user-state episodes. However, the developers criticized that the coding system was not very well suited for generating training material for user-state recognition of facial expressions. Because of the holistic approach, the labels included not only information from the facial expression, but also information from the voice and from the context. This was seen as problematic since a facial expression recognizer obtains information only from the facial expressions and a prosody recognizer obtains information only from the voice. Therefore, we included two additional labeling steps: labeling of the facial expression without the audio information and prosodic labeling without the video channel.

For the facial expression labeling a different labeler group watched the videos without audio. The labelers started with a presegmented file (from the holistic labeling). This presegmentation was derived from the holistic labeling where the names of the categories (apart form "neutral") were deleted, while the borders were retained. We decided to start with the holistic step because humans perceive emotions in a holistic way. Not only do we include visual, auditory and semantic information, we even add assumptions about the probable emotional state of our communication partner and then add these assumptions to our perceptions (Ratner, 1989). For this reason it can be assumed that the judgments about perceived emotions would differ in cases where we have a more natural, holistic situation (watching a video, although in a video, of course, some cues from a real-life situation are missing) and a more unnatural, deprived situation, such as watching only the face without audio (Wallbott, 1988; Munn, 1940; Sherman, 1927). Since we did not use a formal system, but one that had the goal to label subjective categories, the labelers could not be trained to look only for the cues for the category (concentrate only on certain movements, for example). Their task was a holistic one: to label a "positive mood" or a "negative mood." Therefore we assumed that their emotion detection and labeling ability would decrease in a situation with as little information as possible. By putting the holistic step before the nonauditory step we hoped that subtle episodes would not be missed.

Since the facial expression crew had presegmented label material, they knew there should be something and could verify if they actually could decide upon an emotion or not. In some of the cases, they could not perceive an emotion in the segment delivered from the holistic crew. In these cases they recorded "neutral." Conceivably these were the cases where the emotional information was only in the voice or in the context, or the emotion was too weak to be visible in a context-deprived situation. The results of the two steps were linked – the facial expression crew did not start from scratch but got the hint that there "could be something." This generated an expectation that could lead to the perception of an emotion where none would have been seen without this expectation. We decided to rather live with these errors rather than to have a wholly independent video-only labeling step because of practical considerations (the time would have doubled for the user-state labeling), but also for theoretical ones:

1. As already mentioned, without context information judgments about emotions can be blurred (see references above).
2. Because of a low inter-rater reliability for emotional judgments, two wholly independent steps would perhaps have had results that differed so much that no conclusions would have been possible about what is different between holistic and purely visual labeling.

We therefore chose an intermediate step and hoped to get some clues about which material is better suited for training and how the emotional judgments differed with regard to the input information.

Additionally, on request of one of the developer groups, we adopted a formal coding system that was used in Verbmobil (Fischer, 1999) and changed it to suit our needs in SMARTKOM. The labels were chosen according to the requirements for the user-state recognition group in SMARTKOM and were thought to represent prosodic features that are indicative of emotional speech. By the comparison between the holistic labeling and the prosodic labeling we hoped to detect relevant user states in speech. For more information on the usage of prosodic features as indicators of emotional speech, please refer to Batliner et al. (2000). For more information on the development of the user-state label systems, please refer to Steininger et al. (2002a). In the following sections the label procedures for the three different steps are described briefly together with difficulties and proposals for future research.

Fig. 1. Example of the front view that was used for the holistic and the facial expression labeling. The picture was taken from an episode that was labeled as "joy/success" in the holistic labeling step

2.4 Holistic Labeling of User States

The first step during user-state labeling was the so called "holistic labeling". A labeler watched the video of a session and marked each change in the state of the user. The segments that were found in this way were then assigned one of seven labels:

- neutral [Neutral]
- joy/gratification (being successful) [Freude/Erfolg]
- anger/irritation [Ärger/Mißerfolg]
- helplessness [Ratlosigkeit]
- pondering/reflecting [Überlegen/Nachdenken]
- surprise [Überraschung/Verwunderung]
- unidentifiable episodes [Restklasse]

The allocation of the labels was done with regard to the overall, subjective impression and took into account the facial expression, the quality of the voice, the choice of words and the context. However, the sole usage of anger-implicating words that were uttered without any emotional expression were not taken as an indicator for anger.

Additionally the label was given a rating with regard to the intensity of the user-state ("weak" or "strong"). This rating was not absolute, but referred to the general strength of emotion a certain subject showed. An individualized referencing like this poses problems for the training of recognizers. However, there is no way to define weak and strong emotions absolutely. Additionally the impression of weak and strong is dependent on the observer, and he or she also has no absolute sense for the level of a perceived emotion. This has to be taken into account when interpreting results that refer to the weak and strong differentiation. Sequences during which

Fig. 2. Example for an episode that was labeled as "anger/irritation" in the holistic labeling step

the face was partly occluded by the hand/s of the subject were marked, as well as sequences during which the face could be seen only partly in the camera picture. Consistency was enhanced by two correction steps, one of them a final correction by the same corrector for every session. Difficult episodes were discussed.

2.5 Facial Expression Labeling of User States

After the holistic labeling a second labeler group (that was not involved in the holistic labeling) watched the same video without the audio information. The group used the same categories as the holistic crew (without the auditory features). As described above, the labeler was informed about the segment borders and the occurrence of "neutral" segments, which he or she could ignore. Because of the different perception situation the labelers sometimes had a different impression about the borders of an emotional segment. In these cases they could adjust the markers for beginning and end of a segment.

2.6 Categories of Holistic and Facial User-State Labeling

Please note that the categories in the two steps were the same, but in the facial expression labeling the descriptions for features in the voice were not used.

Joy/Gratification (Being Successful)

This label was given if the labeler had the impression that the user was in a positive mood, enjoyed him- or herself, was visibly content, amused or something similar. Amusement that (obviously or probably) stemmed from derision, mockery or

Fig. 3. Example for an episode that was labeled as "helplessness" in the holistic labeling step

something similar was still labeled as joy. We made this decision because sarcasm, derision, etc., are exceedingly hard to detect reliably automatically.

Anger/Irritation

The labeler marked this label if he had the impression that the user was in a negative mood, was visibly not content, was irritated, annoyed, exasperated, angry, disappointed or something similar. Most cases of this category were weak: Full-blown anger almost never showed up. In many cases users showed their anger rather "politely" (sometimes even smiling!).

Helplessness

This label was chosen if the labeler had the impression that the user was helpless, confused or interrogative. The strong cases comprised helplessness and distress: The user did not know what to do and/or had no idea how to go on. The weak cases comprised episodes in the dialogue where the user was puzzled, a bit confused, wanted to know something or had no concrete plan how to go on.

Pondering/Reflecting

If the labeler had the impression that the user was thinking hard, "pondering/reflecting" was chosen. The task of the subject led to the fact that for the most parts of the session he watched the display, concentrated, read or searched the display. This behavior was not labeled as "pondering/reflecting". There had to be visible or audible indicators, like biting the lips.

Fig. 4. Example for an episode that was labeled as "pondering/reflecting" in the holistic labeling step

Surprise

This label was given if the labeler had the impression that the user was surprised. There was a fast movement in the face (in most cases of the eyebrows) or an abrupt vocal reaction in response to an external stimulus.

Neutral

If no emotional or cognitive state in the face or the voice could be detected by the labeler, "neutral" was marked, as well as in cases where an emotional or cognitive state was so faint that the assignment of a label seemed inappropriate.

Unidentifiable Episodes

This label was chosen if the user was neither "neutral" nor any of the other labels could be assigned to the episode. Three cases could be discriminated:

1. grimaces with no emotional content, for example, playing with the tongue in the cheek, twitching muscles, etc. (about 65%)
2. emotional sequences that had no label in our system, for example, "disgust," (about 5%)
3. states that seemed to have an emotional or cognitive meaning, but could not be decided upon by the labelers (about 30%)

For additional information on the user-state labeling please refer to Steininger et al. (2002b). A complete description of the conventions can by found in Steininger et al. (2000).

2.7 Inter-Rater Agreement in Holistic User-State Labeling

To work out the interlabeler agreement, 20 dialogues were labeled twice by different, nonoverlapping groups of labelers. The accuracy and correctness values were calculated as were the single label agreements. The dialogues for the interlabeler agreement probe were chosen by chance. The first labeling was done in the years 2001 and 2002. It took place during the normal processing of the data of the Wizard-of-Oz recordings. The second labeling was done especially for the calculation of the interlabeler agreement. It took place in 2003. The procedure was exactly the same as in the first labeling with the exception that no end correction was done. This was because only one person was responsible for the end correction during normal labeling to keep the consistency high, and this person could not take part in the interlabeler agreement study without influencing the results. All labelers were experienced, with the exception of some of the labelers in the second labeling, who had only medium labeling experience.

Within each category the labels of 20 files (labeling A) were compared to the corresponding labels of their 20 equivalent files (labeling B). Since the labeling was not a discrete process but the segmentation of a continuous data stream, labels from the two files were matched with regard of their time range. The possible hits and mistakes were:

Label A matched label B, same label:	correct
Label A matched label B, different label:	substitution
There was no match for label A in the B-file:	deletion (missing label)
There was no match for label B in the A-file:	insertion (additional label)

The third case could happen if, for example, labeler B marked one long label and labeler A marked the first part of the long label, but inserted a short label afterwards. If the two longer parts matched (with regard to time), they were compared. For the short label a "deletion" was marked, since labeler A perceived a new user state, whereas labeler B did not:

A-File XXXXXX YYY ZZZZZZZ
B-File XXXXXXXX ZZZZZZZZZ

The fourth case could happen, for example, if labeler A marked a long label and labeler B marked a label that matched the long A label (with respect to time) but ended earlier as the label in A. Labeler B then marked a short label that had no match (with respect to time) to a label in file A. In this case an "insertion" was marked:

A-File XXXXXXXX ZZZZZZZZZ
B-File XXXXXX YYY ZZZZZZZ

Note that a "deletion" during the comparison of file A with file B would be counted as an "insertion" during the comparison of file B with file A. Since neither file could be regarded as the "correct" file because no external reference existed, deletions and insertions actually described the same case: One labeler marked an additional label that the other labeler overlooked or did not deem strong enough to warrant an extra label.[4]

The percentage of correctness was based on the number of correct labels divided by the total number of labels, multiplied by one hundred. The percentage of accuracy was based on the total number of labels minus substitutions, minus deletions, minus insertions divided by the total number of labels and multiplied by hundred.

The reliability turned up to be not very good: 55% matches were correct and the accuracy was 1.95%. This was caused by twice as many insertions/deletions and substitutions as concordant labels. The large number of insertions/deletions came from the fact that the second label group set 763 labels in 20 dialogues, whereas the other group set only 494 labels to categorize the user states within the same dialogues. When matching segmentally more than one fourth of the segments were judged differently in the two labelings. The categories "joy/gratification" and "pondering/reflecting" had the largest number of concordant user state labels. A comparison of only the labels that were rated "strong"[5] showed an accuracy of 36.6%, with 51% correct matches. This much higher rate was caused by a similar number of assigned labels during labeling A and labeling B (37 vs. 46) and by a higher number of label matches related to substitutions, insertions and deletions. Again, "joy/gratification" and "pondering/reflecting" showed the highest concordance. For detailed information about the inter-rater analysis, please refer to Jacobi and Schiel (2003).

Correct matches differed strongly with respect to label:

neutral	41%
joy/gratification	47%
anger/irritation	14%
helplessness	25%
pondering/reflecting	34%
surprise	8%
unidentifiable episodes	20%

The results of the inter-rater agreement will be discussed after the next section, which describes the inter-rater agreement results for the nonauditory user-state labeling.

2.8 Inter-Rater Agreement Facial User-State Labeling

The procedure for the analysis of the inter-rater agreement for the labeling without audio information was exactly the same as for the holistic labeling (see above). The results were much better than for the holistic labeling group: 59% of the labels were correct and the accuracy was 45.4%. The concordant labeling of "neutral" segments was not included in the calculation because they were in large part predetermined by

[4] Both files were compared to each other, and the correctness and accuracy were computed for both cases and then divided by two.

[5] A label was chosen for the comparison if the rating "strong" was given in at least one of the labelings.

the holistic labeling group. Apart from "joy/gratification," the variance was overall fairly high. If comparing the labels where at least one was rated "strong," the accuracy climbed to 64.9%. "Joy/gratification" and "pondering/reflecting" were in this case fairly reliable with regard to their hit results. The label "helplessness" showed the highest spreading, respectively the worst reliability.

Correct matches differed strongly with respect to label:

joy/gratification	82%
anger/irritation	59%
helplessness	40%
pondering/reflecting	59%
surprise	33%
unidentifiable episodes	55%

For detailed information about the inter-rater analysis please refer to Jacobi and Schiel (2003).

2.9 Problems and Suggestions for the Future

With regard to one of our goals — labeling training material for recognizers — the inter-rater results described above are disappointing. Only two of the categories were matched correctly relatively often, but even the category that was detected best ("joy/gratification") was labeled as something different more than half of the time by the other group. The missing end correction step certainly increased the number of errors, but it cannot explain the whole range of differing judgments. Apart from the missing end correction we assume, the following main reasons for the large difference between the two labeler groups: First, the emotions in the data mostly were subtle emotions. Obviously, it is very difficult to detect and judge such fine changes in facial expressions and voice. This assumption is strengthened by the finding that the emotions that were rated "strong" showed a higher inter-rater agreement.
Second, the labelers in the holistic group had to make two decisions:

1. Is there an emotional segment, respectively, where does an emotional segment start/end?
2. What is the content of the segment?

This doubly ambiguous situation could lead to more errors. In fact, the second group coded 50% more segments than the first group. The groups therefore differed not only in their judgments with respect to the content of a segment but also with respect to the question if there was an emotion to be labeled. Perhaps the second group used a high criterion (willingness to say "signal there" in an ambiguous situation, compare with signal detection theory (Egan, 1975)) and the first group used a low criterion, which means both labeler groups had a different response bias. Third, the categories with the best results ("joy/gratification" and "pondering/reflecting") were the ones with the highest homogeneity with respect to morphological features. Notably the lips curved upwards ("joy/gratification") and chewing on the lips ("pondering/reflecting") could be observed in most of the cases. The other emotions differed a lot between the subjects (with the notable exception of surprise, which subjectively seemed relatively homogeneous). Probably only emotions that are shown in a relative similar way by different persons evoke a comparable categorization by different observers. Emotions with a wide range of possible realizations (displays), like polite anger, seem to be judged very differently by different observers.

The inter-rater results for the group without auditory information are very interesting if compared to those of the holistic group. Obviously, the task was much easier (assuming that the skill of the two labeler groups was roughly the same, which seems probable because they had equal training). Is the judgment of emotions perhaps easier without (perhaps confusing) context? This would contradict the subjective impression of most labelers that the task was more difficult than with context and also some studies about this subject (Wallbott, 1988; Munn, 1940; Sherman, 1927). We think that the main reason for the better reliability results was that the labelers had to make only one decision: the content decision (is there an emotion, and if yes, which one). The holistic group, as mentioned above, had to make two: one about the boundaries (beginning/end) and one about the content. Actually, many errors in the holistic group were caused by setting differing boundaries ("insertions" and "deletions," see above). In the holistic group 79% of all errors were substitutions/deletions (which actually included two mistakes: differing content and differing segment borders). Substitutions (21% of all errors) included only one mistake: the content. In the facial expression group 57% of all errors were substitutions, 43% were deletions/insertions (which could occur despite of largely predetermined segment boundaries, for example, if a segment was cut in two by the first group and left whole by the second group). Segmentation of a "user-state stream" obviously was as difficult (perhaps even more so) than assigning a label to a segment. Therefore, if a reliable way of segmenting a user-state stream can be found, perhaps even subtle emotions can be labeled relatively reliable.

With these considerations in mind, we propose the following for future attempts to label emotions:

- Try to make sure that all labelers use the same response bias. Since for the labeling of emotions no external validation is possible,[6] this can be achieved only in instructing the labelers to use one of the extremes: "Be certain to label all sequences where you perceive an emotional change," resp. "Only label the sequences where you are absolutely certain that an emotional change has occurred." The second approach would be the more practical, since labelers would ignore many weak emotional changes, which are hard to judge, as we have seen in our data. Of course, the response bias still would not be the same for all labelers, but it can be assumed that the biases would have a smaller variation.

[6] Asking the subjects what they "truly" felt would not help, because first, asking them afterwards results in judgments affected by memory and social response bias, and second, we do not want to label the emotions that the user feels, but the ones that are observed by a communication partner, which is something different.

- Alternatively to the instruction to only label sequences where the labeler is absolutely sure, one could introduce a rating for certainty. In this way data could be collected about the reliability, with respect to the certainty of the judgments.
- With respect to our finding that the labeling of emotions that were rated "strong" had a much higher reliability, one could argue to try to evoke stronger emotions in subjects and then label these. Although this is an interesting approach, it does not solve the general problem that everyday emotions are subtle. If we want to study them we have to accept that they subjectively are categorized very differently and have to look for the reason for this. Another point against evoking stronger emotion is that the validity possibly suffers. Provoked emotions are not the same as strong emotions that show up "by themselves."
- To increase the reliability of segmenting a continuous user-state stream, one could artificially cut the whole stream in small segments of the same length. The task of the labelers then would be to assign a label to each segment. The segments would have to be as small as the shortest possible user state that one wants to include (which can be very short: surprise in our case was the shortest and could comprise as few as five frames). The exact duration should probably be found out empirically.
- A more radical solution to the problem of reliable labels would be to take a step backwards and ask the question if it actually made sense to look for examples of the "big six"[7] like anger or surprise in a human–machine dialogue. The so-called "fundamental" emotions were formulated as categories with regard to evolutionary development (Plutchik, 1980). These emotions are also considered to be fundamental because all other emotions are thought to be somehow derived from them. So in effect they are the "parents" of a huge range of emotional states, which seldom are studied because these substates are even harder to define and to describe than the parents. So does it make sense to look for basic, superordinate concepts that mostly were studied in their "strong" form (that means as "full-blown" emotions) and with the goal to find the categories that exist below social conventions? Perhaps it would be better to look for not-so-well-defined, context-dependent states that serve a certain function within a dialogue. For a computer system it actually is not in the least important to detect if the user is angry, sad, happy or afraid. The computer system has to decide if the user is content (which is not the same as happy), not content (which is not the same as angry) or needs help (for example, other scenarios of course are conceivable). Not only is the state of the user is of interest but at the same time we are interested in the necessary reaction of the communication partner. We observed quite often that subjects were genuinely amused (without sarcasm) but at the same time were not content with the service of the system, so detection of "happiness" or "joy" would not have served our needs. We acknowledged the problem outlined

[7] In the literature on the study of emotions the "big six" are thought by some authors to be the fundamental emotions, probably existing in every culture: happiness, sadness, fear, disgust, anger and surprise (Cornelius, 2000). Authors differ in their opinion about the number of the "basic" emotions, which ranges from six to ten.

above at least a little when we decided not to include "emotion recognition" in SMARTKOM but instead use "user-state recognition." A user state was defined as an emotional or cognitive state, "that is of interest in the communication process." We perhaps did not take the second part of our definition not seriously enough and we concentrated mainly on "standard" emotions, not trying to find out if there were categories that were better suited for our special condition. Of course, it would be great if it was possible to construct universal emotion detectors. But emotional states strongly depend on the context — perhaps not with regard to the feeling (inside) but definitely with regard to the emotional display. And only the emotional display is available for detection. Luckily, emotional displays conform to cultural rules and serve a certain function in human–human interactions. These rules certainly have detectable features since they were developed to help regulate communication. To target these rules and functions for labeling and recognizing user states in human–computer interaction seems to be — after our experience with the SMARTKOM data — a more promising approach than trying to look for consistent features of "basic" emotions. We therefore propose to look for user states that are defined as "emotional or cognitive state of a user that calls for a certain action/reaction by the system and/or contains an important message for the system." One way to do this would be to start with no categories whatsoever and in a first step label only the "functional message" that an observer gets from a displayed user state. In a second step categories could be developed from the segmented material. Another option would be to define which states/messages are important for the system (or to which the system can react). Taking the example above, this could be "content," "not content, but not confused" and "confused/needs help." After that a labeling of the morphological changes in the face, voice quality, body movements, head movements and gestures could be carried out (perhaps realized as modifiers for the functional label). This approach would be similar to the one we chose for the gesture labeling: Do not use existing categories, because you do not know if they make sense in your case or if you lose important information because existing categories are too broad. Look for events with regard to your item of study (gestures, resp. user states) and categorize them according to their "function" in the dialogue. Then mark the observable, formal features of these events.

A lot more could be said with respect to the labeling of emotions/user states. The points above are the the most important considerations from our experience with labeling the data in SMARTKOM. We hope that they will prove themselves useful for future attempts to tackle the task of emotion/user-state labeling and recognition in humanm–machine dialogues and everyday situations.

2.10 Prosodic Annotation of Audio with Formal Criteria

The goal of this labeling step was to capture the information that was contained in the voice: A labeler listened to the audio file and marked if any of the labels below occurred, together with their time of occurrence. A word could carry more than one

prosodic marker. The labels were given with respect to the "normal" speaking habits of a subject: If someone habitually articulated clearly or spoke emphatically only the very clearly articulated words or the words with strong emphasis were marked.
The categories were:

- PAUSE_PHRASE: Irregular pause on a phrasal level/between units of meaning. Pauses between sentences or between main clause and subordinate clauses were not meant, except if the pause was very long.
- PAUSE_WORD: Irregular pauses between words.
- PAUSE_SYLL: Irregular pauses between syllables of a word.
- LENGTH_SYLL: Lengthening of a syllable; could occur at any syllable of a word.
- EMPHASIS: Emphatic accentuation/strong emphasis on a word or syllable.
- STRONG_EMPH: Very strong emphatic accentuation/very strong emphasis on a word or syllable.
- CLEAR_ART: Clearly articulated speech. Clear articulation could be seen as a weak version of hyperarticulated speech. The speaker used (tried to use) less colloquial speech and less dialect, and the speech emulated that of a news reader.
- HYPER_ART: Hyperarticulated speech; very strong increase of the clear articulation.
- LAUGHTER: Speech overlapped by laughter or sighing.

2.11 Inter-Rater Agreement Prosodic Annotation

The procedure for the analysis of the inter-rater agreement for the prosodic annotation was exactly the same as for the holistic labeling (see above). The agreement between the labelers in this group was very low, with 35% correct and an accuracy of -1.14%. The main reason for the negative accuracy was the huge overall amount of insertions/deletions. For detailed information about the inter-rater analysis please refer to Jacobi and Schiel (2003). Here again it seems to be a big problem for the labelers to decide if a label should be placed or not. This is backed up by the observation that the prosodic group was the one that had the most discussions about how to improve the consistency. Some labelers complained that the label instructions were too vague. They found it very hard to decide if a pause was long enough to be classified as an unusual long pause, and to distinguish emphasis and strong emphasis. Since no absolute measure for the length of pauses, the strength of emphasis, etc., exists, the labelers had to compare a possible instance of a label against the rest of the dialogue of the subject. This could be one reason why the judgments varied so widely. One could try to align the response bias of the labelers ("only mark a label if you are absolutely sure that it is there"), but this would reduce the labeled phenomena drastically. Taking everything together we think that the label categories simply were not very well suited for the material we had to label in SmartKom. The coding system was developed for emotional speech. The SmartKom corpus includes subtle everyday emotions but not much speech that actually could be called emotional. We suggest the approach outlined under the holistic section for a new attempt to capture significant aspects of subtle emotions in everyday situations.

3 Development of the Gesture Coding System

To define our categories, we started with a heuristic analysis of a number of recordings of human–machine dialogues. The data we used consisted of 70 sessions of about 4.5-min length each. The subjects had the task to "plan a trip to the cinema this evening." The scenario was Public (for more information on the WOZ recordings please refer to Schiel and Türk (2006)). The recorded sessions were analyzed by the following procedure: Two observers separately listed every episode that they could identify as a "functional unit." Then they tried to assign a label to the unit with respect to the observable meaning of the gesture ("meaning" with respect to the communication process). For obvious reasons, it was not attempted to ascertain the true intent with which the gesture was performed but just the intent that could be identified by an observer. Identifiable units with no observable meaning for the communication process were listed as well. Some questions for the list of units were: What is the best way to categorize the observed units? Is there a reference on the display? If so, which one? How can we define beginning and end of a given unit? The observers made their judgments independently. After completing the list, the identified gestures were discussed. Every label that could not be operationalized satisfactorily was removed. Similar labels were combined. Three broad categories emerged, each with several subcategories, called labels. After this, some additional labels were added that were not observed but were thought probable to appear or had to be added in order to complete the categories. For more information on the development of the gesture coding system please refer to Steininger et al. (2001a). A complete description of the labels can be found in Steininger et al. (2001b). In the next section the categories and labels are outlined.

3.1 Overview and Definitions of Gesture Labeling

The basic procedure for the gesture coding was as follows: All gestures were assigned to one of three broad categories according to the observable intention with which they were used: Is the gesture an interaction with the system (an "interactional gesture"), a preparation of an interaction (a "supporting gesture") or something else (a "residual gesture")? The only (interactional and supporting) gestures that were labeled were the ones that entered the so-called "cubus," the field of the display and the room above the display where the SiViT gesture recognizer recorded data (it corresponded roughly to the border of the display). Additionally, gestures that were interesting (but took place outside the "cubus") were coded in the category of residual gestures. Each label was complemented by several modifiers (see below).

3.2 Gesture Label Categories and Modifiers

Three broad categories were labeled:

1. interactional gesture (I-gesture)

2. supporting gesture (U-gesture)[8]
3. residual gesture (R-gesture)

Interactional Gesture

When a subject used a hand/arm movement to request something from the computer this gesture was called "interactional." A second type of interactional gesture was the confirmation of a question from the system. Only for interactional gestures was the stroke marked. We found that most gestures of this type had a specific acceleration/deceleration curve (whereas other gestures were more erratic with respect to velocity changes), often had an overall higher velocity and were mostly produced vertically to the body and the display (whereas other gestures were mostly produced parallel to the display/body). These last observations were not used as labeling criteria.

Supporting Gesture

The "supporting gesture" occurred in the phase when a request was prepared. It signified the gestural support of a "solo action" of the user (like reading or searching), which means an action that was a request or reaction to a request from the system. To distinguish it from an "interactive gesture," at least one of the following prerequisites had to be confirmed as true:

1. The gesture was accompanied by verbal comments from the user that made the preparational character of the underlying cognitive process obvious (for example, "hmm... where is... here?... well...").
2. The gesture was accompanied by a facial expression that made the preparational character of the underlying cognitive process obvious (like lip movements of silent reading, searching eye movements, frowning).

If a gesture that seemed to be an "interactional gesture" was interrupted (because of a change on the display or speech output from the system) it was marked as "supporting gesture," since it contained no request. In order to be able to differentiate a supporting gesture from a residual gesture at least one (or both) of the following two prerequisites had to be given:

1. A linkage between gesture and speech (both streams referred to the same topic).
2. An identifiable focus, i.e., gaze and gesture fell on the same spot.

Residual Gesture

A "residual gesture" did not prepare a request (at least not obviously) and was not a request or confirmation. A "residual gesture" was either an "emotional gesture" or an "unidentifiable gesture." When an unidentifiable gesture was marked, there existed:

1. no linkage between gesture and speech, and/or
2. no identifiable focus, which means gaze and gesture did not fall on the same spot

[8] "U" for the German "Unterstützende Geste".

Label

Below are listed all labels and modifiers.

I-labels:

- I-circle (+), I-circle (−)
- I-point (long +), I-point (long −)
- I-point (short +), I-point (short −)
- I-free (+), I-free (−)
- not recognizable[9]

U-labels:

- U-continual (read), U-punctual (read)
- U-continual (search)
- U-continual (ponder), U-punctual (ponder)
- U-continual (count)
- not recognizable

R-labels:

- R-emotional (+ cubus), R-emotional (− cubus)
- R-unidentifiable (+ cubus)
- not recognizable

Morphology

This modifier signified which hand and which finger was used to perform the gesture and if a pen was used (in some scenarios the subjects had a pen for the display, in some not. For more information on the scenarios please refer to Schiel and Türk (2006). Some peculiarities (like double pointing) were noted as comment (see below).

Reference Word

This modifier signified the word that was spoken in correspondence to an "interactional gesture." It is known that "hand gestures co-occur with their semantically parallel linguistic units" (McNeill, 1992). "Correspondence" therefore meant that the word occurred shortly (50 ms) before, simultaneously or shortly after the gesture and gesture, and word were linked with regard to the content.

Reference Zone

This modifier signified the region of the display to which an "interactional" or "supporting" gesture referred — the middle, one of the four corners, "whole display" or "not recognizable." "Whole display" was labeled if the hand/finger moved through at least two reference zones or in the case of "I-free" gestures, resp. if the gesture indicated the whole display.

[9] It actually could happen that the category could be decided upon but the exact label was not recognizable because a part of the gesture was occluded. However, this was a rare case.

Object

This modifier signified the "interactional gestures," if the targeted object was hit or missed (and if so, on which side it was missed). It could be "hit," "above," "below," "right," "left" or "not recognizable."

Stroke

This modifier signified the "culmination point" of the gesture, which means the "most energetic" or "most important" part of a gesture that is often aligned with the intonationally most prominent syllable of the accompanying speech segment (McNeill, 1992; Kendon, 1980). For pointing gestures it was the period the finger remained on the indicated object.

Beginning, End

The beginning of a gesture was determined with the help of the following criteria: (1) hand entered cubus, (2) end of previous gesture or (3) hand moved after pausing within the cubus for a certain time. The end of a gesture was determined with the help of similar criteria: (1) hand left cubus, (2) beginning of following gesture, (3) hand stopped moving within the cubus.

Comment

Comments about the stroke, beginning, end or peculiarities that could not be noted anywhere else. Comments referring to the whole file were marked in the header; comments that referred to a single label were marked as a modifier for the label.

3.3 The Gesture Labels

I-Label

- I-circle (+), I-circle (−): A continuous movement with one hand that specified one or more objects on the display. The display was touched (+)/not touched (−).
- I-point (long +), I-point (long −): A selective movement of one hand that specified an object on the display, with the display being touched (+)/not touched (−). The stroke of the movement (which means the pointing of the hand at the object) was of extended duration (20 frames or more).
- I-point (short +), I-point (short −): As I-point (long +/−), but the stroke of the movement (which means the period during which the hand was pointed at the object) was not extended/was very short (up to 19 frames).
- I-free (+), (−): A movement of the hand/the hands which took place above the display (+) or beside the display/between body an display (−) and signified a request of the user (for example, to go to another page). This gesture could vary considerably with respect to its morphology.

U-Label

- U-continual (read): A continuous movement of the hand above or on the display. At least during a part of the movement the subject obviously read, which means the text was read aloud or the lips moved.
- U-point (read): Like U-continual (read), but with a selective movement of the hand (not a continuous one).
- U-continual (count): A continuous movement of the hand above or on the display. At least during a part of the movement the subject obviously counted, which means numbers were uttered or formed with the lips.
- U-continual (search): A continuous movement of the hand above or on the display that moved over a large part of the display. The hand moved through at least two reference zones. From the context it was obvious that the subject was looking for something or preparing a request, that means through the spoken words and/or a following interactional gesture (selection).
- U-point (ponder): A selective movement of the hand to a spot on the display. The hand did not move.
- U-continual (ponder): As U-point (ponder) but with small movements of the hand.

R-Label

- R-emotional (+): A movement of one or both hands within the cubus with an obviously emotional or otherwise interesting content.
- R-emotional (-): Like R-emotional (+), but outside of the cubus.
- R-unidentifiable (+): A movement of one or both hands within the cubus that could not be assigned to any of the other labels, for example, the hand lying on the display.

3.4 Inter-Rater Agreement Gesture Labeling

The procedure for the analysis of the inter-rater agreement for the gesture labeling was exactly the same as for the holistic labeling (see above). Comparing the label categories (I-, U-, R-gestures) the inter-rater reliability was good: 85% correct matches were counted, and the overall accuracy was 52,3%. The most reliable label was "I-gesture." Comparing all labels, not only the categories, the reliability remained good. Here 90% of the matches were correct; however, the accuracy dropped to 45.1%. Since the most important distinction — the one between an interactional gesture (I-gesture) and a supporting gesture (U-gesture) — showed very few confusions, we were content with the results. Most confusions took place between U-gestures and R-gestures, which possibly results from the fact that gestures that help prepare a request or accompany a thought process can sometimes be short, casual and have no distinct acceleration/deceleration curve like the I-gestures. In this they resemble R-gestures that happen without a goal and are mostly subconscious (for example, scratching, impatient drumming). Another reason for the confusions could be that

Table 1. Inter-rater agreement for gesture labeling (%)

Label	I-gesture	U-gesture	R-gesture	None
I-gesture	76	3	3	18
U-gesture	1	54	0	45
R-gesture	9	34	44	13
None	38	38	25	0

an important criterion for distinguishing U- and R-gestures was the gaze: U-gestures were defined as being in the focus of attention, seen in the gaze that followed the gesture (for the most part) or a link between gesture and speech. The last could be asserted with relative ease, but sometimes it was difficult to judge where the g aze of a subject was directed to. Therefore, with more suitable camera angles the reliability of the coding system could perhaps be improved.

The insertions/deletions are unfortunate because the instructions for marking a label or not marking a label are very clear and easily observable (entering the display for the most part). That in spite of this relevant sequences are missed (or irrelevant coded) shows that even a well-defined label system suffers problems with regard to consistency. The insertion/deletion rate could probably reduced with more training and with a more thorough correction.

3.5 Considerations for Future Attempts of Labeling Gestures

The gesture coding system described above was especially designed for labeling gesture in a human–machine dialogue. It categorizes gestures according to their observable function for the user in the communication process with the machine. We think that such coding systems (functional and tailored to the context) are much needed in addition to general taxonomies (McNeill, 1992; Ekman, 1999; Friesen et al., 1979) and structural, formal or micro-coding systems (like “FACS” from Ekman and Friesen (1978)). Additionally, we assume the following two major points to be important for future attempts to label gestures and study gestural aspects of communication:

1. A very important aspect of gestures is their dynamic properties. Until now these properties have largely been ignored in taxonomies (with the exception of notation of dance (Laban, 1956)). This is a pity, since it is very likely that the dynamic properties carry important information about the meaning of a gesture. They likely are good candidates for automatic detection since dynamic properties often are subconscious and therefore are less influenced by social conventions and are less context-dependent. As described above, we found that intentional gestures (“interactional gestures”) often seemed to have a distinct acceleration–deceleration curve, whereas supporting gestures showed relatively erratic velocity changes. To study such differences accurate measurements of the movements would be necessary — which is not too difficult a problem, since there exist many good techniques for measuring trajectories of gestures in space

(Mulder, 1994). However, additionally a wa y to categorize the dynamic properties will have to be found and this could be a tricky problem. Nonetheless, we strongly hope that research along this road will be forthcoming.

2. Taxonomies above the structural level are often very similar (that means they can be translated into each other with relative ease). There seems to exist an underlying assumption that one (functional) taxonomy is enough to describe all kinds of human gestures that show up during communication. We agree that it is desirable to have one "master taxonomy" that is broad enough to cover all relevant gestural phenomena to have a reference and to make studies comparable. However, we think that we additionally need taxonomies that are especially tailored to the different situations where gestures can show up. Not only does speech vary greatly from situation to situation, gestures will look very different during a talk, a phone call, an argument, during talking to a machine like in our data, or during describing an animated cartoon as in one very well known gesture corpus (McNeill, 1992). If one uses a taxonomy that was developed with data from a certain situation, one cannot expect to cover all gestural phenomena adequately. For example, if we had used the category "deictic gesture" (pointing movements) we would have missed the most important aspects of the gestures in our human–machine dialogues: A deictic gesture can have the meaning of a selection ("Ok, I want this"), but in some cases people point at something while they are still making up their minds if they really want the thing indicated ("Hmm, should I go to this cinema") or point to figure something out ("Uh — this restaurant is near my place — or isn't it ?"). These two kinds of deictic gestures not only have a different meaning, they are executed differently as well. If using a standard taxonomy we would have missed these differences. Another example for needing context-dependent studies of gestures are ideographs (sketch a direction of thought). Ideographs normally take place in front of the thorax. We assume that the "supporting gestures" we describe are such "thought-sketching" gestures. Since the subjects think about the things presented on the display, they move their hands in this area (and not in front of their body as in a human–human interaction). We suppose that a display encourages the subjects "to think with the hands" (that means trace a thought with their hands). This can be compared to the "thinking aloud" one sometimes uses to help make up one's mind. In a communication situation where the general physical conditions are different the gestures will change — this is obvious, but until now it has not initiated a gathering of data from such diverse situations.

If we truly want to find out what happens between two speakers on the nonverbal tier, we have to keep looking with a much broader scope than used until now.

4 Conclusion

The challenge we faced at the start of the project was to develop labeling systems for multimodal data that were tailored to the task, that were fast and that were suited for the training of recognizers. We decided not to label on the morphological level but to define the labels with regard to the intent of the user, respectively his observable communication goal (in the case of user-state labeling and gesture labeling). Since we do not yet know which features of the gestures or the user states carry the vital information, we cannot resort to a low-level coding of morphological features, even if this would greatly improve the reliability of categories. For future endeavors of transliterations of spontaneous speech in human–computer interaction, we think that a very important point that has to be solved is the coding of so-called off-talk (Oppermann et al., 2001). For the development of machines that do not need a command word to switch on and off, the discrimination of commands to the computer and talk that is not aimed at the computer such a coding, is essential. Our system allows for such coding but it seems prudent to try to enhance the validity and reliability of off-talk labelings: subjective impression for these judgments plays a greater role than for most other categories of transliteration conventions for spontaneous speech. The system we described for gestural coding is reliable and able to pinpoint the important aspects of gesture events in a dialogue. It can easily be expanded for other contexts and requirements. However, we will need studies that target the dynamical properties of gestures if we want to come to a more complete understanding of the meaning and appearance of the gestural components of communication. In the case of the user-state labeling, the low inter-rater agreement of our nonformal coding system showed that the optimal way of labeling such cloudy categories as emotions or user states still has to be found. We made suggestions how to improve the reliability of a functional coding system without resorting to formal categories. Together with additional collections of emotional speech data — which in itself poses no trivial problem — they should lead to a better understanding of the connections between speech and emotion. Since even human observers have problems in deciding if an emotion is there and if it is, which one (in everyday situations where only subtle emotions are shown, not full-blown emotions like during an argument or in a TV show), feature detection alone will probably not suffice, but will have to be combined with context information and/or information from other recognizers and/or models of the user. If this will be proven as true, one last critical remark about the endeavor of teaching computer systems to detect emotions will have to be made: If it turns out that not only humans need context information to judge emotions but machines need it too, then perhaps they will only get really good at the job if we train them to expect certain emotions in certain situations for certain persons, like humans tend to do. However, we then may have the problem of a system that is projecting emotions onto a communication partner – a well-known phenomenon in human communication. With this option to look forward to, we strongly advise not to enhance the emotion detection capabilities of computers, but the ability to judge the requests from the user correctly (which is not the same), both the ones uttered verbally and the ones given by nonverbal means, be they gestural or through signs of disapproval, confusion or approval. That is to say, let us keep in mind what the computer system should do: fulfill a request. Computer systems that show polite considerateness to my — real or assumed — mood would be a marvel, but interacting with one could get as tedious as conversing with an overpolite human.

Acknowledgments

We give our thanks to the SMARTKOM group of the Institute of Phonetics in Munich that provided the Wizard-of-Oz data. Many thanks to B. Lindemann and M. Pätzold for their contribution to the development of the gesture coding system and to all labelers, who not only coded the data but made important contributions to the development of the label systems.

References

A. Batliner, K. Fischer, R. Huber, J. Spilker, and E. Nöth. Desperately Seeking Emotions Or: Actors, Wizards, and Human Beings. In: *Proc. ISCA Workshop on Speech and Emotion*, Belfast, Irland, 2000.

N. Beringer, S. Burger, and D. Oppermann. Lexikon der Transliterationen. Technical Document SmartKom 2, Bavarian Archive for Speech Signals (BAS), 2000.

R.R. Cornelius. Theoretical Approaches to Emotion. In: *Proc. ISCA Workshop on Speech And Emotion*, Belfast, Ireland, 2000.

J. Egan. *Signal Detection Theory and ROC Analysis*. Academic, New York, 1975.

P. Ekman. Emotional and Conversational Nonverbal Signals. In: L.S. Messing and R. Campbell (eds.), *Gesture, Speech and Sign*, pp. 45–55. Oxford University Press, 1999.

P. Ekman and W.V. Friesen. Facial Action Coding System (FACS). A Technique for the Measurement of Facial Action. *Consulting Psychologists Press*, 1978.

G. Faßnacht. *Systematische Verhaltensbeobachtung*. Reinhardt, Munich, Germany, 1979.

K. Fischer. Annotating Emotional Language Data. Verbmobil Report 236, DFKI, 1999.

W.V. Friesen, P. Ekman, and H.G. Wallbott. Measuring Hand Movements. *Journal of Nonverbal Behavior*, 1:97–112, 1979.

I. Jacobi and F. Schiel. Interlabeller Agreement in SmartKom Multi-Modal Annotations. Technical Document SmartKom 26, Bavarian Archive for Speech Signals (BAS), 2003.

A. Kendon. Gesticulation and Speech: Two Aspects of the Process. In: M.R. Key (ed.), *The Relation Between Verbal and Nonverbal Communication*, The Hague, The Netherlands, 1980. Mouton.

R. Laban. *Principles of Dance and Movement Notation*. Macdonald & Evans Ltd., London, UK, 1956.

D. McNeill. *Hand and Mind: What Gestures Reveal About Thought*. University of Chicago Press, Chicago, IL, 1992.

A. Mulder. Human Movement Tracking Technology. Hand Centered Studies of Human Movement Project 94-1, School of Kinesiology, Simon Fraser University, Burnaby, Canada, 1994.

N. Munn. The Effect of Knowledge of the Situation Upon Judgment of Emotion From Facial Expressions. *Journal of Abnormal and Social Psychology*, 35:324–338, 1940.

D. Oppermann, F. Schiel, S. Steininger, and N. Beringer. Off-Talk — A Problem for Human-Machine-Interaction. In: *Proc. EUROSPEECH-01*, Aalborg, Denmark, 2001.

R. Plutchik. *Emotion: A Psycho-Evolutionary Synthesis*. Harper and Row, New York, 1980.

C. Ratner. A Social Constructionist Critique of Naturalistic Theories of Emotion. *Journal of Mind and Behavior*, 10:211–230, 1989.

F. Schiel and U. Türk. Wizard-of-Oz Recordings, 2006. In this volume.

M. Sherman. The Differentiation of Emotional Responses in Infants. *Journal of Comparative Physiology*, pp. 265–284, 1927.

S. Steininger, O. Dioubina, R. Siepmann, C. Beiras-Cunqueiro, and A. Glesner. Labeling von User-State im Mensch-Maschine Dialog — User-State-Kodierkonventionen SmartKom. Technical Document SmartKom 17, Bavarian Archive for Speech Signals (BAS), 2000.

S. Steininger, B. Lindemann, and T. Paetzold. Labeling of Gestures in SmartKom - The Coding System. In: I. Wachsmuth and T. Sowa (eds.), *Gesture and Sign Languages in Human-Computer Interaction, Int. Gesture Workshop 2001 London*, pp. 215–227, Berlin Heidelberg New York, 2001a. Springer.

S. Steininger, B. Lindemann, and T. Paetzold. Labeling von Gesten im Mensch-Maschine Dialog — Gesten-Kodierkonventionen SmartKom. Technical Document SmartKom 14, Bavarian Archive for Speech Signals (BAS), 2001b.

S. Steininger, F. Schiel, O. Dioubina, and S. Rabold. Development of User-State Conventions for the Multimodal Corpus in SmartKom. In: *Proc. Workshop "Multimodal Resources and Multimodal Systems Evaluation"*, pp. 33–37, Las Palmas, Spain, 2002a.

S. Steininger, F. Schiel, and A. Glesner. Labeling Procedures for the Multi-Modal Data Collection of SmartKom. In: *Proc. 3rd Int. Conf. on Language Resources and Evaluation (LREC 2002)*, pp. 371–377, Las Palmas, Spain, 2002b.

H. Wallbott. Faces in Context: The Relative Importance of Facial Expression and Context Information in Determining Emotion Attributions. In: K. Scherer (ed.), *Facets of Emotion*, Mahwah, NJ, 1988. Lawrence Erlbaum.

Multimodal Emogram, Data Collection and Presentation

Johann Adelhardt, Carmen Frank, Elmar Nöth, Rui Ping Shi, Viktor Zeißler, Heinrich Niemann

Friedrich-Alexander Universität Erlangen-Nürnberg, Germany
{shi,adelhardt,batliner,frank,noeth,zeissler,
niemann}@informatik.uni-erlangen.de,

Summary. There are several characteristics not optimally suited for the user state classification with Wizard-of-Oz (WOZ) data like the nonuniform distribution of emotions in the utterances and the distribution of emotional utterances in speech, facial expression, and gesture. In particular, the fact that most of the data collected in the WOZ experiments are without any emotional expression gives rise to the problem of getting enough representative data for training the classifiers. Because of this problem we collected data in our own database. These data are also relevant for several demonstration sessions, where the functionality of the SMARTKOM system is shown in accordance with the defined use cases.

In the following we first describe the system environment for data collection and then the collected data. At the end we will discuss the tool to demonstrate user states detected in the different modalities.

1 Database with Acted User States

Because of the lack of training data we decided to build our own database and to collect uniformly distributed data containing emotional expression of user state in all three handled modalities — speech, gesture and facial expression (see Streit et al. (2006) and for an online demonstration refer to our website [1]). We collected data of instructed subjects, who should express four user states for recording. Because SMARTKOM is a demonstration system it is sufficient to use instructed data for the training database.

For our study we collected data from 63 naive subjects (41 male/22 female). They were instructed to act as if they had asked the SMARTKOM system for the TV program and felt content, unsatisfied, helpless or neutral with the system feedbacks. Different genres such as news, daily soap and science reports were projected onto the display for selection. The subjects were prompted with an utterance displayed on the screen and were then to indicate their internal state through voice and gesture, and at the same time, through different facial expressions.

[1] http://www5.informatik.uni-erlangen.de/SmartKom/SkBook.html

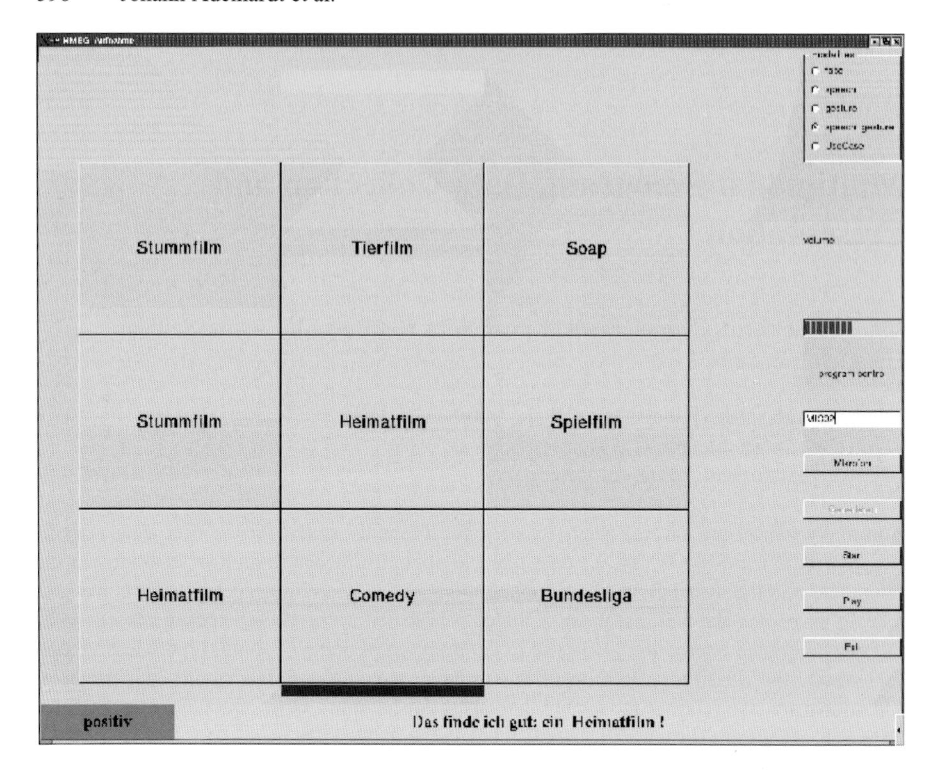

Fig. 1. Screen used for recording the user states in our local environment; several genres are shown together with a state that should be played. The *upper right corner* shows which modalities are currently recorded

1.1 Data Collection Environment

For the data collection we developed a special environment which we used directly with the local SMARTKOM demonstrator of the Institute of Pattern Recognition (LME). The screen of the graphical user interface of our experiment is shown in Fig. 1.

It shows in the center a 3×3 matrix of several genres, from which the subject had to choose one, together with a user state that should be expressed by the subject. This user state is shown in the lower left corner of the screen. At the middle bottom, the utterance, which the user had to speak at the current turn, is shown. Between the bottom of the 3×3 matrix of genres and the generated user utterance a colored bar is shown, which informs the user that the system is recording his speech.

On the right of the screen several features for handling the local data collection environment are shown. In the upper right corner five kinds of data collection can be chosen. Four of them refer directly to the modalities and to the combination of modalities used in SMARTKOM. The fifth one, UseCases, was used to collect data

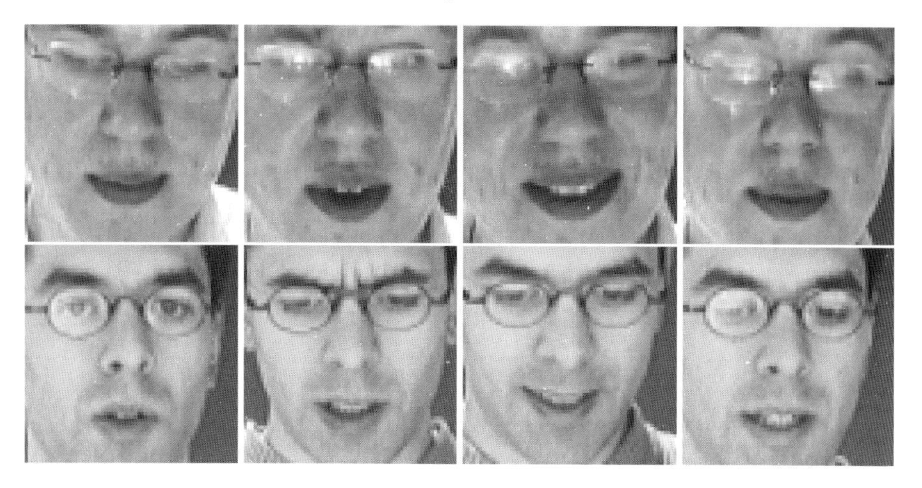

Fig. 2. Sample data; from *left to right*: *hesitant*, *angry*, *joyful*, and *neutral*

related to so-called UseCases, which were used for the demonstration of the SMARTKOM system.

In the middle and the lower right part of the screen control buttons for camera and microphone and the field for the identification label of the experiment are shown. These were handled by the supervisor of the experiment with mouse and keyboard.

Data collection was done separately for all three modalities and for the combination of them. Because of the lack of annotated emotional data, the facial expression was recorded throughout, although the subjects were unaware of it. During the utterance the user had to select the genre by gesture in the centered choice matrix mentioned above.

Data collection was done in several sessions so the users could get familiar with the system and with the way to express their emotions naturally facing the camera and/or microphone in an incremental way. In the first session we recorded only the facial expression as the subject had to express the related emotion by speaking the utterance. In the second session we also recorded the speech of the subject together with facial expression. In the third session gesture and facial expression were recorded, while in the final session all three modalities were recorded as indicated with the tag *"speech_gesture"* (facial expression automatically included as mentioned above) in the upper right corner of the screen.

1.2 Collected Data

Facial expression, gesture and speech were recorded simultaneously in the experiment; this made it possible to combine all three input modalities afterwards. The user states were equally distributed. The test subjects read 20 sentences per user state. The utterances were taken in a random order from a large pool of utterances. About 40% of them were repetitions of TV genres or special expressions, which did not actual depend on the given user state, like *"tolles Programm!"* (*"nice program!"*). In other words, we chose expressions one could produce in each of the given user states. (Note that a prima facie positive statement can be produced in a sarcastic mode and by that, turned into a negative statement.) All the other sentences were multiword expressions, where the user state could be guessed from the semantics of the sentence. The subjects should align to the given text, but minor variations were allowed.

From all collected data we chose 4848 sentences (3.6 hours of speech) with good signal quality and used them for further experiments. For the experiments with prosodic analysis, we randomly chose 4292 sentences for the training set and 556 for the test set.

For the facial analysis video, sequences of ten subjects were used. These subjects where selected because their mouth area was not covered by facial hair or the microphone. As training images, we used image sequences of these subjects without wearing the headset. In the images of the test sequences, there is a headset. Some of the training images can be seen in Fig. 2.

For gesture analysis there are, all in all, 5803 samples of all three user states (note that there are only three user states for gesture as mentioned in Shi et al. (2006)), and 2075 of them are accompanied by speech. As we are interested in the combination of all three modalities, we concentrate on this subset. Of this sample subset, 1891 are used for training and the other 184 are used for testing. Since the samples were recorded according to the user states categories in facial expression and speech, we merged the data of the corresponding user states *neutral* and *joyful* into a general user state category *determined* for gesture. The data we collected for the multimodal emogram (MMEG) are described in Zeißler et al. (2006), Frank et al. (2006) and Shi et al. (2006).

2 Multimodal Emogram (MMEG)

During the system development there always exists a problem if the input and output happen to take place simultanenously. It is also unwise to registrate any kind of output at the end of the processing queue without knowing the steps between input and output. In particular, if there are several steps for processing with statistical methods, as is the case in SmartKom with user states, it is absolutely necessary to know the intermediate results. The *multimodal emogram*[2] is a tool to show the results of user state processed in three modules: prosody, facial expression and gesture analysis.

The advantages of the evaluation tool are the following:

- quick presentation of recognizer results in the running environment
- systematic evaluation
- demonstration of functionality without the full system environment
- possibility to show the user the difficulty to act "as if" he were in the corresponding user state

[2] http://www5.informatik.uni-erlangen.de/SmartKom/SkBook.html

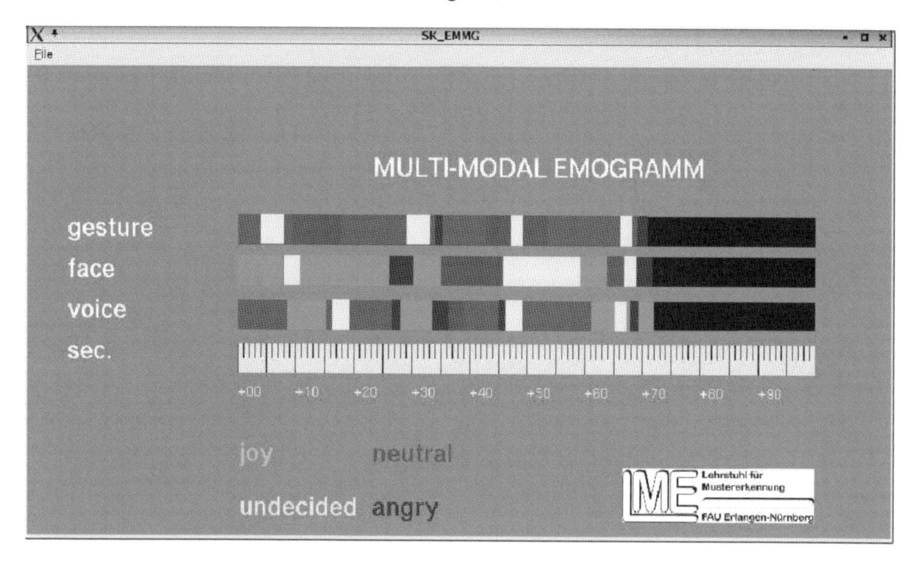

Fig. 3. Presentation of the results of the user state analysis in three modalities: speech, gesture and facial expression

For evaluating the system it is absolutely necessary to know the final result, but it is also important to know the results of the stages between input and output. In the case of error during the early development stage the developers are able to localize system components that produce errors.

In our case this is the presentation of the user states as recognized by the three modules. Their output is taken by the tool and is shown in a presentation where the user states are aligned parallel to the modalities. The presentation shows in this manner the detected user state of each module/modality at a certain time. So the recognition result of each single modality can be analyzed also for itself and at run-time.

The MMEG provides a compressed and transformed presentation of the user utterance in form of user states in quasi–real time. Just as the sonagram shows the spectral energy and the formants of the speech, the multimodal emogram shows user states as they occur in speech, facial expression and gesture. At the moment there are some problems in presentation concerning time alignment, synchronization and real-time behavior. But the idea of presentation generation in real time is realized, and it relies only on technical aspects such as the dependency on the word lattice or the power of the CPU. Another positive side effect of the evaluation tool is the possibility to show the functionality of the components processing the user state without running the full system.

For the robust classification of the user state it is necessary to analyze the input of three different modalities. Therefore, these single results have to be combined in a fusion component, which is performed by the *media fusion module* in the SMART-

KOM environment. By considering the result of the fusion, it is possible to know the contribution of each modality to the final recognized user state. By knowing the results of each system component, it is possible to give a clue to the current user state and thus to increase the system performance in several tuning steps.

Beyond evaluation the multimodal emogram delivers very valuable information of the analysis in the system. So it is possible to adapt it to several conditions:

- A user state is not recognized: the tool shows the results of the different modalities; with the help of these results it is possible to find out at run time why the user state is not recognized,
- Testing of new side conditions like new light conditions or new microphones.
- Training of system conformation behaviour for presentation tasks.

For the tuning task it is helpful to show the contribution and result of each modality. Like the sonagram shows the different frequencies in the speech signal, the multimodal emogram shows the results of user state expressed in speech, facial expression and gesture.

The user states are presented with the help of colored bars as shown in Fig. 3. For each modality there is one bar. We have four colors for the presentation:

- red: angry/anger
- green: joyful/joy
- yellow: hesitant/undecided
- blue: neutral

For the recognition of user states in gesture there is the restriction that only three user states are defined: angry, hesitant, and joyful together with neutral as the third class.

References

C. Frank, J. Adelhardt, A. Batliner, E. Nöth, R.P. Shi, V. Zeißler, and H. Niemann. The Facial Expression Module, 2006. In this volume.

R.P. Shi, J. Adelhardt, A. Batliner, C. Frank, E. Nöth, V. Zeißler, and H. Niemann. The Gesture Interpretation Module, 2006. In this volume.

M. Streit, A. Batliner, and T. Portele. Emotion Analysis and Emotion Handling Subdialogs, 2006. In this volume.

V. Zeißler, J. Adelhardt, A. Batliner, C. Frank, E. Nöth, R.P. Shi, and H. Niemann. The Prosody Module, 2006. In this volume.

Empirical Studies for Intuitive Interaction

Iryna Gurevych* and Robert Porzel

European Media Laboratory GmbH, Heidelberg, Germany
iryna.gurevych@eml-r.villa-bosch.de
robert.porzel@eml-d.villa-bosch.de
* Current affiliation: EML Research gGmbH

Summary. We present three types of data collections and their experimental paradigms. The resulting data were employed to conduct a number of annotation experiments, create evaluation *gold standards* and train statistical models. The data, experiments and their analyses highlight the importance of data-driven empirical laboratory and field work for research on intuitive multimodal human–computer interfaces.

1 Introduction

Research on dialogue systems in the past has focused on engineering the various processing stages involved in dialogical human–computer interaction (HCI), e.g., robust automatic speech recognition, intention recognition, natural language generation or speech synthesis (Allen et al., 1996; Bailly et al., 2003; Cox et al., 2000). Development of such dialogue technologies involves experimental work.

Many issues in the semantics and pragmatics of dialogue can be formulated as empirical topics. Therefore, the analysis of sufficient amounts of data is necessary to study a specific phenomenon or to train statistical models. Given a certain hypothesis, the researcher has to find out whether there is empirical evidence for the phenomenon in question, and how frequent and important it is. Also, rigorous large-scale evaluations of the algorithms are a driving force of the engineering progress. Such evaluations are only possible if sufficient amounts of annotated data are available. In this case, the output of the computer program (an *answer*) is compared to the gold standard (a *key*) defined by the annotated data (Romary et al., 2003).

Alongside these efforts, the characteristics of computer-directed language have also been examined as a general phenomenon (Darves and Oviatt, 2002; Wooffitt et al., 1997; Zoeppritz, 1985). In the following sections we describe new paradigms for collecting dialogue data. These data can be employed for a variety of necessary examinations that improve the performance and reliability of conversational dialogue systems.

Fig. 1. Operator in hidden operator tests

Fig. 2. Subject in hidden operator tests

2 Collection and Usage of Hidden Operator Test Data

The data collected in the first experiment was employed in research on empirical topics, such as semantic postprocessing of the speech recognizer's outputs and assigning domains to speech recognition hypotheses. We designed a set of annotation schemata and performed experiments with human subjects [1] resulting in a set of annotated corpora. The annotated data was, then, employed in order to:

- test whether the annotators are able to annotate the data reliably;
- produce a gold standard for automatic evaluations of the algorithms;
- create data models based on human annotations;
- produce training datasets for statistical classifiers.

2.1 Data Collection Setup

The data collection was conducted following an iterative data collection approach (Rapp and Strube, 2002). This approach has been developed within the EMBASSI research project [2] and employes the *hidden operator test* paradigm. In this experimental setup, the SmartKom system was simulated by an operator with the help

[1] Referred to as "subjects" or "users" in what follows.

[2] http://www.embassi.de

Table 1. Descriptive corpus statistics

	Dataset 1	Dataset 2
Number of dialogues	232	95
Number of utterances	1479	552
Number of SRHs	2.239	1.375
Number of coherent SRHs	1511	867
Number of incoherent SRHs	728	508

Table 2. Multiple domain assignments to a single SRH (in %)

	Dataset 1		Dataset 2	
Number of domains	Annotator 1	Annotator 2	Annotator 1	Annotator 2
1	90.06	87.11	90.77	88.7
2	6.94	11.27	9.11	11.19
3	3.01	1.28	0.16	0
4	0	0.35	0	0

of predefined dialogue scripts (Fig. 1). Twenty-nine subjects were prompted to say certain inputs in 8 dialogues, resulting in 1479 turns (Fig. 2).

Because of the controlled experimental setting and since prompts appeared on a per-turn basis, each user turn in the dialogue corresponded to a single intention, e.g., route request or sights information request. All user turns were recorded in separate audio files. The audio files obtained from the hidden operator tests were then sent to two differently configured versions of the SMARTKOM system. The data describing our corpora is given in Table 1.[3] The first and the second system's runs are referred to as *Dataset 1* and *Dataset 2*, respectively.

The corpora obtained from these experiments were further transformed into a set of annotation files, which can be read into a GUI-based annotation tool, MMAX (Müller and Strube, 2003a,b). This tool can be adopted for annotating different levels of information, e.g., semantic coherence and domains of utterances, the best speech recognition hypothesis in the N-best list, as well as domains of individual concepts. The two annotators were trained with the help of an annotation manual. A reconciled version of both annotations resulted in the gold standard. In the following, we describe a set of annotation experiments performed on the basis of the *Hidden Operator Test* data and present their results. Also, we sketch some applications of the annotated corpora for empirically grounded research.

2.2 Annotation Experiments

2.2.1 Semantic Coherence and Domains of SRHs

In this annotation experiment, the task of annotators was to classify the speech recognition hypotheses (SRHs) as either *coherent* or *incoherent*. We defined semantic coherence as well formedness of an SRH on an abstract semantic level (Gurevych et al., 2002). The SRHs from dataset 1 were randomly mixed in order to prevent them from being annotated in the discourse context. The resulting Kappa statistics (Carletta, 1996) over the annotated data yielded $\kappa = 0.7$. This indicates that human annotators can distinguish between coherent samples and incoherent ones.

In the second step, the coherent SRHs from both dataset 1 and dataset 2 were labeled with respect to their domains. We restricted the annotation to coherent SRHs

[3] We also include the distribution of coherent and incoherent speech recognition hypotheses, which was computed on the basis of annotation experiments described in Sect. 2.2.

Table 3. Class distribution for domain assignments (in %)

	Dataset 1		Dataset 2	
Domain	Annotator 1	Annotator 2	Annotator 1	Annotator 2
Electronic program guide	14.43	14.86	12.13	11.73
Interaction management	15.56	15.17	10.02	9.24
Cinema information	5.32	8.7	4.01	3.43
Personal assistance	0.31	0.3	0	0
Route planning	37.05	36	44.2	46.31
Sights	12.49	12.74	21.94	21.08
Home appliances control	14.12	11.22	7.59	8.1
Off-talk	0.72	1.01	0.11	0.1

only, as automatically assigning domains to SRHs should only be done if the utterance is a coherent one. The annotations resulted in two corpora of coherent speech recognition hypotheses labeled for at least one domain category. The percentage of SRHs with one or more domain attributions can be found in Table 2. It shows that the vast majority of hypotheses (ca. 90%) could be unambiguously classified in a single category. The class distribution is given in Table 3. The majority class is the *Route Planning* domain, while some of the categories, e.g., *Personal Assistance*, *Off Talk*, occur rarely in the data. Table 4 presents the kappa coefficient values computed for individual domain categories in dataset 1.[4] P(A) is the percentage of agreement between annotators. P(E) is the percentage that we would expect them to agree by chance. The annotations are generally considered to be reliable if $\kappa > 0.8$. This is true for all classes except those which are under-represented in this dataset (see Table 3).

2.2.2 Best SRH in the N-Best Lists

For this study a markable, i.e., an expression to be annotated, is a set of SRHs (N-best list) related to a single user's utterance. The number of markables in *Dataset 2* employed here corresponds to the number of utterances 552. The annotators saw the SRHs together with the transcribed user utterances. For each utterance a single SRH had to be labeled as the best one. The guidelines for selecting the best SRH were:

- How well the respective SRH captures the intention contained in the transcribed user's utterance.
- If several SRHs capture the intention equally well, the actual word error rate had to be considered.

The kappa coefficient — often applied to measure the degree of interannotator agreement — is not appropriate in this experiment, as its calculation is class-based. But the number of SRHs underlying the best SRH selection per utterance does vary.

[4] The kappa coefficient was not computed for dataset 2, but we expect the results to be very similar. For dataset 2 we employed a different agreement measure described in a separate experiment.

Table 4. Kappa coefficient for separate domains

	P(A)	P(E)	Kappa
Electronic program guide	0.9743	0.7246	0.9066
Interaction management	0.9836	0.7107	0.9434
Cinema information	0.9661	0.8506	0.7229
Personal assistance	0.9953	0.9930	0.3310
Route planning	0.9777	0.5119	0.9544
Sights	0.9731	0.7629	0.8865
Home appliances control	0.9626	0.7504	0.8501
Off-talk	0.9871	0.9780	0.4145

Therefore, we computed the percentage of utterances, where the annotators agreed on the correct solution resulting in 95.35% of inter-annotator agreement. This number suggests a rather high degree of reliability for identifying the best SRH by humans.

2.2.3 Best Conceptual Representation and Domains of SRHs

The best *conceptual representation* and the domains of coherent SRHs from dataset 2 were annotated. Due to lexical ambiguity of individual words, each SRH can be mapped to a set of possible interpretations called *conceptual representations* (CR). E.g., the German word *kommen* can be mapped either to the ontological concept `MotionDirectedProcess` or to the concept `WatchPerceptualProcess`. The algorithms operating on the ontology convert each utterance to a set of conceptual representations. As a consequence, they must be disambiguated automatically within the running system as well as manually for evaluation purposes.

Eight hundred sixty-seven SRHs used in this experiment are mapped to 2853 CRs, i.e., on average each SRH is mapped to 3.29 conceptual representations. The annotators' agreement on the task of determining the best CR resulted in ca. 88.93%. For the task of domain annotation, we computed two kinds of agreements. The first measure is the absolute agreement $Prec_{mark}$, when the annotators agreed on all domains for a given markable, i.e., an SRH. This resulted in ca. 92.5%. The second measure $Prec_{dom}$ denotes the agreement on individual domain decisions (6936 overall) and amounted to ca. 98.92%.

2.2.4 Domains of Ontological Concepts

In the last experiment, ontological concepts were annotated with zero or more domain categories.[5] The percentage of ambiguous domain attributions is given in Table 5. We extracted 231 concepts from the lexicon, which is a subset of ontological concepts occurring in our data. The annotators were given the textual descriptions of all concepts. These definitions are supplied with the ontology. Again, we computed

[5] Top-level concepts, e.g., Type, Event, are typically not domain-specific. Therefore they will not be assigned any domains.

Table 5. Domain assignments to concepts (in %)

Number of domains	Annotator 1	Annotator 2
0	4.76	16.45
1	60.17	44.16
2	22.51	18.18
3	5.63	10.82
4	1.73	7.79
5	2.16	1.3
6	0.43	0.43
7	0.87	0
8	1.73	0.87

two kinds of interannotator agreement. In the first case, we calculated the percentage of concepts for which the annotators agreed on all domain categories $Prec_{mark}$, resulting in ca. 47.62%. In the second case, the agreement on individual domain decisions $Prec_{dom}$ (1848 overall) was computed, ca. 86.85%.

A comparison between domain annotations of SRHs and ontological concepts (Table 6) suggests that annotating utterances with domains is an easier task for humans than annotating ontological concepts with the same information. A reason for this state of affairs might be that the high-level meaning of the utterance is defined through the context, whereas the meaning of an ontological concept is much less clear in the absence of context.

2.3 Usage of the Annotated Data

In Sect. 2 we discussed the reasons for creating large corpora with annotated data. Such data is necessary to perform empirical research in the automatic language processing. At first, a research hypothesis is formulated. Then, it is translated into an annotation scheme and validated empirically by measuring the reliability of human annotations based on the corresponding annotation scheme. Additionaly, human performance indicates the *upper boundary*, also called *ceiling*, in evaluating computer programs.

The data from hidden operator tests were used for the design and evaluation of some language processing algorithms. Following the common distinction in the evaluation practices, we differentiate between *intrinsic* and *extrinsic* evaluation. For example, the semantic coherence scoring algorithm was evaluated intrinsically for scoring sets of ontological concepts in terms of their semantic coherence (Gurevych et al., 2003). In this kind of evaluation, the output of the program is compared directly with the gold standard produced by the annotators. Extrinsic evaluation measures how a particular program contributes to the overall performance of the system. In our case, we employed the system in order to determine the best SRH in the output of the automatic speech recognizer (Gurevych and Porzel, 2003).

A further application of the annotated data was a set of experiments directed at automatically assigning the domains to speech recognition hypotheses (Rüggenmann

Table 6. Interannotator reliability for domain annotations (in %)

	$Prec_{mark}$		$Prec_{dom}$	
	Concepts	SRHs	Concepts	SRHs
Agreement	47.62	92.5	86.85	98.92

and Gurevych, 2004). The algorithm for determining the domains of conceptual representations and their underlying speech recognition hypotheses employs a knowledge source called the *domain model*. Two kinds of domain models were produced on the basis of the annotated data. The first domain model was derived by means of statistical analysis of speech recognition hypotheses in dataset 1 annotated for their domains. The second domain model was established through the direct annotation of concepts with respect to domains in the corresponding annotation experiments. Finally, the quality of both domain models and the domain recognition algorithm was evaluated intrinsically on the basis of annotated dataset 2.

3 Collection and Usage of the Wizard and Operator Test Data

The main goal of the second data collection and its experimental setup was to enable precise analyses of the differences in the communicative behaviors of the various interlocutors, i.e., human–human, human–computer and computer–human interaction. The setup of the experiment was, therefore, designed to enable the control of various factors. Most important factors were the technical performance (e.g., latency times), the pragmatic performance (e.g., understanding vs. nonunderstanding of the user utterances) and the communicative behavior of the simulated system. They were to be adjustable to resemble the state of the art dialogue systems, such as SMARTKOM. These factors can, of course, also be adjusted to simulate potential future capabilities of dialogue systems and test their effects.

3.1 Data Collection Setup

For conducting the experiments a new paradigm for collecting telephone-based dialogue data, called *Wizard and Operator Test* (WOT) — described by Porzel and Baudis (2004) — was employed. This procedure represents a simplification of classical end-to-end experiments, as it is — much like *Wizard-of-Oz* (WOZ) experiments — conductable without the technically very complex use of a real conversational system. As postexperimental interviews showed, this did not limit the feeling of *authenticity* regarding the simulated conversational system by the human subjects. The WOT setup consists of two major phases that begin after subjects have been given a set of tasks to be solved with the telephone-based dialogue system:

- In `Phase 1` the human assistant is acting as a wizard who is simulating the dialogue system, much like in WOZ experiments, by operating a speech synthesis interface.

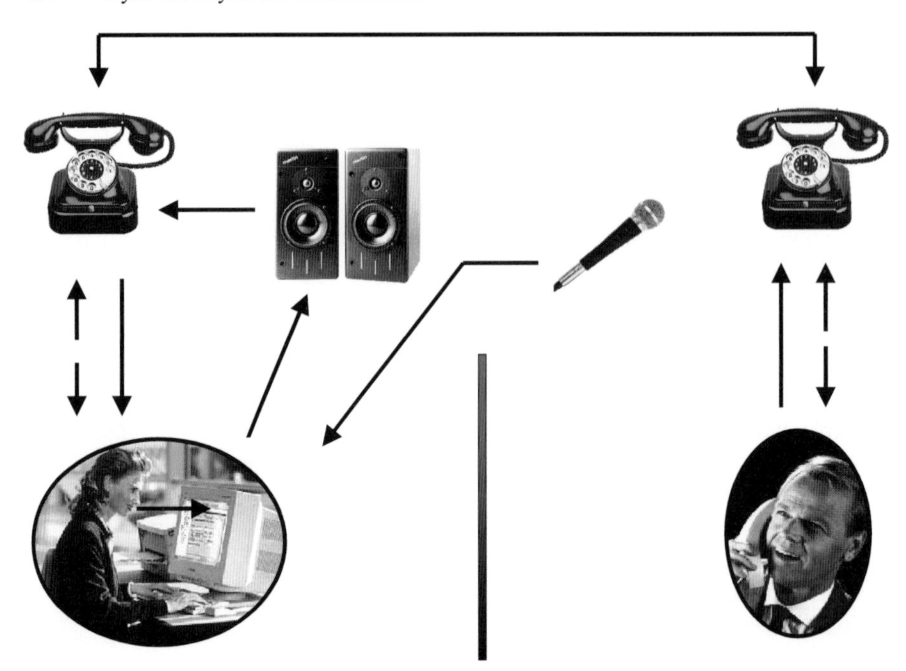

Fig. 3. Communication in `phase 1` goes from synthesized speech out of the loudspeakers into the operator room (*left side*) telephone to the subject room (*right side*), and in `phase 2` directly via the telephone between operator and subject

- In `Phase 2`, which starts immediately after a system breakdown has been simulated by means of beeping noises transmitted via the telephone, the human assistant is acting as a **human** operator asking the subject to continue with the tasks.

During the experiment the subject and the assistant were in separate rooms. Communication between both was conducted via telephone, i.e., for the user only a telephone was visible next to a radio microphone for the recording of the subject's linguistic expressions. As shown in Fig. 3, the assistant/operator room featured a telephone as well as two computers — one for the speech synthesis interface and one for collecting all audio streams. Loudspeakers were also present for feeding the speech synthesis output into the telephone and a microphone for the recording of the synthesis and operator output. With the help of an audio mixer all linguistic data were recorded time-synchronously and stored in one audio file. The assistant/operator acting as the computer system communicated by selecting fitting answers for the subject's request from a prefabricated list, which were returned via speech synthesis through the telephone. Beyond that, it was possible for the assistant/operator to communicate over telephone directly with the subjects when acting as the human operator.

The experiments were conducted with an English setup, with subjects and assistants at the International Computer Science Institute in Berkeley, USA, and with

Table 7. Descriptive corpus statistics of the English data (dataset E) and the German data (dataset G)

	Dataset E	Dataset G
Number of dialogues	22	22
Dialogue average length (min.)	5:53	4:57
Human–operator average length (min.)	2:30	1:52
Human–computer average length (min.)	3:23	2:59

a German setup, with subjects and assistants at the European Media Laboratory in Heidelberg, Germany. Both experiments were otherwise identical, with 22 sessions recorded in each. At the beginning of the WOT, the test manager told the subjects that they were testing a novel telephone-based dialogue system that supplies touristic information on the city of Heidelberg. In order to avoid the usual paraphrases of tasks worded too specifically, the manager gave the subjects an overall list of 20 very general touristic activities, such as *visit museum* or *Museum besuchen*, from which each subject had to pick six tasks that had to be solved in the experiment. The manager then removed the original list, dialed the system's number on the phone and exited from the room after handing over the telephone receiver. The subject was always greeted by the system's standard opening: *Welcome to the Heidelberger tourist information system. How can I help you?* After three tasks were finished — whether successful or not — the assistant simulated the system's breakdown and entered the line by saying *Excuse me, something seems to have happened with our system, may I assist you from here on*, and finishing the remaining three tasks with the subjects. Table 7 gives an overview of the data collected in the WOT experiments. Overall, the subjects featured approximately proportional mixtures of gender (25 male, 18 female), age (12< >71) and computer expertise. The average dialogue (German and English) consisted of 12.3 turns (composed of an average of 8.3 turns per dialogue in `Phase 1` and 16.3 turns in `Phase 2`).

3.2 Measurements on the Data

These data enable analyses of the datasets that are language-specific as well as cross-linguistic analyses on the entire data, such as presented by Porzel and Baudis (2004). In their analyses pauses (i.e., silences over 1 second) not caused by system latency times, overlaps in speech, dialogue structuring signals (e.g., *well*, *yes* or *OK*) and feedback-channeling signals (Yngve, 1970) were in part manually and in part automaticall tagged and measured, yielding the results shown in Table 8.

As the primary effects of the human-directed language exhibited by today's conversational dialogue systems, the data collected in these experiments clearly show that:

- Dialogue efficiency decreases significantly even beyond the effects caused by latency times.

Table 8. Measurements performed on the English data (dataset E) and on the German data (dataset G)

	Dataset E	Dataset G
Dialogue average turns	14,33	10,28
Human–operator average turns	21.25	11.35
Human–computer average turns	7.4	9.2
Pauses	115	89
Pauses human–operator	21	10
Pauses human–computer	94	79
Overlaps	92	56
Overlaps human–operator	88	49
Overlaps human–computer	4	7
Dialogue-structuring signals	292	317
Signals human–operator	202	225
Signals human–computer	90	112
Feedback particles	43	153
Feedback human–operator	43	135
Feedback human–computer	0	18

- The human interlocutor ceases in the production of feedback signals, but still attempts to use his or her turn signals for marking turn boundaries, which, however, remain ignored by the system.
- The increases in the number of pauses is caused by waiting and uncertainty effects, which are also manifested by missing overlaps at turn boundaries.

4 Collection and Usage of the Field Operative Test Data

In an initial data collection we asked American native English speakers to imagine that they were tourists in Heidelberg, Germany, equipped with a small personal computer device that understands them and can answer their questions (Porzel and Gurevych, 2002). Among tasks from hotel and restaurant domains subjects also had to ask for directions to specific places. In the corpus we find 128 instances of instructional requests out of a total of roughly 500 requests from 49 subjects. The types and occurrences of these categories are in Table 9.

As can be seen by looking at category B, *Where interrogatives* are pragmatically ambiguous, i.e., they request either spatial localizations or spatial instructions. In order to perform an empirical study examining this phenomena under realistic conditions we performed the *field operative test* described below.

4.1 Data Collection Setup

Based on the results from these observations we conducted a *field operative test* (FOT) in which field operatives asked people on the street using interrogatives of

Table 9. Request types and occurrences

Type	Example	Occurrences
(A) How interrogatives	*How do I get to the Fischergasse*	38
(B) Where interrogatives	*Where is the Fischergasse*	37
(C) What/which interrogatives	*What is the best way to the castle*	18
(D) Imperatives	*Give me directions to the castle*	12
(E) Declaratives	*I want to go to the castle*	12
(F) Existential interrogatives	*Are there any toilets here*	8
(G) Others	*I do not see any bus stops*	3

Table 10. Request types and occurrences

Factors	Values
The goal object	The castle, city hall, school, discotheque, cinema, bank (ATM), clothing store
The time of day	Morning, afternoon, evening
The proximity to the goal object	Near , medium, far
The approximate age group	Young, middle, old
The gender of the subjects	Male, female
The weather conditions	Warm, cold, rainy, dry
The operative's means of transportation	On foot, bicycle
The baggage situation of the question	With, without luggage
The accessibility of the goal object	Open, closed

type A, B, C, D, E, and F (Table 9). They asked, e.g., `Where interrogatives`, such as *Excuse me, can you tell me where the cinema Europa is*, or `How interrogative`, such as *How do I get to the castle*, or `Existential Interrogatives`, such as *Is there an ATM here*. The passerby's responses were not recorded as that would have required their permission and thwarted a *natural* response. We logged and varied several factors shown in Table 10.

We asked 366 subjects, and their responses were then immediately categorized and logged according to their type, i.e., spatial descriptions, instructions, questions, etc., and their specific features, e.g., suggested means of transport (by bus, foot, etc).

4.2 Usage of the Field Data

This set of collected and categorized data was used to train classifiers and extract decision trees via standard machine learning algorithms. For example, as reported by Porzel and Gurevych (2003), the results of generating and rules applying a c4.5 learning algorithm (Winston, 1992) show that:

- If the object is currently closed, e.g., a discotheque or cinema in the morning, almost 90% of the *Where interrogatives* are answered by means of localizations, a few subjects asked whether we actually wanted to go there now, and one subject gave instructions.

- If the object is currently open, e.g., a store or ATM machine in the morning, people responded with instructions, unless — and this we did not expect — the goal object is near and can be localized by means of a reference object that is within line of sight.

5 Conclusion

We described three experimental paradigms for the collection of data employed in the dialogue system design. We showed that issues in dialogue processing can be formulated as empirical problems, and translated to annotation schemata and annotated corpora. The corpora can be employed as training data for statistical models and as a basis for symbolic models, such as domain models. A further advantage of our approach is a straightforward possibility for thorough evaluations of the developed algorithms. This is especially important for improving the quality and reliability of intuitive dialogue systems.

Acknowledgments

This work was partially funded by the German Federal Ministry of Education and Research (BMBF) as part of the SMARTKOM project under Grant 01 IL 905 C0 and by the Klaus Tschira Foundation. The authors would like to thank Dr. Michael Strube, Nicola Kaiser, Stefani Nellen, Florian Hillenkamp, Klaus Rüggenmann and Christof Müller for their help in designing and conducting the experiments.

References

J.F. Allen, B. Miller, E. Ringger, and T. Sikorski. A Robust System for Natural Spoken Dialogue. In: *Proc. 34th Annual Meeting of the Association for Computational Linguistics*, pp. 62–70, Santa Cruz, CA , June 1996.

G. Bailly, N. Campbell, and B. Mobius. ISCA Special Session: Hot Topics in Speech Synthesis. In: *Proc. EUROSPEECH-03*, pp. 37–40, Geneva, Switzerland, 2003.

J. Carletta. Assessing Agreement on Classification Tasks: The Kappa Statistic. *Computational Linguistics*, 22(2):249–254, 1996.

R.V. Cox, C.A. Kamm, L.R. Rabiner, J. Schroeter, and J.G. Wilpon. Speech and Language Processing for Next-Millenium Communications Services. *Proc. IEEE-2000*, 88(8):1314–1337, 2000.

C. Darves and S. Oviatt. Adaptation of Users' Spoken Dialogue Patterns in a Conversational Interface. In: *Proc. ICSLP-2002*, Denver, CO, 2002.

I. Gurevych, R. Malaka, R. Porzel, and H.P. Zorn. Semantic Coherence Scoring Using an Ontology. In: *Proc. Human Language Technology Conference / North American Chapter of the Association for Computational Linguistics Annual Meeting 2003*, pp. 88–95, Edmonton, Canada, 2003.

I. Gurevych and R. Porzel. Using Knowledge-Based Scores for Identifying Best Speech Recognition Hypotheses. In: *Proc. ISCA Tutorial and Research Workshop on Error Handling in Spoken Dialogue Systems*, pp. 77–81, Chateau-d'Oex, Switzerland, 28–31 August 2003.

I. Gurevych, R. Porzel, and M. Strube. Annotating the Semantic Consistency of Speech Recognition Hypotheses. In: *Proc. 3rd SIGdial Workshop on Discourse and Dialogue*, pp. 46–49, Philadelphia, PA, July 2002.

C. Müller and M. Strube. Multi-Level Annotation in MMAX. In: *Proc. 4th SIGdial Workshop on Discourse and Dialogue*, pp. 198–207, Sapporo, Japan, 4–5 July 2003a.

C. Müller and M. Strube. A Tool for Multi-Level Annotation of Language Data. In: *Proc. Diabruck: 7th Workshop on the Semantics and Pragmatics of Dialogue*, pp. 199–200, Wallerfangen, Germany, 4–6 September 2003b.

R. Porzel and M. Baudis. The Tao of CHI: Towards Felicitous Human-Computer Interaction. In: *Proc. Human Language Technology Conference / North American Chapter of the Association for Computational Linguistics Annual Meeting 2004*, Boston, MA, 2004.

R. Porzel and I. Gurevych. Towards Context-Adaptive Utterance Interpretation. In: *Proc. the 3rd SIGdial Workshop on Discourse and Dialogue*, pp. 90–95, Philadelphia, PA, 2002.

R. Porzel and I. Gurevych. Contextual Coherence in Natural Language Processing. In: P. Blackburn, C. Ghidini, R. Turner, and F. Giunchiglia (eds.), *Modeling and Using Context*, LNAI 2680, Berlin Heidelberg New York, 2003. Springer.

S. Rapp and M. Strube. An Iterative Data Collection Approach for Multimodal Dialogue Systems. In: *Proc. 3rd Int. Conf. on Language Resources and Evaluation (LREC 2002)*, pp. 661–665, Las Palmas, Spain, 2002.

L. Romary, M. Strube, and D. Traum. Best Practice in Empirically-Based Dialogue Research. In: *Proc. Diabruck: 7th Workshop on the Semantics and Pragmatics of Dialogue*, Wallerfangen, Germany, 4–6 September 2003.

K. Rüggenmann and I. Gurevych. Assigning Domains to Speech Recognition Hypotheses. In: S. Bangalore and H.K.J. Kuo (eds.), *Proc. HLT-NAACL 2004 Workshop: Spoken Language Understanding for Conversational Systems and Higher Level Linguistic Information for Speech Processing*, pp. 70–77, Boston, MA, 2–7 May 2004. Association for Computational Linguistics.

P.H. Winston. *Artificial Intelligence*. Addison-Wesley, Reading, MA, 1992.

R. Wooffitt, N. Gilbert, N. Fraser, and S. McGlashan. *Humans, Computers and Wizards: Conversation Analysis and Human (Simulated) Computer Interaction*. Brunner-Routledge, London, UK, 1997.

V. Yngve. On Getting a Word in Edgewise. In: *Papers From the 6th Regional Meeting of the Chicago Linguistic Society*, Chicago, IL, 1970.

M. Zoeppritz. Computer Talk? Technical Report 85.05, IBM Scientific Center Heidelberg, 1985.

Evaluation of Multimodal Dialogue Systems

Florian Schiel

Bavarian Archive for Speech Signals
c/o Institut für Phonetik und Sprachliche Kommunikation
Ludwig-Maximilians-Universität Münchenn, Germany
schiel@bas.uni-muenchen.de

Summary. In this chapter we will give a brief overview about what different methods of evaluation were applied to the SmartKom prototypes and which of them resulted in utilizable results and why others did not. Since there are no established benchmarking methods yet for multimodal HCI systems and very few debatable methods for monomodal dialogue systems our work within the SmartKom prototype evaluations was rather an exploration of new methods than the simple application of standard routines.

1 Introduction

The evaluation of a human–computer interface (HCI) is not a very well defined term. In general, we understand the term "evaluation" as a reproducible experiment or procedure involving human subjects[1] and the HCI which results in some kind of qualitative or quantitative statement about the quality of the HCI system or parts of it. Again, the term "quality" in this context is not very well defined either.

Therefore the "evaluation of a multimodal dialogue system" can have many different realizations with regard to

- which parts (modules, modalities) of the HCI are involved
- type and number of subjects
- which tasks the subjects have to solve with the HCI
- what data are measured during evaluation
- what kind of questionnaire is used for the subjects
- the analysis of the recorded data
- the combination of subjective and objective results

For SmartKom, being a multimodal system with three different technical scenarios[2] and many different task domains, the possible space of evaluations gets even larger.

[1] Referred to as "subjects" or "users" in what follows.
[2] Public, Mobile and Home.

The framework of this book does not allow us to list all the results en detail. The interested reader may refer to some of our publications (Beringer et al., 2002c,b) or contact the authors for the numerous project internal reports. [3]

In the following sections we first will explain the logistical and structural methods of the different SMARTKOM evaluations carried out and then give a short overview of each individual evaluation, and list and discuss the most prominent results or the possible reasons for failure.

2 Methods of Evaluation

The SMARTKOM evaluations can roughly be classified into the broader categories "subjective" and "objective" as well as into "accompanying" and "benchmark" evaluations.

2.1 Accompanying Versus Benchmark Evaluation

In the SMARTKOM project we applied two logistically different types of evaluation while applying the same evaluation techniques to both:

"Accompanying" evaluation denotes an ongoing evaluation during the design process of the prototypes. The motivation for this is quite clear: Flaws in the design should be detected as early as possible and removed. The ideal situation would be to have an ongoing evaluation for all three technical scenarios Public, Home and Mobile for the total period of development. Due to financial and time constraints this was not possible. Therefore only two accompanying evaluations were carried out during the SMARTKOM project.

The results of an accompanying evaluation may theoretically be used to quantify and document the progress of development, because the prototype may change (hopefully for the better) during the evaluation phase. Since we could carry out only two evaluations during development, the only progress documented here is that from development phase to the final prototype.

"Benchmark" evaluation is our term for an evaluation of a "frozen" prototype; a prototype is delivered to the evaluation team and may not be updated during the evaluation phase (which may last for more than a month). In SMARTKOM seven benchmark evaluations were carried out on the subfinal prototypes [4]: three subjective, three objective and one long-term subjective evaluation.

2.2 Subjective Versus Objective

All serious HCI evaluations involve experiments with human subjects as well as the measurement of system parameters. The former usually take place in form of

---

[3] In German only.

[4] After the delivery of the "final" prototype 4.2, another prototype version 5.1 was assembled. The evaluation of version 5.1 in an outdoor setting is described in Sect. 3.3.

questionnaires or interviews before/after the user has tested the system, while the latter are sampled from the system itself during the test runs with the user. The reason for this often-found combination is simply that we do not know which combination of measurable parameters of a HCI system provides an adequate performance score for the system. Or, with other words, the overall performance or quality of a complex HCI system cannot be measured like, for instance, the word accuracy of a speech recognition system.

For example, it is quite easy to measure the delay between the end of a user input and the beginning of the corresponding system output. However, it is not clear a priori that a shorter measured period improves the overall success of the dialogue between user and machine. And if it does, is it an important or less important parameter?

Some people argue that the total is the sum of all components, i.e., if we can objectively measure the performance of each module of a HCI system, then the overall performance should be the sum of all single performances. Since Verbmobil (Burger et al., 2000) we know that this is not even true for a system with 100% optimal harmonized interfaces between the different modules, because one single defective module can crush the entire processing chain of a temporal linear structured information processing system.

Therefore both, the subjective evaluations of subjects and the measurable parameters of a life session, are the basis for the advanced analysis of the quality of the HCI system. In the following we will discuss fundamental techniques for both methods. [5]

2.2.1 Subjective: Speaker Questionnaires/Interviews

Subjective evaluations about a HCI system may be obtained by a questionnaire or — better, but much more time-consuming — by an interview. For all subjective evaluations the rule applies: what you ask, is what you get. We therefore conducted a pretest (Kartal et al., 2002a) in the early phase of SMARTKOM to come up with a reasonable and usable catalogue of questions to the subjects.

In general, we distinguish the following types of questions:

- Questions with categorized answers (*codes*). This includes *multiple choice*, *ranking* (usually on a scale of 7) and answers that allow *defined quantities* only (e.g., the height in centimeters).
- *Cross questions* that are asked in the same way before and after the experiment to detect changes caused by the experience.
- Questions that may be answered in free texts (*free questions*).
- *Subject questions* that are not directly related to the experiment but characterize the background of the subject (e.g., about computer experience).

The subjective evaluation should provide answers to the following general questions (Beringer et al., 2002a):

- How do subjects evaluate the efficiency of SMARTKOM to solve practical tasks?

[5] Other terms often used for "subjective" and "objective" are "ergonomic" and "technical" or "modular."

- How do subjects evaluate input and output modalities and the dialogue?
- What kind of (subjective) problems occurred?
- How do users accept the system?
- How do subjects accept the biometric identification of SMARTKOM?
- Who do the subjects think might use such a system?

For the SMARTKOM evaluation this concept led to a first catalogue of 62 questions (Questionnaire 1 (Kartal et al., 2002a)) in the pretest that was conducted with participants of the Wizard-of-Oz recordings (Schiel and Türk, 2006): [6]

- 2 cross questions about how the user feels about emotion detection
- 7 subject questions (codes)
- 5 PARADISE questions (codes, see below)
- 8 general ergonomic questions (ranking)
- 10 ergonomic questions (free questions)
- 6 + 6 specific evaluations of SMARTKOM techniques (ranking)
- 11 questions about potential user groups (ranking)
- 7 questions about the usability of certain SMARTKOM domains (ranking)

Categorized answers can be easily summarized (mean and median). Free questions require a rather tedious manual processing, often combined with the finding of new categories. The interpretation of results is tricky, especially when the number of subjects is low.

As for the technique, we used HTML-based questionnaires that entered the answers directly into a simple database. We also tested face-to-face interview techniques but did not pursue these due to the costs.

After the analysis of the pretest an improved version was compiled as follows (Questionnaire 2, e.g., Kartal et al. (2002b)):

- 2 cross questions about how the user feels about emotion detection (ranking)
- 13 subject questions (codes)
- 11 general ergonomic questions (ranking)
- 8 evaluations of special SMARTKOM techniques (ranking)
- 11 questions about potential user groups (ranking)
- 5 questions about the usability of certain SMARTKOM domains
- 1 free question about problems and suggestions
- 1 free question about the communication in general
- 1 free question about additional input/output techniques
- 1 free question about the preferred identification technique
- 7 evaluations of the input/output modalities (ranking)
- 1 free question about what is most annoying when using SMARTKOM
- 1 question about how much the subject would pay for a system like that
- 1 free question about the most positive and most negative action of SMARTKOM
- 14 control questions with a different ranking system (ranking)

[6] Note that the participants in the WOZ recordings believed they were communicating with a real technical system.

This results in a catalogue of 78 questions total that was used in all accompanying SMARTKOM evaluations (Sect. 3.1).

For the final benchmark evaluations of all three prototypes (Sect. 3.2) this questionnaire was boiled down to ten free and 31 ranking questions (Questionnaire 3). All questions that were interpreted ambiguously by the majority of the subjects were removed; some additional questions were added. The ten free questions are as follows (abbreviated):

- Describe your expectations of the SMARTKOM system.
- Was the operating of SMARTKOM unambiguous?
- Did you use a functionality that the supervisor did not mention in her instruction?
- How did SMARTKOM help you to operate the system?
- Did you make mistakes? Which ones?
- How did SMARTKOM help you to avoid such mistakes?
- Did SMARTKOM come up with correct answers although you made a mistake?
- How did SMARTKOM fulfil your requests/wishes?
- Was the output (voice and display) always clearly understandable?
- Could you finish your task in an appropriate period of time?

Please refer to Beringer et al. (2003a) for a detailed listing.

2.2.2 Objective: Technical Evaluation

The technical or objective evaluation exploits two sources: the internal parameters of the HCI system itself, e.g., the recognized string of words produced by the speech recognition module, and labelings and segmentations of the recorded modalities during the test which must be produced by a labeler group after the experiment. The latter are needed to provide measurable parameters such as *word accuracy*, *gesture recognition* and *user state recognition*. This implies, of course, that for the evaluation it is not sufficient to run the prototype alone, because the prototype itself will not record, for instance, the video stream of the facial camera. In our setting we augmented the prototype system by the required recording equipment, such as recording of voice input and output, facial video, infrared video (SiViT stream) and system display output. Also, the labeling of the recorded data is a very time-consuming process, which explains that some results of the benchmark evaluations were published long after the termination of the project.

In total, 110 parameters were derived from the running system as well as from the labelings of the recorded data streams. Please refer to Beringer et al. (2002d) for a complete listing. As expected, the majority of the parameters turned out to contribute none or only very little to an overall performance score. Consequently, the number of analyzed parameters was reduced to 30 in the final benchmark evaluations to achieve more utilizable results.

2.2.3 Task Success: End-to-End Evaluation

The subjective and objective measures of an evaluation may be combined into a general score representing the overall performance. However, there are situations where both measures are inadequate to judge the primary success of a human–machine dialogue, namely whether the given task has been solved efficiently or maybe not at all. For lack of a better term, we call this abstract concept "task success." It should be clear that the task success cannot be easily derived from the subjective or objective measures: a dialogue might be successful even if the majority of technical measures scored rather low or the user gave a bad evaluation because she had a different idea on how to solve the task, and vice versa. The task success is therefore judged by objective guidelines based on the recorded input and output information streams only. Because of this "black box" approach[7] we often call this type of evaluation "end-to-end."

In PARADISE (Walker et al., 1997), for instance, this third pillar is the normalized task success κ computed from the attribute value matrix (AVM). In PROMISE (see below) κ was not computable, because SMARTKOM tasks are much more fuzzy than the strictly aimed task of accessing a certain information from a database system.

2.3 PARADISE and PROMISE

As said before, the optimal evaluation strategy combines subjective and objective results as well as the task success of a series of experiments in a way that an overall quality score can be computed from the measured data, which is comparable across different domains and desirably even across different technical scenarios. In Walker et al. (1997), Walker and her colleagues proposed the PARADISE method to calculate comparable performance scores for voice dialogue agents. This method was applied successfully to several different monomodal dialogue systems (Walker et al., 1998). When trying to apply the PARADISE method to SMARTKOM we ran into several problems. The most prominent are:

- SMARTKOM uses multiple asynchronous modalities in combination, and therefore it is not always possible to assign labels of correctness to single recognition results.
- In typical SMARTKOM dialogues the tasks are much more fuzzy than in a simple database access task and also may be solved in very many different ways, simply because SMARTKOM has no strict dialogue control and allows the user to ask almost everything at all times.

In Beringer et al. (2002b) we proposed an adaptation of PARADISE called "Procedure for Multimodal Interactive System Evaluation" (PROMISE) that should work with a system with more freedom such as SMARTKOM. In a nutshell, PROMISE replaces the κ calculated from the AVM by a much broader "task success" score that may have only two possible values $+1$ and -1, while the multilinear regression analysis between subjective judgments and measured cost functions was replaced by

[7] What happens in the "black box" is not relevant, only the outcome of the dialogue.

a simple regression analysis (t-test).[8] By applying PROMISE we could — at least on a formal level — avoid the systematical problems of PARADISE when applied to SMARTKOM.

2.4 Task Success Versus Request Success

The task success of PROMISE is, of course, a rather ambivalent concept, because it is based on a whole recording session (which typically lasts 10 min). Therefore it was not surprising that the labeling in the end-to-end evaluation turned out to be a difficult task: in many cases it was almost impossible to assign the labels "success" (+1) or "failure" (−1) to a session. After the accompanying evaluations were completed we took another step and defined a more fine-grained value called "request success" that is assigned in the end-to-end evaluation to pairs or groups of turns between the subject and the system according to the following rules:

1. Each dialogue is automatically segmented into blocks.
   Each block starts with a regular user request. A regular user request must address the system in a clear and cooperative way. Irregular requests may include incomplete turns, hesitations, off-talk, ungrammatical sentences, inconsistent multimodal inputs, not understandable utterances; these are discarded from the analysis including the possible following system reply. Each block must end with a system reply.[9]
2. Each block is then evaluated with 1.0, 0.5 or 0.0 points.
   - 1.0 points : The user gets a complete and correct reply to her request. For example:
     `USR: ''Show me the TV program!''`
     `SMA: ''Here are the shows running at the moment.''`
     (displaying the correct program of all channels)
   - 0.5 points : The user gets a complete and correct reply to her request, but the system provides *additional and unnecessary* information so that the user has to filter out the requested items. For example:
     `USR: ''Which TV shows are on ARD tomorrow?''`
     `SMA: ''Here are the shows that will run tomorrow.''`
     (displaying the program of all channels)
   - 0.0 points : All other cases. For most SMARTKOM sessions this results in basically three possibilities:
     - no reply at all
     - a wrong or incomplete reply
     - an error message

[8] Of course, this implies that only judgments of the type *ranking* may be used in the PROMISE method, because a t-test correlation is only meaningful across a certain range of possible values.

[9] Please keep in mind that a user request may be just a gesture, while a system reply may also be just a change of display.

3. The total request success per session is calculated.
   Since there exist (rare) cases where in one block more than one user request and system answer take place,[10] there are two possible ways to average the individual judgments: simple average (sum all points and divide by number of requests) or blockwise average (average over requests within each block, then average over all block values). Although both methods show very similar results, it turned out that the blockwise average reached more significant correlations with technical measures. Therefore we decided to use the blockwise averaging.

This procedure results in an average request success value between 0.0 and 1.0 for each session; 0.0 denotes a complete failure, while 1.0 would — theoretically — suggest that all cooperative requests of the subject were satisfied perfectly.

3 SmartKom Evaluations

We are now ready to list the results of the individual SMARTKOM evaluations. In this contribution we will only present the most prominent results.[11]

3.1 Accompanying Evaluations — Home, Mobile

The two accompanying evaluations took place from 1 April 2002 to 31 June 2002 (scenario Home) and from 1 Aug to 31 Aug 2002 (scenario Mobile). Both a full ergonomic and a technical evaluation were carried out.

3.1.1 Technical Setup Home

For the Home scenario the studio was converted into a living room-like environment with sofa, low chair, TV and VCR. The graphical output was displayed on a portable Web pad Fujitsu Siemens, Stylistic 3500 X Pen Tablet (Windows NT, 256 MB) that the subject held on her knees. A pen was used for gesture input. The input microphone was a standard low-cost headset[12] connected to the Web pad. The display and audio modules were running on the Web pad; the Web pad was connected via standard cable network with the remaining prototype cluster (2 x Linux 900/1800 MHz, 1280 MB RAM total).

[10] This happens when the user does not wait for the answer to her first request but utters another request:

```
USR: ''What shows are on TV tonight?''
USR: ''Show me the program of today!''
SMA: ''Sorry, I did not understand you.''
SMA: ''Here are the shows running at the moment.''
```

[11] Please refer to Kartal et al. (2002b); Beringer et al. (2002d); Kartal et al. (2003); Beringer et al. (2003d,a,c,b); Jacobi et al. (2004); Schiel (2004) for a more detailed version of the evaluation results.

[12] Approved by system group.

3.1.2 Ergonomic Accompanying Evaluation Home

Prototype	Subjects	Male/Female	Naive/Experts	Quest.	Remarks
2.1	27	14m / 13f	11n / 10e	2	2 L2 speakers

General Results

The overall picture was rather negative. Subjects criticized long waiting periods between request and answer, nonexistent user guidance and that they felt left alone by the system. The topic of biometric identification seemed to be of high importance to almost all subjects, but with very mixed messages. Nearly all technical aspects of the SmartKom prototype have been evaluated negatively. The only exception was the "persona," although some users doubted that actions like blowing bubble gum or counting fingers were really necessary.

It turned out that naive users were not able to use the system on their own; even experts rarely completed the given task. Very often the long processing periods between request and answer led to delayed answers that were misinterpreted by the subjects.

A detailed discussion of all results can be found in Kartal et al. (2002b).

Most Prominent Problems (Number of Judgments in Brackets)

"SmartKom does not understand me" (13)
"System too slow" (12)
"Pointing gestures, help function, volume control do not work" (8)
"Bad/not working user guidance; actions of 'persona' are hard to understand" (12)
"Did not know what to say" (2)
"System crashed" (1)

Most Prominent Suggestions

"Improve speech recognition" (10) [13]
"Improve speed" (16)
"Improve pen input" (4)
"More functionalities" (2)
"More feedback about what the system is doing" (10)
"Better help" (2)

3.1.3 Technical Accompanying Evaluation Home

Prototype	Subjects	Parameters	Naive/Experts	Remarks
2.1	21	110	11n / 10e	–

[13] This probably did not refer exclusively to the speech recognition module alone, but to the complete "speech understanding."

General Results

The overall picture was rather negative, quite in correspondence to the ergonomic evaluation. Recognizers of all modalities scored rather low; there was not one correct treatment of off-talk and barge-in. The processing time between request and answer was extremely high; as a follow-up it almost uniformly happened that new requests were uttered before the answer to the first request was given. The correct answers to correct recognitions were very few. The dynamic help module was activated very often, but the impact to the dialogue was not effective. The only positive measurements were the output gesture of the persona (75% correct) and the syntax of the speech output (100%). For a detailed listing of all measurements please refer to Beringer et al. (2002d).

The PROMISE value could not be calculated reliably because the correlations between subjective judgments and measurements were in most cases not significant.

3.1.4 Technical Setup Mobile

The Mobile scenario was evaluated in the same studio. An evaluation in the field was not possible because of technical reasons. The subject was asked to stand or walk within a certain area where two video cameras could capture the user actions. The subject was holding the PDA (Compaq IPAQ, embedded Linux, graphical output displayed via VNC) in her left hand and the pen in her right. Input microphone was a headset Sennheiser Evolution Wireless E100 connected directly to the Windows 2000 machine of the prototype cluster (1 x Windows 2000 1800 MHz and 2 x Linux 900/1800 MHz, 1536-MB RAM total).

3.1.5 Ergonomic Accompanying Evaluation Mobile

Prototype	Subjects	Male/Female	Naive/Experts	Quest.	Remarks
3.1	16	7m / 9f	11n / 5e	2	2 L2 speakers

General Results

The overall picture was rather negative; in comparison to the previous evaluation in scenario Home it was even worse. Subjects criticized that in most cases a real interaction was not happening. Processing periods between request and answer were extremely high so that most subjects had doubts whether the system had already crashed. Seven subjects claimed that no other display output happened than the greeting of the persona.

It turned out that naive users were not able to use the system on their own; even experts rarely completed the given task. Very often the long processing periods between request and answer led to delayed answers that were misinterpreted by the subject. A detailed discussion of all results can be found in Kartal et al. (2002b).

Most Prominent Problems

"No communication at all" (6)
"System did not understand me" (6)
"System did not react to anything" (7)
"System too slow" (1)

Most Prominent Suggestions

"Better use single commands instead of sentences" (5)
"Better user guidance: signal when to speak, display should indicate what can be done next" (5)
"System should react faster so that you get a feeling of a real dialogue" (3)
"Input should be improved" (1)
"Better gesture input" (1)

3.1.6 Technical Accompanying Evaluation Mobile

Prototype	Subjects	Parameters	Naive/Experts	Remarks
3.1	18	110	12n / 6e	–

General Results

We found a low performance of all recognizers. Off-talk and barge-in were not handled correctly. Very long processing times occurred between request and answer; this caused double requests that were processed in parallel. Very few correct system replies were encountered; also very few correct input recognitions. For a detailed listing of all measurements please refer to Beringer et al. (2003d).

Unfortunately, in many sessions very few interactions between subject and system took place. Consequently, we had only very few usable measurements; in some cases the experiment was terminated prematurely because the system seemed to have crashed.

The PROMISE value could not be calculated reliably because the correlations between subjective judgments and measurements were in most cases not significant.

3.1.7 Summary Accompanying Evaluations

The results of both accompanying evaluations were fed back into the design process of SMARTKOM. Because of the negative experiences with naive users we decided to recruit only users that were familiar with SMARTKOM in the forthcoming benchmark evaluations. Also, the question arose whether a new and unusual interaction technique such as used in SMARTKOM should not be evaluated during a longer time span, because we learned from the first evaluations that most concepts and interaction patterns of SMARTKOM have to be learned or adapted to. Unfortunately a long-term evaluation clashes with the fundamental claim of every evaluation, namely that the results should be statistically valid across different subjects and therefore needs at least 10–20 subjects per evaluation experiment. Our budget and time constraints did not allow us to blow up the three final benchmark evaluations so that every subject could test the system over a longer period. We therefore decided to stick with the plan and evaluate short sessions of a larger group of subjects while performing one or two long-term evaluations in parallel.

3.2 Benchmark Evaluations — Home, Public, Mobile

The three benchmark evaluations took place 10–20 March 2003 (Home), 11–13 June 2003 (Public) and 4–11 July 2003 (Mobile). A full ergonomic evaluation as described in Sect. 2.2.1 was carried out for each scenario. However, since we recruited only "experts" for the benchmark, evaluations, we changed the instruction phase to the following:

- The supervisor instructed the subject verbally, gave the task description and reminded the subject to wait for a system response before uttering the next request.
- Subjects were asked to start the session after a green light signal given from the operator and to conclude the session when the green light turned off.
- Subjects were shown a short video (5 min) demonstrating the proper usage of SmartKom (scenario Home).
- Subjects could perform a short training dialogue of 5 min.
- After the clarification of last questions the evaluation session of 10-min length was started.

During the session the supervisor did not interact with the subject.

After the negative experiences in the accompanying evaluations, the technical evaluation was reduced from 110 to a set of 30 reliable parameters (Jacobi et al., 2004). The main reason for this was the observation that many desirable parameters could not be extracted reliably from the system's log files and therefore did not show the desired correlations to the subjective judgments.

3.2.1 Technical Setup Home

The studio was equipped in a living room-like fashion with sofa, low chairs, TV and VCR. The subject sat in a low chair directly in front of a low table on which the display output of SmartKom was projected.[14] The facial video camera was active but the subject's face was not visible when leaning back ("lean-back mode," see task). Microphone input was via a wireless headset Sennheiser E100 connected to the Windows 2000 machine of the prototype cluster (1 x Windows 2000 1800 MHz and 2 x Linux 2800 MHz, 4608-MB RAM total). The subject was not supposed to stand up during the session. The task was: "Inform yourself about the TV program (tonight, tomorrow etc.), get details about certain shows, actors. Try to control your TV and VCR. Try deactivating and reactivating the display ("lean-back mode")."

[14] The Web pad was not used in the final benchmark evaluation because the hardware was not available. Therefore the subject could also use her finger to point.

3.2.2 Ergonomic Benchmark Evaluation Home

Prototype	Subjects	Male/Female	Naive/Experts	Quest.	Remarks
4.2	20	10m / 10f	0n / 20e	3	3 L2 speakers

General Results

The overall results were much more promising than in the previous accompanying evaluation. Improvements were judged mainly in speech understanding, gesture recognition, dialogue handling and shorter processing times. Still a problem seemed to be the missing 'user guidance' and some very long processing periods where the subjects were unsure about the state of the system.

Most Prominent Answers to Free Questions

Subjects criticized in unison that SMARTKOM did not help to learn the operation. About half of the subjects were satisfied with the system; others claimed that SMARTKOM did not understand (4), was too slow (2), did not recognize English movie titles (1), had no "back" button (2), had no emotion detection (1). Most of the subjects showed a high level of self-awareness with regard to their own mistakes. They listed 18 cases where they made a mistake. Unfortunately, SMARTKOM didn't seem to help in correcting those mistakes. The output was judged being more or less OK; the meaning of the "red lamp" was unclear to most of the users. Some subjects criticized the pronunciation of foreign words.

Most Prominent Rankings

The following aspects of SMARTKOM were ranked positively: clearly arranged display, speech understanding, speech input, SMARTKOM is easy to understand, task simple to solve. Negative rankings occurred in: speech output, emotion detection, good support, speed, fast reactions, help function, competence.

3.2.3 Technical Benchmark Evaluation Home

Prototype	Subjects	Parameters	Naive/Experts	Remarks
4.2	20	30	0n / 20e	-

As for the recognizer results, the speech recognition module scored much better than in previous prototype versions (47.1% accuracy), but not as good as in Public (57.8%) and Mobile (63.9%). User state recognition is still low (16.2%) but significantly better than in the Public scenario (8.4%). Gesture recognition reached 38.7% correctness. Crosstalk and off-talk was much lower than in previous recordings, probably due to the fact that the subjects were instructed more strictly. The average duration of a user speech input was significantly longer (2070 msec) than in the other two scenarios (approx. 1760 msec). It is unclear why, possibly because the domain TV allows more complex input sentences than other domains.

The request success (Sect. 2.4) scored an average of 0.48 per session, which was right on the total average (0.47). However, request success values varied from 0.8 to 0.27 per session, which shows that the successful usage of SMARTKOM very much depends on the user.

3.2.4 Technical Setup Public

The subject stood in front of the standard Public setup with the display projected on the graphic tableau, SiViT unit and facial camera. Microphone input was via a wireless headset Sennheiser E100 connected to the Windows 2000 machine of the prototype cluster (1 x Windows 2000 1800 MHz and 2 x Linux 2800 MHz, 4608-MB RAM total). The subject was not supposed to leave the range of the facial camera, but due to body and head movements the face of the subject was not always fully visible in the facial video stream. The task was: "Inform yourself about the cinema program in Heidelberg tonight, get details about certain movies, actors. Find the way to the selected cinema (maps). Try to make a reservation for tickets."

3.2.5 Ergonomic Benchmark Evaluation Public

Prototype	Subjects	Male/Female	Naive/Experts	Quest.	Remarks
4.2	11	6m / 5f	0n / 11e	3	2 L2 speakers

General Results

The overall results for the Public benchmark were very similar to the Home benchmark evaluation. Since there has not been an accompanying Public evaluation we cannot give any detailed improvements. Speech understanding, gesture recognition and dialogue handling seemed to work reasonably well. Two subjects claimed that SMARTKOM actually allowed them to solve the task as fast as with traditional techniques (phone, Internet).

Subjects criticized the missing "user guidance," some very long processing periods so that subjects were unsure about the state of the system and that the error message "I did not understand you!" was not very helpful because it did not relay any information about what was hard to understand.

Most prominent answers to free questions

Subject criticized in unison that SMARTKOM did not help to learn the operation. Five of the subjects were satisfied with the system; others claimed that SMARTKOM did not understand (4), was too slow (1), did not recognize gestures (1), had no "back" button (2) and that the reservation of movie tickets did not work at all (1). Most of the subjects showed a high level of self-awareness with regard to their own mistakes. They listed eight cases where they probably made a mistake. Unfortunately, SMARTKOM did not seem to help in correcting those mistakes. The output was judged as being more or less OK. Five subjects criticized the prosody of the speech synthesis; three subjects said that sometimes the meaning of the output was unclear; 1 subject claimed that speech and display were not in sync when the persona was speaking.

Most Prominent Rankings

The following aspects of SMARTKOM were ranked positively: display output, has a clearly arranged display, correct answers in most of the cases, speech input, SMARTKOM is easy to understand, task simple to solve, SMARTKOM is easy to operate. Negative rankings occurred in: emotion detection, gesture recognition, one input sufficient to reach goal, help function, fast reaction to gesture input, competence.

3.2.6 Technical Benchmark Evaluation Public

Prototype	Subjects	Parameters	Naive/Experts	Remarks
4.2	11	30	0n / 11e	–

The speech recognition module scored in the middle range (57.8% accuracy), better than in Home (47.1%), but not as good as in Mobile (63.9%). User state recognition is very low (8.4%) compared to Home (16.2%). This is surprising because one might think that the more rigid hardware setup of the Public scenario (fixed position of user, fixed background texture) would help the user state recognition. Gesture recognition scored 48.6%, which is significantly better than in Home. In four cases the user interrupted his own gesture or speech input caused by a system output (other scenarios: no occurrence). This might indicate a slightly worse dialogue handling in the Public domains. Crosstalk and off-talk was much lower than in previous evaluations, probably due to the fact that the subjects were instructed more strictly.

The request success (Sect. 2.4) scored an average of 0.45 per session which was slightly below the average (0.47). However, request success values varied from 0.62 to 0.23 per session, which shows that the successful usage of SMARTKOM very much depends on the user.

3.2.7 Technical Setup Mobile

For the Mobile scenario the studio was cleared out as much as possible to give subjects room to walk around. Two cameras were mounted to capture user actions. The subjects were asked to stand or walk within a marked area of the studio during the test. No background noise was played back. The subject was holding the PDA (Compaq IPAQ, embedded Linux, graphical output displayed via VNC) in her left hand and the pen in her right. Input microphone was a headset Sennheiser Evolution Wireless E100 connected directly to the Windows 2000 machine of the prototype cluster (1 x Windows 2000 1800 MHz and 2 x Linux 900/1800 MHz, 4608-MB RAM total). No facial video camera was used and therefore no user state recognition was available. The subject was not supposed to give the PDA out of her hands during the test. The task was: "Inform yourself about touristic sites in Heidelberg, get details about them and look at a map. Try to get directions for a walk (navigation). Also get information about parking sites."

3.2.8 Ergonomic Benchmark Evaluation Mobile

Prototype	Subjects	Male/Female	Naive/Experts	Quest.	Remarks
4.2	10	5m / 5f	0n / 10e	3	1 L2 speaker

General Results

The overall results were more promising than in the previous accompanying evaluation. Improvements were judged mainly in speech understanding and faster reaction to gesture input. On the other hand, the display output was judged much lower than in the other two scenarios. A problem still seemed to be the missing "user guidance" and some very long processing periods where the subjects were unsure about the state of the system. In a total comparison of all answers the scenario Mobile showed the lowest user satisfaction.

Most Prominent Answers to Free Questions

Eight subjects criticized that SMARTKOM did not help to learn the operation. About half of the subjects were satisfied with the system; others claimed that the display was confusing (1), SMARTKOM was too slow (2), did not recognize gestures (1), had an unsatisfactory feedback (2), had no emotion detection (1). Six of the subjects showed a high level of self-awareness with regard to their own mistakes. Five subjects claimed that they solved the task with the help of SMARTKOM in the same amount of time as with traditional techniques.

Most Prominent Rankings

The following aspects of SMARTKOM were ranked positively: pen input, speech understanding, speech input, speech output, gesture recognition (pen). Negative rankings occurred in: emotion detection, good support, speed, fast reactions, clear dialogue, help function and competence.

3.2.9 Technical Benchmark Evaluation Mobile

Prototype	Subjects	Parameters	Naive/Experts	Remarks
4.2	10	30	0n / 10e	-

As for the recognizer results, the speech recognition module scored the best result of all evaluated prototypes (63.9%). Crosstalk and off-talk was much lower than in previous recordings, probably because the subjects were instructed more strictly. The number of turns and the number of words [15] were significantly lower than in the other evaluations; it is not quite clear whether this is a good or bad tendency.

The request success (Sect. 2.4) scored an average of 0.49 per session, which was slightly above the total average (0.47). But again request success values varied from 0.70 to 0.25 per session, which shows that the successful usage of SMARTKOM very much depends on the user.

[15] Both user and system.

3.2.10 Pearson Correlation Subjective — Objective

To achieve significant correlations we ran the analysis across all three scenarios. As for the Pearson correlation between the subjective judgments (ranking only) and the measured parameters, we found only three value pairs with an error probability of less than 0.01. On this basis we did not deem the PROMISE approach applicable to this data.

A total of 17 value pairs were found with an error probability of less than 0.05. For a complete listing and discussion of these correlations please refer to Jacobi et al. (2004). Although some of these pairs (4) did not seem to make any sense, there are some interesting observations:

- A larger average processing time in the speech recognition module is not necessarily an indicator for a bad performance. Users who operate SMARTKOM successfully use larger and more complex requests than others. Consequently the speech recognition module needs more time to process them. This does not hurt the subjective positive judgment of the user.
- Some correlation pairs indicate that frequent interrupts of user actions by the system[16] lower overall performance.
- Unspecified error messages lower overall performance. Also, some correlations indicate that error messages should not be given as speech output alone.
- An unstructured display with too much information causes the users to produce off-talk, which in turn may lower the overall performance.
- Complex requests combined of speech and gestures are valued higher than simple requests, granted that they can be answered successfully.
- Frequent off-talk indicates problems in the interaction and dialogue handling.
- A funny acting or strangely designed persona triggers off-talk, which might lower overall performance.

3.2.11 Pearson Correlation Subjective — Request Success

As for the Pearson correlation between the subjective judgments (ranking only) and the request success per session, we found six value pairs with an error probability of less than 0.01. This suggests that the request success is a rather good overall performance indicator even for a multimodal HCI system.

Since the request success cannot determined automatically from the labeling and segmentation of the session data, it would be quite desirable to predict the average request success from the technical parameters. We therefore trained a simple linear model of technical parameters in the previously found correlation ranking, and tested the amount of explained variance (R^2) with increasing number of parameters (see Fig. 1). The measured parameters in the ranking of their descending correlation are as follows:

[16] For instance, when the user is about to point on the display and the display changes simultaneously.

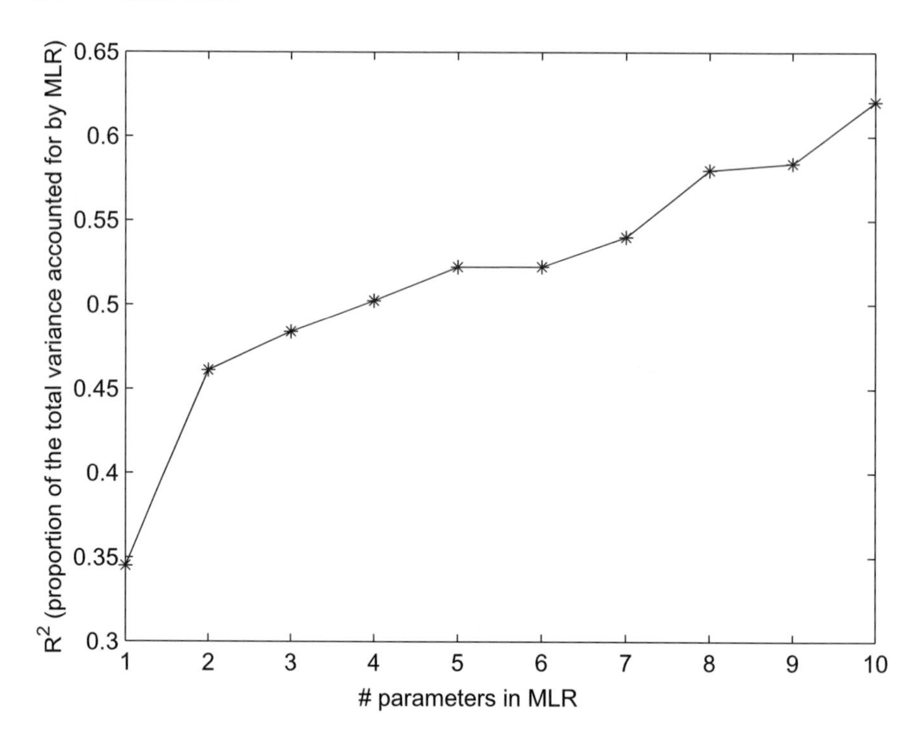

Fig. 1. R^2 explained variance in dependency of measured parameters that were fed into the linear predictor

1.	Number of error messages	* −0.29
2.	Average reaction time of SmartKom	* +0.38
3.	Number of audio turns in dialogue	
4.	Number of user turns in dialogue	
5.	Speech recognition accuracy	* +0.23
6.	Number of cross-talked user words	
7.	Total number of words in dialogue	* +0.84
8.	Total duration of dialogue	* −0.62
9.	Total number of cross-talked words	
10.	Number of SmartKom words in dialogue	* −0.90

Figure 1 shows that 60% of the variance of the request success may be explained with only six measured parameters (marked with "*" in the above listing together with their respective model weight). Although this is not perfect, it might be an alternative solution in cases where neither time nor money are available for the manual labeling and segmentation of request blocks.

3.3 Outdoor Evaluation—Mobile

3.3.1 Introduction

Between December 2003 and March 2004, SMARTKOM-Mobile (version 5.1) was tested in an outdoor setting in Heidelberg.[17] The purpose of this outdoor evaluation was generally similar to that of the benchmark evaluations described in Sect. 3.2, but the specific questions addressed were somewhat different:

1. Can people use SMARTKOM-Mobile effectively in an outdoor setting after a few minutes of instruction?
   Potential problems that could be caused by the outdoor setting include speech recognition problems due to ambient noise and distraction of the users by events and noise in the environment. In accordance with the results of the accompanying evaluations described in Sect. 3.1, it was assumed that previously inexperienced users would require a few minutes of instruction on the tasks that they were about to perform.
2. How do users rate SMARTKOM-Mobile overall, relative to other methods that they might use to accomplish the same tasks?
   Ultimately, users' acceptance of a system like SMARTKOM-Mobile will depend largely on whether they consider it more desirable than other available methods (e.g., conventional tourist guidebooks or more traditional computer-based systems).

Because of the differences in the environment in which the study was conducted and in the nature of the prototype used, the objective and subjective measures collected during this evaluation study were in part different from those of the studies reported in the previous section.

The study was prepared and conducted after the end of the main SMARTKOM project, so that the most advanced prototype from that project could be used as a starting point for the prototype used in the study. Consequently, the analysis of a sizable proportion of the data collected in this study had not been completed at the time of the writing of this book chapter; further results will be reported in later publications.

3.3.2 Method

After indoor pilot testing, the study was conducted in a pedestrian zone in the center of Heidelberg. The system was deployed on two laptops with 1800-GHz processors and with 1-GB RAM each, running Linux and Windows, respectively. A Compaq IPAQ 3630 served as the subjects' handheld device. It communicated with the laptops via a wireless LAN with 2-MBit bandwidth.

[17] The prototype was prepared and the tests were conducted by H. Aras, J. Häußler, M. Jöst and R. Malaka of EML, with consultation from A. Jameson of the Evaluation Center for Language Technology Systems of DFKI's project Collate.

The 8 subjects (6 female, 2 male) were recruited from groups that had no experience with SMARTKOM or other dialogue systems. Their average age was 27.1 years (standard deviation: 4.7 years). Two subjects were not native speakers of German. The subjects had virtually no experience with handheld computing devices (e.g., PDAs or organizers on cell phones) or with automobile navigation systems. Most of them occasionally used mapping systems on the Internet, and all claimed considerable general computer experience.

Subjects used SMARTKOM-Mobile to perform three tasks:

1. Controlling the display of a map: Having all objects of a certain type (e.g., cafés) displayed on the map; zooming in and out
2. Planning a route: Having the system compute and display a walking route between two locations in Heidelberg
3. Requesting information about sights: Having the system display information (a photo plus text) about one or more sights in Heidelberg

In all three tasks, commands to the system were given via speech; in the second and third tasks, pointing gestures could be used as well. The study began with a training phase in which the subject performed one instance of each of these three tasks. For each task, the user was shown a sheet of paper that indicated the actions that could be performed and some typical utterances (some of which were combined with gestures) that the system would understand. The user was asked to perform the actions indicated, using utterances like the ones listed on the sheet of paper.

During the test phase, for each task the subject was once again instructed to perform the actions associated with the task, but they had to recall the appropriate utterances from memory and/or construct similar utterances themselves.

3.3.3 Objective Results

The request success rate for the test phase was computed as described in Sect. 2.4. The overall request success rates for the three tasks were 0.59, 0.64 and 0.78, respectively, the overall rate being 0.65. In some cases, after an unsuccessful request, the subject repeated the request, often in a slightly different way, and ultimately achieved success; in other cases, success was never achieved. Because of the various differences between this study and the previous evaluations (aside from any relevant improvements in the prototype itself), it would not be meaningful to compare these success rates precisely with those found in previous evaluations. But in absolute terms, they do show that reasonable request success rates can be achieved in an outdoor setting with users whose experience with the system has been limited to a few minutes of instruction.

3.3.4 Subjective Ratings and Comments

After performing each task, the subject was asked to name his or her preferred alternative method for performing the same task. For example, for the task "Requesting information about sights," most subjects named the classical tourist guide book; for the task "Controlling the display of a map" they tended to mention digital alternatives such as Web-based mapping systems. For each of five rating scales, the subject was then asked to compare the methods of SMARTKOM Mobile ("speech and pointing gestures") to the preferred alternative, on a 5-point scale ranging from −2 (e.g., "much less convenient") to +2 (e.g., "much more convenient"). The five scales referred to the criteria of convenience, speed,[18] intuitiveness , efficiency and overall preference. Averaging over all criteria and subjects, the average ratings for the three tasks were +0.5, +1.1 and +1.0 for the three tasks, respectively. That is, subjects showed a consistent preference for the methods offered by SMARTKOM-Mobile relative to the best alternative that they knew. This positive result does not imply that the SMARTKOM-Mobile prototype used for the study, with its long response times and far-from-perfect request success rates, would be preferred to the currently available methods; but it does show that the subjects were on the whole convinced of the suitability of SMARTKOM-Mobile's methods for tasks such as the ones that they performed.

After performing all three tasks, the subjects were asked about possible positive aspects of the character Smartakus. Of the eight subjects, four agreed with the statement that he was "entertaining" and four indicated that he enhanced the understanding of internal system states (such as being busy). But only two subjects found him "informative," and none agreed with the statement that he supported the progress of the dialogue. The freely reformulated comments likewise indicated that Smartakus was generally seen as a nice but—if necessary—dispensable extra element.[19]

The most important negative aspects of the system that the subjects mentioned in their freely formulated comments were the low processing speed and the limited accuracy of the speech understanding. In addition, when subjects were asked to specify situations in which they would choose to use a system like SMARTKOM-Mobile, two subjects consistently indicated an unwillingness to use it when other people were present, even though they were happy to use it when they were alone. Even if this attitude turns out to be limited to a small portion of the potential user group, it does point to a serious issue in connection with application domains such as tourism, in which users are often members of groups.

Concerns that could be addressed through relatively straightforward changes in the design and the hardware include: (a) the small size of the display, which was associated with limited legibility and amount of displayed information; (b) the relatively high speed with which textual information about individual sights was scrolled down the screen; (c) and the lack of an opportunity to obtain different information about sights than the information presented by default.

[18] In connection with this criterion, subjects were instructed to ignore the relatively long response times of the prototype used for the study, which ranged from several seconds for simple utterances to 20–30 seconds for complex requests, and to focus on the basic nature of the interaction.

[19] These ratings may have been affected by the fact that, in the prototype, the lip movements of the character were slow and not well synchronized with the speech output.

The positive aspects that were mentioned include: (a) the simplicity and intuitive usability of the interface—in particular, the ability to do without icons and menus; and (b) the speed and accuracy of the interaction via pointing gestures.

3.4 Long-Term Evaluations–Public, Home

The motivation for a long-term evaluation was the trivial observation that the operating humans need to be trained to new technologies. This is true, for instance, for the basic desktop paradigm, where certain actions are always initiated with the left or right mouse button, but also true for simple interactions such as using a cellular phone. On the other hand, in the SMARTKOM evaluations we often found that subjects complained that the recording time for the session was limited to 10 min. They claimed that they were "... just getting the grip on it ...," etc. Therefore we wanted to see how judgments and dialogue efficiency changed for the better when the same user was able to use SMARTKOM more often and without any time limit.

The only long-term evaluation of the SMARTKOM prototype took place in Sept 2003 with only one person.[20] It comprised ten experiments with differing length, differing domains (tasks) but always the same technical setup Public.

Sex	Experience	Age	Smoker	Glasses	Beard	Weight	Height
male	high	40	no	no	no	75kg	186cm

Jewelry	Piercing	Handedness	Education	L1	L2	Bilingual	Culture
no	no	right	academic	German	English	no	German

3.4.1 Technical Setup Long-Term Evaluation

The technical setup of the prototype was identical to that of the Public benchmark evaluation (Sect. 3.2.4). Total recording time was 6.14 hours. Eight sessions were transcribed in a basic transliteration, which is a subset of the official SMARTKOM transliteration standard resulting in 810 turns. Gestures and user states were not labeled and segmented, because without the time limit of 10 min it was not possible to record the corresponding video signals. There was a minimum interval of 2 hours between each session. In most sessions no supervisor was present during the recording. No technical evaluation was carried out for the long-term evaluation.

3.4.2 Ergonomic Long Term Evaluation

The subject started the session and decided when to finish. Right after each session the user completed a questionnaire with seven free general questions, one coded

[20] Since this activity was not covered by the SMARTKOM budget we could not afford more than one study.

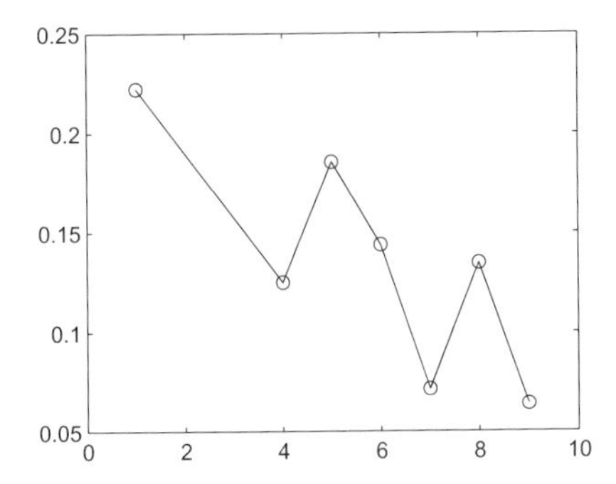

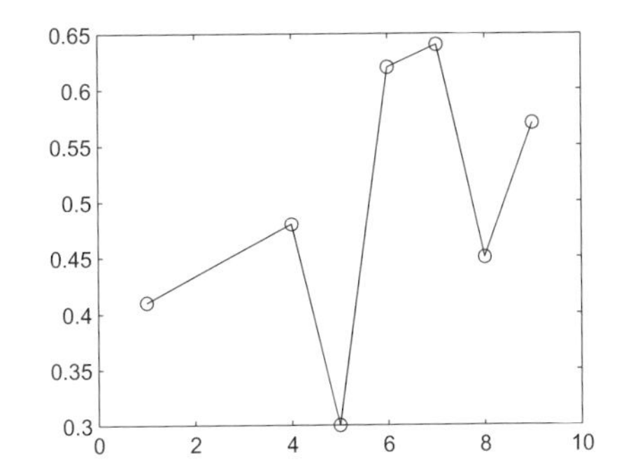

Fig. 2. Number of error messages per turn (*left*) and average request success (*right*) over long-term evaluation sessions.

question (domains) and five free questions per domain. For instance, in a session that comprised the topic TV+VCR, Cinema and Navigation this resulted in $7+3*15=52$ questions. A full listing of all answers as well as the transcripts can be found in Schiel (2004).

Adaptation

As expected, the user adapted his behavior quickly to the specific SMARTKOM dialogue technique. In Fig. 2 we show the average number of error messages of the system (left) and the average request success per session (right) against the total time the subject spent communicating with SMARTKOM. If we ignore the last session for the time being,[21] there is definitely a positive tendency over time, because the subject learned to operate SMARTKOM efficiently (error messages decrease, request success increases).

Most Prominent Positive Judgments

- Dialogue flow was structured if the user patiently waited for each answer.
- The subject was never bored.
- Reaction times were judged adequately.
- Speech understanding was very good if the headset level was adjusted carefully.
- Speech output was very good, but not with names.
- Locations on display and spoken input were associated correctly.
- Time concepts were recognized correctly (e.g., “right now,” “tonight, “Friday night”, etc.)
- Complex spoken inputs with three different types of information were recognized.

[21] In this last session the subject insistently tried to program the VCR although this functionality was not implemented in the testbed. Since it did not work, the subject tried about four different approaches to get the programming done. Naturally this trial-and-error technique resulted in a very high percentage of error messages.

- Switch on and off of VCR and TV worked.
- Change channel by pointing to a logo worked.
- Dialing the phone by using short names worked.
- Navigating on the maps worked by speech commands.

Most Prominent Negative Judgments

- Help function was not effective. It is not clear what the system calls "alternatives" and how the user should react to them.
- Anaphora were not resolved correctly ("Switch to that channel!")
- No explicit help function.
- Some system outputs came in two separated blocks with a silence interval in between; this can lead to cross-talk with the next user request.
- SMARTKOM seems to have a very limited dialogue memory. After asking for the TV program of Friday, the request: "What about the evening?" was answered by showing the program of tonight.
- Level of microphone was critical; only the slightest change in position may cause the speech recognition to break down.
- Reservation for cinema tickets did not work because the pointing finger always triggered more than six seats on the display.
- Sometimes SMARTKOM misunderstood an input ("ProSieben") even if the subject pointed to a logo of ProSieben at the same time.
- Using the phone was tricky because
  - the input of numbers by pointing was too slow
  - input of a number by voice did not work
  - after scanning a fax the user couldn't input the addressee
- Navigating in Heidelberg worked only if both the starting point and the end point of the desired route were entered in one sentence.

Most Prominent Suggestions for Improvement

- Context-dependent help function that can be called any time. After the help the dialogue should return to the point where help was called.
- Level dynamics of the microphone input more robust; it would be better to use speech input with directed microphone instead of headset.
- Output of the system should never come in more than one block with a silence interval in between.
- User should be able to interrupt system output any time.
- Dialogue handling: If the system gives a wrong answer, the user should be able to repeat the input. Currently this is only possible when the system issues an error message.
- Error messages should contain a diagnosis about the nature of the error (especially if the error is caused by an external device).

4 Conclusion

During the late design phase of SMARTKOM two accompanying ergonomic and technical evaluations were carried out, partly with naive subjects. Although the results provided valuable input to improve the prototypes, the results were statistically invalid and the PROMISE approach could not be applied.

Then the prefinal prototype 4.2 was evaluated in all three technical scenarios using an improved subjective and objective evaluation scheme. These benchmark evaluations showed a progress in almost every aspect of the prototype functionalities. The changes in the ergonomic and technical evaluation techniques led to more substantial results, which were easier to interpret than in the previous accompanying evaluations. In most sessions real dialogues took place between the user and the system, which allowed us to make constructive suggestions for further improvement.

The subjective judgments showed that the basic combined multimodal interaction between human and machine was working successfully. However, it is still a long way to an absolutely smoothly running interaction. The users made many recurring suggestions about how to better the dialogue flow. The overall impression was that the technical quality of the single modules for multimodal input and output was sufficient; the problem now lies more in the dialogue handling, error treatment and the ergonomic user guidance than in the recognizer techniques.

The PROMISE approach was not applied to the data, because the number of significant correlations was too small. On the other hand, the proposed request success value seemed to be in good correlation with the summarized subjective judgments and had more significant correlations with technical parameters. The request success was in the same range for all three scenarios (0,47 on a scale from 0 to 1).

The measured technical parameters of the system show an ambivalent picture. Although some values (e.g., speech recognition accuracy) seem to be low at first glance — especially when compared to accuracies from speech recognition tasks which range between 70–80% for a comparable task (Burger et al., 2000) — this is obviously not a severe problem when the module runs in a well-defined framework together with other recognizers (as shown in the correlation values between technical parameters and user judgments). Insofar one of the claims of SMARTKOM namely that multimodal interaction will be more robust than monomodal seems to be true.

Finally, a long-term evaluation was carried out with a single user. The results confirmed our hypothesis that a new technology like SMARTKOM needs to be adapted to. The user quickly learned how to operate SMARTKOM successfully and managed to use all capabilities of the system while decreasing the error rate over time. We therefore strongly recommend to use long-term evaluations for complex HCI systems in the future.

Acknowledgments

This contribution would not have been possible without the work of the following members of the SMARTKOM team (in alphabetical order): H. Aras, C. Beiras-Cunqueiro, N. Beringer, S. Hans, J. Häußler, I. Jacobi, A. Jameson, M. Jöst, U.

Kartal, G. Kouam-Wotchung, M. Libossek, R. Malaka, H. Mögele, M. Mozul, D. Sonntag, S. Steininger, J. Tang, S.V. Tiedemann, U. Türk and L. Wang.

References

N. Beringer, I. Jacobi, and S. von Tiedemann. Auswertung der ergonomischen Evaluation des SmartKom Prototypen -Szenario Home- Befragungszeitraum 10. Mrz 03 bis 20. Mrz 03. Memo SmartKom 16, Bavarian Archive for Speech Signals (BAS), 2003a.

N. Beringer, I. Jacobi, and S. von Tiedemann. Auswertung der ergonomischen Evaluation des SmartKom Prototypen -Szenario Mobil- Befragungszeitraum 04. Juli 03 bis 11. Juli 03. Memo SmartKom 18, Bavarian Archive for Speech Signals (BAS), 2003b.

N. Beringer, I. Jacobi, and S. von Tiedemann. Auswertung der ergonomischen Evaluation des SmartKom Prototypen -Szenario Public- Befragungszeitraum 11. Juni 03 bis 13. Juni 03. Memo SmartKom 17, Bavarian Archive for Speech Signals (BAS), 2003c.

N. Beringer, U. Kartal, M. Libossek, and S. Steininger. Gestaltung der End-to-End Evaluation in SmartKom 2.0. Technical Document SmartKom 19, Bavarian Archive for Speech Signals (BAS), 2002a.

N. Beringer, U. Kartal, K. Louka, F. Schiel, and U. Türk. PROMISE: A Procedure for Multimodal Interactive System Evaluation. In: *Proc. Workshop "Multimodal Resources and Multimodal Systems Evaluation"*, pp. 77–80, Las Palmas, Spain, 2002b.

N. Beringer, K. Louka, V. Penide-Lopez, and U. Türk. End-To-End Evaluation of Multimodal Dialogue Systems — Can We Transfer Established Methods? In: *Proc. 3rd Int. Conf. on Language Resources and Evaluation (LREC 2002)*, pp. 558–563, Las Palmas, Spain, 2002c.

N. Beringer, H. Mögele, I. Jacobi, C. Beiras-Cunqueiro, L. Wang, U. Türk, G. Kouam-Wotchung, M. Mozul, and U. Kartal. Auswertung der End-to-End Evaluation des SmartKom Prototypen (technische Evaluation, Evaluationstool, PROMISE) -Szenario Home-, Aufnahmezeitraum 01. April 02 bis 31. Juni 02. Memo SmartKom 13, Bavarian Archive for Speech Signals (BAS), 2002d.

N. Beringer, H. Mögele, I. Jacobi, C. Beiras-Cunqueiro, L. Wang, U. Türk, G. Kouam-Wotchung, M. Mozul, and U. Kartal. Auswertung der End-to-End Evaluation des SmartKom Prototypen (technische Evaluation, Evaluationstool, PROMISE) -Szenario Mobil-, Aufnahmezeitraum 01.August 02 bis 31. August 02. Memo SmartKom 15, Bavarian Archive for Speech Signals (BAS), 2003d.

S. Burger, K. Weilhammer, F. Schiel, and H.G. Tillmann. Verbmobil Data Collection and Annotation. In: W. Wahlster (ed.), *Verbmobil: Foundations of Speech-to-Speech Translation*, pp. 537–549, Berlin Heidelberg Germany, 2000. Springer.

I. Jacobi, H. Mögele, S. von Tiedemann, and U. Türk. Auswertung der End-to-end Evaluation des SmartKom Prototypen (technische Evaluation, PROMISE)

- Szenario Home, Public, Mobil. Memo SmartKom 19, Bavarian Archive for Speech Signals (BAS), 2004.

U. Kartal, N. Beringer, and S. Steininger. Auswertung der ergonomischen Evaluation der WOZ-Aufnahmen im Projekt SmartKom. Memo SmartKom 11, Bavarian Archive for Speech Signals (BAS), 2002a.

U. Kartal, N. Beringer, S. Steininger, and D. Sonntag. Auswertung der ergonomischen Evaluation des SmartKom Prototypen -Szenario Home-, Befragungszeitraum 01. April 02 bis 31. Juni 02. Memo SmartKom 12, Bavarian Archive for Speech Signals (BAS), 2002b.

U. Kartal, N. Beringer, S. Steininger, and D. Sonntag. Auswertung der ergonomischen Evaluation des SmartKom Prototypen -Szenario Mobil-, Befragungszeitraum 01.August 02 bis 31. August 02. Memo SmartKom 14, Bavarian Archive for Speech Signals (BAS), 2003.

F. Schiel. SmartKom – Langzeitevaluation. Memo SmartKom 23, Bavarian Archive for Speech Signals (BAS), 2004.

F. Schiel and U. Türk. Wizard-of-Oz Recordings, 2006. In this volume.

M.A. Walker, D.J. Litman, C.A. Kamm, and A. Abella. PARADISE: A Framework for Evaluating Spoken Dialogue Agents. In: *Proc. 35th ACL*, Madrid, Spain, 1997.

M.A. Walker, D.J. Litman, C.A. Kamm, and A. Abella. Evaluating Spoken Dialogue Agents With Paradise: Two Case Studies. *Computer, Speech and Language*, 12 (3), 1998.

Cognitive Technologies

T. Frühwirth, S. Abdennadher:
Essentials of Constraint Programming.
IX, 144 pages. 2003

J. W. Lloyd:
Logic for Learning.
Learning Comprehensive Theories from Structured Data.
X, 256 pages. 2003

S. K. Pal, L. Polkowski, A. Skowron (Eds.):
Rough-Neural Computing.
Techniques for Computing with Words.
XXV, 734 pages. 2004

H. Prendinger, M. Ishizuka (Eds.):
Life-Like Characters.
Tools, Affective Functions, and Applications.
IX, 477 pages. 2004

H. Helbig:
Knowledge Representation and the Semantics of Natural Language.
XVIII, 646 pages. 2006

P. M. Nugues:
An Introduction to Language Processing with Perl and Prolog.
An Outline of Theories, Implementation, and Application with Special Consideration of English, French, and German.
XX, 513 pages. 2006

W. Wahlster (Ed.):
SmartKom: Foundations of Multimodal Dialogue Systems
XVIII, 644 pages. 2006